Sampling: Design and Analysis

Second Edition

Sharon L. Lohr
Arizona State University

BROOKS/COLE
CENGAGE Learning™

Australia • Brazil • Japan • Korea • Mexico • Singapore • Spain • United Kingdom • United States

BROOKS/COLE
CENGAGE Learning™

Sampling: Design and Analysis, Second Edition
Sharon L. Lohr

to Doug

Editor in Chief: Michelle Julet

Publisher: Richard Stratton

Senior Sponsoring Editor: Molly Taylor

Associate Editor: Daniel Seibert

Editorial Assistant: Shaylin Walsh

Associate Media Editor: Catie Ronquillo

Senior Marketing Manager: Greta Kleinert

Marketing Coordinator: Erica O'Connell

Marketing Communications Manager: Mary Anne Payumo

Associate Content Project Manager: Jill Clark

Art Director: Linda Helcher

Senior Manufacturing Buyer: Diane Gibbons

Senior Text Rights Account Manager: Bob Kauser

Production Service: MPS Limited, A Macmillan Company

Cover Designer: Denise Davidson

Compositor: MPS Limited, A Macmillan Company

© 2010, 1999 Brooks/Cole, Cengage Learning

ALL RIGHTS RESERVED. No part of this work covered by the copyright herein may be reproduced, transmitted, stored or used in any form or by any means graphic, electronic, or mechanical, including but not limited to photocopying, recording, scanning, digitizing, taping, Web distribution, information networks, or information storage and retrieval systems, except as permitted under Section 107 or 108 of the 1976 United States Copyright Act, without the prior written permission of the publisher.

For product information and technology assistance, contact us at
Cengage Learning Customer & Sales Support, 1-800-354-9706

For permission to use material from this text or product, submit all requests online at www.cengage.com/permissions. Further permissions questions can be emailed to **permissionrequest@cengage.com**.

Library of Congress Control Number: 2009938648

ISBN-13: 978-0-495-11084-2

ISBN-10: 0-495-11084-1

Brooks/Cole
20 Channel Center Street
Boston, MA 02210
USA

Cengage Learning is a leading provider of customized learning solutions with office locations around the globe, including Singapore, the United Kingdom, Australia, Mexico, Brazil, and Japan. Locate your local office at: **international.cengage.com/region**

Cengage Learning products are represented in Canada by Nelson Education, Ltd.

For your course and learning solutions, visit **www.cengage.com**

Purchase any of our products at your local college store or at our preferred online store **www.ichapters.com**

Computer output in this book was produced using SAS software, version 9.2. Copyright © (2009), SAS Institute Inc. SAS and all other SAS Institute Inc. product or service names are registered trademarks or trademarks of SAS Institute Inc. in the USA and other countries; ® indicates USA registration.

Printed in the United States of America
1 2 3 4 5 6 7 13 12 11 10 09

Contents

Preface

Surveys and samples sometimes seem to surround you. Many give valuable information; some, unfortunately, are so poorly conceived and implemented that it would be better for science and society if they were simply not done. This book concentrates on the statistical aspects of taking and analyzing a sample. It gives you guidance on how to tell when a sample is valid or not, and how to design and analyze many different forms of sample surveys.

Much research has been done on theoretical and applied aspects of survey sampling since the publication of the first edition of this book. The second edition incorporates some of this recent research, contains new topics such as total survey design and statistical issues in Internet surveys, and expands coverage of weighting, calibration, two-phase sampling, and sampling for rare events. The order of topics has been streamlined to be more intuitive, chapter summaries have been added for quick review, and exercise sets are now better categorized by problem type. SAS® software is now used for calculations, with downloadable SAS code provided on the book's companion website.

Six main features distinguish this book from other texts about sampling methods.

- The book is accessible to students with a wide range of statistical backgrounds, and is flexible for content and level. By appropriate choice of sections, this book can be used for a first-year graduate course for statistics students or for a class with students from business, sociology, psychology, or biology who want to learn about designing and analyzing data from sample surveys. It is also useful for a person doing survey research who wants to learn more about the statistical aspects of surveys and recent developments.

- I have tried to use real data as much as possible—the Acme Widget Company never appears in this book. The examples and exercises come from social sciences, engineering, agriculture, ecology, medicine, and a variety of other disciplines, and are selected to illustrate the wide applicability of sampling methods. A number of data sets have extra variables not specifically referenced in the text; an instructor can use these for additional exercises or variations. The exercises also give the instructor much flexibility for course level. Some emphasize mastering the mechanics, but many encourage the student to think about the sampling issues involved and to understand the structure of sample designs at a deeper level, while others are open-ended and encourage further exploration of the ideas.

- I have incorporated model-based as well as randomization-based theory into the text, with the goal of placing sampling methods within the framework used in other areas of statistics. Many of the important results in the last twenty-plus years of sampling research have involved models, and an understanding of both approaches is essential for the survey practitioner. The model-based approach is introduced in Section 2.9 and further developed in successive chapters; however, those sections could be discussed at any time later in the course.

- The book covers many topics not found in other textbooks at this level. Chapters 7 through 15 discuss how to analyze complex surveys such as those administered by the United States Census Bureau or Statistics Canada, computer-intensive methods for estimating variances in complex surveys, what to do if there is nonresponse, and how to perform chi-squared tests and regression analyses using data from complex surveys.

- This book emphasizes the importance of graphing the data. Graphical analysis of survey data is challenging because of the large sizes and complexity of survey data sets but graphs can provide insight into the data structure.

- Design of surveys is emphasized throughout, and is related to methods for analyzing the data from a survey. The book presents the philosophy that the design is by far the most important aspect of any survey: no amount of statistical analysis can compensate for a badly designed survey. Models are used to motivate designs, and graphs are presented to check the sensitivity of the design to model assumptions.

Chapters 1 through 6 cover the building blocks of simple random, stratified, and cluster sampling, as well as ratio and regression estimation. To read them requires familiarity with basic ideas of expectation, sampling distributions, confidence intervals, and linear regression—material covered in most introductory statistics classes. Optional sections on the statistical theory for designs are marked with asterisks—these require you to be familiar with calculus and mathematical statistics. Along with Chapters 7 and 8, these chapters form the foundation of a one-quarter or one-semester course. The material in Chapters 9 through 15 can be covered in almost any order, with topics chosen to fit the needs of the students. Appendix A reviews probability concepts.

The second edition introduces some organizational changes to the chapters. The central concept of sampling weights is now introduced in Chapter 2. Stratified sampling has been moved earlier to Chapter 3, preceding ratio and regression estimation. This allows students to become more familiar with the use of weights to account for inclusion probabilities before they are exposed to adjusting the weights for calibration. Chapter 6 contains more intuition and theory on the Horvitz–Thompson estimator, and Chapter 7 provides additional methods for graphing survey data. Chapter 9 has expanded treatment of computer-intensive methods such as jackknife and bootstrap. Material in Chapter 12 of the first edition has been expanded in Chapters 12 to 14 of the second edition. Chapter 15 on total survey design is completely new, and ties together much of the material in the earlier chapters. Each chapter now concludes with a chapter summary, including key terms and references for further exploration.

The exercises in the second edition have been reordered into four categories in each chapter, with many new exercises added to the book's already extensive problem sets.

- *Introductory Exercises* give more routine problems intended to develop skills at the basic ideas in the book. Many of these would be suitable for hand calculations.

- *Working with Survey Data* exercises ask students to analyze data from real surveys. Most require use of statistical software such as SAS. Data sets and SAS code for dealing with special problems in reading the data are available for download from the book's companion website.

- *Working with Theory* exercises are intended for a more mathematically oriented class, allowing students to work through proofs of results in a step-by-step manner and explore the theory of sampling in more depth. They also include presentations of some results about survey sampling that may be of interest to more advanced students. Many of these exercises require students to know calculus and material from an undergraduate probability class.

- *Projects and Activities*—new for the second edition—contain activities suitable for classroom use or for assignment as projects. Many of these activities ask the student to design, collect, and analyze a sample selected from a population of real data provided on the book's companion website. The activities continue from chapter to chapter, allowing students to build on their knowledge and compare various sampling designs. I always assign Exercise 31 from Chapter 7 and its continuation in subsequent chapters as a course project. This exercise asks students to download data from a survey on a topic of their choice from the Internet and analyze the data. Along the way, the students read and translate the survey design descriptions into the design features studied in class, develop skills in analyzing survey data, and gain experience in dealing with nonresponse and other challenges.

You must know how to use a statistical computer package to be able to do the problems in this book. The second edition uses SAS software for computing estimates and graphing data, with selected output presented in the book and annotated code used for the examples available for download on the book's companion website. Other software packages that calculate estimates for survey data can also be used with the book; the website *www.hcp.med.harvard.edu/statistics/survey-soft/* provides an up-to-date overview of these programs.

The book's companion website, *www.cengage.com/statistics/lohr*, contains the SAS code and data sets referenced in the book, as well as the exercises and appendix from the first edition for the SURVEY computer program. Additionally, worked solutions to the exercises in the book are provided online to instructors who sign up for an account with Cengage's Solution Builder service at *www.cengage.com/solutionbuilder*.

Many people have been generous with their encouragement and suggestions for this book. I am grateful to J.N.K. Rao for his permission to adapt material that he and I presented at the 2004 Joint Statistical Meetings for inclusion in the second edition, and for his suggestions and unfailing support. The following persons reviewed or used various versions of the manuscript, providing valuable suggestions for improvement: Elizabeth Stasny, Fritz Scheuren, Nancy Heckman, Ted Chang, Steve MacEachern, Mark Conaway, Ron Christensen, Michael Hamada, Partha Lahiri, Dale Everson, James Gentle, Ruth Mickey, Sarah Nusser, N.G.N. Prasad, Deborah Rumsey, Fritz Scheuren, David Bellhouse, David Marker, Tim Johnson, Stas Kolenikov, Serge Alalouf, Trent Buskirk, and Jae-kwang Kim. In addition, Anders Lundqvist, Imbi Traat, Andrew Gelman, Ron Christensen, Paul Biemer, Ron Fecso, Steve Fienberg, Pierre Lavallée, Mike Hidiroglou, Dave Chapman, Mike Brick, Thomas P. Ryan, Kinley Larntz, Shap Wolf, and Burke Grandjean provided much helpful advice and encouragement. Ted Chang first encouraged me to turn my class notes into a book, and generously allowed use of the SURVEY program. Alastair Scott's inspiring class on sampling at the University of Wisconsin introduced me to the joys of the subject.

Finally, thanks to my wonderful husband Doug for his patience and cheerful encouragement.

Sharon L. Lohr

1

Introduction

When statistics are not based on strictly accurate calculations, they mislead instead of guide. The mind easily lets itself be taken in by the false appearance of exactitude which statistics retain in their mistakes, and confidently adopts errors clothed in the form of mathematical truth.

—Alexis de Tocqueville, *Democracy in America*

1.1

A Sample Controversy

Shere Hite's book *Women and Love: A Cultural Revolution in Progress* (1987) had a number of widely quoted results:

- 84% of women are "not satisfied emotionally with their relationships" (p. 804).
- 70% of all women "married five or more years are having sex outside of their marriages" (p. 856).
- 95% of women "report forms of emotional and psychological harassment from men with whom they are in love relationships" (p. 810).
- 84% of women report forms of condescension from the men in their love relationships (p. 809).

The book was widely criticized in newspaper and magazine articles throughout the United States. The *Time* magazine cover story "Back Off, Buddy" (October 12, 1987), for example, called the conclusions of Hite's study "dubious" and "of limited value."

Why was Hite's study so roundly criticized? Was it wrong for Hite to report the quotes from women who feel that the men in their lives refuse to treat them as equals, who perhaps have never been given the chance to speak out before? Was it wrong to report the percentages of these women who are unhappy in their relationships with men?

Of course not. Hite's research allowed women to discuss how they viewed their experiences, and reflected the richness of these women's experience in a way that a

1

multiple choice questionnaire could not. Hite erred in generalizing these results to all women, whether they participated in the survey or not, and in claiming that the percentages above applied to all women. The following characteristics of the survey make it unsuitable for generalizing the results to all women.

- The sample was self-selected—that is, recipients of questionnaires decided whether they would be in the sample or not. Hite mailed 100,000 questionnaires; of these, 4.5% were returned.

- The questionnaires were mailed to such organizations as professional women's groups, counseling centers, church societies, and senior citizens' centers. The members may differ in political views, but many have joined an "all-women" group, and their viewpoints may differ from other women in the United States.

- The survey has 127 essay questions, and most of the questions have several parts. Who will tend to return such a survey?

- Many of the questions are vague, using words such as "love." The concept of love probably has as many interpretations as there are people, making it impossible to attach a single interpretation to any statistic purporting to state how many women are "in love." Such question wording works well for eliciting the rich individual vignettes that comprise most of the book, but makes interpreting percentages difficult.

- Many of the questions are leading—they suggest to the respondent which response she should make. For instance: "Does your husband/lover see you as an equal? Or are there times when he seems to treat you as an inferior? Leave you out of the decisions? Act superior?" (p. 795)

Hite writes "Does research that is not based on a probability or random sample give one the right to generalize from the results of the study to the population at large? If a study is large enough and the sample broad enough, and if one generalizes carefully, yes" (p. 778). Most survey statisticians would answer Hite's question with a resounding "no." In Hite's survey, because the women sent questionnaires were purposefully chosen and an extremely small percentage of those women returned the questionnaires, statistics calculated from these data cannot be used to indicate attitudes of all women in the United States. The final sample is not *representative* of women in the United States, and the statistics can only be used to describe women who would have responded to the survey.

Hite claims that results from the sample could be generalized because characteristics such as the age, educational, and occupational profiles of women in the sample matched those for the population of women in the United States. But the women in the sample differed on one important aspect—they were willing to take the time to fill out a long questionnaire dealing with harassment by men, and to provide intensely personal information to a researcher. We would expect that in every age group and socioeconomic class, women who choose to report such information would in general have had different experiences than women who choose not to participate in the survey.

1.2
Requirements of a Good Sample

In the movie "Magic Town," the public opinion researcher played by James Stewart discovered a town that had exactly the same characteristics as the whole United States: Grandview had exactly the same proportion of people who voted Republican, the same proportion of people under the poverty line, the same proportion of auto mechanics, and so on, as the United States taken as a whole. All that Stewart's character had to do was to interview the people of Grandview, and he would know what public opinion was in the United States.

A perfect sample would be like Grandview: a "scaled-down" version of the population, mirroring every characteristic of the whole population. Of course, no such perfect sample can exist for complicated populations (even if it did exist, we would not know it was a perfect sample without measuring the whole population). But a good sample will be **representative** in the sense that characteristics of interest in the population can be estimated from the sample with a known degree of accuracy.

Some definitions are needed to make the notion of a good sample more precise.

Observation unit An object on which a measurement is taken. This is the basic unit of observation, sometimes called an **element**. In studying human populations, observation units are often individuals.

Target population The complete collection of observations we want to study. Defining the target population is an important and often difficult part of the study. For example, in a political poll, should the target population be all adults eligible to vote? All registered voters? All persons who voted in the last election? The choice of target population will profoundly affect the statistics that result.

Sample A subset of a population.

Sampled population The collection of all possible observation units that might have been chosen in a sample; the population from which the sample was taken.

Sampling unit A unit that can be selected for a sample. We may want to study individuals, but do not have a list of all individuals in the target population. Instead, households serve as the sampling units, and the observation units are the individuals living in the households.

Sampling frame A list, map, or other specification of sampling units in the population from which a sample may be selected. For a telephone survey, the sampling frame might be a list of all residential telephone numbers in the city. For a survey using in-person interviews, the sampling frame might be a list of all street addresses. For an agricultural survey, a sampling frame might be a list of all farms, or a map of areas containing farms.

In an ideal survey, the sampled population will be identical to the target population, but this ideal is rarely met exactly. In surveys of people, the sampled population is usually smaller than the target population: as illustrated in Figure 1.1, not all persons in the target population are included in the sampling frame, and a number of persons will not respond to the survey.

FIGURE **1.1**

Target population and sampled population in a telephone survey of likely voters. Not all households have telephones, so a number of persons in the target population of likely voters will not be associated with a telephone number in the sampling frame. In some households with telephones, the residents are not registered to vote and hence are not eligible for the survey. Some eligible persons in the sampling frame population do not respond because they cannot be contacted, some refuse to respond to the survey, and some may be ill and incapable of responding.

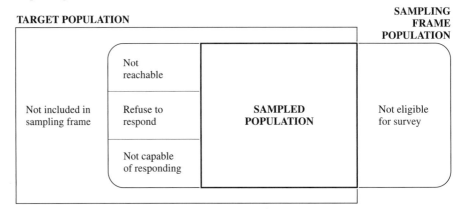

In the Hite (1987) study, one characteristic of interest was the percentage of women who are harassed in their relationship. An individual woman was an element. The target population was all adult women in the United States. Hite's sampled population was women belonging to women's organizations who would return the questionnaire. Consequently, inferences can only be made to the sampled population, not to the population of all adult women in the United States.

The National Crime Victimization Survey (NCVS) is an ongoing survey to study victimization rates, administered by the U.S. Census Bureau and the Bureau of Justice Statistics. If the characteristic of interest is the total number of households in the United States that were victimized by crime last year, the elements are households, the target population consists of all households in the United States, and the sampled population consists of households in the sampling frame, constructed from U.S. Census information and building permits, that are "at home" and agree to answer questions.

The goal of the National Pesticide Survey, conducted by the U.S. Environmental Protection Agency, was to study pesticides and nitrate in drinking water wells nationwide. The target population was all community water systems and rural domestic wells in the United States. The sampled population was all community water systems (all are listed in the Federal Reporting Data System) and all identifiable domestic wells outside of government reservations that belonged to households willing to cooperate with the survey.

Public opinion polls are often taken to predict which candidate will win the next election. The target population is persons who will vote in the next election; the sampled population is often persons who can be reached by telephone and who are judged to be likely to vote in the next election. Few national polls in the

United States include persons in hospitals, dormitories, or jails; they, and persons without telephones, are not part of the sampling frame or of the sampled population.

1.3
Selection Bias

A good sample will be as free from selection bias as possible. **Selection bias** occurs when some part of the target population is not in the sampled population, or, more generally, when some population units are sampled at a different rate than intended by the investigator. If a survey designed to study household income omits transient persons, the estimates from the survey of the average or median household income are likely to be too large. A **sample of convenience** is often biased, since the units that are easiest to select or that are most likely to respond are usually not representative of the harder-to-select or nonresponding units. The following examples indicate some ways in which selection bias can occur.

- Using a sample selection procedure that, unknown to the investigators, depends on some characteristic associated with the properties of interest. For example, investigators took a convenience sample of adolescents to study how frequently adolescents talk to their parents and teachers about AIDS. But adolescents willing to talk to the investigators about AIDS are probably also more likely to talk to other authority figures about AIDS. The investigators, who simply averaged the amounts of time that adolescents in the sample said they spent talking with their parents and teachers, probably overestimated the amount of communication occurring between parents and adolescents in the population.

- Deliberately or purposively selecting a "representative" sample. If we want to estimate the average amount a shopper spends at the Mall of America in a shopping trip, and we sample shoppers who look like they have spent an "average" amount, we have deliberately selected a sample to confirm our prior opinion. This type of sample is sometimes called a **judgment sample**—the investigator uses his or her judgment to select the specific units to be included in the sample.

- Misspecifying the target population. For instance, all the polls in the 1994 Democratic gubernatorial primary election in Arizona predicted that candidate Eddie Basha would trail the front-runner in the polls by at least nine percentage points. In the election, Basha won 37% of the vote; the other two candidates won 35% and 28%, respectively. One problem is that many voters were undecided at the time the polls were taken. Another is that the target population for the polls was registered voters who had voted in previous primary elections and were interested in this one. In the primary election, however, Basha had heavy support in rural areas from demographic groups that had not voted before and hence were not targeted in the surveys.

- Failing to include all of the target population in the sampling frame, called **undercoverage**. The U.S. Behavioral Risk Factor Surveillance System survey, described at www.cdc.gov, illustrates some of the coverage problems that may occur in a household telephone survey. The target population for this survey on preventive health practices and risk behaviors is adults aged 18 and older in the United States. Some undercoverage occurs because persons in institutions such as nursing homes or prisons are excluded. Additional undercoverage occurs because the survey is conducted by telephone. Some households do not have telephones and telephone coverage varies across states. Households in the southern part of the United States, minority households, and low-income households are less likely to have telephones, so those households are likely to be underrepresented in the sample because of the undercoverage. Households that have only a cellular telephone are also not included in the sampling frame at this writing.

- Including population units in the sampling frame that are not in the target population, called **overcoverage**. Overcoverage can occur when persons not in the target population are not screened out of the sample, or when data collectors are not given specific instructions on sample eligibility. The target population for a telephone survey on radio listening habits might be persons aged 18 and over, but some interviewers might include persons under age 18 when taking the sample, and children and teenagers may well listen to different radio stations than adults.

- Having multiplicity of listings in the sampling frame, without adjusting for the multiplicity in the analysis. In its simplest form, random digit dialing prescribes selecting a random sample of 10-digit numbers. Households with more than one telephone line then have a higher chance of being selected in the sample. This multiplicity can be compensated in the estimation (we'll discuss this in Section 6.5); if it is ignored, bias can result. One might expect households with more telephone lines to be larger or more affluent, so if no adjustment is made for those households having a higher probability of being selected for the sample, estimates of average income or household size may be too large.

- Substituting a convenient member of a population for a designated member who is not readily available. For example, if no one is at home in the designated household, a field representative might try next door. In a wildlife survey, the investigator might substitute an area next to a road for a less accessible area. In each case, the sampled units most likely differ on a number of characteristics from units not in the sample. The substituted household may be more likely to have a member who does not work outside of the house than the originally selected household. The area by the road may have fewer frogs than the area that is harder to reach.

- Failing to obtain responses from all of the chosen sample. **Nonresponse** distorts the results of many surveys, even surveys that are carefully designed to minimize other sources of selection bias. Often, nonrespondents differ critically from the respondents, but the extent of that difference is unknown unless you can later obtain information about the nonrespondents. Many surveys reported in newspapers or research journals have dismal response rates—in some, the response rate is as low as 10%. It is difficult to see how results can be generalized to the population when 90% of the targeted sample cannot be reached or refuses to participate.

The Adolescent Health Database Survey was designed to obtain a representative sample of Minnesota junior and senior high school students in public schools (Remafedi et al., 1992). Overall, 49% of the school districts that were invited to participate in the survey agreed to participate. The response rate varied with the size of the school district:

Type of School District	Participation Rate (%)
Urban	100
Metropolitan suburban	25
Nonmetropolitan with more than 2000 students	62
Nonmetropolitan with 1000–1999 students	27
Nonmetropolitan with 500–999 students	61
Nonmetropolitan with fewer than 500 students	53

In each of the school districts that participated, surveys were distributed to students and students' participation was voluntary. Of the 52,553 surveys distributed to students, 36,741 were completed and returned, resulting in a student response rate of 69%. The survey asked questions about health habits, religious affiliation, psychosocial status, and sexual orientation. It seems likely that responding and nonresponding school districts have different levels of health and activity. It seems even more likely that students who respond to the survey will, on average, have a different health profile than students who do not respond to the survey.

Many studies comparing respondents and nonrespondents have found differences in the two groups. In the Iowa Women's Health Study, 41,836 women responded to a mailed questionnaire in 1986. Bisgard et al. (1994) compared those respondents to the 55,323 nonrespondents by checking records in the State Health Registry; they found that the age-adjusted mortality rate and the cancer attack rate were significantly higher for the nonrespondents than for the respondents.

■ Allowing the sample to consist entirely of volunteers. Such is the case in radio and television call-in polls, and in most online surveys. The statistics from such surveys cannot be trusted. At best, they are entertainment; at worst, they mislead, particularly when statistics from polls with self-selected respondents are cited in policy debates without any mention of their unscientific nature. CNN.com's daily QuickVote, which invites site visitors to vote on an issue of the day, carefully states that "This QuickVote is not scientific and reflects the opinions of only those Internet users who have chosen to participate. The results cannot be assumed to represent the opinions of Internet users in general, nor the public as a whole" (Cable News Network, 2002). Yet statistics from QuickVote and other online surveys are frequently quoted by independent research institutes, policy organizations, and scholarly journals. For example, Christian and Kinney (1999) cited a 1999 Internet poll on CNN.com, where 98% of the 17,000 visitors to a website linked to the science and technology reports voted "yes" to a question on whether the Hubble Space Telescope was worth the investment, as an indication of "a great improvement in public opinion." In fact, all that can be concluded from the Internet poll is that nearly 17,000 people who visited a website voted "yes" on the question; nothing can be inferred about the rest of the population without making heroic assumptions. Some individuals or organizations may respond multiple times to a voluntary survey, and a determined organization may skew the results.

EXAMPLE 1.1   Many surveys have more than one of these problems. *The Literary Digest* (1932, 1936a, b, c) began taking polls to forecast the outcome of the U.S. presidential election in 1912, and their polls attained a reputation for accuracy because they forecast the correct winner in every election between 1912 and 1932. In 1932, for example, the poll predicted that Roosevelt would receive 56% of the popular vote and 474 votes in the Electoral College; in the actual election, Roosevelt received 58% of the popular vote and 472 votes in the Electoral College.

With such a strong record of accuracy, it is not surprising that the editors of *The Literary Digest* had a great deal of confidence in their polling methods by 1936. Launching the 1936 poll, they said:

> The Poll represents thirty years' constant evolution and perfection. Based on the "commercial sampling" methods used for more than a century by publishing houses to push book sales, the present mailing list is drawn from every telephone book in the United States, from the rosters of clubs and associations, from city directories, lists of registered voters, classified mail-order and occupational data. (1936a, p. 3)

On October 31, the poll predicted that Republican Alf Landon would receive 55% of the popular vote, compared with 41% for President Roosevelt. The article "Landon, 1,293,669; Roosevelt, 972,897: Final Returns in The Digest's Poll of Ten Million Voters" contained the statement "We make no claim to infallibility. We did not coin the phrase 'uncanny accuracy' which has been so freely applied to our Polls" (1936b). It is a good thing they made no claim to infallibility: In the election, Roosevelt received 61% of the vote; Landon, 37%.

What went wrong? One problem may have been undercoverage in the sampling frame, which relied heavily on telephone directories and automobile registration lists—the frame was used for advertising purposes, as well as for the poll. Households with a telephone or automobile in 1936 were generally more affluent than other households, and opinion of Roosevelt's economic policies was generally related to the economic class of the respondent. But sampling frame bias does not explain all the discrepancy. Postmortem analyses of the poll by Squire (1988) and Calahan (1989) indicate that even persons with both a car and a telephone tended to favor Roosevelt, though not to the degree that persons with neither car nor telephone supported him.

The low response rate to the survey was likely the source of much of the error. *Ten million* questionnaires were mailed out, and 2.3 million were returned—an enormous sample, but a response rate of less than 25%. In Allentown, Pennsylvania, for example, the survey was mailed to every registered voter, but the survey results for Allentown were still incorrect because only one-third of the ballots were returned. Squire (1988) reports that persons supporting Landon were much more likely to have returned the survey; in fact, many Roosevelt supporters did not even remember receiving a survey even though they were on the mailing list.

One lesson to be learned from *The Literary Digest* poll is that the sheer size of a sample is no guarantee of its accuracy. The *Digest* editors became complacent because they sent out questionnaires to more than one quarter of all registered voters and obtained a huge sample of 2.3 million people. But large unrepresentative samples can perform as badly as small unrepresentative samples. A large unrepresentative sample may do more damage than a small one because many people think that large samples are always better than small ones. The design of the survey is far more important than the absolute size of the sample. ■

What good are samples with selection bias?   We prefer to have samples with no selection bias, that serve as a microcosm of the population. When the primary interest is in estimating the total number of victims of violent crime in the United States, or the percentage of likely voters in the United Kingdom who intend to vote for the Labour Party in the next election, serious selection bias can cause the sample estimates to be invalid.

Purposive or judgment samples can provide valuable information, though, particularly in the early stages of an investigation. Teichman et al. (1993) took soil samples along Interstate 880 in Alameda County, California, to determine the amount of lead in yards of homes and in parks close to the freeway. In taking the samples, they concentrated on areas where they thought children were likely to play and areas where soil might easily be tracked into homes. The purposive sampling scheme worked well for justifying the conclusion of the study, that "lead contamination of urban soil in the east bay area of the San Francisco metropolitan area is high and exceeds hazardous waste levels at many sites." A sampling scheme that avoided selection bias would be needed for this study if the investigators wanted to generalize the estimated percentage of contaminated sites to the entire area.

1.4
Measurement Error

A good sample has accurate responses to the items of interest. When a response in the survey differs from the true value, **measurement error** has occurred. **Measurement bias** occurs when the response has a tendency to differ from the true value in one direction. As with selection bias, measurement error and bias must be considered and minimized in the design stage of the survey; no amount of statistical analysis will disclose that the scale erroneously added 5 kilograms to the weight of every person in the health survey.

Measurement error is a concern in all surveys and can be insidious. In many surveys of vegetation, for example, areas to be sampled are divided into smaller plots. A sample of plots is selected, and the number of plants in each plot is recorded. When a plant is near the boundary of the region, the field researcher needs to decide whether to include the plant in the tally. A person who includes all plants near or on the boundary in the count is likely to produce an estimate of the total number of plants in the area that is too high because some plants may be counted twice. Duce et al. (1972) report concentrations of trace metals, lipids, and chlorinated hydrocarbons in the top 100 micrometers of Narragansett Bay that are 1.5 to 50 times as great as those in the water 20 cm below the surface. If studying the transport of pollutants from coastal waters to the deeper waters of the ocean, a sampling scheme that ignores this boundary effect may underestimate the amount transported.

Sometimes measurement bias is unavoidable. In the North American Breeding Bird Survey, observers stop every one-half mile on designated routes and count all birds heard or seen during a 3-minute period within a quarter-mile radius

(Sauer et al., 1997). The count of birds for that point is almost always an under-estimate of the number of birds in the area; statistical models may possibly be used to adjust for the measurement bias. If data are collected with the same procedure and with similarly skilled observers from year to year, the survey can be used to estimate trends in the population of different species—the biases from different years are expected to be similar, and may cancel when year-to-year differences are calculated.

Obtaining accurate responses is challenging in all types of surveys, but particularly so in surveys of people:

- People sometimes do not tell the truth. In an agricultural survey, farmers in an area with food-aid programs may underreport crop yields, hoping for more food aid. Obtaining truthful responses is a particular challenge in surveys involving sensitive subject matter, such as surveys about drug use.

- People do not always understand the questions. Many persons in the United States were shocked by the results of a 1993 Roper poll reporting that 25% of Americans did not believe the Holocaust really happened. When the double-negative structure of the question was eliminated, and the question reworded, only 1% thought it was "possible … the Nazi extermination of the Jews never happened."

- People forget. One problem faced in the design of the NCVS is that of **telescoping**: Persons are asked about experiences as a crime victim in the last six months, but some include victimizations that occurred more than six months ago.

- People give different answers to different interviewers. Schuman and Converse (1971) employed white and black interviewers to interview black residents of Detroit. In response to the question "Do you personally feel that you can trust most white people, some white people, or none at all?" 35% of the respondents interviewed by a white person said they could trust most white people. The percentage was 7% for those interviewed by a black person.

- People may say what they think an interviewer wants to hear or what they think will impress the interviewer. In experiments done with questions beginning "Do you agree or disagree with the following statement" it has been found that a subset of the population tends to agree with any statement regardless of its content. Lenski and Leggett (1960) found that about one-tenth of their sample agreed with both of the following statements:

    It is hardly fair to bring children into the world, the way things look for the future.

    Children born today have a wonderful future to look forward to.

    Some responses are perceived as being more socially desirable than others, so that persons may overreport behaviors such as exercising and donating to charities, and underreport behaviors such as smoking or drinking.

- A particular interviewer may affect the accuracy of the response, by misreading questions, recording responses inaccurately, or antagonizing the respondent. In a survey on abortion, a poorly trained interviewer with strong feelings about abortion may encourage the respondent to provide one answer rather than another. In extreme cases, an interviewer may change the answers given by the respondent, or simply make up data and not contact the respondent at all.

- Certain words mean different things to different people. A simple question such as "Do you own a car?" may be answered yes or no depending on the respondent's interpretation of "you" (does it refer to just the individual, or to the household?), "own" (does it count as ownership if you are making payments to a finance company?), or "car" (are pickup trucks included?).

- Question wording and question order have a large effect on the responses obtained. Two surveys were taken in late 1993/early 1994 about Elvis Presley. One survey asked, "In the past few years, there have been a lot of rumors and stories about whether Elvis Presley is really dead. How do you feel about this? Do you think there is any possibility that these rumors are true and that Elvis Presley is still alive, or don't you think so?" The other survey asked, "A recent television show examined various theories about Elvis Presley's death. Do you think it is possible that Elvis is alive or not?" Eight percent of the respondents to the first question said it is possible that Elvis is still alive; 16% of respondents to the second question said it is possible that Elvis is still alive.

Excellent discussions of these problems can be found in Groves et al. (2009) and Tourangeau et al. (2000). In some cases, accuracy can be increased by careful questionnaire design.

1.5
Questionnaire Design

This section gives a very brief introduction to writing and testing questions. It provides some general guidelines and examples, but if you are writing a questionnaire, you should consult one of the more comprehensive references on questionnaire design listed at the end of this chapter.

The most important step in writing a questionnaire is to decide what you want to find out. Write down the goals of your survey, and be precise. "I want to learn something about the homeless" won't do. Instead, you should write down specific questions, such as "What percentage of persons using homeless shelters in Chicago between January and March 1996 are under 16 years old?" Then, write or select questions that will elicit accurate answers to the research questions, and that will encourage persons in the sample to respond to the questions.

- *Always test your questions before taking the survey.* Ideally, the questions would be tested on a small sample of members of the target population. Try different versions for the questions, and ask respondents in your pretest how they interpret the questions.

  The NCVS was tested for several years before it was conducted on a national scale (Lehnen and Skogan, 1981). The pretests were used to help decide on a recall period (it was decided to ask respondents about victimizations that had occurred in the previous six months), test interviewing procedures and questions, and compare information from selected interviews with information found in the police report about the victimization. As a result of the pretests, some of the long and repetitive questions were shortened and more specific wording introduced.

The questionnaire was revised in 1985 and again in 1991 to make use of recent research in cognitive psychology and to include topics, such as victim and bystander behavior, that were not found in the earlier versions. All revisions are tested extensively in the field before being used (Taylor, 1989). In the past, for example, the NCVS has been criticized for underreporting the crime of rape; when the questionnaire was designed in the early 1970s, there was worry that asking about rape directly would be perceived as insensitive and embarrassing, and would provoke congressional outrage. The original NCVS questionnaire asked a series of specific questions intended to prompt the memory of respondents. These included questions such as "Did anyone take something directly from you by using force, such as by a stickup, mugging or threat?" The last question in the violent crime screening section of the questionnaire was "Did anyone try to attack you in some other way?" If the respondent mentioned in response that he or she was raped, then a rape was reported. Not surprisingly, the victimization rate for rape reported for the 1990 and earlier NCVS is very low: It is reported that about 1 per 1000 females aged 12 and older were raped in 1990. The current version of the NCVS questionnaire asks about rape directly.

You will not necessarily catch misinterpretations of questions by trying them out on friends or colleagues; your friends and colleagues may have backgrounds similar to yours, and may not have the same understanding of words as persons in your target population. Belson (1981) demonstrated that each of 29 questions about television viewing was misinterpreted by some respondents. The question "Do you think that the television news programmes are impartial about politics?" was tested on 56 people. Of these, 13 interpreted the question as intended, 18 respondents narrowed the term *news programmes* to mean "news bulletins," 21 narrowed it to "political programmes," and 1 interpreted it as "newspapers." Only 25 persons interpreted "impartial" as intended; 5 inferred the opposite meaning, "partial"; 11, as "giving too much or too little attention to"; and the others were simply unfamiliar with the word. Suessbrick et al. (2000) found that the concepts in a seemingly clear question such as "Have you smoked at least 100 cigarettes in your entire life?" were commonly interpreted in a different way than the authors intended: Some respondents included marijuana cigarettes or cigars, while others excluded cigarettes that were only partially smoked or hand-rolled cigarettes.

- *Keep it simple and clear.* Questions that seem clear to you may not be clear to someone listening to the whole question over the telephone, or to a person with a different native language. Belson (1981, p. 240) tested the question "What proportion of your evening viewing time do you spend watching news programmes?" on 53 people. Only 14 people correctly interpreted the word "proportion" as "percentage," "part," or "fraction." Others interpreted it as "how long do you watch" or "which news programmes do you watch."

- *Use specific questions instead of general ones, if possible.* Strunk and White advised writers to "Prefer the specific to the general, the definite to the vague, the concrete to the abstract" (1959, p. 15). Good questions result from good writing.

  Instead of asking "Did anyone attack you in the last six months," the NCVS asks a series of specific questions detailing how one might be attacked. The NCVS question is "Has anyone attacked or threatened you in any of these ways: (a) With any weapon, for instance, a gun or knife, (b) With anything like a baseball bat, frying pan, scissors, or stick ....."

■ *Relate your questions to the concept of interest.* This seems obvious but is forgotten or ignored in many surveys. In some disciplines, a standard set of questions has been developed and tested, and these are then used by subsequent researchers. Often, use of a common survey instrument allows results from different studies to be compared. In some cases, however, the standard questions are inappropriate for addressing the research hypotheses.

Pincus (1993) criticizes early research that concluded that persons with arthritis were more likely to have psychological problems than persons without arthritis. In those studies, persons with arthritis were given the Minnesota Multiphasic Personality Inventory, a test of 566 true/false questions commonly used in psychological research. Patients with rheumatoid arthritis tended to have high scores on the scales of hypochondriasis, depression, and hysteria. Part of the reason they scored highly on those scales is clear when the actual questions are examined. A person with arthritis can truthfully answer false to questions such as "I am about as able to work as I ever was," "I am in just as good physical health as most of my friends," and "I have few or no pains" without being either hysterical or a hypochondriac.

■ *Decide whether to use open or closed questions.* An **open question** allows respondents to form their own response categories; in a **closed question** (multiple choice), the respondent chooses from a set of categories read or displayed. Each has advantages. A closed question may prompt the respondent to remember responses that might otherwise be forgotten, and is in accordance with the principle that specific questions are better than general ones. If the subject matter has been thoroughly pretested and responses of interest are known, a well-written closed question will usually elicit more accurate responses, as in the NCVS question "Has anyone attacked or threatened you with anything like a baseball bat, frying pan, scissors, or stick?" If the survey is exploratory or questions are sensitive, though, it is often better to use an open question: Bradburn and Sudman (1979) note that respondents reported higher frequency of drinking alcoholic beverages when asked an open question than a closed question with categories "never" through "daily."

Schuman and Scott (1987) conclude that, depending on the context, either open or closed questions can limit the types of responses received. In one experiment, the most common responses to the open question "What do you think is the most important problem facing this country today?" were "unemployment" (17%) and "general economic problems" (17%). The closed version asked, "Which of the following do you think is the most important problem facing this country today—the energy shortage, the quality of public schools, legalized abortion, or pollution—or if you prefer, you may name a different problem as most important"; 32% or respondents chose "the quality of public schools." In this case, the limited options in the closed question guided respondents to one of the listed responses. In another experiment, Schuman and Scott (1987) asked respondents to name one or two of the most important national world events or changes during the last 50 years. Persons asked the open question most frequently gave responses such as World War II or the Vietnam War; they typically did not mention events such as the invention of the computer, which was the most prevalent response to the closed question including this option.

If using a closed question, always have an "other" category. In one study of sexual activity among adolescents, adolescents were asked from whom they felt the most pressure to have sex. Categories for the closed question were "friends of same sex," "boyfriend/girlfriend," "friends of opposite sex," "TV or radio," "don't feel pressure," and "other." The response "parents" or "father" was written in by a number of the adolescent respondents, a response that had not been anticipated by the researchers.

- *Report the actual question asked.* Public opinion is complex, and you inevitably leave a distorted impression of it when you compress the results of your careful research into a summary statement "*x%* of Americans favor affirmative action."

  The results of three surveys in Spring 1995, all purportedly about affirmative action, emphasize the importance of reporting the question. A *Newsweek* poll asked "Should there be special consideration for each of the following groups to increase their opportunities for getting into college and getting jobs or promotions?" and asked about these groups: blacks, women, Hispanics, Asians, and Native Americans. The poll found that 62% of blacks but only 25% of whites answered "yes" to the question about blacks. A *USA Today*–CNN–Gallup poll asked the question "What is your opinion on affirmative action programs for women and minorities: do you favor them or oppose them?" and reported that 55% of respondents favored such programs. A Harris poll asking "Would you favor or oppose a law limiting affirmative action programs in your state?" reported 51% of respondents favoring such a law. These questions are clearly addressing different concepts because the differences in percentages obtained are too great to be ascribed to the different samples of people taken by the three organizations. Yet all three polls' results were described in newspapers in terms of percentages of persons who support affirmative action.

- *Avoid questions that prompt or motivate the respondent to say what you would like to hear.* These are often called **leading,** or **loaded, questions.** The May 17, 1994 issue of *The Wall Street Journal* reported the following question asked by the Gallup Organization in a survey commissioned by the American Paper Institute: "It is estimated that disposable diapers account for less than 2 percent of the trash in today's landfills. In contrast, beverage containers, third-class mail and yard waste are estimated to account for about 21 percent of trash in landfills. Given this, in your opinion, would it be fair to tax or ban disposable diapers?"

- *Consider the social desirability of responses to questions, and write questions that elicit honest responses.* Abelson et al. (1992) review several studies that find many people say they voted in the last election when they actually did not vote. They argue that voting is a socially desirable behavior, and many respondents do not want to admit that they did not vote; respondents need to be prompted to report their actual behavior.

- *Avoid double negatives.* Double negatives needlessly confuse the respondent. A question such as "Do you favor or oppose not allowing drivers to use cell phones while driving?" might elicit either "favor" or "oppose" from a respondent who thinks persons should not use cell phones while driving.

- *Use forced-choice, rather than agree/disagree questions.* As noted earlier, some persons will agree with almost any statement. Schuman and Presser (1981, p. 223) report the following differences from an experiment comparing agree/disagree with forced-choice versions:

  Q1: Do you agree or disagree with this statement: Most men are better suited emotionally for politics than are most women.

  Q2: Would you say that most men are better suited emotionally for politics than are most women, that men and women are equally suited, or that women are better suited than men in this area?

	Years of schooling		
	0–11	12	13+
Q1: Percent "agree"	57	44	39
Q2: Percent "men better suited"	33	38	28

- *Ask only one concept per question.* In particular, avoid what are sometimes called **double-barreled** questions, so named because if one barrel of the shotgun does not get you, the other one will.

  The question "Do you agree with Bill Clinton's $50 billion bailout of Mexico?" appeared on a survey distributed by a member of the U.S. House of Representatives to his constituents. The question is really confusing two opinions of the respondent: the opinion of Bill Clinton, and the opinion of the Mexico policy. Disapproval of either one will lead to a "disagree" answer to the question. Note also the loaded content of the word *bailout,* which will almost certainly elicit more negative responses than the term *aid package* would.

- *Pay attention to question order effects.* If you ask more than one question on a topic, it is usually (but not always) better to ask the more general question first and follow it by the specific questions. McFarland (1981) conducted an experiment in which half of the respondents were given general questions (for example, "How interested would you say you are in religion: very interested, somewhat interested, or not very interested") first, followed by specific questions on the subject ("Did you, yourself, happen to attend church in the last seven days?"); the other half were asked the specific questions first and then asked the general questions. When the general question was asked first, 56% reported that they were "very interested in religion"; the percentage rose to 64% when the specific question was asked first.

  Serdula et al. (1995) found that in the years in which a respondent of a health survey was asked to report his or her weight and then immediately asked "Are you trying to lose weight?" 28.8% of men and 48.0% of women reported that they were trying to lose weight. When "Are you trying to lose weight?" was asked in the middle of the survey and the self-report question on weight at the end of the survey, 26.5% of the men and 40.9% of the women reported that they were trying to lose weight. The authors speculate that respondents who are reminded of their weight status may overreport trying to lose weight.

The 2000 U.S. Census had separate questions for race (with categories white, black, American Indian or Alaskan Native, Asian Indian, Chinese, and Filipino, among others) and ethnicity (with categories non-Hispanic, Mexican-American, Puerto Rican, Cuban, and other Hispanic). These are considered separate classifications; an individual can, for instance, be white Cuban or black non-Hispanic. The Census Bureau has done a great deal of experimental research to determine effects of alternate wordings and orderings of these questions on responses (Bates et al., 1995). Martin et al. (2005) report results of experiments comparing the questions used for the 1990 Census with those used for the 2000 Census. In 1990, race was question 4 and ethnicity was question 7; in 2000, ethnicity was question 7 and race was question 8. When the question on race occurred first, as in the 1990 Census, some Hispanic respondents looked for a Hispanic category, did not find it, and checked the "Other Race" category. After answering the race question, some persons skipped the ethnicity question so that there was substantial nonresponse on the ethnicity question. The reversed question order and other changes in the 2000 Census led to less missing data on both the race and ethnicity questions.

1.6
Sampling and Nonsampling Errors

Most opinion polls that you see report a "margin of error." Many merely say that the margin of error is 3 percentage points. Others give more detail, as in this excerpt from a *New York Times* poll: "In theory, in 19 cases out of 20 the results based on such samples will differ by no more than three percentage points in either direction from what would have been obtained by interviewing all Americans." The margin of error given in polls is an expression of **sampling error**, the error that results from taking one sample instead of examining the whole population. If we took a different sample, we would most likely obtain a different sample percentage of persons who visited the public library last week. Sampling errors are usually reported in probabilistic terms. We discuss the calculation of sampling errors for different survey designs in Chapters 2 through 7.

Selection bias and measurement error are examples of **nonsampling errors**, which are any errors that cannot be attributed to the sample-to-sample variability. In many surveys, the sampling error that is reported for the survey may be negligible compared to the nonsampling errors; you often see surveys with a 30% response rate proudly proclaiming their 3% margin of error, while ignoring the tremendous selection bias in their results.

The goal of this chapter was to sensitize you to various forms of selection bias and inaccurate responses. We can reduce some forms of selection bias by using probability sampling methods, as described in the next chapter. Accurate responses can often be achieved through careful design and testing of the survey instrument, thorough training of interviewers, and pretesting the survey. We shall return to nonsampling errors in Chapter 8, where we discuss methods that have been proposed for trying to reduce nonresponse error after the survey has been collected (sneak preview: none of the methods is as good as obtaining a high response rate to begin with), and Chapter 15, where we present a unified approach to survey design that attempts to minimize both sampling and nonsampling error.

Why sample at all?   With the abundance of poorly done surveys, it is not surprising that some people are skeptical of *all* surveys. "After all," some say, "my opinion has never been asked, so how can the survey results claim to represent me?" Public questioning of the validity of surveys intensifies after a survey makes a large mistake in predicting the results of an election, such as in the *Literary Digest* survey of 1936 or in the 1948 U.S. presidential election in which most pollsters predicted that Dewey would defeat Truman. A public backlash against survey research occurred again after the British general election of 1992, when the Conservative government won reelection despite the predictions from all but one of the major polling organizations that it would be a dead heat or that Labour would win. One member of Parliament expressed his opinion that "extrapolating what tens of millions are thinking from a tiny sample of opinions affronts human intelligence and negates true freedom of thought."

Some people insist that only a complete census, in which every element of the population is measured, will be satisfactory. This objection to sampling has a long history. When Anders Kiaer, director of Norwegian statistics, proposed using sampling for collecting official governmental statistics (Kiaer, 1897), his proposal was by no means universally well received. Opponents of sampling argued that sampling was dangerous, and that samples would never be able to replace a census. Within a few years, however, the international statistical community was largely persuaded that representative samples are a good thing, although probability samples were not widely used until the 1930s and 1940s.

For small populations, a census may of course be practical. If you want to know about the employment history of 2005 Arizona State University graduates who majored in mathematics, it would be sensible to try to contact all of them. If all of the graduates respond, then estimates from the survey will have no sampling error. The estimates will have nonsampling errors, however, if the questions are poor or if respondents give inaccurate information. If some of the graduates do not return the questionnaire, then the estimates will likely be biased because of nonresponse.

In general, taking a complete census of a population uses a great deal of time and money, and does not eliminate error. The biggest causes of error in a survey are often undercoverage, nonresponse, and sloppiness in data collection. Most of you have kept a paper or electronic checkbook register at some time, and essentially keep a census of all of the check and deposit amounts. How many of you can say that you have never made an error in your checkbook? It is usually much better to take a high-quality sample and allocate resources elsewhere, for instance, by being more careful in collecting or recording data, or doing follow-up studies, or measuring more variables.

After all, the *Literary Digest* poll discussed in Example 1.1 predicted the vote wrong even in some counties in which it attempted to take a census. The U.S. Decennial Census, which attempts to enumerate every resident of the country, misses segments of the population. Citro et al. (2004) document the coverage error in the 2000 Census, reporting that black males had greater undercoverage than other demographic groups.

There are three main justifications for using sampling:

■ Sampling can provide reliable information at far less cost than a census. With probability samples (described in the next chapter), you can quantify the sampling error from a survey. In some instances, an observation unit must be destroyed to be measured, as when a cookie must be pulverized to determine the fat content. In such a case, a sample provides reliable information about the population; a census destroys the population and, with it, the need for information about it.

■ Data can be collected more quickly, so estimates can be published in a timely fashion. An estimate of the unemployment rate for 2005 is not very helpful if it takes until 2015 to interview every household.

■ Finally, and less well known, estimates based on sample surveys are often more accurate than those based on a census because investigators can be more careful when collecting data. A complete census often requires a large administrative organization, and involves many persons in the data collection. With the administrative complexity and the pressure to produce timely estimates, many types of errors can be injected into the census. In a sample, more attention can be devoted to data quality through training personnel and following up on nonrespondents. It is far better to have good measurements on a representative sample than unreliable or biased measurements on the whole population.

Deming says, "Sampling is not mere substitution of a partial coverage for a total coverage. Sampling is the science and art of controlling and measuring the reliability of useful statistical information through the theory of probability" (1950, p. 2). In the remaining chapters of this book, we explore this science and art in detail.

Key Terms

Census: A survey in which the entire population is measured.

Coverage: The percentage of the population of interest that is included in the sampling frame.

Measurement error: The difference between the response coded and the true value of the characteristic being studied for a respondent.

Nonresponse: Failure of some units in the sample to provide responses to the survey.

Nonsampling error: An error from any source other than sampling error. Examples include nonresponse and measurement error.

Sampling error: Error in estimation due to taking a sample instead of measuring every unit in the population.

Sampling frame: A list, map, or other specification of units in the population from which a sample may be selected. Examples include a list of all university students, a telephone directory, or a map of geographic segments.

Selection bias: Bias that occurs because the actual probabilities with which units are sampled differ from the selection probabilities specified by the investigator.

For Further Reading

The American Statistical Association series "What is a Survey?" provides an introduction to survey sampling, with examples of many of the concepts discussed in Chapter 1. In particular, see the chapter "Judging the Quality of a Survey." This series is available on the American Statistical Association Survey Research Methods Section website at www.amstat.org/sections/srms/. The American Association of Public Opinion Research website, www.aapor.org, contains many resources for the sampling practitioner, including a guide to Standards and Best Practices.

The following three books are recommended for further reading about general issues for taking surveys. Groves et al. (2009) discuss statistical and nonstatistical issues in survey sampling, with examples from large-scale surveys. Biemer and Lyberg (2003) provide a thorough treatment of issues in survey quality. Dillman et al. (2009) give practical, research-supported guidance on everything from questionnaire design to choice of survey mode to timing of follow-up letters.

If you are interested in more information on questionnaire design and or on procedures for taking social surveys, start with the books by Presser et al. (2004), Fowler (1995), Converse and Presser (1986), Schuman and Presser (1981), and Sudman and Bradburn (1982). Much recent research has been done in the area of using results from cognitive psychology when writing questionnaires: Tanur (1992), Sudman et al. (1995), Schwarz and Sudman (1996), Tourangeau et al. (2000), and Bradburn (2004) are useful references on the topic. All are clearly written and list other references. In addition, many issues of the journal *Public Opinion Quarterly* have articles dealing with questionnaire design.

1.7
Exercises

A. Introductory Exercises

For each survey in Exercises 1–20, describe the target population, sampling frame, sampling unit, and observation unit. Discuss any possible sources of selection bias or inaccuracy of responses.

1 The article "What Readers Say about Marijuana" (*Parade*, July 31, 1994, p. 16) reported "More than 75% of the readers who took part in an informal PARADE telephone poll say marijuana should be as legal as alcoholic beverages." The telephone poll was announced on page 5 of the June 12 issue; readers were instructed to "Call 1-900-773-1200, at 75 cents a call, if you would like to answer the following questions. Use touch-tone phones only. To participate, call between 8 a.m. EDT [Eastern Daylight Time] on Saturday, June 11, and midnight EDT on Wednesday, June 15."

2 A student wants to estimate the percentage of mutual funds whose shares went up in price last week. She selects every tenth fund listing in the Mutual Fund pages of the newspaper, and calculates the percentage of those in which the share price increased.

3 Amazon books (www.amazon.com) summarizes reader reviews of the books it sells. Persons who want to review a book can submit a review online; Amazon then reports the average rating from all reader reviews on its website.

4   Potential jurors in some jurisdictions are chosen from a list of county residents who are registered voters or licensed drivers over age 18. In the fourth quarter of 1994, 100,300 jury summons were mailed to Maricopa County, Arizona, residents. Approximately 23,000 of those were returned from the post office as undeliverable. Approximately 7000 persons were unqualified for service because they were not citizens, were under 18, were convicted felons, or other reason that disqualified them from serving on a jury. An additional 22,000 were excused from jury service because of illness, financial hardship, military service, or other acceptable reason. The final sample consists of persons who appear for jury duty; some unexcused jurors fail to appear.

5   Many scholars and policy makers are interested in the proportion of homeless people who are mentally ill. Wright (1988) estimates that 33% of all homeless people are mentally ill, by sampling homeless persons who received medical attention from one of the clinics in the Health Care for the Homeless (HCH) project. He argues that selection bias is not a serious problem because the clinics were easily accessible to the homeless and because the demographic profiles of HCH clients were close to those of the general homeless population in each city in the sample. Do you agree?

6   Approximately 16,500 women returned the Healthy Women Survey that appeared in the September 1992 issue of *Prevention*. The May 1993 issue, reporting on the survey, stated that "Ninety-two percent of our readers rated their health as excellent, very good or good."

7   A survey is conducted to find the average weight of cows in a region. A list of all farms is available for the region, and 50 farms are selected at random. Then the weight of each cow at the 50 selected farms is recorded.

8   To study nutrient content of menus in boarding homes for the elderly in Washington State, Goren et al. (1993) mailed surveys to all 184 licensed homes in Washington State, directed to the administrator and food service manager. Of those, 43 were returned by the deadline and included menus.

9   Entries in the online encyclopedia *Wikipedia* can be written or edited by anyone with Internet access. This has given rise to concern about the accuracy of the information. Giles (2005) reports on a *Nature* study assessing the accuracy of *Wikipedia* science articles. Fifty subjects were chosen "on a broad range of scientific disciplines." For each subject, the entries from *Wikipedia* and *Encyclopaedia Brittanica* were sent to a relevant expert; 42 sets of usable reviews were returned. The editors of *Nature* then tallied the number of errors reported for each encyclopedia.

10   The December 2003 issue of *PC World* reported the results from a survey of over 32,000 subscribers asking about reliability and service for personal computers and other electronic equipment. The magazine "invited subscribers to take the Web-based survey from April 1 through June 30, 2003" and received 32,051 responses. Survey respondents were entered in a drawing to win prizes. They reported that 46% of desktop PCs had at least one significant malfunction.

11   Karras (2008) reports on a survey conducted by *SELF* magazine on prevalence of eating disorders in women. The survey, posted online at self.com, obtained responses from 4000 women. Based on these responses, the article reports that 27% of women in the survey "say they would be 'extremely upset' if they gained just 5 pounds"; it is estimated that 10% of women have eating disorders such as anorexia or bulimia.

12   Shen and Hsieh (1999) took a purposive sample of 29 higher education institutions; the institutions were "representative in terms of institutional type, geographic and demographic diversity, religious/nonreligious affiliation, and the public/private dimension" (p. 318). They then mailed the survey to 2042 faculty members in the institutions, of whom 1219 returned the survey.

13   The American Statistical Association sent the following e-mail with subject line "Joint Statistical Meetings 2005 Participants Survey" to a sample of persons who attended the 2005 Joint Statistical Meetings: "Thank you for attending the 2005 Joint Statistical Meetings (JSM) in Minneapolis, Minnesota. We need your help to complete an online survey about the JSM. Because the quality of the JSM is very important, a survey is being conducted to find out how we might improve future meetings. We would like to get your opinion about various aspects of the 2005 meeting your preferences for 2006 and beyond.

    You are part of a small sample of conference registrants who have been selected randomly to participate in the survey. We hope you will take the time to complete this short questionnaire online at www.amstat.org/meetings/jsm/2005/survey. In order to tabulate and analyze the data, please submit your response by mid-September 2005."

14   Fark and Johnson (1997) report on a survey of professors of education taken in summer of 1997 and conclude that there is a large disparity between the views of education professors and those of the general public. A sample of 5324 education professors was drawn from a population of about 34,000 education professors in colleges and universities across the country. A letter was mailed to each professor in the sample in May 1997, inviting him or her to participate and to provide a number where he or she could be reached during the summer for a telephone interview. During the summer, a total of 778 interviews were completed by telephone. An additional 122 interviews were obtained by calling professors in the sample at work in August and September. To attempt to minimize question order effects, the survey was pretested and some questions were asked in random order.

    Respondents were asked which in a series of qualities were "absolutely essential" to be imparted to prospective teachers: 84% of the respondents selected having teachers who are "life-long learners and constantly updating their skills"; 41%, having teachers "trained in pragmatic issues of running a classroom such as managing time and preparing lesson plans"; 19%, for teachers to "stress correct spelling, grammar, and punctuation"; and 12%, for teachers to "expect students to be neat, on time, and polite" (p. 30).

15   Kripke et al. (2002) claim that persons who sleep 8 or more hours per night have a higher mortality risk than persons who sleep 6 or 7 hours. They analyzed data from the 1982 Cancer Prevention Study II of the American Cancer Society, a national survey taken by about 1.1 million people. The survival or date of death was determined for about 98% of the sample six years later. Most of the respondents were friends and relatives of American Cancer Society volunteers; the purpose of the original survey was to explore factors associated with the development of cancer, but the survey also contained a few questions about sleep and insomnia.

16   In lawsuits about trademarks, a plaintiff claiming that another company is infringing on its trademarks must often show that the marks have a "secondary meaning" in the marketplace—that is, potential users of the product associate the trademarks with the plaintiff even when the company's name is missing. In the court case *Harlequin Enterprises Ltd v. Gulf & Western Corporation* (503 F. Supp. 647, 1980), the publisher of Harlequin Romances persuaded the court that the cover design for "Harlequin Presents" novels had acquired secondary meaning. Part of the evidence presented was a survey of 500 women from three cities who identified themselves as readers of romance fiction. They were shown copies of unpublished "Harlequin Presents" novels with the Harlequin name hidden; over 50% identified the novel as a Harlequin product.

17   Theoharakis and Skordia (2003) asked statisticians who responded to their survey to rank statistics journals in terms of prestige, importance, and usefulness. They gathered e-mail addresses for 12,053 statisticians from the online directories of statistical organizations and sent an e-mail invitation to each to participate in the online survey at www.alba.edu.gr/statsurvey/. A total of 2190 responses were obtained. The authors suggest that the results of their survey could help universities making promotion and tenure decisions about statistics faculty by providing information about the perceived quality of statistics journals.

18   Ann Landers (1976) asked readers of her column to respond to the question: "If you had it to do over again, would you have children?" About 70% of the readers who responded said "No." She received over 10,000 responses, 80% of those from women.

19   The August, 1996 issue of *Consumer Reports* contained satisfaction ratings for various health maintenance organizations used by readers of the magazine. Describing the survey, the editors say on p. 40, "The Ratings are based on more than 20,000 responses to our 1995 Annual Questionnaire about experiences in HMOs between May 1994 and April 1995. Those results reflect experiences of *Consumer Reports* subscribers, who are a more affluent and educated cross-section of the U.S. population." Answer the general questions about target population, sampling frame, and units for this survey. Also, do you think that this survey provides valuable information for comparing health plans? If you were selecting an HMO for yourself, which information would you rather have: results from this survey, or results from customer satisfaction surveys conducted by the individual HMOs?

20   Ebersole (2000) studied how students in selected public schools describe their use of the Internet. Five school districts in a Western state were selected to give a cross-section of urban and rural schools that have Internet access. A survey, administered electronically, was installed as the home page in middle and high school media centers in the districts "for a period of time to gather approximately 100 responses from each school." Students who had parental permission to access the Internet were permitted to access the computer-administered survey. Participation in the survey was voluntary.

21   The following questions, quoted in Kinsley (1981), were from a survey conducted by Cambridge Reports, Inc., and financed by Union Carbide Corporation. Critique these questions.

> Some people say that granting companies tax credits for the taxes they actually pay to foreign nations could increase these companies' international competitiveness. If you

knew for a fact that the tax credits for taxes paid to foreign countries would increase the money available to US companies to expand and modernize their plants and create more jobs, would you favor or oppose such a tax policy?

Do you favor or oppose changing environmental regulations so that while they still protect the public, they cost American businesses less and lower product costs?

22 Frankovic (2008) reported that in 1970, a poll conducted by the Harris organization for Virginia Slims, a brand of cigarettes marketed primarily to women, had the following question: "There won't be a woman president of the U.S. for a long time and that's probably just as well." Sixty-seven percent of female respondents agreed with the statement. Critique this question.

23 On March 21, 1993, NBC televised "The First National Referendum—Government Reform Presented by Ross Perot." During the show, 1992 U.S. presidential candidate Perot asked viewers to express their opinions by mailing in *The National Referendum on Government Reform*, printed in the March 20 issue of *TV Guide*. Some of the questions on the survey were the following:

Do you believe that for every dollar of tax increase there should be $2.00 in spending cuts with the savings earmarked for deficit and debt reduction?

Should the President present an overall plan including spending cuts, spending increases, and tax increases and present the net result of the overall plan, so that the people can know the net result before paying more taxes?

Should the electoral college be replaced with a popular vote for the Presidential election?

Was this TV forum worthwhile? Do you wish to continue participating as a voting member of United We Stand America?

Critique these questions.

D. Projects and Activities

24 Read the article by Roush (1996), which describes a proposal for using sampling in the 2000 U.S. Census. What are the main arguments for using sampling in 2000? Against? What do you think? You may also want to read Holden (2009), about issues in the 2010 Census.

25 (For students of U.S. history.) Eighty-five letters appeared in New York City newspapers in 1787 and 1788, with the purpose of drawing support in the state for the newly drafted Constitution. Collectively, these letters are known as *The Federalist*. Read Number 54 of *The Federalist*, in which the author (widely thought to be James Madison) discusses using a population census to apportion elected representatives and taxes among the states. This article explains part of Article I, Section II of the United States Constitution.

Write a short paper discussing Madison's view of a population census. What is the target population and sampling frame? What sources of bias does Madison mention, and how does he propose to reduce bias? What is your reaction to Madison's plan, from a statistical point of view? Where do you think Madison would stand today on the issue of using sampling versus complete enumeration to obtain population estimates?

26   Read the article by Horvitz et al. (1995) on Self-Selected Opinion Polls, which they term SLOPs. Find an example of a SLOP and explain why results from the poll should not be generalized to the population of interest.

27   Find a recent survey reported in a newspaper, academic journal, or popular magazine. Describe the survey. What are the target population and sampled population? What conclusions are drawn about the survey in the article? Do you think those conclusions are justified? What are possible sources of bias for the survey?

28   Find a survey on the Internet. For example, Survey.Net (www.survey.net) allows you to participate in surveys on a variety of subjects; you can find other surveys by searching online for "survey" or "take survey." Participate in one of the surveys yourself, and write a paragraph or two describing the survey and its results (most online surveys allow you to see the statistics from all the persons who have taken the survey). What are the target population and sampled population? What biases do you think might occur in the results?

29   Some polling organizations recruit volunteers for an Internet panel and then take samples from the panel to measure public opinion. Volunteer to be in an Internet panel for a survey organization (search for "online poll" to find one). What information are you asked to provide? Report how the organization produces estimates.

30   At about the same time Hite (1987) conducted the survey described in Section 1.1, a survey on similar topics was taken in the United Kingdom. Read the article by Wadsworth et al. (1993) describing the National Survey of Sexual Attitudes and Lifestyles. How do the authors describe potential biases and sources of error in the survey? Contrast the possible errors in this survey with those in Hite (1987).

2

Simple Probability Samples

[Kennedy] read every fiftieth letter of the thirty thousand coming weekly to the White House, as well as a statistical summary of the entire batch, but he knew that these were often as organized and unrepresentative as the pickets on Pennsylvania Avenue.

—Theodore Sorensen, *Kennedy*

The examples of bad surveys in Chapter 1—for example, the *Literary Digest* survey in Example 1.1—had major flaws that resulted in unrepresentative samples. In this chapter, we discuss how to use **probability sampling** to conduct surveys. In a probability sample, each unit in the population has a known probability of selection, and a random number table or other randomization mechanism is used to choose the specific units to be included in the sample. If a probability sampling design is implemented well, an investigator can use a relatively small sample to make inferences about an arbitrarily large population.

In Chapters 2 through 6, we explore survey design and properties of estimators for the three major design components used in a probability sample: simple random sampling, stratified sampling, and cluster sampling. We shall integrate all these ideas in Chapter 7, and show how they are combined in complex surveys such as the U.S. National Crime Victimization Survey. To simplify presentation of the concepts, we assume for now that the sampled population is the target population, that the sampling frame is complete, that there is no nonresponse or missing data, and that there is no measurement error. We return to nonsampling errors in Chapter 8.

As you might suppose, you need to know some probability to be able to understand probability sampling. You may want to review the material in Sections A.1 and A.2 of Appendix A while reading this chapter.

2.1
Types of Probability Samples

The terms *simple random sample*, *stratified sample*, *cluster sample*, and *systematic sample* are basic to any discussion of sample surveys, so let's define them now.

- A **simple random sample** (SRS) is the simplest form of probability sample. An SRS of size n is taken when every possible subset of n units in the population has the same chance of being the sample. SRSs are the focus of this chapter and the foundation for more complex sampling designs. In taking a random sample, the investigator is in effect mixing up the population before grabbing n units. The investigator does not need to examine every member of the population for the same reason that a medical technician does not need to drain you of blood to measure your red blood cell count: Your blood is sufficiently well mixed that any sample should be representative. SRSs are discussed in Section 2.3, after we present the basic framework for probability samples in Section 2.2.

- In a **stratified random sample**, the population is divided into subgroups called **strata**. Then an SRS is selected from each stratum, and the SRSs in the strata are selected independently. The strata are often subgroups of interest to the investigator—for example, the strata might be different regions of the country in a survey of people, different types of terrain in an ecological survey, or sizes of firms in a business survey. Elements in the same stratum often tend to be more similar than randomly selected elements from the whole population, so stratification often increases precision, as we shall see in Chapter 3.

- In a **cluster sample**, observation units in the population are aggregated into larger sampling units, called *clusters*. Suppose you want to survey Lutheran church members in Minneapolis but do not have a list of all church members in the city, so you cannot take an SRS of Lutheran church members. However, you do have a list of all the Lutheran churches. You can then take an SRS of the churches and then subsample all or some church members in the selected churches. In this case, the churches form the clusters, and the church members are the observation units. It is more convenient to sample at the church level; however, members of the same church may have more similarities than Lutherans selected at random in Minneapolis, so a cluster sample of 500 Lutherans may not provide as much information as an SRS of 500 Lutherans. We shall explore this idea further in Chapter 5.

- In a **systematic sample**, a starting point is chosen from a list of population members using a random number. That unit, and every kth unit thereafter, is chosen to be in the sample. A systematic sample thus consists of units that are equally spaced in the list. Systematic samples will be discussed in more detail in Sections 2.7 and 5.5.

Suppose you want to estimate the average amount of time that professors at your university say they spent grading homework in a specific week. To take an SRS, construct a list of all professors and randomly select n of them to be your sample. Now ask each professor in your sample how much time he or she spent grading homework that week—you would of course have to define the words *homework* and *grading* carefully in your questionnaire. In a stratified sample, you might classify faculty by college: engineering, liberal arts and sciences, business, nursing, and fine arts. You would then take an SRS of faculty in the engineering college, a separate SRS of faculty in liberal arts and sciences, and so on. For a cluster sample, you might randomly select 10 of the 60 academic departments in the university and ask

FIGURE **2.1**

Examples of a simple random sample, stratified random sample, cluster sample, and systematic sample of 20 integers from the population $\{1, 2, \ldots, 100\}$.

Simple random sample of 20 numbers from population of 100 numbers

Stratified random sample of 20 numbers from population of 100 numbers

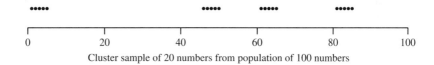

Cluster sample of 20 numbers from population of 100 numbers

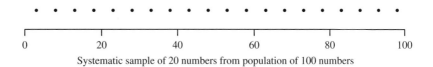

Systematic sample of 20 numbers from population of 100 numbers

each faculty member in those departments how much time he or she spent grading homework. A systematic sample could be chosen by selecting an integer at random between 1 and 20; if the random integer is 16, say, then you would include professors in positions 16, 36, 56, and so on, in the list.

EXAMPLE **2.1**   Figure 2.1 illustrates the differences among simple random, stratified, cluster, and systematic sampling for selecting a sample of 20 integers from the population $\{1, 2, \ldots, 100\}$. For the stratified sample, the population was divided into the 10 strata $\{1, 2, \ldots, 10\}$, $\{11, 12, \ldots, 20\}, \ldots, \{91, 92, \ldots, 100\}$, and an SRS of 2 numbers was drawn from each of the 10 strata. This ensures that each stratum is represented in the sample. For the cluster sample, the population was divided into 20 clusters $\{1, 2, 3, 4, 5\}$, $\{6, 7, 8, 9, 10\}, \ldots, \{96, 97, 98, 99, 100\}$; an SRS of 4 of these clusters was selected. For the systematic sample, the random starting point was 3, so the sample contains units 3, 8, 13, 18, and so on. ■

All of these methods—simple random sampling, stratified random sampling, cluster sampling, and systematic sampling—involve random selection of units to be in the sample. In an SRS, the observation units themselves are selected at random from the population of observation units; in a stratified random sample, observation units within each stratum are randomly selected; in a cluster sample, the clusters are randomly selected from the population of all clusters. Each method is a form of probability sampling, which we discuss in the next section.

2.2
Framework for Probability Sampling

To show how probability sampling works, we need to be able to list the N units in the finite population. The finite **population,** or **universe,** of N units is denoted by the index set

$$\mathcal{U} = \{1, 2, \ldots, N\}. \tag{2.1}$$

Out of this population we can choose various samples, which are subsets of $\mathcal{U}$. The particular sample chosen is denoted by $\mathcal{S}$, a subset consisting of n of the units in $\mathcal{U}$.

Suppose the population has four units: $\mathcal{U} = \{1, 2, 3, 4\}$. Six different samples of size 2 could be chosen from this population:

$$\begin{aligned}
\mathcal{S}_1 &= \{1, 2\} & \mathcal{S}_4 &= \{2, 3\} \\
\mathcal{S}_2 &= \{1, 3\} & \mathcal{S}_5 &= \{2, 4\} \\
\mathcal{S}_3 &= \{1, 4\} & \mathcal{S}_6 &= \{3, 4\}
\end{aligned}$$

In probability sampling, each possible sample $\mathcal{S}$ from the population has a known probability $P(\mathcal{S})$ of being chosen, and the probabilities of the possible samples sum to 1. One possible sample design for a probability sample of size 2 would have $P(\mathcal{S}_1) = 1/3$, $P(\mathcal{S}_2) = 1/6$, and $P(\mathcal{S}_6) = 1/2$, and $P(\mathcal{S}_3) = P(\mathcal{S}_4) = P(\mathcal{S}_5) = 0$. The probabilities $P(\mathcal{S}_1)$, $P(\mathcal{S}_2)$, and $P(\mathcal{S}_6)$ of the possible samples are known before the sample is drawn. One way to select the sample would be to place six labeled balls in a box; two of the balls are labeled 1, one is labeled 2, and three are labeled 6. Now choose one ball at random; if a ball labeled 6 is chosen, then $\mathcal{S}_6$ is the sample.

In a probability sample, since each possible sample has a known probability of being the chosen sample, each unit in the population has a known probability of appearing in our selected sample. We calculate

$$\pi_i = P(\text{unit } i \text{ in sample}) \tag{2.2}$$

by summing the probabilities of all possible samples that contain unit i. In probability sampling, the π_i are known before the survey commences, and we assume that $\pi_i > 0$ for every unit in the population. For the sample design described above, $\pi_1 = P(\mathcal{S}_1) + P(\mathcal{S}_2) + P(\mathcal{S}_3) = 1/2$, $\pi_2 = P(\mathcal{S}_1) + P(\mathcal{S}_4) + P(\mathcal{S}_5) = 1/3$, $\pi_3 = P(\mathcal{S}_2) + P(\mathcal{S}_4) + P(\mathcal{S}_6) = 2/3$, and $\pi_4 = P(\mathcal{S}_3) + P(\mathcal{S}_5) + P(\mathcal{S}_6) = 1/2$.

Of course, we never write all possible samples down and calculate the probability with which we would choose every possible sample—this would take far too long. But such enumeration underlies all of probability sampling. Investigators using a probability sample have much less discretion about which units are included in the sample, so using probability samples helps us avoid some of the selection biases described in Chapter 1. In a probability sample, the interviewer cannot choose to substitute a friendly looking person for the grumpy person selected to be in the sample by the random selection method. A forester taking a probability sample of trees cannot simply measure the trees near the road but must measure the trees designated for inclusion in the sample. Taking a probability sample is much harder than taking a convenience sample, but a probability sampling procedure guarantees that each unit in the population could appear in the sample and provides information that can be used to assess the precision of statistics calculated from the sample.

Within the framework of probability sampling, we can quantify how likely it is that our sample is a "good" one. A single probability sample is not guaranteed to be representative of the population with regard to the characteristics of interest, but we can quantify how often samples will meet some criterion of representativeness. The notion is the same as that of confidence intervals: We do not know whether the particular 95% confidence interval we construct for the mean contains the true value of the mean. We do know, however, that if the assumptions for the confidence interval procedure are valid and if we repeat the procedure over and over again, we can expect 95% of the resulting confidence intervals to contain the true value of the mean.

Let y_i be a characteristic associated with the ith unit in the population. We consider y_i to be a fixed quantity; if Farm 723 is included in the sample, then the amount of corn produced on Farm 723, y_{723}, is known exactly.

E X A M P L E 2.2 To illustrate these concepts, let's look at an artificial situation in which we know the value of y_i for each of the $N = 8$ units in the whole population. The index set for the population is

$$\mathcal{U} = \{1, 2, 3, 4, 5, 6, 7, 8\}.$$

The values of y_i are

i	1	2	3	4	5	6	7	8
y_i	1	2	4	4	7	7	7	8

There are 70 possible samples of size 4 that may be drawn without replacement from this population; the samples are listed in file samples.dat on the website. If the sample consisting of units $\{1, 2, 3, 4\}$ were chosen, the corresponding values of y_i would be 1, 2, 4, and 4. The values of y_i for the sample $\{2, 3, 6, 7\}$ are 2, 4, 7, and 7. Define $P(\mathcal{S}) = 1/70$ for each distinct subset of size four from $\mathcal{U}$. As you will see after you read Section 2.3, this design is an SRS without replacement. Each unit is in exactly 35 of the possible samples, so $\pi_i = 1/2$ for $i = 1, 2, \ldots, 8$.

A random mechanism is used to select one of the 70 possible samples. One possible mechanism for this example, because we have listed all possible samples, is to generate a random number between 1 and 70 and select the corresponding sample. With large populations, however, the number of samples is so great that it is impractical to list all possible samples—instead, another method is used to select the sample. Methods that will give an SRS will be described in Section 2.3. ∎

Most results in sampling rely on the **sampling distribution** of a statistic, the distribution of different values of the statistic obtained by the process of taking all possible samples from the population. A sampling distribution is an example of a discrete probability distribution.

Suppose we want to use a sample to estimate a population quantity, say the population total $t = \sum_{i=1}^{N} y_i$. One estimator we might use for t is $\hat{t}_S = N\bar{y}_S$, where $\bar{y}_S$ is the average of the y_i's in S, the chosen sample. In our example, $t = 40$. If the sample S consists of units 1, 3, 5, and 6, then $\hat{t}_S = 8 \times (1 + 4 + 7 + 7)/4 = 38$. Since we know the whole population here, we can find $\hat{t}_S$ for each of the 70 possible samples. The probabilities of selection for the samples give the sampling distribution of $\hat{t}$:

$$P\{\hat{t} = k\} = \sum_{S:\hat{t}_S = k} P(S).$$

The summation is over all samples S for which $\hat{t}_S = k$. We know the probability $P(S)$ with which we select a sample S because we take a probability sample.

EXAMPLE **2.3**   The sampling distribution of $\hat{t}$ for the population and sampling design in Example 2.2 derives entirely from the probabilities of selection for the various samples. Four samples ({3,4,5,6}, {3,4,5,7}, {3,4,6,7}, and {1,5,6,7}) result in the estimate $\hat{t} = 44$, so $P\{\hat{t} = 44\} = 4/70$. For this example, we can write out the sampling distribution of $\hat{t}$ because we know the values for the entire population.

k	22	28	30	32	34	36	38	40	42	44	46	48	50	52	58
$P\{\hat{t}=k\}$	$\frac{1}{70}$	$\frac{6}{70}$	$\frac{2}{70}$	$\frac{3}{70}$	$\frac{7}{70}$	$\frac{4}{70}$	$\frac{6}{70}$	$\frac{12}{70}$	$\frac{6}{70}$	$\frac{4}{70}$	$\frac{7}{70}$	$\frac{3}{70}$	$\frac{2}{70}$	$\frac{6}{70}$	$\frac{1}{70}$

Figure 2.2 displays the sampling distribution. ∎

The **expected value** of $\hat{t}$, $E[\hat{t}]$, is the mean of the sampling distribution of $\hat{t}$:

$$E[\hat{t}] = \sum_{S} \hat{t}_S P(S) \tag{2.3}$$

$$= \sum_{k} k P(\hat{t} = k).$$

FIGURE   **2.2**

Sampling distribution of the sample total in Example 2.3.

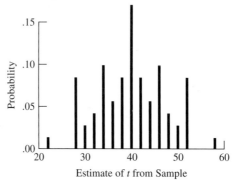

The expected value of the statistic is the weighted average of the possible sample values of the statistic, weighted by the probability that particular value of the statistic would occur.

The **estimation bias** of the estimator $\hat{t}$ is

$$\text{Bias}\,[\hat{t}] = E[\hat{t}] - t. \tag{2.4}$$

If Bias$[\hat{t}] = 0$, we say that the estimator $\hat{t}$ is **unbiased** for t. For the data in Example 2.2 the expected value of $\hat{t}$ is

$$E[\hat{t}] = \frac{1}{70}(22) + \frac{6}{70}(28) + \cdots + \frac{1}{70}(58) = 40.$$

Thus, the estimator is unbiased.

Note that the mathematical definition of bias in (2.4) is *not* the same thing as the selection or measurement bias described in Chapter 1. All indicate a systematic deviation from the population value, but from different sources. Selection bias is due to the method of selecting the sample—often, the investigator acts as though every possible sample S has the same probability of being selected, but some subsets of the population actually have a different probability of selection. With undercoverage, for example, the probability of including a unit not in the sampling frame is zero. Measurement bias means that the y_i's are not really the quantities of interest, so although $\hat{t}$ may be unbiased in the sense of (2.4) for $t = \sum_{i=1}^{N} y_i$, t itself would not be the true total of interest. Estimation bias means that the estimator chosen results in bias—for example, if we used $\hat{t}_S = \sum_{i \in S} y_i$ and did not take a census, $\hat{t}$ would be biased. To illustrate these distinctions, suppose you wanted to estimate the average height of male actors belonging to the Screen Actors Guild. Selection bias would occur if you took a convenience sample of actors on the set—perhaps taller actors are more or less likely to be working. Measurement bias would occur if your tape measure inaccurately added 3 centimeters (cm) to each actor's height. Estimation bias would occur if you took an SRS from the list of all actors in the Guild, but estimated mean height by the average height of the six shortest men in the sample—the sampling procedure is good, but the estimator is bad.

The **variance** of the sampling distribution of $\hat{t}$ is

$$V(\hat{t}) = E[(\hat{t} - E[\hat{t}])^2] = \sum_{\substack{\text{all possible} \\ \text{samples } S}} P(S)\,[\hat{t}_S - E(\hat{t})]^2. \tag{2.5}$$

For the data in Example 2.2,

$$V(\hat{t}) = \frac{1}{70}(22 - 40)^2 + \cdots + \frac{1}{70}(58 - 40)^2 = \frac{3840}{70} = 54.86.$$

Because we sometimes use biased estimators, we often use the **mean squared error** (MSE) rather than variance to measure the accuracy of an estimator.

$$
\begin{aligned}
\text{MSE}[\hat{t}] &= E[(\hat{t} - t)^2] \\
&= E[(\hat{t} - E[\hat{t}] + E[\hat{t}] - t)^2] \\
&= E[(\hat{t} - E[\hat{t}])^2] + (E[\hat{t}] - t)^2 + 2\,E[(\hat{t} - E[\hat{t}])(E[\hat{t}] - t)] \\
&= V(\hat{t}) + \left[\text{Bias}(\hat{t})\right]^2.
\end{aligned}
$$

FIGURE   **2.3**

Unbiased, precise, and accurate archers. Archer A is unbiased—the average position of all arrows is at the bull's-eye. Archer B is precise but not unbiased—all arrows are close together but systematically away from the bull's-eye. Archer C is accurate—all arrows are close together and near the center of the target.

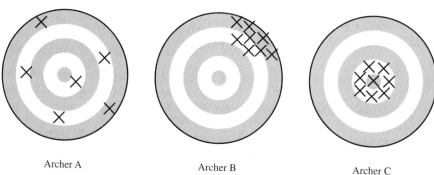

Archer A   Archer B   Archer C

Thus, an estimator $\hat{t}$ of t is **unbiased** if $E(\hat{t}) = t$, **precise** if $V(\hat{t}) = E[(\hat{t} - E[\hat{t}])^2]$ is small, and **accurate** if $\mathrm{MSE}[\hat{t}] = E[(\hat{t} - t)^2]$ is small. A badly biased estimator may be precise but it will not be accurate; accuracy (MSE) is how close the estimate is to the true value, while precision (variance) measures how close estimates from different samples are to each other. Figure 2.3 illustrates these concepts.

In summary, the finite population $\mathcal{U}$ consists of units $\{1, 2, \ldots, N\}$ whose measured values are $\{y_1, y_2, \ldots, y_N\}$. We select a sample $\mathcal{S}$ of n units from $\mathcal{U}$ using the probabilities of selection that define the sampling design. The y_i's are fixed but unknown quantities—unknown unless that unit happens to appear in our sample $\mathcal{S}$. Unless we make additional assumptions, the only information we have about the set of y_i's in the population is in the set $\{y_i : i \in \mathcal{S}\}$.

You may be interested in many different population quantities from your population. Historically, however, the main impetus for developing theory for sample surveys has been estimating population means and totals. Suppose we want to estimate the total number of persons in Canada who have diabetes, or the average number of oranges produced per orange tree. The population total is

$$t = \sum_{i=1}^{N} y_i,$$

and the mean of the population is

$$\bar{y}_U = \frac{1}{N} \sum_{i=1}^{N} y_i.$$

Almost all populations exhibit some variability; for example, households have different incomes and trees have different diameters. Define the **variance** of the population values about the mean as

$$S^2 = \frac{1}{N-1} \sum_{i=1}^{N} (y_i - \bar{y}_U)^2. \qquad (2.6)$$

The population **standard deviation** is $S = \sqrt{S^2}$.

It is sometimes helpful to have a special notation for proportions. The proportion of units having a characteristic is simply a special case of the mean, obtained by letting $y_i = 1$ if unit i has the characteristic of interest, and $y_i = 0$ if unit i does not have the characteristic. Let

$$p = \frac{\text{number of units with the characteristic in the population}}{N}.$$

EXAMPLE **2.4** For the population in Example 2.2, let

$$y_i = \begin{cases} 1 \text{ if unit } i \text{ has the value 7} \\ 0 \text{ if unit } i \text{ does not have the value 7} \end{cases}$$

Let $\hat{p}_S = \sum_{i \in S} y_i / 4$, the proportion of 7s in the sample. The list of all possible samples in the data file samples.dat has 5 samples with no 7s, 30 samples with exactly one 7, 30 samples with exactly two 7s, and 5 samples with three 7s. Since one of the possible samples is selected with probability 1/70, the sampling distribution of $\hat{p}$ is[1]:

k	0	$\dfrac{1}{4}$	$\dfrac{1}{2}$	$\dfrac{3}{4}$
$P\{\hat{p} = k\}$	$\dfrac{5}{70}$	$\dfrac{30}{70}$	$\dfrac{30}{70}$	$\dfrac{5}{70}$

∎

2.3
Simple Random Sampling

Simple random sampling is the most basic form of probability sampling, and provides the theoretical basis for the more complicated forms. There are two ways of taking a simple random sample: with replacement, in which the same unit may be included more than once in the sample, and without replacement, in which all units in the sample are distinct.

A **simple random sample with replacement** (SRSWR) of size n from a population of N units can be thought of as drawing n independent samples of size 1. One unit is randomly selected from the population to be the first sampled unit, with probability $1/N$. Then the sampled unit is replaced in the population, and a second unit is randomly selected with probability $1/N$. This procedure is repeated until the sample has n units, which may include duplicates from the population.

In finite population sampling, however, sampling the same person twice provides no additional information. We usually prefer to sample without replacement, so that the sample contains no duplicates. A **simple random sample without replacement** (SRS) of size n is selected so that every possible subset of n distinct units in the population has the same probability of being selected as the sample. There are $\binom{N}{n}$ possible samples (see Appendix A), and each is equally likely, so the probability of

---

[1]An alternative derivation of the sampling distribution is in Exercise A.2 in Appendix A.

selecting any individual sample S of n units is

$$P(S) = \frac{1}{\binom{N}{n}} = \frac{n!(N-n)!}{N!}. \qquad (2.7)$$

As a consequence of this definition, the probability that the ith unit appears in the sample is $\pi_i = n/N$, as shown in Section 2.8.

To take an SRS, you need a list of all observation units in the population; this list is the sampling frame. In an SRS, the sampling unit and observation unit coincide. Each unit is assigned a number, and a sample is selected so that each possible sample of size n has the same chance of being the sample actually selected. This can be thought of as drawing numbers out of a hat; in practice, computer-generated pseudo-random numbers are usually used to select a sample.

One method for selecting an SRS of size n from a population of size N is to generate N random numbers between 0 and 1, then select the units corresponding to the n smallest random numbers to be the sample. For example, if $N = 10$ and $n = 4$, we generate 10 numbers between 0 and 1:

unit i	1	2	3	4	5	6	7	8	9	10
random number	0.837	0.636	0.465	0.609	0.154	0.766	0.821	0.713	0.987	0.469

The smallest 4 of the random numbers are 0.154, 0.465, 0.469 and 0.609, leading to the sample with units {3, 4, 5, 10}. Other methods that might be used to select an SRS are described in Example 2.5 and Exercises 21 and 29. Several survey software packages will select an SRS from a list of N units; the file srsselect.sas on the website gives code for selecting an SRS using SAS PROC SURVEYSELECT.

EXAMPLE 2.5   The U.S. government conducts a Census of Agriculture every five years, collecting data on all farms (defined as any place from which $1000 or more of agricultural products were produced and sold) in the 50 states.[2] The Census of Agriculture provides data on number of farms, the total acreage devoted to farms, farm size, yield of different crops, and a wide variety of other agricultural measures for each of the $N = 3078$ counties and county-equivalents in the United States. The file agpop.dat contains the 1982, 1987, and 1992 information on the number of farms, acreage devoted to farms, number of farms with fewer than 9 acres, and number of farms with more than 1000 acres for the population.

To take an SRS of size 300 from this population, I generated 300 random numbers between 0 and 1 on the computer, multiplied each by 3078, and rounded the result up to the next highest integer. This procedure generates an SRSWR. If the population is large relative to the sample, it is likely that each unit in the sample only occurs once in the list. In this case, however, 13 of the 300 numbers were duplicates. The duplicates were discarded, and replaced with new randomly generated numbers between 1 and

---

[2]The Census of Agriculture was formerly conducted by the U.S. Census Bureau; it is currently conducted by the U.S. National Agricultural Statistics Service (NASS). More information about the census and selected data are available on the web through the NASS material on www.fedstats.gov; also see www.agcensus.usda.gov.

FIGURE   **2.4**

Histogram: number of acres devoted to farms in 1992, for an SRS of 300 counties. Note the skewness of the data. Most of the counties have fewer than 500,000 acres in farms; some counties, however, have more than 1.5 million acres in farms.

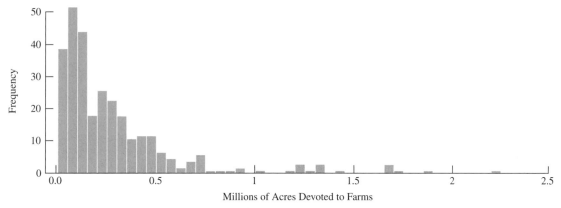

3078 until all 300 numbers were distinct; the set of random numbers generated is in file selectrs.dat, and the data set for the SRS is in agsrs.dat.

The counties selected to be in the sample may not "feel" very random at first glance. For example, counties 2840, 2841, and 2842 are all in the sample while none of the counties between 2740 and 2787 appear. The sample contains 18% of Virginia counties, but no counties in Alaska, Arizona, Connecticut, Delaware, Hawaii, Rhode Island, Utah, or Wyoming. There is a quite natural temptation to want to "adjust" the random number list, to spread it out a bit more. If you want a random sample, you must resist this temptation. Research, beginning with Neyman (1934), has repeatedly demonstrated that purposive samples often do not represent the population on key variables. If you deliberately substitute other counties for those in the randomly generated sample, you may be able match the population on one particular characteristic such as geographic distribution; however, you will likely fail to match the population on characteristics of interests such as number of farms or average farm size. If you want to ensure that all states are represented, do not adjust your randomly selected sample purposively but take a stratified sample (to be discussed in Chapter 3).

Let's look at the variable *acres92,* the number of acres devoted to farms in 1992. A small number of counties in the population are missing that value—in some cases, the data are withheld to prevent disclosing data on individual farms. Thus we first check to see the extent of the missing data in our sample. Fortunately, our sample has no missing data (Exercise 23 tells how likely such an occurrence is). Figure 2.4 displays a histogram of the acreage devoted to farms in each of the 300 counties. ∎

For estimating the population mean $\bar{y}_U$ from an SRS, we use the sample mean

$$\bar{y}_S = \frac{1}{n} \sum_{i \in S} y_i. \tag{2.8}$$

In the following, we use $\bar{y}$ to refer to the sample mean and drop the subscript $\mathcal{S}$ unless it is needed for clarity. As will be shown in Section 2.8, $\bar{y}$ is an unbiased estimator of the population mean $\bar{y}_U$, and the variance of $\bar{y}$ is

$$V(\bar{y}) = \frac{S^2}{n}\left(1 - \frac{n}{N}\right) \tag{2.9}$$

for S^2 defined in (2.6). The variance $V(\bar{y})$ measures the variability among estimates of $\bar{y}_U$ from different samples.

The factor $(1 - n/N)$ is called the **finite population correction** (fpc). Intuitively, we make this correction because with small populations the greater our **sampling fraction** n/N, the more information we have about the population and thus the smaller the variance. If $N = 10$ and we sample all 10 observations, we would expect the variance of $\bar{y}$ to be 0 (which it is). If $N = 10$, there is only one possible sample $\mathcal{S}$ of size 10 without replacement, with $\bar{y}_\mathcal{S} = \bar{y}_U$, so there is no variability due to taking a sample. For a census, the fpc, and hence $V(\bar{y})$, is 0. When the sampling fraction n/N is large in an SRS without replacement, the sample is closer to a census, which has no sampling variability.

For most samples that are taken from extremely large populations, the fpc is approximately 1. For large populations it is the size of the sample taken, not the percentage of the population sampled, that determines the precision of the estimator: If your soup is well stirred, you need to taste only one or two spoonfuls to check the seasoning, whether you have made 1 liter or 20 liters of soup. A sample of size 100 from a population of 100,000 units has almost the same precision as a sample of size 100 from a population of 100 million units:

$$V[\bar{y}] = \frac{S^2}{100}\frac{99,900}{100,000} = \frac{S^2}{100}(0.999) \qquad \text{for } N = 100,000$$

$$V[\bar{y}] = \frac{S^2}{100}\frac{99,999,900}{100,000,000} = \frac{S^2}{100}(0.999999) \qquad \text{for } N = 100,000,000$$

The population variance S^2, which depends on the values for the entire population, is in general unknown. We estimate it by the sample variance:

$$s^2 = \frac{1}{n-1}\sum_{i \in \mathcal{S}}(y_i - \bar{y})^2. \tag{2.10}$$

An unbiased estimator of the variance of $\bar{y}$ is (see Section 2.8)

$$\hat{V}(\bar{y}) = \left(1 - \frac{n}{N}\right)\frac{s^2}{n}. \tag{2.11}$$

The **standard error** (SE) is the square root of the estimated variance of $\bar{y}$:

$$\text{SE}(\bar{y}) = \sqrt{\left(1 - \frac{n}{N}\right)\frac{s^2}{n}}. \tag{2.12}$$

The population standard deviation is often related to the mean. A population of trees might have a mean height of 10 meters (m) and standard deviation of one m. A population of small cacti, however, with a mean height of 10 cm, might have a standard deviation of 1 cm. The **coefficient of variation** (CV) of the estimator $\bar{y}$ is a measure of relative variability, which may be defined when $\bar{y}_U \neq 0$ as:

$$CV(\bar{y}) = \frac{\sqrt{V(\bar{y})}}{E(\bar{y})} = \sqrt{1 - \frac{n}{N}} \frac{S}{\sqrt{n}\bar{y}_U}. \tag{2.13}$$

If tree height is measured in meters, then $\bar{y}_U$ and S are also in meters. The CV does not depend on the unit of measurement: In this example, the trees and the cacti have the same CV. We can estimate the CV of an estimator using the standard error divided by the mean (defined only when the mean is nonzero): In an SRS,

$$\widehat{CV}(\bar{y}) = \frac{\text{SE}[\bar{y}]}{\bar{y}} = \sqrt{1 - \frac{n}{N}} \frac{s}{\sqrt{n}\bar{y}}. \tag{2.14}$$

The estimated CV is thus the standard error expressed as a percentage of the mean. All these results apply to the estimation of a population total, t, since

$$t = \sum_{i=1}^{N} y_i = N\bar{y}_U.$$

To estimate t, we use the unbiased estimator

$$\hat{t} = N\bar{y}. \tag{2.15}$$

Then, from (2.9),

$$V(\hat{t}) = N^2 V(\bar{y}) = N^2 \left(1 - \frac{n}{N}\right) \frac{S^2}{n} \tag{2.16}$$

and

$$\hat{V}(\hat{t}) = N^2 \left(1 - \frac{n}{N}\right) \frac{s^2}{n}. \tag{2.17}$$

Note that $CV(\hat{t}) = \sqrt{V(\hat{t})}/E(\hat{t})$ is the same as $CV(\bar{y})$.

EXAMPLE 2.6 For the data in Example 2.5, $N = 3078$ and $n = 300$, so the sampling fraction is $300/3078 = 0.097$. The sample statistics are $\bar{y} = 297,897$, $s = 344,551.9$, and $\hat{t} = N\bar{y} = 916,927,110$. Standard errors are

$$\text{SE}[\bar{y}] = \sqrt{\frac{s^2}{n}\left(1 - \frac{300}{3078}\right)} = 18,898.434428$$

and

$$\text{SE}[\hat{t}] = (3078)(18,898.434428) = 58,169,381,$$

and the estimated coefficient of variation is

$$\widehat{CV}\,[\hat{t}] = \widehat{CV}\,[\bar{y}]$$

$$= \frac{SE\,[\bar{y}]}{\bar{y}}$$

$$= \frac{18{,}898.434428}{297{,}897}$$

$$= 0.06344.$$

Since these data are so highly skewed, we should also report the median number of farm acres in a county, which is 196,717. ∎

We might also want to estimate the proportion of counties in Example 2.5 with fewer than 200,000 acres in farms. Since estimating a proportion is a special case of estimating a mean, the results in (2.8)–(2.14) hold for proportions as well, and they take a simple form. Suppose we want to estimate the proportion of units in the population that have some characteristic—call this proportion p. Define y_i to be 1 if the unit has the characteristic and to be 0 if the unit does not have that characteristic. Then $p = \sum_{i=1}^{N} y_i/N = \bar{y}_U$, and p is estimated by $\hat{p} = \bar{y}$. Consequently, $\hat{p}$ is an unbiased estimator of p. For the response y_i, taking on values 0 or 1,

$$S^2 = \frac{\sum_{i=1}^{N}(y_i - p)^2}{N-1} = \frac{\sum_{i=1}^{N} y_i^2 - 2p\sum_{i=1}^{N} y_i + Np^2}{N-1} = \frac{N}{N-1}p(1-p).$$

Thus, (2.9) implies that

$$V(\hat{p}) = \left(\frac{N-n}{N-1}\right)\frac{p(1-p)}{n}. \qquad (2.18)$$

Also,

$$s^2 = \frac{1}{n-1}\sum_{i \in \mathcal{S}}(y_i - \hat{p})^2 = \frac{n}{n-1}\hat{p}(1-\hat{p}),$$

so from (2.11),

$$\hat{V}(\hat{p}) = \left(1 - \frac{n}{N}\right)\frac{\hat{p}(1-\hat{p})}{n-1}. \qquad (2.19)$$

EXAMPLE 2.7 For the sample described in Example 2.5, the estimated proportion of counties with fewer than 200,000 acres in farms is

$$\hat{p} = \frac{153}{300} = 0.51$$

with standard error

$$SE(\hat{p}) = \sqrt{\left(1 - \frac{300}{3078}\right)\frac{(0.51)(0.49)}{299}} = 0.0275. \quad ∎$$

Note: In an SRS, $\pi_i = n/N$ for all units $i = 1, \ldots, N$. However, many other probability sampling designs also have $\pi_i = n/N$ for all units but are not SRSs. To have an SRS, it is not sufficient for every individual to have the same probability of being in the sample; in addition, every possible *sample* of size n must have the same probability $1/\binom{N}{n}$ of being the sample selected, as defined in (2.7). The cluster sampling design in Example 2.1, in which the population of 100 integers is divided into 20 clusters $\{1, 2, 3, 4, 5\}, \{6, 7, 8, 9, 10\}, \ldots, \{96, 97, 98, 99, 100\}$ and an SRS of 4 of these clusters selected, has $\pi_i = 20/100$ for each unit in the population but is *not* an SRS of size 20 because different possible samples of size 20 have different probabilities of being selected. To see this, let's look at two particular subsets of $\{1, 2, \ldots, 100\}$. Let $\mathcal{S}_1$ be the cluster sample depicted in the third panel of Figure 2.1, with

$$\mathcal{S}_1 = \{1, 2, 3, 4, 5, 46, 47, 48, 49, 50, 61, 62, 63, 64, 65, 81, 82, 83, 84, 85\},$$

and let

$$\mathcal{S}_2 = \{1, 6, 11, 16, 21, 26, 31, 36, 41, 46, 51, 56, 61, 66, 71, 76, 81, 86, 91, 96\}.$$

The cluster sampling design specifies taking an SRS of 4 of the 20 clusters, so $P(\mathcal{S}_1) = 1/\binom{20}{4} = 4!(20-4)!/20! = 1/4845$. Sample $\mathcal{S}_2$ cannot occur under this design, however, so $P(\mathcal{S}_2) = 0$. An SRS with $n = 20$ from a population with $N = 100$ would have

$$P(\mathcal{S}) = \frac{1}{\binom{100}{20}} = \frac{20!(100-20)!}{100!} = \frac{1}{5.359834 \times 10^{20}}$$

for *every* subset $\mathcal{S}$ of size 20 from the population $\{1, 2, \ldots, 100\}$.

2.4
Sampling Weights

In (2.2), we defined π_i to be the probability that unit i is included in the sample. In probability sampling, these inclusion probabilities are used to calculate point estimates such as $\hat{t}$ and $\bar{y}$. Define the **sampling weight**, for any sampling design, to be the reciprocal of the inclusion probability:

$$w_i = \frac{1}{\pi_i}. \tag{2.20}$$

The sampling weight of unit i can be interpreted as the number of population units represented by unit i.

In an SRS, each unit has inclusion probability $\pi_i = n/N$; consequently, all sampling weights are the same with $w_i = 1/\pi_i = N/n$. We can thus think of every unit in the sample as representing itself plus $N/n - 1$ of the unsampled units in the population. Note that for an SRS,

$$\sum_{i \in \mathcal{S}} w_i = \sum_{i \in \mathcal{S}} \frac{N}{n} = N,$$

$$\sum_{i \in S} w_i y_i = \sum_{i \in S} \frac{N}{n} y_i = \hat{t},$$

and

$$\frac{\displaystyle\sum_{i \in S} w_i y_i}{\displaystyle\sum_{i \in S} w_i} = \frac{\hat{t}}{N} = \bar{y}.$$

All weights are the same in an SRS—that is, every unit in the sample represents the same number of units, N/n, in the population. We call such a sample, in which every unit has the same sampling weight, a **self-weighting** sample.

EXAMPLE **2.8**   Let's look at the sampling weights for the sample described in Example 2.5. Here, $N = 3078$ and $n = 300$, so the sampling weight is $w_i = 3078/300 = 10.26$ for each unit in the sample. The first county in the data file agsrs.dat, Coffee County, Alabama, thus represents itself and 9.26 counties from the 2778 counties not included in the sample. We can create a column of sampling weights as follows:

A County	B State	C *acres92*	D *weight*	E *weight*acres92*
Coffee County	AL	175,209	10.26	1,797,644.34
Colbert County	AL	138,135	10.26	1417,265.10
Lamar County	AL	56,102	10.26	575,606.52
Marengo County	AL	199,117	10.26	2,042,940.42
Marion County	AL	89,228	10.26	915,479.28
Tuscaloosa County	AL	96,194	10.26	986,950.44
Columbia County	AR	57,253	10.26	587,415.78
⋮	⋮	⋮	⋮	⋮
Pleasants County	WV	15,650	10.26	160,569.00
Putnam County	WV	55,827	10.26	572,785.02
Sum		89,369,114	3078	916,927,109.60

The last column is formed by multiplying columns C and D, so the entries are $w_i y_i$. We see that $\sum_{i \in S} w_i y_i = 916,927,110$, which is the same value we obtained for the estimated population total in Example 2.5. ∎

2.5
Confidence Intervals

When you take a sample survey, it is not sufficient to simply report the average height of trees or the sample proportion of voters who intend to vote for Candidate B in the next election. You also need to give an indication of how accurate your estimates are. In statistics, **confidence intervals** (CIs) are used to indicate the accuracy of an estimate.

A 95% confidence interval is often explained heuristically: If we take samples from our population over and over again, and construct a confidence interval using our procedure for each possible sample, we expect 95% of the resulting intervals to include the true value of the population parameter.

In probability sampling from a finite population, only a finite number of possible samples exist and we know the probability with which each will be chosen; if we were able to generate all possible samples from the population, we would be able to calculate the exact confidence level for a confidence interval procedure.

EXAMPLE 2.9   Return to Example 2.2, in which the entire population is known. Let's choose an arbitrary procedure for calculating a confidence interval, constructing interval estimates for t as

$$\text{CI}(\mathcal{S}) = [\hat{t}_{\mathcal{S}} - 4s_{\mathcal{S}}, \hat{t}_{\mathcal{S}} + 4s_{\mathcal{S}}].$$

There is no theoretical reason to choose this procedure, but it will illustrate the concept of a confidence interval. Define $u(\mathcal{S})$ to be 1 if $\text{CI}(\mathcal{S})$ contains the true population value 40, and 0 if $\text{CI}(\mathcal{S})$ does not contain 40. Since we know the population, we can calculate the confidence interval $\text{CI}(\mathcal{S})$ and the value of $u(\mathcal{S})$ for each possible sample $\mathcal{S}$. Some of the 70 confidence intervals are shown in Table 2.1 (all entries in the table are rounded to two decimals).

Each individual confidence interval either does or does not contain the population total 40. The probability statement in the confidence interval is made about the collection of all possible samples; for this confidence interval procedure and population, the confidence level is

$$\sum_{\mathcal{S}} P(\mathcal{S})u(\mathcal{S}) = 0.77.$$

That means that if we take an SRS of four elements without replacement from this population of eight elements, there is a 77% chance that our sample is one of the "good" ones whose confidence interval contains the true value 40. This procedure thus creates a 77% confidence interval.

Of course, in real life, we only take one sample and do not know the value of the population total t. Without further investigation, we have no way of knowing whether the sample we obtained is one of the "good" ones, such as $\mathcal{S} = \{2, 3, 5, 6\}$, or one of the "bad" ones such as $\mathcal{S} = \{4, 6, 7, 8\}$. The confidence interval gives us only a probabilistic statement of how often we expect to be right. ∎

In practice, we do not know the values of statistics from all possible samples, so we cannot calculate the exact confidence coefficient for a procedure as done in Example 2.9. In your introductory statistics class, you relied largely on **asymptotic** (as the sample size goes to infinity) results to construct confidence intervals for an unknown mean μ. The central limit theorem says that if we have a random sample with replacement, then the probability distribution of $\sqrt{n}(\bar{y} - \mu)$ converges to a normal distribution as the sample size n approaches infinity.

In most sample surveys, though, we only have a finite population. To use asymptotic results in finite population sampling, we pretend that our population is itself part of a larger **superpopulation**; the superpopulation is itself a subset of a larger

TABLE 2.1

Confidence intervals for possible samples from small population

Sample $\mathcal{S}$	$y_i, i \in \mathcal{S}$	$\hat{t}_\mathcal{S}$	$s_\mathcal{S}$	CI($\mathcal{S}$)	$u(\mathcal{S})$
{1, 2, 3, 4}	1,2,4,4	22	1.50	[16.00, 28.00]	0
{1, 2, 3, 5}	1,2,4,7	28	2.65	[17.42, 38.58]	0
{1, 2, 3, 6}	1,2,4,7	28	2.65	[17.42, 38.58]	0
{1, 2, 3, 7}	1,2,4,7	28	2.65	[17.42, 38.58]	0
{1, 2, 3, 8}	1,2,4,8	30	3.10	[17.62, 42.38]	1
{1, 2, 4, 5}	1,2,4,7	28	2.65	[17.42, 38.58]	0
{1, 2, 4, 6}	1,2,4,7	28	2.65	[17.42, 38.58]	0
{1, 2, 4, 7}	1,2,4,7	28	2.65	[17.42, 38.58]	0
{1, 2, 4, 8}	1,2,4,8	30	3.10	[17.62, 42.38]	1
{1, 2, 5, 6}	1,2,7,7	34	3.20	[21.19, 46.81]	1
⋮	⋮	⋮	⋮	⋮	⋮
{2, 3, 4, 8}	2,4,4,8	36	2.52	[25.93, 46.07]	1
{2, 3, 5, 6}	2,4,7,7	40	2.45	[30.20, 49.80]	1
{2, 3, 5, 7}	2,4,7,7	40	2.45	[30.20, 49.80]	1
{2, 3, 5, 8}	2,4,7,8	42	2.75	[30.98, 53.02]	1
{2, 3, 6, 7}	2,4,7,7	40	2.45	[30.20, 49.80]	1
⋮	⋮	⋮	⋮	⋮	⋮
{4, 5, 6, 8}	4,7,7,8	52	1.73	[45.07, 58.93]	0
{4, 5, 7, 8}	4,7,7,8	52	1.73	[45.07, 58.93]	0
{4, 6, 7, 8}	4,7,7,8	52	1.73	[45.07, 58.93]	0
{5, 6, 7, 8}	7,7,7,8	58	0.50	[56.00, 60.00]	0

superpopulation, and so on until the superpopulations are as large as we could wish. Our population is embedded in a series of increasing finite populations. This embedding can give us properties such as consistency and asymptotic normality. One can imagine the superpopulations as "alternative universes" in a science fiction sense—what might have happened if circumstances were slightly different.

Hájek (1960) proves a central limit theorem for simple random sampling without replacement (also see Lehmann, 1999, Sections 2.8 and 4.4, for a derivation). In practical terms, Hájek's theorem says that if certain technical conditions hold and if n, N, and $N - n$ are all "sufficiently large," then the sampling distribution of

$$\frac{\bar{y} - \bar{y}_U}{\sqrt{\left(1 - \frac{n}{N}\right)} \frac{S}{\sqrt{n}}}$$

is approximately normal (Gaussian) with mean 0 and variance 1. A large-sample $100(1 - \alpha)\%$ CI for the population mean is

$$\left[\bar{y} - z_{\alpha/2}\sqrt{1 - \frac{n}{N}} \frac{S}{\sqrt{n}}, \bar{y} + z_{\alpha/2}\sqrt{1 - \frac{n}{N}} \frac{S}{\sqrt{n}}\right], \tag{2.21}$$

where $z_{\alpha/2}$ is the $(1 - \alpha/2)$th percentile of the standard normal distribution. In simple random sampling without replacement, 95% of the possible samples that could be chosen will give a 95% CI for $\bar{y}_U$ that contains the true value of $\bar{y}_U$. Usually, S is unknown, so in large samples s is substituted for S with little change in the approximation; the large-sample CI is

$$[\bar{y} - z_{\alpha/2}\mathrm{SE}(\bar{y}), \bar{y} + z_{\alpha/2}\mathrm{SE}(\bar{y})].$$

In practice, we often substitute $t_{\alpha/2,n-1}$, the $(1-\alpha/2)^{th}$ percentile of a t distribution with $n - 1$ degrees of freedom, for $z_{\alpha/2}$. For large samples, $t_{\alpha/2,n-1} \approx z_{\alpha/2}$. In smaller samples, using $t_{\alpha/2,n-1}$ instead of $z_{\alpha/2}$ produces a wider CI. Most software packages use the following CI for the population mean from an SRS:

$$\left[\bar{y} - t_{\alpha/2,n-1}\sqrt{1 - \frac{n}{N}}\frac{s}{\sqrt{n}}, \bar{y} + t_{\alpha/2,n-1}\sqrt{1 - \frac{n}{N}}\frac{s}{\sqrt{n}}\right], \qquad (2.22)$$

The imprecise term "sufficiently large" occurs in the central limit theorem because the adequacy of the normal approximation depends on n and on how closely the population $\{y_i, i = 1, \ldots, N\}$ resembles a population generated from the normal distribution. The "magic number" of $n = 30$, often cited in introductory statistics books as a sample size that is "sufficiently large" for the central limit theorem to apply, often does not suffice in finite population sampling problems. Many populations we sample are highly skewed—we may measure income, number of acres on a farm that are devoted to corn, or the concentration of mercury in Minnesota lakes. For all of these examples, we expect most of the observations to be relatively small, but a few to be very, very large, so that a smoothed histogram of the entire population would look like this:

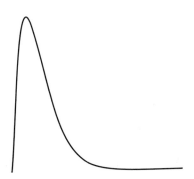

Thinking of observations as generated from some distribution is useful in deciding whether it is safe to use the central limit theorem. If you can think of the generating distribution as being somewhat close to normal, it is probably safe to use the central limit theorem with a sample size as small as 50. If the sample size is too small and the sampling distribution of $\bar{y}$ not approximately normal, we need to use another method, relying on distributional assumptions, to obtain a CI for $\bar{y}_U$. Such methods rely on a model-based perspective for sampling (Section 2.9).

Sugden et al. (2000) discuss and extend Cochran's rule (Cochran, 1977, p. 42) for the sample size needed for the normal approximation to be adequate. They recommend a minimum sample size of

$$n_{\min} = 28 + 25 \left(\frac{\sum_{i=1}^{N} (y_i - \bar{y}_U)^3}{NS^3} \right)^2 \tag{2.23}$$

for the CI in (2.22) to have confidence level approximately equal to $1 - \alpha$. The quantity $\sum_{i=1}^{N} (y_i - \bar{y}_U)^3 / (NS^3)$ is the skewness of the population; if the skewness is large, a large sample size is needed for the normal approximation to be valid. Another approach for considering whether the sample size is adequate for a normal approximation to be used is to look at a bootstrap approximation to the sampling distribution (see Exercise 25).

EXAMPLE 2.10   The histogram in Figure 2.4 exhibited an underlying distribution for farm acreage that was far from normal. Is the sample size large enough to apply the central limit theorem? We substitute the sample values $s = 344{,}551.9$ and $\sum_{i \in \mathcal{S}} (y_i - \bar{y})^3 / n = 1.05036 \times 10^{17}$ for the population quantities S and $\sum_{i=1}^{N} (y_i - \bar{y}_U)^3 / N$ in (2.23), giving an estimated minimum sample size of

$$n_{\min} = 28 + 25 \left[\frac{1.05036 \times 10^{17}}{(344{,}551.9)^3} \right]^2 \approx 193.$$

For this example, our sample of size 300 appears to be sufficiently large for the sampling distribution of $\bar{y}$ to be approximately normal.

For the data in Example 2.5, an approximate 95% CI for $\bar{y}_U$, using $t_{\alpha/2,299} = 1.968$, is

$$[297{,}897 - (1.968)(18{,}898.434),\ 297{,}897 + (1.968)(18{,}898.434)]$$
$$= [260{,}706,\ 335{,}088].$$

For the population total t, an approximate 95% CI is

$$[916{,}927{,}110 - 1.968(58{,}169{,}381),\ 916{,}927{,}110 + 1.968(58{,}169{,}381)]$$
$$= [8.02 \times 10^8,\ 1.03 \times 10^9].$$

For estimating proportions, the usual criterion that the sample size is large enough to use the normal distribution if both $np \geq 5$ and $n(1 - p) \geq 5$ is a useful guideline. A 95% CI for the proportion of counties with fewer than 200,000 acres in farms is

$$0.51 \pm 1.968(0.0275),\ \text{or } [0.456, 0.564].$$

To find a 95% CI for the total number of counties with fewer than 200,000 acres in farms, we only need to multiply all quantities by N, so the point estimate is $3078(0.51) = 1570$, with standard error $3078 \times \text{SE}(\hat{p}) = 84.65$ and 95% CI [1403, 1736].

Software packages such as SAS that calculate estimates for survey samples use the weight variable to find point estimates of means and totals. Here is output from SAS PROC SURVEYMEANS. The variable *acres92* is the number of acres devoted to farms in 1992, and the variable *lt200k* takes on the value 1 if the county has less than 200,000 acres in farms and takes on the value 1 if the county has greater than 200,000 acres in farms. The SAS code used to produce this output is given on the website in file example0210.sas.

```
The SURVEYMEANS Procedure

Data Summary

Number of Observations          300
Sum of Weights                  3078


Class Level Information

Class
Variable          Levels      Values

lt200k              2          0 1


Statistics
```

Variable	Mean	Std Error of Mean	Lower 95% CL for Mean	Upper 95% CL for Mean	Sum
acres92	297897	18898	260706	335088	916927110
lt200k=0	0.490000	0.027465	0.435951	0.544049	1508.220000
lt200k=1	0.510000	0.027465	0.455951	0.564049	1569.780000

```
Statistics
```

Variable	Std Dev	Lower 95% CL for Sum	Upper 95% CL for Sum
acres92	58169381	802453859	1031400361
lt200k=0	84.537220	1341.856696	1674.583304
lt200k=1	84.537220	1403.416696	1736.143304

The weight for every observation in this sample is $w_i = 3078/300$; note that the sum of the weights is 3078 ($= N$). ∎

2.6
Sample Size Estimation

An investigator often measures several variables and has a number of goals for a survey. Anyone designing an SRS must decide what amount of sampling error in the estimates is tolerable and must balance the precision of the estimates with the cost of the survey. Even though many variables may be measured, an investigator can often focus on one or two responses that are of primary interest in the survey, and use these for estimating a sample size.

For a single response, follow these steps to estimate the sample size:

1 Ask "What is expected of the sample, and how much precision do I need?" What are the consequences of the sample results? How much error is tolerable? If your survey measures the unemployment rate every month, you would like your estimates to be very precise indeed so that you can detect changes in unemployment rates from month to month. A preliminary investigation, however, often needs less precision than an ongoing survey.

Instead of asking about required precision, many people ask, "What percentage of the population should I include in my sample?" This is usually the wrong question to be asking. Except in very small populations, precision is obtained through the absolute size of the sample, not the proportion of the population covered. We saw in Section 2.3 that the fpc, which is the only place that the population size N occurs in the variance formula, has little effect on the variance of the estimator in large populations.

2 Find an equation relating the sample size n and your expectations of the sample.

3 Estimate any unknown quantities and solve for n.

4 If you are relatively new at designing surveys, you will find at this point that the sample size you calculated in step 3 is much larger than you can afford. Go back and adjust some of your expectations for the survey and try again. In some cases, you will find that you cannot even come close to the precision you need with the resources you have available; in that case, perhaps you should consider whether you should even conduct your study.

Specify the Tolerable Error   Only the investigators in the study can say how much precision is needed. The desired precision is often expressed in absolute terms, as

$$P(|\bar{y} - \bar{y}_U| \leq e) = 1 - \alpha.$$

The investigator must decide on reasonable values for α and e; e is called the **margin of error** in many surveys. For many surveys of people in which a proportion is measured, $e = 0.03$ and $\alpha = 0.05$.

Sometimes you would like to achieve a desired relative precision, controlling the CV in (2.13) rather than the absolute error. In that case, if $\bar{y}_U \neq 0$ the precision may be expressed as

$$P\left(\left|\frac{\bar{y} - \bar{y}_U}{\bar{y}_U}\right| \leq r\right) = 1 - \alpha.$$

Find an Equation   The simplest equation relating the precision and sample size comes from the confidence intervals in the previous section. To obtain absolute precision e, find a value of n that satisfies

$$e = z_{\alpha/2}\sqrt{\left(1 - \frac{n}{N}\right)}\frac{S}{\sqrt{n}}.$$

To solve this equation for n, we first find the sample size n_0 that we would use for an SRSWR:

$$n_0 = \left(\frac{z_{\alpha/2}S}{e}\right)^2. \tag{2.24}$$

Then (see Exercise 9) the desired sample size is

$$n = \frac{n_0}{1 + \dfrac{n_0}{N}} = \frac{z_{\alpha/2}^2 S^2}{e^2 + \dfrac{z_{\alpha/2}^2 S^2}{N}}. \tag{2.25}$$

Of course, if $n_0 \geq N$ we simply take a census with $n = N$.

In surveys in which one of the main responses of interest is a proportion, it is often easiest to use that response in setting the sample size. For large populations, $S^2 \approx p(1 - p)$, which attains its maximal value when $p = 1/2$. So using $n_0 = 1.96^2/(4e^2)$ will result in a 95% CI with width at most $2e$.

To calculate a sample size to obtain a specified relative precision, substitute $r\bar{y}_U$ for e in (2.24) and (2.25). This results in sample size

$$n = \frac{z_{\alpha/2}^2 S^2}{(r\bar{y}_U)^2 + \dfrac{z_{\alpha/2}^2 S^2}{N}}. \tag{2.26}$$

To achieve a specified relative precision, the sample size may be determined using only the ratio $S/\bar{y}_U$, the CV for a sample of size 1.

EXAMPLE 2.11   Suppose we want to estimate the proportion of recipes in the *Better Homes & Gardens New Cook Book* that do not involve animal products. We plan to take an SRS of the $N = 1251$ test kitchen-tested recipes, and want to use a 95% CI with margin of error 0.03. Then,

$$n_0 = \frac{(1.96)^2\left(\dfrac{1}{2}\right)\left(1 - \dfrac{1}{2}\right)}{(0.03)^2} \approx 1067.$$

The sample size ignoring the fpc is large compared with the population size, so in this case we would make the fpc adjustment and use

$$n = \frac{n_0}{1 + \dfrac{n_0}{N}} = \frac{1067}{1 + \dfrac{1067}{1251}} = 576. \quad \blacksquare$$

In Example 2.11, the fpc makes a difference in the sample size because N is only 1251. If N is large, however, typically n_0/N will be very small so that for large populations we usually have $n \approx n_0$. Thus, we need approximately the same sample size for any large population—whether that population has 10 million or 1 billion or 100 billion units.

EXAMPLE **2.12**   Many public opinion polls specify using a sample size of about 1100. That number comes from rounding the value of n_0 in Example 2.11 up to the next hundred, and then noting that the population size is so large relative to the sample that the fpc should be ignored. For large populations, it is the size of the sample, not the proportion of the population that is sampled, that determines the precision. $\quad \blacksquare$

Estimate unknown quantities.   When interested in a proportion, we can use 1/4 as an upper bound for S^2. For other quantities, S^2 must be estimated or guessed at. Some methods for estimating S^2 include:

1 Use sample quantities obtained when pretesting your survey. This is probably the best method, as your pretest should be similar to the survey you take. A **pilot sample**, a small sample taken to provide information and guidance for the design of the main survey, can be used to estimate quantities needed for setting the sample size.

2 Use previous studies or data available in the literature. You are rarely the first person in the world to study anything related to your investigation. You may be able to find estimates of variances that have been published in related studies, and use these as a starting point for estimating your sample size. But you have no control over the quality or design of those studies, and their estimates may be unreliable or may not apply for your study. In addition, estimates may change over time and vary in different geographic locations.

Sometimes you can use the CV for a sample of size 1, the ratio of the standard deviation to the mean, in obtaining estimates of variability. The CV of a quantity is a measure of relative error, and tends to be more stable over time and location than the variance. If we take a random sample of houses for sale in the United States today, we will find that the variability in price will be much greater than if we had taken a similar survey in 1930. But the average price of a house has also increased from 1930 to today. We would probably find that the CV today is close to the CV in 1930.

3 If nothing else is available, guess the variance. Sometimes a hypothesized distribution of the data will give us information about the variance. For example, if you believe the population to be normally distributed, you may not know what the variance is, but you may have an idea of the range of the data. You could then estimate

FIGURE   **2.5**

Plot of $t_{0.025,n-1}s/\sqrt{n}$ vs. n, for two possible values of the standard deviation s.

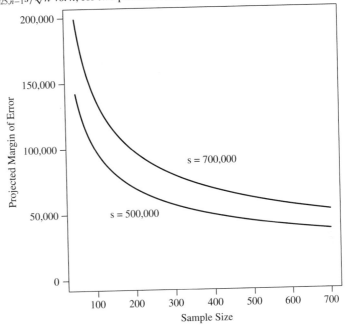

S by range/4 or range/6, as approximately 95% of values from a normal population are within 2 standard deviations of the mean, and 99.7% of the values are within 3 standard deviations of the mean.

EXAMPLE 2.13   Before taking the sample of size 300 in Example 2.5, we took a pilot sample of size 30 from the population. One county in the pilot sample of size 30 was missing the value of *acres92*; the sample standard deviation of the remaining 29 observations was 519,085. Using this value, and a desired margin of error of 60,000,

$$n_0 = (1.96)^2 \frac{519,085^2}{60,000^2} = 288.$$

We took a sample of size 300 in case the estimated standard deviation from the pilot sample is too low. Also, we ignored the fpc in the sample size calculations; in most populations, the fpc will have little effect on the sample size.

You may also view possible consequences of different sample sizes graphically. Figure 2.5 shows the value of $t_{0.025,n-1} s/\sqrt{n}$, for a range of sample sizes between 50 and 700, and for two possible values of the standard deviation s. The plot shows that if we ignore the fpc and if the standard deviation is about 500,000, a sample of size 300 will give a margin of error of about 60,000. ∎

Determining the sample size is one of the early steps that must be taken in an investigation, and no magic formula will tell you the perfect sample size for your investigation (you only know that in hindsight, after you have completed the study!). Choosing a sample size is somewhat like deciding how much food to take on a picnic. You have a rough idea of how many people will attend, but do not know how much food you should have brought until after the picnic is over. You also need to bring extra food to allow for unexpected happenings, such as 2-year-old Freddie feeding a bowl of potato salad to the ducks or cousin Ted bringing along some extra guests. But you do not want to bring too much extra food, or it will spoil and you will have wasted money. Of course, the more picnics you have organized, and the better acquainted you are with the picnic guests, the better you become at bringing the right amount of food. It is comforting to know that the same is true of determining sample sizes—experience and knowledge about the population make you much better at designing surveys.

The results in this section can give you some guidance in choosing the size of the sample, but the final decision is up to you. In general, the larger the sample, the smaller the sampling error. Remember, though, that in most surveys you also need to worry about nonsampling errors, and need to budget resources to control selection and measurement bias. In many cases, nonsampling errors are greater when a larger sample is taken—with a large sample, it is easy to introduce additional sources of error (for example, it becomes more difficult to control the quality of the interviewers or to follow up on nonrespondents) or to become more relaxed about selection bias. In Chapter 15, we shall revisit the issue of designing a sample with the aim of reducing nonsampling as well as sampling error.

2.7
Systematic Sampling

Sometimes **systematic sampling** is used as a proxy for simple random sampling, when no list of the population exists or when the list is in roughly random order. To obtain a systematic sample, choose a sample size n. If N/n is an integer, let $k = N/n$; otherwise, let k be the next integer after N/n. Then find a random integer R between 1 and k, which determines the sample to be the units numbered $R, R + k, R + 2k$, etc. For example, to select a systematic sample of 45 students from the list of 45,000 students at a university, the sampling interval k is 1000. Suppose the random integer we choose is 597. Then the students numbered 597, 1597, 2597, . . . , 44,957 would be in the sample.

If the list of students is ordered by randomly generated student identification numbers, we shall probably obtain a sample that will behave much like an SRS—it is unlikely that a person's position in the list is associated with the characteristic of interest. However, systematic sampling is not the same as simple random sampling; it does not have the property that every possible group of n units has the same probability of being the sample. In the example above, it is impossible to have students 345 and 346 both appear in the sample. Systematic sampling is technically a form of cluster sampling, as will be discussed in Chapter 5.

If the population is in random order, the systematic sample will behave much like an SRS, and SRS methods can be used in the analysis. The population itself can be thought of as being mixed. In the quote at the beginning of this chapter, Sorensen reports that President Kennedy used to read a systematic sample of letters written to him at the White House. This systematic sample most likely behaved much like a random sample. Note that Kennedy was well aware that the letters he read, while representative of letters written to the White House, were not at all representative of public opinion.

Systematic sampling does not necessarily give a representative sample, though, if the listing of population units is in some periodic or cyclical order. If male and female names alternate in the list, for example, and k is even, the systematic sample will contain either all men or all women—this cannot be considered a representative sample. In ecological surveys done on agricultural land, a ridge-and-furrow topography may be present that would lead to a periodic pattern of vegetation. If a systematic sampling scheme follows the same cycle, the sample will not behave like an SRS.

On the other hand, some populations are in increasing or decreasing order. A list of accounts receivable may be ordered from largest amount to smallest amount. In this case, estimates from the systematic sample may have smaller (but unestimable) variance than comparable estimates from the SRS. A systematic sample from an ordered list of accounts receivable is forced to contain some large amounts and some small amounts. It is possible for an SRS to contain all small amounts or all large amounts, so there may be more variability among the sample means of all possible SRSs than there is among the sample means of all possible systematic samples.

In systematic sampling, we must still have a sampling frame and be careful when defining the target population. Sampling every 20th student to enter the library will not give a representative sample of the student body. Sampling every 10th person exiting an airplane, though, will probably give a representative sample of the persons on that flight. The sampling frame for the airplane passengers is not written down, but it exists all the same.

2.8
Randomization Theory Results for Simple Random Sampling*[3]

In this section, we show that $\bar{y}$ is an unbiased estimator of $\bar{y}_U$. We also calculate the variance of $\bar{y}$ given in (2.9), and show that the estimator in (2.11) is unbiased over repeated sampling.

No distributional assumptions are made about the y_i's in order to ascertain that $\bar{y}$ is unbiased for estimating $\bar{y}_U$. We do not, for instance, assume that the y_i's are normally distributed with mean μ. In the **randomization theory** (also called **design-based**) approach to sampling, the y_i's are considered to be fixed but unknown numbers—the random variables used in randomization theory inference indicate which population units are in the sample.

Let's see how the randomization theory works for deriving properties of the sample mean in simple random sampling. As in Cornfield (1944), define

$$Z_i = \begin{cases} 1 & \text{if unit } i \text{ is in the sample} \\ 0 & \text{otherwise} \end{cases}. \qquad (2.27)$$

---

[3]An asterisk (*) indicates a section, chapter, or exercise that requires more mathematical background.

Then

$$\bar{y} = \sum_{i \in S} \frac{y_i}{n} = \sum_{i=1}^{N} Z_i \frac{y_i}{n}. \tag{2.28}$$

The Z_i's are the *only* random variables in (2.28) because, according to randomization theory, the y_i's are fixed quantities. When we choose an SRS of n units out of the N units in the population, $\{Z_1, \ldots, Z_N\}$ are identically distributed Bernoulli random variables with

$$\pi_i = P(Z_i = 1) = P(\text{select unit } i \text{ in sample}) = \frac{n}{N} \tag{2.29}$$

and

$$P(Z_i = 0) = 1 - \pi_i = 1 - \frac{n}{N}.$$

The probability in (2.29) follows from the definition of an SRS. To see this, note that if unit i is in the sample, then the other $n-1$ units in the sample must be chosen from the other $N-1$ units in the population. A total of $\binom{N-1}{n-1}$ possible samples of size $n-1$ may be drawn from a population of size $N-1$, so

$$P(Z_i = 1) = \frac{\text{number of samples including unit } i}{\text{number of possible samples}} = \frac{\binom{N-1}{n-1}}{\binom{N}{n}} = \frac{n}{N}.$$

As a consequence of (2.29),

$$E[Z_i] = E[Z_i^2] = \frac{n}{N}$$

and

$$E[\bar{y}] = E\left[\sum_{i=1}^{N} Z_i \frac{y_i}{n}\right] = \sum_{i=1}^{N} E[Z_i]\frac{y_i}{n} = \sum_{i=1}^{N} \frac{n}{N}\frac{y_i}{n} = \sum_{i=1}^{N} \frac{y_i}{N} = \bar{y}_U. \tag{2.30}$$

This shows that $\bar{y}$ is an unbiased estimator of $\bar{y}_U$. Note that in (2.30), the random variables are $Z_1, \ldots, Z_N$; $y_1, \ldots, y_N$ are treated as constants.

The variance of $\bar{y}$ is also calculated using properties of the random variables $Z_1, \ldots Z_N$. Note that

$$V(Z_i) = E[Z_i^2] - (E[Z_i])^2 = \frac{n}{N} - \left(\frac{n}{N}\right)^2 = \frac{n}{N}\left(1 - \frac{n}{N}\right).$$

For $i \neq j$,

$$E[Z_i Z_j] = P(Z_i = 1 \text{ and } Z_j = 1)$$
$$= P(Z_j = 1 \mid Z_i = 1)P(Z_i = 1)$$
$$= \left(\frac{n-1}{N-1}\right)\left(\frac{n}{N}\right).$$

Because the population is finite, the Z_i's are not quite independent—if we know that unit i is in the sample, we do have a small amount of information about whether unit j is in the sample, reflected in the conditional probability $P(Z_j = 1 \mid Z_i = 1)$. Consequently, for $i \neq j$, the covariance of Z_i and Z_j is:

$$\text{Cov}(Z_i, Z_j) = E[Z_i Z_j] - E[Z_i]E[Z_j]$$

$$= \frac{n-1}{N-1}\frac{n}{N} - \left(\frac{n}{N}\right)^2$$

$$= -\frac{1}{N-1}\left(1 - \frac{n}{N}\right)\left(\frac{n}{N}\right).$$

The negative covariance of Z_i and Z_j is the source of the fpc. The following derivation shows how we can use the random variables $Z_1, \ldots, Z_N$ and the properties of covariances given in Appendix A to find $V(\bar{y})$:

$$V(\bar{y}) = \frac{1}{n^2}V\left(\sum_{i=1}^{N} Z_i y_i\right)$$

$$= \frac{1}{n^2}\text{Cov}\left(\sum_{i=1}^{N} Z_i y_i, \sum_{j=1}^{N} Z_j y_j\right)$$

$$= \frac{1}{n^2}\sum_{i=1}^{N}\sum_{j=1}^{N} y_i y_j \,\text{Cov}(Z_i, Z_j)$$

$$= \frac{1}{n^2}\left[\sum_{i=1}^{N} y_i^2 V(Z_i) + \sum_{i=1}^{N}\sum_{j\neq i}^{N} y_i y_j \,\text{Cov}(Z_i, Z_j)\right]$$

$$= \frac{1}{n^2}\left[\frac{n}{N}\left(1 - \frac{n}{N}\right)\sum_{i=1}^{N} y_i^2 - \sum_{i=1}^{N}\sum_{j\neq i}^{N} y_i y_j \frac{1}{N-1}\left(1 - \frac{n}{N}\right)\left(\frac{n}{N}\right)\right]$$

$$= \frac{1}{n^2}\frac{n}{N}\left(1 - \frac{n}{N}\right)\left[\sum_{i=1}^{N} y_i^2 - \frac{1}{N-1}\sum_{i=1}^{N}\sum_{j\neq i}^{N} y_i y_j\right]$$

$$= \frac{1}{n}\left(1 - \frac{n}{N}\right)\frac{1}{N(N-1)}\left[(N-1)\sum_{i=1}^{N} y_i^2 - \left(\sum_{i=1}^{N} y_i\right)^2 + \sum_{i=1}^{N} y_i^2\right]$$

$$= \frac{1}{n}\left(1 - \frac{n}{N}\right)\frac{1}{N(N-1)}\left[N\sum_{i=1}^{N} y_i^2 - \left(\sum_{i=1}^{N} y_i\right)^2\right]$$

$$= \left(1 - \frac{n}{N}\right)\frac{S^2}{n}.$$

To show that the estimator in (2.11) is an unbiased estimator of the variance, we need to show that $E[s^2] = S^2$. The argument proceeds much like the previous one. Since $S^2 = \sum_{i=1}^{N}(y_i - \bar{y}_U)^2/(N-1)$, it makes sense when trying to find an unbiased estimator to find the expected value of $\sum_{i\in S}(y_i - \bar{y})^2$, and then find the multiplicative constant that will give the unbiasedness:

$$E\left[\sum_{i\in S}(y_i - \bar{y})^2\right] = E\left[\sum_{i\in S}\{(y_i - \bar{y}_U) - (\bar{y} - \bar{y}_U)\}^2\right]$$

$$= E\left[\sum_{i\in S}(y_i - \bar{y}_U)^2 - n(\bar{y} - \bar{y}_U)^2\right]$$

$$= E\left[\sum_{i=1}^{N} Z_i(y_i - \bar{y}_U)^2\right] - n\,V(\bar{y})$$

$$= \frac{n}{N}\sum_{i=1}^{N}(y_i - \bar{y}_U)^2 - \left(1 - \frac{n}{N}\right)S^2$$

$$= \frac{n(N-1)}{N}S^2 - \frac{N-n}{N}S^2$$

$$= (n-1)S^2.$$

Thus,

$$E\left[\frac{1}{n-1}\sum_{i\in S}(y_i - \bar{y})^2\right] = E[s^2] = S^2.$$

2.9
A Prediction Approach for Simple Random Sampling*

Unless you have studied randomization theory in the design of experiments, the proofs in the preceding section probably seemed strange to you. The random variables in randomization theory are not concerned with the responses y_i. The random variables $Z_1, \ldots, Z_N$ are indicator variables that tell us whether the ith unit is in the sample or not. In a design-based, or randomization-theory, approach to sampling inference, the only relationship between units sampled and units not sampled is that the nonsampled units could have been sampled had we used a different starting value for the random number generator.

In Section 2.8 we found properties of the sample mean $\bar{y}$ using randomization theory: $y_1, y_2, \ldots, y_N$ were considered to be fixed values, and $\bar{y}$ is unbiased because $\bar{y} = (1/N)\sum_{i=1}^{N} Z_i y_i$ and $E[Z_i] = P(Z_i = 1) = n/N$. The only probabilities used in finding the expected value and variance of $\bar{y}$ are the probabilities that subsets of units are included in the sample. The quantity measured on unit i, y_i can be anything: Whether y_i is number of television sets owned, systolic blood pressure, or acreage devoted to soybeans, the properties of estimators depend exclusively on the joint distribution of the random variables $\{Z_1, \ldots, Z_N\}$.

In your other statistics classes, you most likely learned a different approach to inference, an approach explained in Chapter 5 of Casella and Berger (2002). There, you had random variables $\{Y_i\}$ that followed some probability distribution, and the actual sample values were realizations of those random variables. Thus you assumed, for example, that $Y_1, Y_2, \ldots, Y_n$ were independent and identically distributed from a normal distribution with mean μ and variance σ^2, and used properties of independent random variables and the normal distribution to find expected values of various statistics.

We can extend this approach to sampling by thinking of random variables Y_1, $Y_2, \ldots, Y_N$ generated from some model. The actual values for the finite population, $y_1, y_2, \ldots, y_N$, are one realization of the random variables. The joint probability distribution of $Y_1, Y_2, \ldots, Y_N$ supplies the link between units in the sample and units not in the sample in this **model-based** approach—a link that is missing in the randomization approach. Here, we sample $\{y_i, i \in S\}$ and use these data to predict the unobserved values $\{y_i, i \notin S\}$. Thus, problems in finite population sampling may be thought of as prediction problems.

In an SRS, a simple model to adopt is:

$$Y_1, Y_2, \ldots, Y_N \text{ independent with } E_M[Y_j] = \mu \text{ and } V_M[Y_j] = \sigma^2. \quad (2.31)$$

The subscript M indicates that the expectation uses the model, not the randomization distribution used in Section 2.8. Here, μ and σ^2 represent unknown infinite population parameters, not the finite population quantities in Section 2.8. This model makes assumptions about the observations not in the sample; namely, that they have the same mean and variance as observations that are in the sample. We take a sample S and observe the values y_i for $i \in S$. That is, we see realizations of the random variables Y_i for $i \in S$. The other observations in the population $\{y_i, i \notin S\}$ are also realizations of random variables, but we do not see those. The finite population total t for our sample can be written as

$$t = \sum_{i=1}^{N} y_i = \sum_{i \in S} y_i + \sum_{i \notin S} y_i$$

and is one possible value that can be taken on by the random variable

$$T = \sum_{i=1}^{N} Y_i = \sum_{i \in S} Y_i + \sum_{i \notin S} Y_i.$$

We know the values $\{y_i, i \in S\}$. To estimate t for our sample, we need to predict values for the y_i's not in the sample. This is where our model of the common mean μ comes in. The least squares estimator of μ from the sample is $\overline{Y}_S = \sum_{i \in S} Y_i / n$, and this is the best linear unbiased predictor (under the model) of each unobserved random variable, so that

$$\hat{T} = \sum_{i \in S} Y_i + \sum_{i \notin S} \hat{Y}_i = \sum_{i \in S} Y_i + \frac{N-n}{n} \sum_{i \in S} Y_i = \frac{N}{n} \sum_{i \in S} Y_i.$$

The estimator $\hat{T}$ is *model-unbiased*: if the model is correct, then the average of $\hat{T} - T$ over repeated realizations of the population is

$$E_M[\hat{T} - T] = \frac{N}{n} \sum_{i \in S} E_M[Y_i] - \sum_{i=1}^{N} E_M[Y_i] = 0.$$

(Notice the difference between finding expectations under the model-based approach and under the design-based approach. In the model-based approach, the Y_i's are the random variables, and the sample has no information for calculating expected values, so we can take the sum $\sum_{i \in S}$ outside of the expected value. In the design-based approach, the random variables are contained in the sample S.)

The mean squared error is also calculated as the average squared deviation between the estimate and the finite population total. For any given realization of the random variables, the squared error is

$$(\hat{t} - t)^2 = \left[\frac{N}{n} \sum_{i \in S} y_i - \sum_{i=1}^{N} y_i \right]^2.$$

Averaging this quantity over all possible realizations of the random variables gives the mean squared error under the model assumptions:

$$E_M[(\hat{T} - T)^2] = E_M \left[\left(\frac{N}{n} \sum_{i \in S} Y_i - \sum_{i=1}^{N} Y_i \right)^2 \right]$$

$$= E_M \left[\left\{ \left(\frac{N}{n} - 1 \right) \sum_{i \in S} Y_i - \sum_{i \notin S} Y_i \right\}^2 \right]$$

$$= E_M \left[\left\{ \left(\frac{N}{n} - 1 \right) \sum_{i \in S} Y_i - \sum_{i \notin S} Y_i - \left(\frac{N}{n} - 1 \right) n\mu + (N - n)\mu \right\}^2 \right]$$

$$= E_M \left[\left(\frac{N}{n} - 1 \right)^2 \left(\sum_{i \in S} Y_i - n\mu \right)^2 + \left(\sum_{i \notin S} Y_i - (N - n)\mu \right)^2 \right]$$

$$= \left(\frac{N}{n} - 1 \right)^2 n\sigma^2 + (N - n)\sigma^2$$

$$= N^2 \left(1 - \frac{n}{N} \right) \frac{\sigma^2}{n}.$$

In practice, if the model in (2.31) were adopted, you would estimate σ^2 by the sample variance s^2. Thus the design-based approach and the model-based approach—with the model in (2.31)—lead to the same estimator of the population total and the same variance estimator. If a different model were adopted, however, the estimators might differ. We shall see in Chapter 4 how a design-based approach and a model-based approach can lead to different inferences.

The design-based approach and the model-based approach with model (2.31) also lead to the same CI for the mean. These CIs have different interpretations, however. The design-based CI for $\bar{y}_U$ may be interpreted as follows: If we take all possible SRSs of size n from the finite population of size N, and construct a 95% CI for each sample, we expect 95% of all the CIs constructed to include the true population value $\bar{y}_U$. Thus, the design-based CI has a *repeated sampling* interpretation. Statistical inference is based on repeated sampling from the finite population.

To construct CIs in the model-based approach, we rely on a central limit theorem that states that if n/N is small, the standardized prediction error

$$\frac{\hat{T} - T}{\sqrt{E_M[(\hat{T} - T)^2]}}$$

converges to a standard normal distribution (Valliant et al., 2000, Section 2.5). For the model in (2.31), with $E_M[(\hat{T} - T)^2] = N^2 (1 - n/N)\, \sigma^2/n$, this central limit theorem says that for sufficiently large sample sizes,

$$P\left[\hat{T} - z_{\alpha/2}N\sqrt{\left(1 - \frac{n}{N}\right)\frac{\sigma^2}{n}} \leq T \leq \hat{T} + z_{\alpha/2}N\sqrt{\left(1 - \frac{n}{N}\right)\frac{\sigma^2}{n}}\right] \approx 1 - \alpha.$$

Substituting the sample standard deviation s for σ, we get the large sample CI

$$\hat{T} \pm z_{\alpha/2}N\sqrt{\left(1 - \frac{n}{N}\right)\frac{s^2}{n}},$$

which has the same form as the design-based CI for the population total t. This model-based CI is also interpreted using repeated sampling ideas, but in a different way than in Section 2.8. The design-based confidence level gives the expected proportion of CIs that will include the true finite population total $t = \sum_{i=1}^{N} y_i$, from the set of all CIs that could be constructed by taking an SRS of size n from the finite population of fixed values $\{y_1, y_2, \ldots, y_N\}$. The model-based confidence level gives the expected proportion of CIs that will include the realization of the population total, from the set of all samples that could be generated from the model in (2.31).

In the model-based approach, the probability model is proposed for all population units, whether in the sample or not. If the model assumptions are valid, model-based inference does not require random sampling—it is assumed that *all* units in the population follow the assumed model, so it makes no difference which ones are chosen for the sample. Thus, model-based analyses can be used for nonprobability samples. The assumptions for model-based analysis are strong—for the model in (2.31), it is assumed that all random variables for the response of interest in the population are independent and have mean μ and variance σ^2. But we only observe the units in the sample, and cannot examine the assumption of whether the model holds for units not in the sample. If you take a sample of your friends to estimate the average amount of time students at your university spend studying, there is no reason to believe that the students not in your sample spend the same average amount of time studying as your friends do. As Box (1979, p. 202) said, "All models are wrong but some are useful." If your model is deficient, inferences made using a model-based analysis may be seriously flawed.

A Note on Notation. Many books (Cochran, 1977, for example) and journal articles use Y to represent the population total (t in this book), and $\overline{Y}$ to represent the finite population mean (our $\bar{y}_U$). In this book, we reserve Y and T to represent random variables in a model-based approach. Our usage is consistent with other areas of statistics, in which capital letters near the end of the alphabet represent random variables. However, you should be aware that notation in the survey sampling literature is not uniform.

2.10
When Should a Simple Random Sample Be Used?

Simple random samples are usually easy to design and easy to analyze. But they are not the best design to use in the following situations:

- Before taking an SRS, you should consider whether a survey sample is the best method for studying your research question. If you want to study whether a certain brand of bath oil is an effective mosquito repellent, you should perform a controlled experiment, not take a survey. You should take a survey if you want to estimate how many people use the bath oil as a mosquito repellent, or if you want to estimate how many mosquitoes are in an area.

- You may not have a list of the observation units, or it may be expensive in terms of travel time to take an SRS. If interested in the proportion of mosquitoes in southwestern Wisconsin that carry an encephalitis virus, you cannot construct a sampling frame of the individual mosquitoes. You would need to sample different areas, and then examine some or all of the mosquitoes found in those areas, using a form of cluster sampling. Cluster sampling will be discussed in Chapters 5 and 6.

- You may have additional information that can be used to design a more cost-effective sampling scheme. In a survey to estimate the total number of mosquitoes in an area, an entomologist would know what terrain would be likely to have high mosquito density, and what areas would be likely to have low mosquito density, before any samples were taken. You would save effort in sampling by dividing the area into *strata*, groups of similar units, and then sampling plots within each stratum. (Stratified sampling will be discussed in Chapter 3.)

You should use an SRS in these situations:

- Little extra information is available that can be used when designing the survey. If your sampling frame is merely a list of university students' names in alphabetical order and you have no additional information such as major or year, simple random or systematic sampling is probably the best probability sampling strategy.

- Persons using the data insist on using SRS formulas, whether they are appropriate or not. Some persons will not be swayed from the belief that one should only estimate the mean by taking the average of the sample values—in that case, you should design a sample in which averaging the sample values is the right thing to do. SRSs are often recommended when sample evidence is used in legal actions; sometimes, when a more complicated sampling scheme is used, an opposing counsel will try to persuade the jury that the sample results are not valid.

- The primary interest is in multivariate relationships such as regression equations that hold for the whole population, and there are no compelling reasons to take a stratified or cluster sample. Multivariate analyses can be done in complex samples, but they are much easier to perform and interpret in an SRS.

2.11
Chapter Summary

In probability sampling, every possible subset from the population has a known probability of being selected as the sample. These probabilities provide a basis for inference to the finite population.

Simple random sampling without replacement is the simplest of all probability sampling methods. In an SRS, each subset of the population of size n has the same probability of being chosen as the sample. The probability that unit i of the population appears in the sample is

$$\pi_i = \frac{n}{N}.$$

The sampling weight for each unit in the sample is

$$w_i = \frac{1}{\pi_i} = \frac{N}{n};$$

each unit in the sample can be thought of as representing N/n units in the population.

Estimators for an SRS are similar to those in your introductory statistics class, using $\bar{y} = \sum_{i \in \mathcal{S}} y_i/n$ and $s^2 = \sum_{i \in \mathcal{S}} (y_i - \bar{y})^2/(n-1)$:

Population Quantity	Estimator	Standard Error of Estimator
Population total, $t = \sum_{i=1}^{N} y_i$	$\hat{t} = \sum_{i \in \mathcal{S}} w_i y_i = N\bar{y}$	$N\sqrt{\left(1 - \dfrac{n}{N}\right)\dfrac{s^2}{n}}$
Population mean, $\bar{y}_U = \dfrac{t}{N}$	$\dfrac{\hat{t}}{N} = \dfrac{\sum_{i \in \mathcal{S}} w_i y_i}{\sum_{i \in \mathcal{S}} w_i} = \bar{y}$	$\sqrt{\left(1 - \dfrac{n}{N}\right)\dfrac{s^2}{n}}$
Population proportion, p	$\hat{p}$	$\sqrt{\left(1 - \dfrac{n}{N}\right)\dfrac{\hat{p}(1 - \hat{p})}{n - 1}}$

The only feature found in the estimators for without-replacement random samples that does not occur in with-replacement random samples is the finite population correction, $(1 - n/N)$, which decreases the standard error when the sample size is large relative to the population size. In most surveys done in practice, the fpc is so close to one that it can be ignored.

For "sufficiently large" sample sizes, an approximate 95% CI is given by

$$\text{estimate} \pm z_{\alpha/2} \text{ SE (estimate)}.$$

The margin of error of an estimate is the half-width of the CI, that is, $z_{\alpha/2} \times$ SE (estimate).

Key Terms

Cluster sample: A probability sample in which each population unit belongs to a group, or cluster, and the clusters are sampled according to the sampling design.

Coefficient of variation (CV): The CV of a statistic $\hat{\theta}$, where with $E(\hat{\theta}) \neq 0$, is $\mathrm{CV}(\hat{\theta}) = \sqrt{V(\hat{\theta})}/E(\hat{\theta})$.

Confidence interval (CI): An interval estimate for a population quantity, for which the probability that the random interval contains the true value of the population quantity is known.

Design-based inference: Inference for finite population characteristics based on the survey design, also called **randomization inference**.

Finite population correction (fpc): A correction factor which, when multiplied by the with-replacement variance, gives the without-replacement variance. For an SRS of size n from a population of size N, the fpc is $1 - n/N$.

Inclusion probability: $\pi_i =$ probability that unit i is included in the sample.

Margin of error: Half of the width of a 95% CI.

Model-based inference: Inference for finite population characteristics based on a model for the population, also called **prediction inference**.

Probability sampling: Method of sampling in which every subset of the population has a known probability of being included in the sample.

Sampling distribution: The probability distribution of a statistic generated by the sampling design.

Sampling weight: Reciprocal of the inclusion probability; $w_i = 1/\pi_i$.

Self-weighting sample: A sample in which all probabilities of inclusion π_i are equal, so that all sampling weights w_i are the same.

Simple random sample with replacement (SRSWR): A probability sample in which the first unit is selected from the population with probability $1/N$; then the unit is replaced and the second unit is selected from the set of N units with probability $1/N$, and so on until n units are selected.

Simple random sample without replacement (SRS): An SRS of size n is a probability sample in which any possible subset of n units from the population has the same probability ($= n!(N-n)!/N!$) of being the sample selected.

Standard error (SE): The square root of the estimated variance of a statistic.

Stratified sample: A probability sample in which population units are partitioned into strata, and then a probability sample of units is taken from each stratum.

Systematic sample: A probability sample in which every kth unit in the population is selected to be in the sample, starting with a randomly chosen value R. Systematic sampling is a special case of cluster sampling.

For Further Reading

Stuart's (1984) book *The Ideas of Sampling* is an intuitive, nontechnical introduction into the structure of probability sampling. He gives simple examples to illustrate the difference among the different probability sampling designs. For mathematical treatments of simple random sampling, see Raj (1968) and Cochran (1977). Levy and Lemeshow (2008) and S.K. Thompson (2002) are general references on sampling. M.E. Thompson (1997) and Fuller (2009) develop the mathematical theory of survey sampling and prove central limit theorems. Prediction- (model-) based inference is developed in Valliant et al. (2000) and Brewer (2002).

2.12
Exercises

Data files referenced in the exercises are provided and described on the website.

A. Introductory Exercises

1 Let $N = 6$ and $n = 3$. For purposes of studying sampling distributions, assume that all population values are known.

$$y_1 = 98 \qquad y_2 = 102 \qquad y_3 = 154$$
$$y_4 = 133 \qquad y_5 = 190 \qquad y_6 = 175$$

We are interested in $\bar{y}_U$, the population mean. Two sampling plans are proposed.

- Plan 1. Eight possible samples may be chosen.

Sample Number	Sample, $\mathcal{S}$	$P(\mathcal{S})$
1	{1,3,5}	1/8
2	{1,3,6}	1/8
3	{1,4,5}	1/8
4	{1,4,6}	1/8
5	{2,3,5}	1/8
6	{2,3,6}	1/8
7	{2,4,5}	1/8
8	{2,4,6}	1/8

- Plan 2. Three possible samples may be chosen.

Sample Number	Sample, $\mathcal{S}$	$P(\mathcal{S})$
1	{1,4,6}	1/4
2	{2,3,6}	1/2
3	{1,3,5}	1/4

a What is the value of $\bar{y}_U$?

b Let $\bar{y}$ be the mean of the sample values. For each sampling plan, find

(i) $E[\bar{y}]$;   (ii) $V[\bar{y}]$;   (iii) Bias($\bar{y}$);   (iv) MSE($\bar{y}$).

c Which sampling plan do you think is better? Why?

2   For the population in Example 2.2, consider the following sampling scheme:

S	$P(S)$
{1,3,5,6}	1/8
{2,3,7,8}	1/4
{1,4,6,8}	1/8
{2,4,6,8}	3/8
{4,5,7,8}	1/8

   a   Find the probability of selection π_i for each unit i.

   b   What is the sampling distribution of $\hat{t} = 8\bar{y}$?

3   Each of the 10,000 shelves in a certain library is 300 cm long. To estimate how many books in the library need rebinding, a librarian takes a sample of 50 books using the following procedure: He first generates a random integer between 1 and 10,000 to select a shelf, and then generates a random number between 0 and 300 to select a location on that shelf. Thus, the pair of random numbers (2531, 25.4) would tell the librarian to include the book that is above the location 25.4 cm from the left end of shelf number 2531 in the sample. Does this procedure generate an SRS of the books in the library? Explain why, or why not.

4   For the population in Example 2.2, find the sampling distribution of $\bar{y}$ for

   a   an SRS of size 3 (without replacement)

   b   an SRSWR of size 3 (with replacement).

For each, draw the histogram of the sampling distribution of $\bar{y}$. Which sampling distribution has the smaller variance, and why?

5   An SRS of size 30 is taken from a population of size 100. The sample values are given below, and in the data file srs30.dat.

8 5 2 6 6 3 8 6 10 7 15 9 15 3 5 6 7 10 14 3 4 17 10 6 14 12 7 8 12 9

   a   What is the sampling weight for each unit in the sample?

   b   Use the sampling weights to estimate the population total, t.

   c   Give a 95% CI for t. Does the fpc make a difference for this sample?

6   A university has 807 faculty members. For each faculty member, the number of refereed publications was recorded. This number is not directly available on the database, so requires the investigator to examine each record separately. A frequency table for number of refereed publications is given below for an SRS of 50 faculty members.

Refereed Publications	0	1	2	3	4	5	6	7	8	9	10
Faculty Members	28	4	3	4	4	2	1	0	2	1	1

   a   Plot the data using a histogram. Describe the shape of the data.

   b   Estimate the mean number of publications per faculty member, and give the SE for your estimate.

   c   Do you think that $\bar{y}$ from (b) will be approximately normally distributed? Why or why not?

d Estimate the proportion of faculty members with no publications and give a 95% CI.

7 A letter in the December 1995 issue of *Dell Champion Variety Puzzles* stated: "I've noticed over the last several issues there have been no winners from the South in your contests. You always say that winners are picked at random, so does this mean you're getting fewer entries from the South?" In response, the editors took a random sample of 1,000 entries from the last few contests, and found that 175 of those came from the South.

    a Find a 95% CI for the percentage of entries that come from the South.

    b According to *Statistical Abstract of the United States*, 30.9% of the U.S. population live in states that the editors considered to be in the South. Is there evidence from your CI that the percentage of entries from the South differs from the percentage of persons living in the South?

8 Discuss whether an SRS would be appropriate for the following situations. What other designs might be used?

    a For an e-mail survey of students, you have a sampling frame that contains a list of e-mail addresses for all students.

    b You want to take a sample of patients of board-certified allergists.

    c You want to estimate the percentage of topics in a medical website that have errors.

    d A county election official wants to assess the accuracy of the machine that counts the ballots by taking a sample of the paper ballots and comparing the estimated vote tallies for candidates from the sample to the machine counts.

9 Show that if $n_0/N \leq 1$, the value of n in (2.25) satisfies

$$e = z_{\alpha/2}\sqrt{\left(1 - \frac{n}{N}\right)}\frac{S}{\sqrt{n}}.$$

10 Which of the following SRS designs will give the most precision for estimating a population mean? Assume that each population has the same value of the population variance S^2.

    1. An SRS of size 400 from a population of size 4000

    2. An SRS of size 30 from a population of size 300

    3. An SRS of size 3000 from a population of size 300,000,000

B. Working with Survey Data

11 Mayr et al. (1994) took an SRS of 240 children who visited their pediatric outpatient clinic. They found the following frequency distribution for the age (in months) of free (unassisted) walking among the children:

Age (months)	9	10	11	12	13	14	15	16	17	18	19	20
Number of Children	13	35	44	69	36	24	7	3	2	5	1	1

a   Construct a histogram of the distribution of age at walking. Is the shape normally distributed? Do you think the sampling distribution of the sample average will be normally distributed? Why, or why not?

b   Find the mean, SE, and a 95% CI for the average age for onset of free walking.

c   Suppose the researchers wanted to do another study in a different region, and wanted a 95% CI for the mean age of onset of walking to have margin of error 0.5. Using the estimated standard deviation for these data, what sample size would they need to take?

12   The percentage of patients overdue for a vaccination is often of interest for a medical clinic Some clinics examine every record to determine that percentage; in a large practice, though, taking a census of the records can be time-consuming. Cullen (1994) took a sample of the 580 children served by an Auckland family practice to estimate the proportion of interest.

a   What sample size in an SRS (without replacement) would be necessary to estimate the proportion with 95% confidence and margin of error 0.10?

b   Cullen actually took an SRS *with* replacement of size 120, of whom 27 were *not* overdue for vaccination. Give a 95% CI for the proportion of children not overdue for vaccination.

13   Einarsen et al. (1998) selected an SRS of 935 assistant nurses from a Norwegian county with 2700 assistant nurses. A total of 745 assistant nurses (80%) responded to the survey.

a   20% of the 745 respondents reported that bullying occurred in their department. Using these respondents as the sample, give a 95% CI for the total number of nurses in the county who would report bullying in their department.

b   What assumptions must you make about the nonrespondents for the analysis in (a)?

14   In 2005, the Statistical Society of Canada (SSC) had 864 members listed in the online directory. An SRS of 150 of the members was selected; the sex and employment category (industry, academic, government) was ascertained for each person in the SRS, with results in file ssc.dat.

a   What are the possible causes of selection bias in this sample?

b   Estimate the percentage of members who are female, and give a 95% CI for your estimate.

c   Assuming that all members are listed in the online directory, estimate the total number of SSC members who are female, along with a 95% CI.

15   The data set agsrs.dat also contains information on other variables. For each of the following quantities, plot the data, and estimate the population mean for that variable along with a 95% CI.

a   Number of acres devoted to farms in 1987

b   Number of farms, 1992

c   Number of farms with 1000 acres or more, 1992

d   Number of farms with 9 acres or fewer, 1992

16 The Internet site www.golfcourse.com listed 14,938 golf courses by state. It gave a variety of information about each course, including greens fees, course rating, par for the course, and facilities. Data from an SRS of 120 of the golf courses is in file golfsrs.dat.

    a Display the data in a histogram for the weekday greens fees for nine holes of golf. How would you describe the shape of the data?

    b Find the average weekday greens fee to play nine holes of golf, and give the SE for your estimate.

17 Repeat Exercise 16 for the back tee yardage.

18 For the data in golfsrs.dat, estimate the proportion of golf courses that have 18 holes, and give a 95% CI for the population proportion.

19 The Special Census of Maricopa County, Arizona, gave 1995 populations for the following cities:

City	Population
Buckeye	4,857
Gilbert	59,338
Gila Bend	1,724
Phoenix	1,149,417
Tempe	153,821

Suppose that you are interested in estimating the percentage of persons who have been immunized against polio in each city and can take an SRS of persons. What should your sample size be in each of the 5 cities if you want the estimate from each city to have margin of error of 4 percentage points? For which cities does the finite population correction make a difference?

C. Working with Theory

20 Define the confidence interval procedure by

$$\text{CI}(\mathcal{S}) = [\hat{t}_{\mathcal{S}} - 1.96 \text{ SE}(\hat{t}_{\mathcal{S}}), \hat{t}_{\mathcal{S}} + 1.96 \text{ SE}(\hat{t}_{\mathcal{S}})].$$

Using the method illustrated in Example 2.9, find the exact confidence level for a CI based on an SRS (without replacement) of size 4 from the population in Example 2.2. Does your confidence level equal 95%?

21 One way of selecting an SRS is to assign a number to every unit in the population, then use a random number table to select units from the list. A page from a random number table is given in file rnt.dat. Explain why each of the following methods will or will not result in a simple random sample.

    a The population has 742 units, and we want to take an SRS of size 30. Divide the random digits into segments of size 3 and throw out any sequences of three digits not between 001 and 742. If a number occurs that has already been included in the sample, ignore it. If we used this method with the first line of random numbers in rnt.dat, the sequence of three-digit numbers would be

749   700   699   611   136   ...

We would include units 700, 699, 611, and 136 in the sample.

b For the situation in (a), when a random three-digit number is larger than 742, eliminate only the first digit and start the sequence with the next digit. With this procedure, the first five numbers would be 497, 006, 611, 136, and 264.

c Now suppose the population has 170 items. If we used the procedures described in (a) or (b), we would throw away many of the numbers from the list. To avoid this waste, divide every random three-digit number by 170 and use the rounded remainder as the unit in the sample. If the remainder is 0, use unit 170. For the sequence in the first row of the random number table, the numbers generated would be

69   20   19   101   136   ...

d Suppose the population has 200 items. Take two-digit sequences of random numbers and put a decimal point in front of each to obtain the sequence

0.74   0.97   0.00   0.69   0.96   ...

Then multiply each decimal by 200 to get the units for the sample (convert 0.00 to 200):

148   194   200   138   192   ...

e A school has 20 homeroom classes; each homeroom class contains between 20 and 40 students. To select a student for the sample, draw a random number between 1 and 20; then select a student at random from the chosen class. Do not include duplicates in your sample.

f For the situation in the preceding question, select a random number between 1 and 20 to choose a class. Then select a second random number between 1 and 40. If the number corresponds to a student in the class then select that student; if the second random number is larger than the class size, then ignore this pair of random numbers and start again. As usual, eliminate duplicates from your list.

22 Suppose we are interested in estimating the proportion p of a population that has a certain disease. As in Section 2.3 let $y_i = 1$ if person i has the disease, and $y_i = 0$ if person i does not have the disease. Then $\hat{p} = \bar{y}$.

a Show, using the definition in (2.13), that

$$CV(\hat{p}) = \sqrt{\frac{N-n}{N-1}\frac{1-p}{np}}.$$

If the population is large and the sampling fraction is small, so that $\dfrac{N-n}{N-1} \approx 1$, write (2.26) in terms of the CV for a sample of size 1.

FIGURE **2.6**
Histogram of the means of 1000 samples of size 300, taken with replacement from the data in Example 2.5.

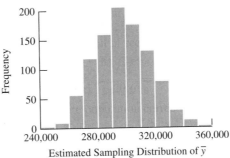

Estimated Sampling Distribution of $\bar{y}$

b Suppose that the fpc ≈ 1. Consider populations with p taking the successive values

$$0.001, 0.005, 0.01, 0.05, 0.10, 0.30, 0.50, 0.70, 0.90, 0.95, 0.99, 0.995, 0.999.$$

For each value of p, find the sample size needed to estimate the population proportion (a) with fixed margin of error 0.03, using (2.25), and (b) with relative error $0.03p$, using (2.26). What happens to the sample sizes for small values of p?

23 (Requires probability.) In the population used in Example 2.5, 19 of the 3078 counties in the population are missing the value of *acres92*. What is the probability that an SRS of size 300 would have no missing data for that variable?

24 *Decision theoretic approach for sample size estimation.* (Requires calculus.) In a decision theoretic approach, two functions are specified:

$$L(n) = \text{ Loss or "cost" of a bad estimate}$$
$$C(n) = \text{ Cost of taking the sample}$$

Suppose that for some constants c_0, c_1, and k,

$$L(n) = k\, V(\bar{y}_S) = k \left(1 - \frac{n}{N}\right) \frac{S^2}{n}$$
$$C(n) = c_0 + c_1 n.$$

What sample size n minimizes the total cost $L(n) + C(n)$?

25 (Requires computing.) If you have a large SRS, you can estimate the sampling distribution of $\bar{y}_S$ by repeatedly taking samples of size n with replacement from the list of sample values. A histogram of the means from 1000 samples of size 300 with replacement from the data in Example 2.5 is displayed in Figure 2.6; the shape may be slightly skewed, but still appears approximately normal. Would a sample of size 100 from this population be sufficiently large to use the central limit theorem? Take 500 samples with replacement of size 100 from the variable *acres92* in agsrs.dat, and draw a histogram of the 500 means. The approach described in this exercise is known as the **bootstrap** (see Efron and Tibshirani, 1993); we discuss the bootstrap further in Section 9.3.

26 (Requires probability.) In an SRS, each possible subset of n units has probability $1/\binom{N}{n}$ of being chosen as the sample; in this chapter, we showed that this definition implies that each unit has probability n/N of appearing in the sample. The converse is not true, however. Show that the inclusion probability π_i for each unit in a systematic sample is n/N, but that condition (2.7) is not met.

27 (Requires probability.) A typical opinion poll surveys about 1000 adults. Suppose that the sampling frame contains 100 million adults including yourself, and that an SRS of 1000 adults is chosen from the frame.

    a What is the probability that you are selected to be in the sample?

    b Now suppose that 2000 such samples are selected, each sample selected independently of the others. What is the probability that you will *not* be in any of the samples?

    c How many samples must be selected for you to have a 0.5 probability of being in at least one sample?

28 (Requires probability.) In an SRSWR, a population unit can appear in the sample anywhere between 0 and n times. Let

$$Q_i = \text{number of times unit } i \text{ appears in the sample,}$$

and

$$\hat{t} = \frac{N}{n} \sum_{i=1}^{N} Q_i y_i.$$

    a Argue that the joint distribution of $Q_1, Q_2, \ldots, Q_N$ is multinomial with n trials and $p_1 = p_2 = \cdots = p_N = 1/N$.

    b Using (a) and properties of the multinomial distribution, show that $E[\hat{t}] = t$.

    c Using (a) and properties of the multinomial distribution, find $V[\hat{t}]$.

29 (Requires probability.) Suppose you would like to take an SRS of size n from a list of N units, but do not know the population size N in advance. Consider the following procedure:

    a Set $S_0 = \{1, 2, \ldots, n\}$, so that the initial sample for consideration consists of the first n units on the list.

    b For $k = 1, 2, \ldots$, generate a random number u_k between 0 and 1. If $u_k > n/(n+k)$, then set S_k equal to S_{k-1}. If $u_k \leq n/(n+k)$, then select one of the units in S_{k-1} at random, and replace it by unit $(n+k)$ to form S_k.

    Show that S_{N-n} from this procedure is an SRS of size n. HINT: Use induction.

D. Projects and Activities

30 *Rectangles.* This activity was suggested by Gnanadesikan et al. (1997). Figure 2.7 contains a population of 100 rectangles. Your goal is to estimate the total area of all the rectangles by taking a sample of 10 rectangles. Keep your results from this exercise; you will use them again in later chapters.

FIGURE 2.7
Population of 100 rectangles

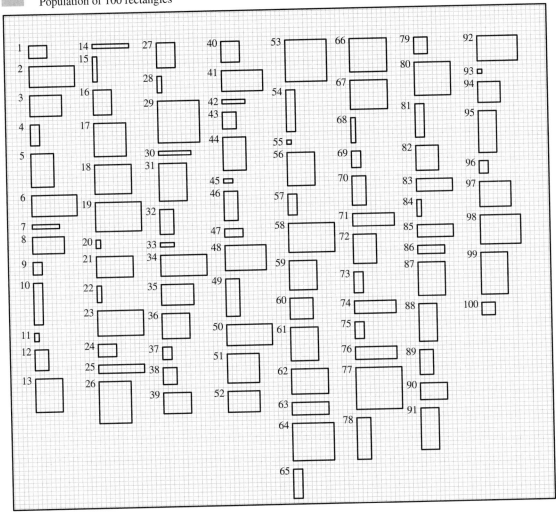

a   Select a purposive sample of 10 rectangles that you think will be representative of the population of 100 rectangles. Record the area (number of small squares) for each rectangle in your sample. Use your sample to estimate the total area. How did you choose the rectangles for your sample?

b   Find the sample variance for your purposive sample of 10 rectangles from part (a), and use (2.22) to form an interval estimate for the total area t.

c   Now take an SRS of 10 rectangles. Use your SRS to estimate the total area of all 100 rectangles, and find a 95% CI for the total area.

d   Compare your intervals with those of other students in the class. What percentage of the intervals from part (b) include the true total area of 3079? What about the CIs from part (c)?

31   *Mutual funds.* The websites of companies such as Fidelity (www.fidelity.com), Vanguard (www.vanguard.com), and T. Rowe Price (www.troweprice.com) list the mutual funds of those companies, along with some statistics about the performance of those funds. Take an SRS of 25 mutual funds from one of these companies. Describe how you selected the SRS. Find the mean and a 95% CI for the mean of a variable you are interested in, such as daily percentage change, or 1-year performance, or length of time the fund has existed.

32   *Baseball data.* This activity is due to Jenifer Boshes, who also compiled the data from Forman (2004) and publicly available salary information. The data file baseball.dat contains statistics on 797 baseball players from the rosters of all major league teams in November, 2004. In this exercise (which will be continued in later chapters), you will treat the file baseball.dat as a population and draw samples from it using different sampling designs.

    a   Take an SRS of 150 players from the file. Describe how you selected the SRS. Save your data set for use in future exercises (if you are selecting it using SAS PROC SURVEYSELECT, you can recreate the data set by using seed = *number*).

    b   Calculate logsal = ln(salary). Construct a histogram of the variables *salary* and *logsal* from your SRS. Does the distribution of *salary* appear approximately normal? What about *logsal*?

    c   Find the mean of the variable *logsal*, and give a 95% CI.

    d   Estimate the proportion of players in the data set who are pitchers, and give a 95% CI.

    e   Since you have the full data file for the population, you can find the true mean and proportion for the population. Do your CIs in (c) and (d) contain the true population values?

33   *Online bookstore.* The website amazon.com can be used to obtain populations of books, CDs, and other wares.

    a   In the books search window, type in a genre you like, such as mystery or sports; you may want to narrow your search by selecting a subcategory since an upper bound is placed on the number of books that can be displayed. Choose a genre with at least 20 pages of listings. The list of books forms your population.

    b   What is your target population? What is the population size, N?

    c   Take an SRS of 50 books from your population. Describe how you selected the SRS, and record the amount of time you spent taking the sample and collecting the data.

    d   Record the following information for each book in your SRS: price, number of pages, and whether the book is paperback or hardback.

    e   Give a point estimate and a 95% CI for the mean price of books in the genre you selected.

    f   Give a point estimate and a 95% CI for the mean number of pages for books in the genre you selected.

    g   Explain, to a person who knows no statistics, what your estimates and CIs mean.

34 Take a small SRS of something you're interested in. Explain what it is you decide to study and carefully describe how you chose your SRS (give the random numbers generated and explain how you translated them into observations), report your data, and give a point estimate and the SE for the quantity or quantities of interest.

The data collection for this exercise should not take a great deal of effort, as you are surrounded by things waiting to be sampled. Some examples: mutual fund data in the financial section of today's newspaper, actual weights of 1-pound bags of carrots at the supermarket, or the cost of a used dining room table from an online classified advertisement site.

35 *Estimating the size of an audience.* A common method for estimating the size of an audience is to take an SRS of n of the N rows in an auditorium, count the number of people in each of the selected rows, then multiply the total number of people in your sample by N/n.

a Why is it important to take an SRS instead of a convenience sample of the first 10 rows?

b Go to a performance or a lecture, and count the number of rows in the auditorium. Take an SRS of 10 or 20 rows, count the number of people in each row, and estimate the number of people in the audience using this method. Give a 95% CI.

36 *Forest data.* The data in file forest.dat are from kdd.ics.uci.edu/databases/covertype/ covertype.data.html (Blackard, 1998). They consist of a subset of the measurements from 581,012 30×30 m cells from Region 2 of the U.S. Forest Service Resource Information System. The original data were used in a data mining application, predicting forest cover type from covariates. Data-mining methods are often used to explore relationships in very large data sets; in many cases, the data sets are so large that statistical software packages cannot analyze them. Many data-mining problems, however, can be alternatively approached by analyzing probability samples from the population. In these exercises, we treat forest.dat as a population.

a Select an SRS of size 2000 from the 581,012 records. Keep this sample, or the random number seed you used to generate the sample, for later use in Chapter 4.

b Using your SRS, estimate the percentage of cells in each of the 7 forest cover types, along with 95% CIs.

c Estimate the average elevation in the population, with 95% CI.

37 *IPUMS data.* This exercise is designed for the Integrated Public Use Microdata Series (IPUMS), available online at www.ipums.org/usa/ (Ruggles et al., 2004). The IPUMS site hosts a collection of samples from the U.S. Decennial Census and American Community Survey. In the following exercises, we use a self-weighting sample selected from the 1980 Decennial Census sample, selected using the "Small Sample Density" option in the data extract tool. The data are in file ipums.dat. We treat these data as a population.

a The variable *inctot* is total personal income from all sources. Note from the documentation for the variable that it is "topcoded" at $75,000 to protect the confidentiality of the respondents. What effect does the topcoding have on estimates from the file?

b   Draw a pilot sample (SRS) of size 50 from the IPUMS population. Use the sample variance you get for *inctot* to determine the sample size you need to estimate the average of *inctot* with a margin of error of 700 or less.

c   Take an SRS of your desired sample size from the population. Estimate the total income for the population, and give a 95% CI. Make sure you save the seed number you use in SAS PROC SURVEYSELECT or other software so you can recreate this sample in later chapters.

E. SURVEY Exercises

The program SURVEY (Chang et al., 1992) allows you to draw samples from a hypothetical population to learn about cable TV practices. The website contains the programs and a description of the population, and gives exercises and activities for using the population. The SURVEY exercises continue in subsequent chapters.

3

Stratified Sampling

One of the things she [Mama] taught me should be obvious to everyone, but I still find a lot of cooks who haven't figured it out yet. Put the food on first that takes the longest to cook.

—Pearl Bailey, *Pearl's Kitchen*

3.1
What Is Stratified Sampling?

EXAMPLE 3.1 The Federal Deposit Insurance Corporation (FDIC) was created in 1933 by the U.S. Congress to supervise banks; it insures deposits at member banks up to a specified limit. When a bank fails, the FDIC acquires the assets from that bank and uses them to help pay the insured depositors. Valuing the assets is time-consuming, so the FDIC selects a sample of the assets in order to estimate the total amount recovered from financial institutions (Chapman, 2005). The assets from failed institutions fall into several types: (1) consumer loans, (2) commercial loans, (3) securities, (4) real estate mortgages, (5) other owned real estate, (6) other assets, and (7) net investments in subsidiaries. A simple random sample (SRS) of assets may result in an imprecise estimate of the total amount recovered. Consumer loans tend to be much smaller on average than assets in the other classes, so the sample variance from an SRS can be very large. In addition, an SRS might contain no assets from one or more of the asset types; if category (2) assets tend to have the most monetary value and the sample chosen has no assets from category (2), that sample may result in an estimate of total assets that is too small. It would be desirable to have a method for sampling that prevents samples that we know would produce bad estimates, and that increases the precision of the estimators. **Stratified sampling** can accomplish these goals. ■

Often, we have supplementary information that can help us design our sample. For example, we would know before undertaking an income survey that men generally earn more than women, that New York City residents pay more for housing than residents of Des Moines, or that rural residents shop for groceries less frequently than urban residents. The FDIC has information on the type of each asset, which is related to the value of the asset.

If the variable we are interested in takes on different mean values in different subpopulations, we may be able to obtain more precise estimates of population quantities by taking a **stratified** random sample. The word *stratify* comes from Latin words meaning "to make layers"; we divide the population into H subpopulations, called **strata**. The strata do not overlap, and they constitute the whole population so that each sampling unit belongs to exactly one stratum. We draw an independent probability sample from each stratum, then pool the information to obtain overall population estimates.

We use stratified sampling for one or more of the following reasons:

1 We want to be protected from the possibility of obtaining a really bad sample. When taking an SRS of size 100 from a population of 1000 male and 1000 female students, obtaining a sample with no or very few males is theoretically possible, although such a sample is not likely to occur. Most people would not consider such a sample to be representative of the population and would worry that men and women might respond differently on the item of interest. In a stratified sample, you can take an SRS of 50 males and an independent SRS of 50 females, guaranteeing that the proportion of males in the sample is the same as that in the population. With this design, a sample with no or few males cannot be selected.

2 We may want data of known precision for subgroups of the population. These subgroups should be the strata. McIlwee and Robinson (1992) sampled graduates from electrical and mechanical engineering programs at public universities in southern California. They were interested in comparing the educational and workforce experiences of male and female graduates, so they stratified their sampling frame by gender and took separate random samples of male and female graduates. Because there were many more male than female graduates, they sampled a higher fraction of female graduates than male graduates in order to obtain comparable precisions for the two groups.

3 A stratified sample may be more convenient to administer and may result in a lower cost for the survey. For example, sampling frames may be constructed differently in different strata, or different sampling designs or field procedures may be used. In a survey of businesses, an Internet survey might be used for large firms while a mail or telephone survey is used for small firms. In other surveys, a different procedure may be used for sampling households in urban strata than in rural strata.

4 Stratified sampling often gives more precise (having lower variance) estimates for population means and totals. Persons of different ages tend to have different blood pressures, so in a blood pressure study it would be helpful to stratify by age groups. If studying the concentration of plants in an area, one would stratify by type of terrain; marshes would have different plants than woodlands. Stratification works for lowering the variance because the variance within each stratum is often lower than the variance in the whole population. Prior knowledge can be used to save money in the sampling procedure.

EXAMPLE **3.2** Refer to Example 2.5, in which we took an SRS to estimate the average number of farm acres per county. In Example 2.5, we noted that even though we scrupulously generated a random sample, some areas were overrepresented, and others not represented at all. Taking a stratified sample can provide some balance in the sample on the stratifying variable.

The SRS in Example 2.5 exhibited a wide range of values for y_i, the number of acres devoted to farms in county i in 1992. You might conjecture that part of the large variability arises because counties in the western United States are larger, and thus tend to have larger values of y, than counties in the eastern United States.

For this example, we use the four census regions of the United States—Northeast, North Central, South, and West—as strata. The SRS in Example 2.5 sampled about 10% of the population; to be able to compare the results of the stratified sample with the SRS, we also sample about 10% of the counties in each stratum. (We discuss other stratified sampling designs later in the chapter.)

Stratum	Number of Counties in Stratum	Number of Counties in Sample
Northeast	220	21
North Central	1054	103
South	1382	135
West	422	41
Total	3078	300

We select four separate SRSs, one from each of the four strata. To select the SRS from the Northeast stratum, we number the counties in that stratum from 1 to 220, and select 21 numbers randomly from $\{1, \ldots, 220\}$. We follow a similar procedure for the other three strata, selecting 103 counties at random from the 1054 in the North Central region, 135 counties from the 1382 in the South, and 41 counties from the 422 in the West. The four SRSs are selected independently: Knowing which counties are in the sample from the Northeast tells us nothing about which counties are in the sample from the South.

The data sampled from all four strata are in data file agstrat.dat. A boxplot, showing the data for each stratum, is in Figure 3.1. Summary statistics for each stratum are given below:

Region	Sample Size	Average	Variance
Northeast	21	97,629.8	7,647,472,708
North Central	103	300,504.2	29,618,183,543
South	135	211,315.0	53,587,487,856
West	41	662,295.5	396,185,950,266

Since we took an SRS in each stratum, we can use (2.15) and (2.17) to estimate the population quantities for each stratum. We use

$$(220)(97,629.81) = 21,478,558.2.$$

FIGURE   3.1

The boxplot of data from Example 3.2. The thick line for each region is the median of the sample data from that region; the other horizontal lines in the boxes are the 25th and 75th percentiles. The Northeast region has a relatively small median and small variance; the West region, however, has a much higher median and variance. The distribution of farm acreage appears to be positively skewed in each of the regions.

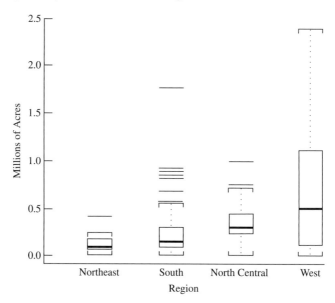

to estimate the total number of acres devoted to farms in the Northeast, with estimated variance

$$(220)^2 \left(1 - \frac{21}{220}\right) \frac{7,647,472,708}{21} = 1.594316 \times 10^{13}.$$

The following table gives estimates of the total number of farm acres and estimated variance of the total for each of the four strata:

Stratum	Estimated Total of Farm Acres	Estimated Variance of Total
Northeast	21,478,558.2	1.59432×10^{13}
North Central	316,731,379.4	2.88232×10^{14}
South	292,037,390.8	6.84076×10^{14}
West	279,488,706.1	1.55365×10^{15}
Total	909,736,034.4	2.5419×10^{15}

We can estimate the total number of acres devoted to farming in the United States in 1992 by adding the totals for each stratum; as sampling was done independently in each stratum, the variance of the U.S. total is the sum of the variances of the stratum totals. Thus we estimate the total number of acres devoted to farming as 909,736,034, with standard error (SE) $\sqrt{2.5419 \times 10^{15}} = 50,417,248$. We would estimate the average number of acres devoted to farming per county as 909,736,034/3078 = 295,560.7649, with standard error 50,417,248/3078 = 16,379.87.

For comparison, the estimate of the population total in Example 2.5, using an SRS of size 300, was 916,927,110, with standard error 58,169,381. For this example, stratified sampling ensures that each region of the United States is represented in the sample, and produces an estimate with smaller standard error than an SRS with the same number of observations. The sample variance in Example 2.5 was $s^2 = 1.1872 \times 10^{11}$. Only the West had sample variance larger than s^2; the sample variance in the Northeast was only 7.647×10^9.

Observations within many strata tend to be more homogeneous than observations in the population as a whole, and the reduction in variance in the individual strata often leads to a reduced variance for the population estimate. In this example, the relative gain from stratification can be estimated by the ratio

$$\frac{\text{estimated variance from stratified sample, with } n = 300}{\text{estimated variance from SRS, with } n = 300} = \frac{2.5419 \times 10^{15}}{3.3837 \times 10^{15}} = 0.75.$$

If these figures were the population variances, we would expect that we would need only (300)(0.75) = 225 observations with a stratified sample to obtain the same precision as from an SRS of 300 observations.

Of course, no law says that you must sample the same fraction of observations in every stratum. In this example, there is far more variability from county to county in the western region; if acres devoted to farming were the primary variable of interest, you would reduce the variance of the estimated total even further by taking a higher sampling fraction in the western region than in the other regions. You will explore an alternative sampling design in Exercise 12. ∎

3.2
Theory of Stratified Sampling

We divide the population of N sampling units into H "layers" or strata, with N_h sampling units in stratum h. For stratified sampling to work, we must know the values of $N_1, N_2, \ldots, N_H$, and must have

$$N_1 + N_2 + \cdots + N_H = N,$$

where N is the total number of units in the entire population.

In **stratified random sampling**, the simplest form of stratified sampling, we independently take an SRS from each stratum, so that n_h observations are randomly selected from the N_h population units in stratum h. Define $\mathcal{S}_h$ to be the set of n_h units in the SRS for stratum h. The total sample size is $n = n_1 + n_2 + \cdots + n_H$.

Notation for Stratification: The population quantities are:

$$y_{hj} = \text{value of } j\text{th unit in stratum } h$$

$$t_h = \sum_{j=1}^{N_h} y_{hj} = \text{population total in stratum } h$$

$$t = \sum_{h=1}^{H} t_h = \text{population total}$$

$$\bar{y}_{hU} = \frac{\sum_{j=1}^{N_h} y_{hj}}{N_h} = \text{population mean in stratum } h$$

$$\bar{y}_U = \frac{t}{N} = \frac{\sum_{h=1}^{H} \sum_{j=1}^{N_h} y_{hj}}{N} = \text{overall population mean}$$

$$S_h^2 = \sum_{j=1}^{N_h} \frac{(y_{hj} - \bar{y}_{hU})^2}{N_h - 1} = \text{population variance in stratum } h$$

Corresponding quantities for the sample, using SRS estimators within each stratum, are:

$$\bar{y}_h = \frac{1}{n_h} \sum_{j \in \mathcal{S}_h} y_{hj}$$

$$\hat{t}_h = \frac{N_h}{n_h} \sum_{j \in \mathcal{S}_h} y_{hj} = N_h \bar{y}_h$$

$$s_h^2 = \sum_{j \in \mathcal{S}_h} \frac{(y_{hj} - \bar{y}_h)^2}{n_h - 1}$$

Suppose we only sampled the hth stratum. In effect, we have a population of N_h units and take an SRS of n_h units. Then we would estimate $\bar{y}_{hU}$ by $\bar{y}_h$, and t_h by $\hat{t}_h = N_h \bar{y}_h$. The population total is $t = \sum_{h=1}^{H} t_h$, so we estimate t by

$$\hat{t}_{\text{str}} = \sum_{h=1}^{H} \hat{t}_h = \sum_{h=1}^{H} N_h \bar{y}_h. \tag{3.1}$$

To estimate $\bar{y}_U$, then, we use

$$\bar{y}_{\text{str}} = \frac{\hat{t}_{\text{str}}}{N} = \sum_{h=1}^{H} \frac{N_h}{N} \bar{y}_h. \tag{3.2}$$

This is a weighted average of the sample stratum averages; $\bar{y}_h$ is multiplied by N_h/N, the proportion of the population units in stratum h. To use stratified sampling, the sizes or relative sizes of the strata must be known.

The properties of these estimators follow directly from the properties of SRS estimators:

- **Unbiasedness.** $\bar{y}_{str}$ and $\hat{t}_{str}$ are unbiased estimators of $\bar{y}_U$ and t. An SRS is taken in each stratum, so (2.30) implies that $E[\bar{y}_h] = \bar{y}_{hU}$ and consequently

$$E\left[\sum_{h=1}^{H} \frac{N_h}{N} \bar{y}_h\right] = \sum_{h=1}^{H} \frac{N_h}{N} E[\bar{y}_h] = \sum_{h=1}^{H} \frac{N_h}{N} \bar{y}_{hU} = \bar{y}_U.$$

- **Variance of the estimators.** Since we are sampling independently from the strata, and we know $V(\hat{t}_h)$ from the SRS theory, the properties of expected value in Section A.2 and (2.16) imply that

$$V(\hat{t}_{str}) = \sum_{h=1}^{H} V(\hat{t}_h) = \sum_{h=1}^{H} \left(1 - \frac{n_h}{N_h}\right) N_h^2 \frac{S_h^2}{n_h}. \tag{3.3}$$

- **Standard errors for stratified samples.** We can obtain an unbiased estimator of $V(\hat{t}_{str})$ by substituting the sample estimators s_h^2 for the population parameters S_h^2. Note that in order to estimate the variances, we need to sample at least two units from each stratum.

$$\hat{V}(\hat{t}_{str}) = \sum_{h=1}^{H} \left(1 - \frac{n_h}{N_h}\right) N_h^2 \frac{s_h^2}{n_h} \tag{3.4}$$

$$\hat{V}(\bar{y}_{str}) = \frac{1}{N^2} \hat{V}(\hat{t}_{str}) = \sum_{h=1}^{H} \left(1 - \frac{n_h}{N_h}\right) \left(\frac{N_h}{N}\right)^2 \frac{s_h^2}{n_h}. \tag{3.5}$$

As always, the standard error of an estimator is the square root of the estimated variance: $\text{SE}(\bar{y}_{str}) = \sqrt{\hat{V}(\bar{y}_{str})}$.

- **Confidence intervals for stratified samples.** If either (1) the sample sizes within each stratum are large, or (2) the sampling design has a large number of strata, an approximate $100(1-\alpha)\%$ confidence interval (CI) for the population mean $\bar{y}_U$ is

$$\bar{y}_{str} \pm z_{\alpha/2} \, \text{SE}\,(\bar{y}_{str}).$$

The central limit theorem used for constructing this CI is stated in Krewski and Rao (1981). Some survey software packages use the percentile of a t distribution with $n - H$ degrees of freedom (df) rather than the percentile of the normal distribution.

EXAMPLE **3.3** Siniff and Skoog (1964) used stratified random sampling to estimate the size of the Nelchina herd of Alaska caribou in February of 1962. In January and early February, several sampling techniques were field-tested. The field tests told the investigators that several of the proposed sampling units, such as equal-flying-time sampling units, were difficult to implement in practice, and that an equal-area sampling unit of 4 square miles (mi^2) would work well for the survey. The biologists used preliminary estimates of caribou densities to divide the area of interest into six strata; each stratum was then divided into a grid of 4-mi^2 sampling units. Stratum A, for example, contained

TABLE   3.1

Spreadsheet for Calculations in Example 3.3

	A	B	C	D	E	F	G
1	Stratum	N_h	n_h	$\bar{y}_h$	s_h^2	$\hat{t}_h = N_h\bar{y}_h$	$\left(1 - \dfrac{n_h}{N_h}\right)N_h^2\dfrac{s_h^2}{n_h}$
2	A	400	98	24.1	5,575	9,640	6,872,040.82
3	B	30	10	25.6	4,064	768	243,840.00
4	C	61	37	267.6	347,556	16,323.6	13,751,945.51
5	D	18	6	179.0	22,798	3,222	820,728.00
6	E	70	39	293.7	123,578	20,559	6,876,006.67
7	F	120	21	33.2	9,795	3,984	5,541,171.43
8	total		211			54,496.6	34,105,732.43
9	$\sqrt{\text{total}}$						5,840.01

$N_1 = 400$ sampling units; $n_1 = 98$ of these were randomly selected to be in the survey. The following data were reported:

Stratum	N_h	n_h	$\bar{y}_h$	s_h^2
A	400	98	24.1	5,575
B	30	10	25.6	4,064
C	61	37	267.6	347,556
D	18	6	179.0	22,798
E	70	39	293.7	123,578
F	120	21	33.2	9,795

The spreadsheet shown in Table 3.1 displays the calculations for finding the stratified sampling estimates. The estimated total number of caribou is 54,497 with standard error 5,840. An approximate 95% CI for the total number of caribou is

$$54{,}497 \pm 1.96(5840) = [43{,}051,\ 65{,}943].$$

Of course, this CI only reflects the uncertainty due to sampling error; if the field procedure for counting caribou tends to miss animals, then the entire CI will be too low. ∎

Stratified Sampling for Proportions   As we observed in Section 2.3, a proportion is a mean of a variable that takes on values 0 and 1. To make inferences about proportions, we simply use the results in (3.1)–(3.5), with $\bar{y}_h = \hat{p}_h$ and $s_h^2 = \dfrac{n_h}{n_h - 1}\hat{p}_h(1 - \hat{p}_h)$. Then,

$$\hat{p}_{str} = \sum_{h=1}^{H} \frac{N_h}{N}\hat{p}_h \tag{3.6}$$

and

$$\hat{V}(\hat{p}_{\text{str}}) = \sum_{h=1}^{H} \left(1 - \frac{n_h}{N_h}\right) \left(\frac{N_h}{N}\right)^2 \frac{\hat{p}_h(1 - \hat{p}_h)}{n_h - 1}. \tag{3.7}$$

Estimating the total number of population units having a specified characteristic is similar:

$$\hat{t}_{\text{str}} = \sum_{h=1}^{H} N_h \hat{p}_h,$$

so the estimated total number of population units with the characteristic is the sum of the estimated totals in each stratum. Similarly, $\hat{V}(\hat{t}_{\text{str}}) = N^2 \hat{V}(\hat{p}_{\text{str}})$.

EXAMPLE 3.4   The American Council of Learned Societies (ACLS) used a stratified random sample of selected ACLS societies in seven disciplines to study publication patterns and computer and library use among scholars who belong to one of the member organizations of the ACLS (Morton and Price, 1989). The data are shown in Table 3.2.

Ignoring the nonresponse for now (we'll return to the nonresponse in Exercise 7 of Chapter 8) and supposing there are no duplicate memberships, let's use the stratified sample to estimate the percentage and number of respondents of the major societies in those seven disciplines that are female. Here, let N_h be the membership figures, and let n_h be the number of valid surveys. Thus,

$$\hat{p}_{\text{str}} = \sum_{h=1}^{7} \frac{N_h}{N} \hat{p}_h = \frac{9100}{44{,}000} 0.38 + \ldots + \frac{9000}{44{,}000} 0.26 = 0.2465$$

and

$$\text{SE}(\hat{p}_{\text{str}}) = \sqrt{\sum_{h=1}^{7} \left(1 - \frac{n_h}{N_h}\right) \left(\frac{N_h}{N}\right)^2 \frac{\hat{p}_h(1 - \hat{p}_h)}{n_h - 1}} = 0.0071.$$

TABLE 3.2
Data from ACLS Survey

Discipline	Membership	Number Mailed	Valid Returns	Female Members (%)
Literature	9,100	915	636	38
Classics	1,950	633	451	27
Philosophy	5,500	658	481	18
History	10,850	855	611	19
Linguistics	2,100	667	493	36
Political Science	5,500	833	575	13
Sociology	9,000	824	588	26
Totals	44,000	5,385	3,835	

The estimated total number of female members in the societies is $\hat{t}_{\text{str}} = 44{,}000 \times (0.2465) = 10{,}847$, with $\text{SE}(\hat{t}_{\text{str}}) = 44{,}000 \times (0.0071) = 312$. ∎

3.3
Sampling Weights in Stratified Random Sampling

We introduced the notion of sampling weight, $w_i = 1/\pi_i$, in Section 2.4. For an SRS, the sampling weight for each observation is the same since all of the inclusion probabilities π_i are the same. In stratified sampling, however we may have different inclusion probabilities in different strata so that the weights may be unequal for some stratified sampling designs.

The stratified sampling estimator $\hat{t}_{\text{str}}$ can be expressed as a weighted sum of the individual sampling units: Using (3.1),

$$\hat{t}_{\text{str}} = \sum_{h=1}^{H} N_h \bar{y}_h = \sum_{h=1}^{H} \sum_{j \in \mathcal{S}_h} \frac{N_h}{n_h} y_{hj}.$$

The estimator of the population total in stratified sampling may thus be written as

$$\hat{t}_{\text{str}} = \sum_{h=1}^{H} \sum_{j \in \mathcal{S}_h} w_{hj} y_{hj}, \tag{3.8}$$

where the sampling weight for unit j of stratum h is $w_{hj} = (N_h/n_h)$. The sampling weight can again be thought of as the number of units in the population represented by the sample member y_{hj}. If the population has 1600 men and 400 women, and the stratified sample design specifies sampling 200 men and 200 women, then each man in the sample has weight 8 and each woman has weight 2. Each woman in the sample represents herself and another woman not selected to be in the sample, and each man represents himself and seven other men not in the sample. Note that the probability of including unit j of stratum h in the sample is $\pi_{hj} = n_h/N_h$, the sampling fraction in stratum h. Thus, as before, the sampling weight is simply the reciprocal of the inclusion probability:

$$w_{hj} = \frac{1}{\pi_{hj}}. \tag{3.9}$$

The sum of the sampling weights in stratified random sampling equals the population size N; each sampled unit "represents" a certain number of units in the population, so the whole sample "represents" the whole population. In a stratified random sample, the population mean is thus estimated by

$$\bar{y}_{\text{str}} = \frac{\displaystyle\sum_{h=1}^{H} \sum_{j \in \mathcal{S}_h} w_{hj} y_{hj}}{\displaystyle\sum_{h=1}^{H} \sum_{j \in \mathcal{S}_h} w_{hj}}. \tag{3.10}$$

FIGURE   3.2

A stratified random sample from a population with $N = 500$. The top row is Stratum 1; rows 2–4 comprise Stratum 2; the bottom 21 rows are Stratum 3. Units in the sample are shaded. Stratum 1 has $N_1 = 20$ and $n_1 = 10$, so the sampling weight for each unit in Stratum 1 is 2. For Stratum 2, $N_2 = 60$, $n_2 = 12$, and the sampling weight for each unit in Stratum 2 is 5. For Stratum 3, $N_3 = 420$, $n_3 = 20$, and the sampling weight for each unit in Stratum 3 is 21.

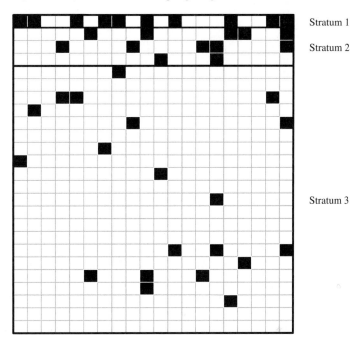

Stratum 1

Stratum 2

Stratum 3

Figure 3.2 illustrates a stratified random sample for a population with 3 strata. The sampling weights are smallest in Stratum 1, where half of the stratum population units are sampled.

A stratified sample is self-weighting if the sampling fraction n_h/N_h is the same for each stratum. In that case, the sampling weight for each observation is N/n, exactly the same as in an SRS. The variance of a stratified random sample, however, depends on the stratification, so (3.4) must be used to estimate the variance of $\hat{t}_{\text{str}}$. Equation (3.4) requires that you calculate the variance separately within each stratum; the weights do not tell you the stratum membership of the observations.

EXAMPLE 3.5   For the caribou survey in Example 3.3, the weights are

Stratum	N_h	n_h	w_{hj}
A	400	98	4.08
B	30	10	3.00
C	61	37	1.65
D	18	6	3.00
E	70	39	1.79
F	120	21	5.71

In stratum A, each sampling unit of 4 mi^2 represents 4.08 sampling units in the stratum (including itself); in stratum B, a sampling unit in the sample represents itself and 2 other sampling units that are not in the sample. To estimate the population total, then, a new variable of weights could be constructed. This variable would contain the value 4.08 for every observation in stratum A, 3.00 for every observation in stratum B, and so on. ∎

EXAMPLE 3.6   The sample in Example 3.2 was designed so that each county in the United States would have approximately the same probability of appearing in the sample. To estimate the total number of acres devoted to agriculture in the United States, we create the variable *strwt* in file agstrat.dat with the sampling weights; it contains the value 220/21 for counties in the Northeast stratum, 1054/103 for the North Central counties, 1382/135 for the South counties, and 422/41 for the West counties. We can use (3.8) to estimate the population total by forming a new column containing the product of variables *strwt* and *acres92*, then calculating the sum of the new column. In doing so, we calculate $\hat{t}_{str} = 909,736,035$, the same (up to roundoff error) estimate as obtained in Example 3.2. Note that even though this sample is approximately self-weighting, it is not exactly self-weighting because the stratum sample sizes must be integers. When calculating estimates, use the exact weights from each stratum.

The variable *strwt* can be used to estimate population means or totals for every variable measured in the sample, and most computer packages for surveys use the weight variable to calculate point estimates. Note, however, that you cannot calculate the standard error of $\hat{t}_{str}$ unless you know the stratification—you need to use (3.4) to estimate the variance. Partial output from SAS PROC SURVEYMEANS for the variable *acres92* is given in below.

```
                        Data Summary

                Number of Strata               4
                Number of Observations       300
                Sum of Weights              3078

                        Statistics

                              Std Error
    Variable   N     DF    Mean   of Mean   95% CL for Mean
    ----------------------------------------------------------------
    acres92   300   296   295561   16380   263325.000 327796.530
    ----------------------------------------------------------------


                        Statistics

        Variable        Sum     Std Dev        95% CL for Sum
        ----------------------------------------------------------------
        acres92   909736035   50417248   810514350 1008957721
        ----------------------------------------------------------------
```

The SAS code on the website also plots the data and finds estimates for other variables. ∎

3.4
Allocating Observations to Strata

So far we have simply analyzed data from a survey that someone else has designed. Designing the survey is the most important part of using a survey in research: If the survey is badly designed, then no amount of analysis will yield the needed information. Survey design includes methods for controlling nonsampling as well as sampling error. We discuss design issues for nonsampling error in Chapters 8 and 15. In this chapter, we discuss design features that affect the sampling error. Simple random sampling involved one design feature: the sample size (Section 2.6). For stratified random sampling, we need to design what the strata should be, then decide how many observations to sample in each stratum. It is somewhat easier to look at these in reverse order. In this section, we assume that the strata have already been fixed, and we discuss methods of allocating observations to the strata. File stratselect.sas gives sample SAS code that can be used to select stratified samples using the allocation methods in this section.

3.4.1   Proportional Allocation

If you are taking a stratified sample in order to ensure that the sample reflects the population with respect to the stratification variable and you would like your sample to be a miniature version of the population, you should use proportional allocation when designing the sample.

In **proportional allocation**, so called because the number of sampled units in each stratum is proportional to the size of the stratum, the inclusion probability $\pi_{hj} = n_h/N_h$ is the same ($= n/N$) for all strata; in a population of 2400 men and 1600 women, proportional allocation with a 10% sample would mean sampling 240 men and 160 women. Thus the probability that an individual will be selected to be in the sample, n/N, is the same as in an SRS, but many of the "bad" samples that could occur in an SRS (for example, a sample in which all 400 persons are men) cannot be selected in a stratified sample with proportional allocation.

If proportional allocation is used, each unit in the sample represents the same number of units in the population: In our example, each man in the sample represents 10 men in the population, and each woman represents 10 women in the population. The sampling weight for every unit in the sample thus equals 10, and the stratified sampling estimator of the population mean is simply the average of all of the observations. Proportional allocation thus results in a self-weighting sample. The sample in Example 3.2 was designed to be approximately self-weighting. In a self-weighting sample, $\bar{y}_{str}$ is the average of all observations in the sample.

When the strata are large enough, the variance of $\bar{y}_{str}$ under proportional allocation is usually at most as large as the variance of the sample mean from an SRS with the same number of observations. This is true no matter how silly the stratification scheme may be. To see why this might be so, let's display the variances between strata and within strata, for proportional allocation, in an ANOVA table for the population (Table 3.3).

In a stratified sample of size n with proportional allocation, since $n_h/N_h = n/N$, Equation (3.3) implies that

$$V_{\text{prop}}(\hat{t}_{\text{str}}) = \sum_{h=1}^{H} \left(1 - \frac{n_h}{N_h}\right) N_h^2 \frac{S_h^2}{n_h}$$

$$= \left(1 - \frac{n}{N}\right) \frac{N}{n} \sum_{h=1}^{H} N_h S_h^2$$

$$= \left(1 - \frac{n}{N}\right) \frac{N}{n} \left(\text{SSW} + \sum_{h=1}^{H} S_h^2\right).$$

The sums of squares add up, with SSTO = SSW + SSB, so the variance of the estimated population total from an SRS of size n is

$$V_{\text{SRS}}(\hat{t}) = \left(1 - \frac{n}{N}\right) N^2 \frac{S^2}{n}$$

$$= \left(1 - \frac{n}{N}\right) \frac{N^2}{n} \frac{\text{SSTO}}{N-1}$$

$$= \left(1 - \frac{n}{N}\right) \frac{N^2}{n(N-1)}(\text{SSW} + \text{SSB})$$

$$= V_{\text{prop}}(\hat{t}_{\text{str}}) + \left(1 - \frac{n}{N}\right) \frac{N}{n(N-1)} \left[N(\text{SSB}) - \sum_{h=1}^{H}(N - N_h)S_h^2\right].$$

The above result shows us that proportional allocation with stratification always gives smaller variance than SRS *unless*

$$\text{SSB} < \sum_{h=1}^{H} \left(1 - \frac{N_h}{N}\right) S_h^2. \tag{3.11}$$

TABLE 3.3

Population ANOVA Table

Source	df	Sum of Squares
Between strata	$H - 1$	$\text{SSB} = \displaystyle\sum_{h=1}^{H}\sum_{j=1}^{N_h} (\bar{y}_{hU} - \bar{y}_U)^2 = \sum_{h=1}^{H} N_h(\bar{y}_{hU} - \bar{y}_U)^2$
Within strata	$N - H$	$\text{SSW} = \displaystyle\sum_{h=1}^{H}\sum_{j=1}^{N_h} (y_{hj} - \bar{y}_{hU})^2 = \sum_{h=1}^{H} (N_h - 1)S_h^2$
Total, about $\bar{y}_U$	$N - 1$	$\text{SSTO} = \displaystyle\sum_{h=1}^{H}\sum_{j=1}^{N_h} (y_{hj} - \bar{y}_U)^2 = (N - 1)S^2$

This rarely happens when the N_h are large; generally, the large population sizes of the strata will force $N_h(\bar{y}_{hU} - \bar{y}_U)^2 > S_h^2$. In general, the variance of the estimator of t from a stratified sample with proportional allocation will be smaller than the variance of the estimator of t from an SRS with the same number of observations. The more unequal the stratum means $\bar{y}_{hU}$, the more precision you will gain by using proportional allocation. The variance of $\hat{t}_{str}$ depends primarily on SSW; since SSTO is a fixed value for the finite population, SSW is smaller when SSB is larger. Of course, this result only holds for population variances; it is possible for a variance estimate from proportional allocation to be larger than that from an SRS merely because the sample selected had large within-stratum sample variances.

EXAMPLE 3.7 Index mutual funds attempt to mimic the performance of one of the indices of overall stock or bond market performance. The Dow Jones Wilshire 5000 Composite Index[SM] includes all U.S. equity securities with readily available price information. The stocks are weighted by market capitalization to form the index. Total stock market index funds attempt to have the same performance as the Wilshire 5000 Index; however, buying all of the stocks in the index and adjusting the holdings every time the index is revised would lead to excessive transaction costs. As a result, mutual fund companies often use stratified sampling to select stocks from the index to include in their index funds. The largest 500 companies in the index make up more than 70% of its value; most total stock market index funds include all of the companies in this stratum. The remaining stocks in the index mutual fund are then sampled from the remaining strata constructed using factors including market capitalization, industry exposures, dividend yield, price/earnings (P/E) ratio, price/book (P/B) ratio, and earnings growth. Not all index funds use stratified random sampling. The prospectuses for some funds state that the fund manager selects representative stocks from the different strata. In some cases, the fund manager uses a proprietary computer program to select stocks within the strata, with the goal of obtaining a better match to the index or reducing transaction costs. ∎

3.4.2 Optimal Allocation

If the variances S_h^2 are more or less equal across all the strata, proportional allocation is probably the best allocation for increasing precision. In cases where the S_h^2 vary greatly, **optimal allocation** can result in smaller costs. In practice, when we are sampling units of different sizes, the larger units are likely to be more variable than the smaller units, and we would sample them with a higher sampling fraction. For example, if we were to take a sample of American corporations and our goal was to estimate the amount of trade with Europe, the variation among large corporations would be greater than the variation among small ones. As a result, we would sample a higher percentage of the large corporations. Optimal allocation works well for sampling units such as corporations, cities, and hospitals, which vary greatly in size. It is also effective when some strata are much more expensive to sample than others.

Neter (1978) tells of a study done by the Chesapeake and Ohio (C&O) Railroad Company to determine how much revenue they should get from interline freight shipments, since the total freight from a shipment that traveled along several railroads was divided among the different railroads. The C&O took a stratified sample of waybills, the documents that detailed the goods, route, and charges for the shipments. The waybills were stratified by the total freight charges, and all of the waybills with charges of over $40 were sampled, whereas only 1% of the waybills with charges less than $5 were sampled. The justification was that there was little variability among the amounts due the C&O in the stratum of the smallest total freight charges, whereas the variability in the stratum with charges of over $40 was much higher.

EXAMPLE 3.8   How are musicians paid when their compositions are performed? In the United States, many composers are affiliated with the American Society of Composers, Authors and Publishers (ASCAP). Television networks, local television and radio stations, services such as Muzak, symphony orchestras, restaurants, nightclubs, and other operations pay ASCAP an annual license fee, based largely on the size of the audience, that allows them to play compositions in the ASCAP catalog. ASCAP then distributes royalties to composers whose works are played.

Theoretically, an ASCAP member should get royalties every time one of his or her compositions is played. Taking a census of every piece of music played in the United States, however, would be impractical; to estimate the amount of royalties due to members, ASCAP uses sampling. According to Dobishinski (1991), Krasilovsky and Shemel (2003, p. 139), and www.ascap.com, ASCAP relies on television producers' cue sheets, which provide details on the music used in a program, to identify and tabulate musical pieces played on network television and major cable channels. About 60,000 hours of tapes are made from radio broadcasts each year, and experts identify the musical compositions aired in these broadcasts.

Stratified sampling is used to sample radio stations for the survey. Radio stations are grouped into strata based on the license fee paid to ASCAP, the type of community the station is in, and the geographic region. As stations paying higher license fees contribute more money for royalties, they are more likely to be sampled; once in the sample, high-fee stations are taped more often than low-fee stations. ASCAP thus uses a form of optimal allocation in taping: Strata with the highest radio fees, and thus with the highest variability in royalty amounts, have larger sampling fractions than strata containing radio stations that pay small fees.   ■

The objective in optimal allocation is to gain the most information for the least cost. A simple cost function is given below: Let C represent total cost, c_0 represent overhead costs such as maintaining an office, and c_h represent the cost of taking an observation in stratum h, so that

$$C = c_0 + \sum_{h=1}^{H} c_h n_h. \qquad (3.12)$$

We want to allocate observations to strata so as to minimize $V(\bar{y}_{\text{str}})$ for a given total cost C, or, equivalently, to minimize C for a fixed $V(\bar{y}_{\text{str}})$. Suppose that the costs $c_1, c_2, \ldots, c_H$ are known. To minimize the total cost for a fixed variance, we can prove using calculus that the optimal allocation has n_h proportional to

$$\frac{N_h S_h}{\sqrt{c_h}} \qquad (3.13)$$

for each h (see Exercise 24). Thus, the optimal sample size in stratum h is

$$n_h = \left(\frac{\dfrac{N_h S_h}{\sqrt{c_h}}}{\displaystyle\sum_{l=1}^{H} \dfrac{N_l S_l}{\sqrt{c_l}}} \right) n. \tag{3.14}$$

We shall then sample heavily within a stratum if

- The stratum accounts for a large part of the population.
- The variance within the stratum is large; we sample more heavily to compensate for the heterogeneity.
- Sampling in the stratum is inexpensive.

Sometimes applying the optimal allocation formula in Equation (3.14) results in one or more of the "optimal" n_h's being larger than the population size N_h in those strata. In that case, take a sample size of N_h in those strata, and then apply (3.14) again with the remaining strata.

EXAMPLE **3.9**   **Dollar stratification** is often used in accounting. The recorded book amounts are used to stratify the population. If you are auditing the loan amounts for a financial institution, stratum 1 might consist of all loans of more than \$1 million, stratum 2 might consist of loans between \$500,000 and \$999,999, and so on down to the smallest stratum of loans less than \$10,000. Optimal allocation is often an efficient strategy for such a stratification: S_h^2 will be much larger in the strata with the large loan amounts, so optimal allocation will prescribe a higher sampling fraction for those strata. If the goal of the audit is to estimate the dollar discrepancy between the audited amounts and the amounts in the institution's books, an error in the recorded amount of one of the \$3,000,000 loans is likely to contribute more to the audited difference than an error in the recorded amount of one of the \$3,000 loans. In a survey such as this, you may even want to use sample size N_1 in stratum 1, so that each population unit in stratum 1 has probability 1 of appearing in the sample. ∎

If all variances and costs are equal, proportional allocation is the same as optimal allocation. If we know the variances within each stratum and they differ, optimal allocation gives a smaller variance for the estimator of $\bar{y}_U$ than proportional allocation. But optimal allocation is a more complicated scheme; often the simplicity and self-weighting property of proportional allocation are worth the extra variance. In addition, the optimal allocation will differ for each variable being measured, whereas the proportional allocation depends only on the number of population units in each stratum. Stokes and Plummer (2002) describe linear programming methods that can be used to determine optimal allocations when more than one variable is of interest.

Neyman allocation is a special case of optimal allocation, used when the costs in the strata (but not the variances) are approximately equal. Under Neyman allocation, n_h is proportional to $N_h S_h$. If the variances S_h^2 are specified correctly, Neyman allocation will give an estimator with smaller variance than proportional allocation (see Exercise 25).

TABLE   3.4

Quantities Used for Designing the Caribou Survey in Example 3.10

	A	B	C	D	E	F
1	Stratum	N_h	s_h	$N_h s_h$	n_h	Sample size
2				B*C	D*225/D9	
3	A	400	3,000	1,200,000	96.26	98
4	B	30	2,000	60,000	4.81	10
5	C	61	9,000	549,000	44.04	37
6	D	18	2,000	36,000	2.89	6
7	E	70	12,000	840,000	67.38	39
8	F	120	1,000	120,000	9.63	21
9	total	699		2,805,000	225	211

EXAMPLE 3.10   The caribou survey in Example 3.3 used a form of optimal allocation to determine the n_h. Before taking the survey, the investigators obtained approximations of the caribou densities and distribution, and constructed strata to be relatively homogeneous in terms of population density. They set the total sample size as $n = 225$. They then used the estimated count in each stratum as a rough estimate of the standard deviation, with the result shown in Table 3.4. The first row contains the names of the spreadsheet columns, and the second row contains the formulas used to calculate the table. The investigators wanted the sampling fraction to be at least 1/3 in smaller strata, so they used the optimal allocation sample sizes in column E as a guideline for determining the sample sizes they actually used, in column F.   ∎

When the stratum variances S_h^2 are approximately known, Neyman allocation gives higher precision than proportional allocation. If the information about the stratum variances is of poor quality, however, disproportional allocation can result in a higher variance than simple random sampling. Proportional allocation, on the other hand, almost always has smaller variance than simple random sampling.

3.4.3 Allocation for Specified Precision within Strata

Sometimes you are less interested in the precision of the estimate of the population total or mean for the whole population than in comparing means or totals among different strata. In that case, you would determine the sample size needed for the individual strata using the guidelines in Section 2.6.

EXAMPLE 3.11   The U.S. Postal Service often conducts surveys asking postal customers about their perceptions of the quality of mail service. The population of residential postal service customers is stratified by geographic area, and it is desired that the precision be ±3 percentage points, at a 95% confidence level, within each area. If there were no nonresponse, such a requirement would lead to sampling at least 1067 households in each stratum, as calculated in Example 2.11. Such an allocation is neither proportional, as the number of residential households in the population vary a great deal from stratum to stratum, nor optimal in the sense of providing the greatest efficiency for estimating percentages for the whole population. It does, however, provide the desired precision within each stratum. ∎

3.4.4 Determining Sample Sizes

The different methods of allocating observations to strata give the relative sample sizes n_h/n. After strata are constructed (see Section 3.5) and observations allocated to strata, Equation (3.3) can be used to determine the sample size necessary to achieve a prespecified margin of error. Recall that

$$V(\bar{y}_{str}) \leq \frac{1}{n} \sum_{h=1}^{H} \frac{n}{n_h} \left(\frac{N_h}{N}\right)^2 S_h^2 = \frac{v}{n},$$

where $v = \sum_{h=1}^{H} (n/n_h) (N_h/N)^2 S_h^2$. Thus, if the fpcs can be ignored and if the normal approximation is valid, an approximate 95% CI for the population mean will be $\bar{y}_{str} \pm z_{\alpha/2} \sqrt{v/n}$. Set $n = z_{\alpha/2}^2 v/e^2$ to achieve a desired margin of error e.

The quantity v depends on the stratum population sizes N_h and variances S_h^2, and on the relative sample sizes n_h/n. If we took an SRS of size n instead of a stratified random sample, the variance of $\bar{y}_{SRS}$ would be (again, ignoring the fpc) S^2/n. Thus, S^2 can be thought of as the variability per observation unit in an SRS, and v can be thought of as the "average" variability per observation unit in a stratified random sample with the specified allocation. We substitute v for S^2 in the sample size (without fpc) formula (2.24) to obtain the necessary sample size for stratified sampling. If sampling fractions in the strata are high, this sample size can be adjusted for the finite population corrections. In Section 7.5, we shall use a similar method to find the necessary sample size for any survey design.

3.5
Defining Strata

One might wonder, since stratified sampling almost always gives higher precision than simple random sampling, why anyone would ever take a sample that is not stratified. The answer is that stratification adds complexity to the survey, and the added complexity may not be worth a small gain in precision. In addition, we need to have information in the sampling frame that can be used to form the strata. For each stratum, we need to know how many and which members of the population belong to that stratum. When such information is available, however, stratification is often well worth the effort.

Remember, stratification is most efficient when the stratum means differ widely; then the between sum of squares is large, and the variability within strata will be smaller. Consequently, when constructing strata we want the strata means to be as different as possible. Ideally, we would stratify by the values of y; if our survey is to estimate total business expenditures on advertising, we would like to put businesses that spent the most on advertising in stratum 1, businesses with the next highest level of advertising expenditures in stratum 2, and so on, until the last stratum contained businesses that spent nothing on advertising. The problem with this scheme is that we do not know the advertising expenditures for all the businesses while designing the survey—if we did, we would not need to do a survey at all! Instead, we try to find some variable closely related to y. For estimating total business expenditures on advertising, we might stratify by number of employees or size of the business and by the type of product or service. For farm income, we might use the size of the farm as a stratifying variable, since we expect that larger farms would have higher incomes.

Most surveys measure more than one variable, so any stratification variable should be related to many characteristics of interest. The U.S. Current Population Survey, which measures characteristics relating to employment, stratifies the areas that form the primary sampling units by geographic region, population density, racial composition, principal industry, and similar variables. In the Canadian Survey of Employment, Payrolls, and Hours, business establishments are stratified by industry, province, and estimated number of employees. The Nielsen television ratings stratify by geographic region, county size, and cable penetration, among other variables. If several stratification variables are available, use the variables associated with the most important responses.

The number of strata you choose depends upon many factors such as the difficulty in constructing a sampling frame with stratifying information, and the cost of stratifying. A general rule to keep in mind is: The more information, the more strata you should use. Thus, you should use an SRS when little prior information about the target population is available.

You can often collect preliminary data that can be used to stratify your design. If you are taking a survey to estimate the number of fish in a region, you can use physical features of the area that are related to fish density, such as depth, salinity, and water temperature. Or you can use survey information from previous years, or data from a preliminary cruise to aid in constructing strata. In this situation, according to Saville (1977, p. 10): "Usually there will be no point in designing a sampling scheme with more than 2 or 3 strata, because our knowledge of the distribution of fish will be rather imprecise. Strata may be of different size, and each stratum may be composed of several distinct areas in different parts of the total survey area." In a survey with more precise prior information, we will want to use more strata—many surveys are stratified to the point that only two sampling units are observed in each stratum.

For many surveys, stratification can increase precision dramatically, and often well repays the effort used in constructing the strata. Example 3.12 tells how strata were constructed in one large-scale survey, the National Pesticide Survey.

EXAMPLE 3.12   Between 1988 and 1990, the U.S. Environmental Protection Agency (1990a, b) sampled drinking water wells to estimate the prevalence of pesticides and nitrate. When designing the National Pesticide Survey (NPS), the EPA scientists wanted a sample that was representative of drinking water wells in the United States. In particular, they wanted to guarantee that wells in the sample would have a wide range of levels of pesticide use and susceptibility to ground-water pollution. They also wanted to study two categories of wells: *community water systems* (CWSs), defined as "systems of piped drinking water with at least 15 connections and/or 25 or more permanent residents of the service area that have at least one working well used to obtain drinking water"; and *rural domestic wells*, "drinking water wells supplying occupied housing units located in rural areas of the United States, except for wells located on government reservations."

The following selections from the EPA describe how it chose the strata for the survey:

In order to determine how many wells to visit for data collection, EPA first needed to identify approximately how many drinking water wells exist in the United States. This process was easier for community water systems than for rural domestic wells because a list of all public water systems, with their addresses, is contained in the Federal Reporting Data System (FRDS), which is maintained by EPA. From FRDS, EPA estimated that there were approximately 51,000 CWSs with wells in the United States. EPA did not have a comprehensive list of rural domestic wells to serve as the foundation for well selection, as it did for CWSs. Using data from the Census Bureau for 1980, EPA estimated that there were approximately 13 million rural domestic wells in the country, but the specific owners and addresses of these rural domestic wells were not known.

EPA chose a survey design technique called "stratification" to ensure that survey data would meet its objectives. This technique was used to improve the precision of the estimates by selecting extra wells from areas with substantial agricultural activity and high susceptibility to ground-water pollution (vulnerability). EPA developed criteria for separating the population of CWS wells and rural domestic wells into four categories of pesticide use and three relative ground-water vulnerability measures. This design ensures that the range of variability that exists nationally with respect to the agricultural use of pesticides and ground-water vulnerability is reflected in the sample of wells.

EPA identified five subgroups of wells for which it was interested in obtaining information. These subgroups were community water system wells in counties with relatively high average ground-water vulnerability; rural domestic wells in counties with relatively high average ground-water vulnerability; rural domestic wells in counties with high pesticide use; rural domestic wells in counties with both high pesticide use and relatively high average ground-water vulnerability; and rural domestic wells in "cropped and vulnerable" parts of counties (high pesticide use and relatively high ground-water vulnerability).

Two of the most difficult design questions were determining how many wells to include in the Survey and determining the level of precision that would be sought for the NPS national estimates. These two questions were connected, because greater precision is usually obtained by collecting more data. Resolving these questions would have been simpler if the Survey designers had known in advance what proportion of wells in the nation contained pesticides, but answering that question was one of the purposes of the Survey. Although many State studies have been conducted for specific pesticides, no reliable national estimates of well water contamination existed. EPA evaluated alternative precision requirements and costs for collecting data from different numbers of wells to determine the Survey size that would meet EPA's requirements and budget.

The Survey designers ultimately selected wells for data collection so that the Survey provided a 90 percent probability of detecting the presence of pesticides in the CWS wells sampled, assuming 0.5 percent of all community water system wells in the country contained pesticides. The rural domestic well Survey design was structured with different probabilities of detection for the several subgroups of interest, with the greatest emphasis placed on the cropped and vulnerable subcounty areas, where EPA was interested in obtaining very precise estimates of pesticide occurrence. EPA assumed that 1 percent of rural domestic wells in these areas would contain pesticides and designed the Survey to have about a 97 percent probability of detection in "cropped and vulnerable" areas if the assumption proved accurate. EPA concluded that sampling approximately 1,300 wells (564 public wells and 734 private wells) would meet the Survey's accuracy specifications and provide a representative national assessment of the number of wells containing pesticides.

Selecting Wells for the Survey. Because the exact number and location of rural domestic wells was unknown, EPA chose a survey design composed of several steps (stages) for those wells. The design began with a sampling of counties, and then characterized pesticide use and ground-water vulnerability for subcounty areas. This eventually allowed small enough geographic areas to be delineated to enable the sampling of individual rural domestic wells. This procedure was not needed for community water system wells, because their number and location were known.

The first step in well selection was common to both CWS wells and rural domestic wells. Each of the 3,137 counties or county equivalents in the U.S. was characterized according to pesticide use and ground-water vulnerability to ensure that the variability in agricultural pesticide use and ground-water vulnerability was reflected in the Survey. EPA used data on agricultural pesticide use obtained from a marketing research source and information on the proportion of the county area that was in agricultural production to rank agricultural pesticide use for each county as high, medium, low, or uncommon. Ground-water vulnerability of each county was estimated using a numerical classification system called Agricultural DRASTIC, which assesses seven factors: (depth of water, recharge, aquifer media, soil media, topography, impact of unsaturated zone, conductivity of the aquifer). The model was modified for the Survey to evaluate the vulnerability of aquifers to pesticide and nitrate contamination, and one of the subsidiary purposes of the Survey was to assess the effectiveness of the DRASTIC classification. Each area was evaluated and received a score of high, moderate, or low, based on information obtained from U.S. Geological Survey maps, U.S. Department of Agriculture soil survey maps and other resources from State agencies, associations, and universities. (1990a)

The procedure resulted in 12 strata for counties, as given in Table 3.5.

Stratification provides several advantages in this survey. It allows for more precise estimates of pesticide and nitrate concentrations in the United States as a whole, as it is expected that the wells within a stratum are more homogeneous than the entire population of wells. Stratification ensures that wells for each level of pesticide use and ground-water vulnerability are included in the sample, and allows estimation of pesticide concentration with prespecified sample size in each stratum. The factorial design, with four levels of the factor *pesticide use* and three levels of the factor *groundwater*

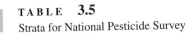

TABLE 3.5
Strata for National Pesticide Survey

Stratum	Pesticide Use	Groundwater Vulnerability (as Estimated by DRASTIC)	Number of Counties
1	High	High	106
2	High	Moderate	234
3	High	Low	129
4	Moderate	High	110
5	Moderate	Moderate	204
6	Moderate	Low	267
7	Low	High	193
8	Low	Moderate	375
9	Low	Low	404
10	Uncommon	High	186
11	Uncommon	Moderate	513
12	Uncommon	Low	416

SOURCE: Adapted from U.S. EPA (1990a), p. 3.

vulnerability, allows investigation of possible effects of each factor separately, and the interaction of the factors, on pesticide concentrations. ∎

3.6
Model-Based Inference for Stratified Sampling*

The one-way ANOVA model with fixed effects provides an underlying structure for stratified sampling. Here,

$$Y_{hj} = \mu_h + \varepsilon_{hj}, \tag{3.15}$$

where the ε_{hj}'s are independent with mean 0 and variance σ_h^2. Then the least squares estimator of μ_h is $\bar{Y}_h$, the average in stratum h.

Let the random variable

$$T_h = \sum_{j=1}^{N_h} Y_{hj}$$

represent the total in stratum h and the random variable

$$T = \sum_{h=1}^{H} T_h$$

represent the overall total. From Section 2.9, the best linear unbiased estimator for T_h is

$$\hat{T}_h = \frac{N_h}{n_h} \sum_{j \in \mathcal{S}_h} Y_{hj}.$$

Then, from the results shown for simple random sampling in Section 2.9,

$$E_M[\hat{T}_h - T_h] = 0$$

and

$$E_M[(\hat{T}_h - T_h)^2] = N_h^2 \left(1 - \frac{n_h}{N_h}\right) \frac{\sigma_h^2}{n_h}.$$

Since observations in different strata are independent under the model in (3.15),

$$E_M[(\hat{T} - T)^2] = E_M\left[\left\{\sum_{h=1}^{H} (\hat{T}_h - T_h)\right\}^2\right]$$

$$= E_M\left[\sum_{h=1}^{H} (\hat{T}_h - T_h)^2 + \sum_{h=1}^{H}\sum_{k \neq h} (\hat{T}_h - T_h)(\hat{T}_k - T_k)\right]$$

$$= E_M\left[\sum_{h=1}^{H} (\hat{T}_h - T_h)^2\right]$$

$$= \sum_{h=1}^{H} N_h^2 \left(1 - \frac{n_h}{N_h}\right) \frac{\sigma_h^2}{n_h}.$$

The theoretical variance σ_h^2 can be estimated by s_h^2. Adopting the model in (3.15) results in the same estimators for t and its standard error as found under randomization theory in (3.5). If a different model is used, however, then different estimators are obtained.

3.7
Quota Sampling

Many samples that masquerade as stratified random samples are actually quota samples. In **quota sampling**, the population is divided into different subpopulations just as in stratified random sampling, but with one important difference: Probability sampling is not used to choose individuals in the subpopulation for the sample. In extreme versions of quota sampling, choice of units in the sample is entirely at the discretion of the interviewer, so that a sample of convenience is chosen within each subpopulation.

In quota sampling, specified numbers (quotas) of particular types of population units are required in the final sample. For example, to obtain a quota sample with $n = 3000$, you might specify that the sample contain 1000 white males, 1000 white females, 500 men of color, and 500 women of color, but you might give no further instructions about how these quotas are to be filled. Thus, quota sampling is not a form of probability sampling—we do not know the probabilities with which each individual is included in the sample. It is often used when probability sampling is impractical, overly costly, or considered unnecessary, or when the persons designing the sample just do not know any better.

The big drawback of quota sampling is that we do not know if the units chosen for the sample exhibit selection bias. If selection of units is totally up to the interviewer, she or he is likely to choose the most accessible members of the population—for instance, persons who are easily reached by telephone, households without menacing dogs, or areas of the forest close to the road. The most accessible members of a population are likely to differ in a systematic way from less accessible members. Thus, unlike in stratified random sampling, we cannot say that the estimator of the population total from quota sampling is unbiased over repeated sampling—one of our usual criteria of goodness in probability samples. In fact, in quota samples, we cannot measure sampling error over repeated samples and we have no way of estimating the bias from the sample data. Since selection of units is up to the individual interviewer, we cannot expect that repeating the sample will give similar results. Thus, anyone drawing inferences from a quota sample must necessarily take a model-based approach.

EXAMPLE 3.13   The 1945 survey on reading habits taken for the Book Manufacturer's Institute (Link and Hopf, 1946), like many surveys in the 1940s and 1950s, used a quota sample. Some of the classifications used to define the quota classes were area, city size, age, sex, and socioeconomic status; a local supervising psychologist in each city determined the blocks of the city in which interviewers were to interview people from a specified socioeconomic group. The interviewers were then allowed to choose the specific households to be interviewed in the designated city blocks.

The quota procedure followed in the survey did not result in a sample that reflected demographic characteristics of the 1945 U.S. population. The following table compares the educational background of the survey respondents with figures from the 1940 U.S. Census, adjusted to reflect the wartime changes in population.

Distribution by Educational Levels	4,000 People Interviewed (%)	U.S. Census, Urban and Rural Nonfarm (%)
8th grade or less	28	48
1–3 years high school	18	19
4 years high school	25	21
1–3 years college	15	7
4 or more years college	13	5

SOURCE: Link and Hopf (1946).

The oversampling of better-educated persons casts doubt on many of the statistics given in the book. The study concluded that 31% of "active readers" (those who had read at least one book in the past month) had bought the last book they read, and that 25% of all last books read by active readers cost $1 or less. Who knows whether a stratified random sample would have given the same results?   ■

In the 1948 U.S. presidential elections, all of the major polls printed just a few days before the election predicted that Dewey would defeat Truman handily. In fact, of course, Truman won the election. According to Mosteller et al. (1949), one of the problems of those polls was that they all used quota sampling, not a probability-based method—the polling debacle in 1948 spurred many survey organizations in the United States to turn away from quota sampling, at least for a few years. The polls that erred in predicting the winner in the British general election of 1992 all used quota methods in selecting persons to interview in their homes or in the street; the primary quota classes used were sex, age, socio-economic class, and employment status. Although we may never know exactly what went wrong in those polls (see Crewe, 1992, for some other explanations), the use of quota samples may have played a part—if interviewing persons "in the street," it is certainly plausible that persons from a quota class that are accessible differ from persons that are less accessible.

While quota sampling is not as good as probability sampling under ideal conditions, it may give better results than a completely haphazard sample because it at least forces the inclusion of members of the different quota groups. Quota samples have the advantage of being less expensive than probability samples. The quality of the data from quota samples can be improved by allowing the interviewer less discretion in the choice of persons or households to be included in the sample. Many survey organizations use probability sampling along with quotas; they use probability sampling to select small blocks of potential respondents, and then take a quota sample within each block, using variables such as age, sex, and race.

Because we do not know the probabilities with which units were sampled, we must take a model-based approach, and make strong assumptions about the data structure, when analyzing data from a quota sample. The model generally adopted is that of Section 3.6—within each subclass the random variables generating the subpopulation are assumed independent and identically distributed. Such a model implies that any selection of units from the quota class will give a representative sample; if the model holds, then quota sampling will likely give good estimates of the population quantity. If the model does not hold, then the estimates from quota sampling may be badly biased.

EXAMPLE **3.14** Sanzo et al. (1993) used a combination of stratified random sampling and quota sampling for estimating the prevalence of *Coxiella burnetii* infection within the Basque country in northern Spain. *Coxiella burnetii* can cause Q fever, which can lead to complications such as heart and nerve damage. Reviews of Q fever patient records from Basque hospitals showed that about three-fourths of the victims were male, about half were between 16 and 30 years old, and victims were disproportionately likely to be from areas with low population density.

The authors stratified the target population by population density and then randomly selected health care centers from the three strata. In selecting persons for blood testing, however, "a probabilistic approach was rejected as we considered that the refusal rate of blood testing would be high" (p. 1185). Instead, they used quota sampling to balance the sample by age and gender; physicians asked patients who needed laboratory tests whether they would participate in the study, and recruited subjects for the study until the desired sample sizes in the six quota groups were reached for each stratum.

Because a quota sample was taken instead of a probability sample, persons analyzing the data must make strong assumptions about the representativeness of the sample in order to apply the results to the general population of the Basque country. First, the assumption must be made that persons attending a health clinic for laboratory tests (the sampled population of the study) are neither more nor less likely to be infected than persons who would not be visiting the clinic. Second, one must assume that persons who are requested and agree to do the study are similar in terms of the infection to persons in the same quota class having laboratory tests that do not participate in the study. These are strong assumptions: the authors of the article argue that the assumptions are justified, but of course they cannot prove that the assumptions hold unless follow-up investigations are done.

If they had taken a probability sample of persons instead of the quota sample, they would not have had to make these strong assumptions. A probability sample of persons, however, would have been exhorbitantly expensive when compared with the quota sampling scheme used, and a probability sample would also have taken longer to design and implement. With the quota sample, the authors were able to collect information about the public health problem; it is unclear whether the results can be generalized to the entire population, but the data do provide a great deal of quick information on the prevalence of infection that can be used in future investigation of who is likely to be infected, and why. ■

Deville (1991, p. 177) argues that quota samples may be useful for market research, when the organization requesting the survey is aware of the model being used. Persons collecting official statistics about crime, unemployment, or other matters that are used for setting public policy should use probability samples, however.

Quota samples, while easier to collect than a probability sample, suffer from the same disadvantages as other convenience samples. Some survey organizations now use quota sampling to recruit volunteers for online surveys; they accumulate respondents until they have specified sample sizes in the desired demographic classes. In such online surveys, the respondents in each quota class are self-selected—if, as argued by Couper (2000), Internet users who volunteer for such surveys differ from members of the target population in those quota classes, results will be biased.

3.8
Chapter Summary

Stratification uses additional information about a population in the survey design. In the simplest form, stratified random sampling, we take an SRS of size n_h in stratum h, for each of the H strata in the population. To use stratification, we must know the population size N_h for each stratum; we must also know the stratum membership for every unit in the population. The inclusion probability for unit i in stratum h is $\pi_{hi} = n_h/N_h$; consequently, the sampling weight for that unit is $w_{hi} = N_h/n_h$.

To estimate the population total t using a stratified random sample, let $\hat{t}_h$ estimate the population total in stratum h. Then

$$\hat{t}_{\text{str}} = \sum_{h=1}^{H} \hat{t}_h = \sum_{h=1}^{H} \sum_{j \in \mathcal{S}_h} w_{hj} y_{hj}$$

and

$$\hat{V}(\hat{t}_{\text{str}}) = \sum_{h=1}^{H} \hat{V}(\hat{t}_h) = \sum_{h=1}^{H} \left(1 - \frac{n_h}{N_h}\right) N_h^2 \frac{s_h^2}{n_h}.$$

The population mean $\bar{y}_U = t/N$ is estimated by

$$\bar{y}_{\text{str}} = \frac{\hat{t}_{\text{str}}}{N} = \sum_{h=1}^{H} \frac{N_h}{N} \bar{y}_h = \frac{\sum_{h=1}^{H} \sum_{j \in \mathcal{S}_h} w_{hj} y_{hj}}{\sum_{h=1}^{H} \sum_{j \in \mathcal{S}_h} w_{hj}}$$

with $\hat{V}(\bar{y}_{\text{str}}) = \hat{V}(\hat{t}_{\text{str}})/N^2$.

Stratified sampling has three major design issues: defining the strata, choosing the total sample size, and allocating the observations to the defined strata. With proportional allocation, the same sampling fraction is used in each stratum. Proportional allocation almost always results in smaller variances for estimated means and totals than simple random sampling. Disproportional allocation may be preferred if some strata should have higher sampling fractions than others, for example, if it is desired to have larger sample sizes for strata with minority populations or for strata with large companies. Optimal allocation specifies taking larger sampling fractions in strata that have larger variances.

Key Terms

Disproportional allocation: Allocation of sampling units to strata so that the sampling fractions n_h/N_h are unequal.

Optimal allocation: Allocation of sampling units to strata so that the variance of the estimator is minimized for a given total cost.

Proportional allocation: Allocation of sampling units to strata so that $n_h/N_h = n/N$ for each stratum. Proportional allocation results in a self-weighting sample.

Quota sampling: A nonprobability sampling method which many persons confuse with stratified sampling. In quota sampling, quota classes are formed that serve the role of strata, but the survey taker uses a nonprobability sampling method such as convenience sampling to reach the desired sample size in each quota class.

Stratified random sampling: Probability sampling method in which population units are partitioned into strata, and then an SRS is taken from each stratum.

Stratum: One of the subpopulations or classes that make up the entire population. Every unit in the population is in exactly one stratum.

For Further Reading

The references in Chapter 2 also describe stratified sampling. Chapter 4 of Raj (1968) gives a rigorous and concise treatment of stratified sampling theory. Cochran (1977) has further results on allocation and construction of strata, and uses ANOVA tables to compare precisions of sampling designs (first described in Cochran, 1939).

Neyman (1934) wrote one of the most important papers in the historical development of survey sampling. He presented a framework for stratified sampling and demonstrated its superiority to purposive selection methods. Neyman's paper pretty much finished off the idea that results from purposive samples could be generalized to the population. He presented an example of a sample of 29 districts, purposely chosen to give the averages of all 214 districts in the 1921 Italian Census on a dozen variables. But Neyman showed that "all statistics other than the average values of the controls showed a violent contrast between the sample and the whole population."

3.9
Exercises

A. Introductory Exercises

1 What stratification variable(s) would you use for each of the following situations:

a A political poll to estimate the percentage of registered voters in Arizona that approve of the governor's performance.

b An e-mail survey of students at your university, to estimate the total amount of money students spend on textbooks in a term.

c A sample of high schools in New York City to estimate what percentage of high schools offer one or more classes in computer programming.

d A sample of public libraries in California to study the availability of computer resources, and the per capita expenditures.

e A survey of anglers visiting a freshwater lake, to learn about which species of fish are preferred.

f An aerial survey to estimate the number of walrus in the pack ice near Alaska between 173° East and 154° West longitude.

g A sample of prime-time (7–10 pm, Monday through Saturday; 6–10 pm, Sunday) TV programs on CBS to estimate the average number of promotional announcements (ads for other programming on the station) per hour of broadcast.

2 Consider the hypothetical population below (this population is also used in Example 2.2). Consider the stratification below, with $N_1 = N_2 = 4$. The population is:

Unit number	Stratum	y
1	1	1
2	1	2
3	1	4
8	1	8
4	2	4
5	2	7
6	2	7
7	2	7

Consider the stratified sampling design in which $n_1 = n_2 = 2$.

a Write out all possible SRSs of size 2 from stratum 1, and find the probability of each sample. Do the same for stratum 2.

b Using your work in (a), find the sampling distribution of $\hat{t}_{str}$.

c Find the mean and variance of the sampling distribution of $\hat{t}_{str}$. How do these compare to the mean and variance in Example 2.2?

3 Consider a population of 6 students. Suppose we know the test scores of the students to be

Student	1	2	3	4	5	6
Score	66	59	70	83	82	71

a Find the mean $\bar{y}_U$ and variance S^2 of the population.

b How many SRS's of size 4 are possible?

c List the possible SRS's. For each, find the sample mean. Using Equation (2.9), find $V(\bar{y})$.

d Now let stratum 1 consist of students 1–3, and stratum 2 consist of students 4–6. How many stratified random samples of size 4 are possible in which 2 students are selected from each stratum?

e List the possible stratified random samples. Which of the samples from (c) cannot occur with the stratified design?

f Find $\bar{y}_{str}$ for each possible stratified random sample. Find $V(\bar{y}_{str})$, and compare it to $V(\bar{y})$.

4 For Example 3.4, construct a data set with 3835 observations. Include three columns: column 1 is the stratum number (from 1 to 7), column 2 contains the response variable of gender (0 for males and 1 for females), and column 3 contains the sampling weight N_h/n_h for each observation. Using columns 2 and 3 along with (3.10), calculate $\hat{p}_{str}$. Is it possible to calculate $SE(\hat{p}_{str})$ by using only columns 2 and 3, with no additional information? Explain.

5 The survey in Example 3.4 collected much other data on the subjects. Another of the survey's questions asked whether the respondent agreed with the following statement: "When I look at a new issue of my discipline's major journal, I rarely find an article that interests me." The results are as follows:

Discipline	Agree (%)
Literature	37
Classics	23
Philosophy	23
History	29
Linguistics	19
Political Science	43
Sociology	41

a What is the sampled population in this survey?

b Find an estimate of the percentage of persons in the sampled population that agree with the statement, and give the standard error of your estimate.

6 Suppose that a city has 90,000 dwelling units, of which 35,000 are houses, 45,000 are apartments, and 10,000 are condominiums.

a You believe that the mean electricity usage is about twice as much for houses as for apartments or condominiums, and that the standard deviation is proportional to the mean so that $S_1 = 2S_2 = 2S_3$. How would you allocate a stratified sample of 900 observations if you wanted to estimate the mean electricity consumption for all households in the city?

b Now suppose that you take a stratified random sample with proportional allocation and want to estimate the overall proportion of households in which energy conservation is practiced. If 45% of house dwellers, 25% of apartment dwellers, and 3% of condomium residents practice energy conservation, what is p for the population? What gain would the stratified sample with proportional allocation offer over an SRS, that is, what is $V_{prop}(\hat{p}_{str})/V_{SRS}(\hat{p}_{SRS})$?

7 In Exercise 6 of Chapter 2, data on numbers of publications were given for an SRS of 50 faculty members. Not all departments were represented, however, in the SRS. The SRS contained several faculty members from psychology and from chemistry, but none from foreign languages. The following data are from a stratified sample of faculty, using the areas biological sciences, physical sciences, social sciences, and humanities as the strata.

Stratum	Number of Faculty Members in Stratum	Number of Faculty Members in Sample
Biological Sciences	102	7
Physical Sciences	310	19
Social Sciences	217	13
Humanities	178	11
Total	807	50

The frequency table for number of publications in the strata is given below.

Number of Refereed Publications	Number of Faculty Members			
	Biological	Physical	Social	Humanities
0	1	10	9	8
1	2	2	0	2
2	0	0	1	0
3	1	1	0	1
4	0	2	2	0
5	2	1	0	0
6	0	1	1	0
7	1	0	0	0
8	0	2	0	0

a   Estimate the total number of refereed publications by faculty members in the college, and give the standard error.

b   How does your result from (a) compare with the result from the SRS in Exercise 6 of Chapter 2?

c   Estimate the proportion of faculty with no refereed publications, and give the standard error.

d   Did stratification increase precision in this example? Explain why you think it did or did not.

8   A public opinion researcher has a budget of $20,000 for taking a survey. She knows that 90% of all households have telephones. Telephone interviews cost $10 per household; in-person interviews cost $30 each if all interviews are conducted in person, and $40 each if only nonphone households are interviewed in person (because there will be extra travel costs). Assume that the variances in the phone and nonphone groups are similar, and that the fixed costs are $c_0 = \$5000$. How many households should be interviewed in each group if

a   all households are interviewed in person

b   households with a phone are contacted by telephone and households without a phone are contacted in person.

B. Working with Survey Data

9   The data file agstrat.dat also contains information on other variables. For each of the following quantities, plot the data, and estimate the population mean for that variable along with its standard error and a 95% CI. Compare your answers with those from the SRS in Exercise 15 of Chapter 2.

a   Number of acres devoted to farms, 1987

b   Number of farms, 1992

c   Number of farms with 1000 acres or more, 1992

d   Number of farms with 9 acres or fewer, 1992

10   Hard shell clams may be sampled by using a dredge. Clams do not tend to be uniformly distributed in a body of water, however, as some areas provide better habitat than others. Thus, taking a simple random sample is likely to result in a large estimated variance for the number of clams in an area. Russell (1972) used stratified random sampling to estimate the total number of bushels of hard shell clams (*Mercenaria mercenaria*) in Narragansett Bay, Rhode Island. The area of interest was divided into four strata based on preliminary surveys that identified areas in which clams were abundant. Then n_h dredge tows were made in stratum h, for $h = 1, 2, 3, 4$. The acreage for each stratum was known, and Russell calculated that the area fished during a standard dredge tow was 0.039 acres, so that we may use $N_h = 25.6 \times \text{Area}_h$.

a   Here are the results from the survey taken before the commercial season. Estimate the total number of bushels of clams in the area, and give the standard error of your estimate.

Stratum	Area (Acres)	Number of Tows Made	Average Number of Bushels per Tow	Sample Variance for Stratum
1	222.81	4	0.44	0.068
2	49.61	6	1.17	0.042
3	50.25	3	3.92	2.146
4	197.81	5	1.80	0.794

b Another survey was performed at the end of the commercial season. In this survey, strata 1, 2, and 3 were collapsed into a single stratum, called stratum 1 below. Estimate the total number of bushels of clams (with standard error) at the end of the season.

Stratum	Area (Acres)	Number of Tows Made	Average Number of Bushels per Tow	Sample Variance for Stratum
1	322.67	8	0.63	0.083
4	197.81	5	0.40	0.046

11 Lydersen and Ryg (1991) used stratification techniques to estimate ringed seal populations in Svalbard fjords. The 200 km^2 study area was divided into three zones: Zone 1, outer Sassenfjorden, was covered with relatively new ice during the study period in March, 1990, and had little snow cover; Zone 3, Tempelfjorden, had a stable ice cover throughout the year; Zone 2, inner Sassenfjorden, was intermediate between the stable Zone 3 and the unstable Zone 1. Ringed seals need good ice to establish territories with breathing holes, and snow cover enables females to dig out birth lairs. Thus, it was thought that the three zones would have different seal densities. The investigators took a stratified random sample of 20% of the 200 1-km^2 areas. The following table gives the number of plots, and the number of plots sampled, in each zone:

Zone	Number of Plots	Plots Sampled
1	68	17
2	84	12
3	48	11
Total	200	40

In each sampled area, Imjak the Siberian husky tracked seal structures by scent; the number of breathing holes in sampled square was recorded. A total of 199 breathing holes were located in zones 1, 2, and 3 altogether. The data (reconstructed from information given in the paper) are in the file seals.dat.

a Estimate the total number of breathing holes in the study region, along with its standard error.

b   If you were designing this survey, how would you allocate observations to strata if the goal is to estimate the total number of breathing holes? If the goal is to compare the density of breathing holes in the three zones?

12   Proportional allocation was used in the stratified sample in Example 3.2. It was noted, however, that variability was much higher in the West than in the other regions. Using the estimated variances in Example 3.2, and assuming that the sampling costs are the same in each stratum, find an optimal allocation for a stratified sample of size 300.

13   Select a stratified random sample of size 300 from the data in the file agpop.dat, using your allocation in Exercise 12. Estimate the total number of acres devoted to farming in the United States, and give the standard error of your estimate. How does this standard error compare with that found in Example 3.2?

14   Burnard (1992) sent a questionnaire to a stratified sample of nursing tutors and students in Wales, to study what the tutors and students understood by the term *experiential learning*. The population size and sample size obtained for each of the four strata are given below:

Stratum	Population Size	Sample Size
General nursing tutors (GT)	150	109
Psychiatric nursing tutors (PT)	34	26
General nursing students (GS)	2680	222
Psychiatric nursing students (PS)	570	40
Total	3434	397

Respondents were asked which of the following techniques could be identified as experiential learning methods; the number of students in each group who identified the method as an experiential learning method are given below:

Method	GS	PS	PT	GT
Role play	213	38	26	104
Problem solving activities	182	33	22	95
Simulations	95	20	22	64
Empathy-building exercises	89	25	20	54
Gestalt exercises	24	4	5	12

Estimate the overall percentage of nursing students and tutors who identify each of these techniques as "experiential learning." Be sure to give standard errors for your estimates.

15   Hayes (2000) took a stratified sample of New York City food stores. The sampling frame consisted of 1408 food stores with at least 4000 square feet of retail space. The population of stores was stratified into three strata using median household income within the zip code. The prices of a "market basket" of goods were determined for each store; the goal of the survey was to investigate whether prices differ among the three strata. Hayes used the logarithm of total price for the basket as the response y. Results are given in the following table:

Stratum, h	N_h	n_h	$\bar{y}_h$	s_h
1 Low income	190	21	3.925	0.037
2 Middle income	407	14	3.938	0.052
3 Upper income	811	22	3.942	0.070

a The planned sample size was 30 in each stratum; this was not achieved because some stores went out of business while the data were being collected. What are the advantages and disadvantages of sampling the same number of stores in each stratum?

b Estimate $\bar{y}_U$ for these data and give a 95% CI.

c Is there evidence that prices are different in the three strata?

16 Kruuk et al. (1989) used a stratified sample to estimate the number of otter (*Lutra lutra*) dens along the 1400-km coastline of Shetland, UK. The coastline was divided into 242 (237 that were not predominantly buildings) 5-km sections, and each section was assigned to the stratum whose terrain type predominated. Then sections were chosen randomly from the sections in each stratum. In each section chosen, the investigators counted the total number of dens in a 110-m-wide strip along the coast. The data are in file otters.dat. The population sizes for the strata are as follows:

Stratum	Total Sections	Sections Counted
1 Cliffs over 10m	89	19
2 Agriculture	61	20
3 Not 1 or 2, peat	40	22
4 Not 1 or 2, non-peat	47	21

a Estimate the total number of otter dens in Shetland, along with a standard error for your estimate.

b Discuss possible sources of bias in this study. Do you think it is possible to avoid all selection and measurement bias?

17 Marriage and divorce statistics are compiled by the National Center for Health Statistics and published in volumes of *Vital Statistics of the United States*. State and local officials provide NCHS with annual counts of marriages and divorces in each county. In addition, some states send computer tapes of additional data, or microfilm copies of marriage or divorce certificates to NCHS. These additional data are used to calculate statistics about age at marriage or divorce, previous marital status of marrying couples, and children involved in divorce. In 1987, if a state sent a computer tape, all records were included in the divorce statistics; if a state sent microfilm copies, a specified fraction of the divorce certificates was randomly sampled and data recorded. The sampling rates (probabilities of selection) and number of records sampled in each state in the divorce registration area for 1987 are in file divorce.dat.

a How many divorces were there in the divorce registration area in 1987? HINT: Construct and use the sampling weights.

b Why did NCHS use different sampling rates in different states?

c Estimate the total number of divorces granted to men aged 24 or less. To women aged 24 or less. Give 95% CIs for your estimates.

d In what proportion of all divorces is the husband between 40 and 50 years old? In what proportion is the wife between 40 and 50? Give 95% CIs for your estimates.

18 Wilk et al. (1977) reported data on the number and types of fishes and environmental data for the area of the Atlantic continental shelf between eastern Long Island, New York and Cape May, New Jersey. The ocean survey area was divided into strata based on depth. Sampling was done at a higher rate close to shore than farther away from shore: "In-shore strata (0–28 m) were sampled at a rate of approximately one station per 515 km^2 and off-shore strata (29–366 m) were sampled at a rate of approximately one station per 1,030 km^2" (p. 1). Thus each record in strata 3–6 represents twice as much area as each record in strata 1 and 2. In calculating average numbers of fish caught and numbers of species, we may use a relative sampling weight of 1 for strata 1 and 2, and weight 2 for strata 3–6.

Stratum	Depth (m)	Relative Sampling Weight
1	0–19	1
2	20–28	1
3	29–55	2
4	56–100	2
5	111–183	2
6	184–366	2

The data file nybight.dat contains data on the total catch for sampling stations visited in June 1974 and June 1975.

a Construct side-by-side boxplots of the number of fish caught in the trawls in June, 1974. Does there appear to be a large variation among the strata?

b Calculate estimates of the average number and average weight of fish caught per haul in June 1974, along with the standard error.

c Calculate estimates of the average number and average weight of fish caught per haul in June 1975, along with the standard error.

d Is there evidence that the average weight of fish caught per haul differ between June 1974 and June 1975? Answer using an appropriate hypothesis test.

19 In January 1995, the Office of University Evaluation at Arizona State University surveyed faculty and staff members to find out their reaction to the closure of the university during Winter Break, 1994. Faculty and staff in academic units that were closed during the winter break were divided into four strata and subsampled.

Stratum Number	Employee Type	Population Size (N_h)	Sample Size (n_h)
1	Faculty	1374	500
2	Classified staff	1960	653
3	Administrative staff	252	74
4	Academic professional	95	95

Questionnaires were sent through campus mail to persons in strata 1 through 4; the sample size in the above table is the number of questionnaires mailed in each stratum. We'll come back to the issue of nonresponse in this survey in Chapter 8; for now, just analyze the respondents in the stratified sample of employees in closed units; the data are in the file "winter.dat." For this exercise, look at the answers to the question "Would you want to have Winter Break Closure again?" (variable *breakaga*).

a   Not all persons in the survey responded to the question. Find the number of persons that responded to the question in each of the four strata. For this exercise, use these values as the n_h.

b   Use (3.6) and (3.7) to estimate the proportion of faculty and staff that would answer yes to the question "Would you want to have Winter Break Closure again" and give the standard error.

c   Create a new variable, in which persons who respond "yes" to the question take on the value 1, persons who respond "no" to the question take on the value 0, and persons who do not respond are either left blank (if you are using a spreadsheet) or assigned the missing value code (if you are using statistical software). Construct a column of sampling weights N_h/n_h for the observations in the sample. (The sampling weight will be 0 or missing for nonrespondents.) Now use (3.10) to estimate the proportion of faculty and staff that would answer yes to the question "Would you want to have Winter Break Closure again?"

d   Using the column of 0s and 1s you constructed in the previous question, find s_h^2 for each stratum by calculating the sample variance of the observations in that stratum. Now use (3.5) to calculate the standard error of your estimate of the proportion. Why is your answer the same as you calculated in (b)?

e   Stratification is sometimes used as a method of dealing with nonresponse. Calculate the response rates (the number of persons who responded divided by the number of questionnaires mailed) for each stratum. Which stratum has the lowest response rate for this question? How does stratification treat the nonrespondents?

20   The data in the file radon.dat were collected from 1003 homes in Minnesota in 1987 (Tate, 1988) in order to estimate the prevalence and distribution of households with high indoor radon concentrations. The data are adapted from www.stat.berkeley.edu/users/statlabs/labs.html, the website for Nolan and Speed (2000). Since the investigators were interested in how radon levels varied across counties, each of the 87 counties in Minnesota served as a stratum. An SRS of telephone numbers from county telephone directories was selected in each county. When a household could not be contacted or was unwilling to participate in the study, an alternate

telephone number was used, until the desired sample size in the stratum was reached.

a Discuss possible sources of nonsampling error in this survey.

b Calculate the sampling weight for each observation, using the values for N_h and n_h in the data file.

c Treating the sample as a stratified random sample, estimate the average radon level for Minnesota homes, along with a 95% CI. Do the same for the response log(*radon*).

d Estimate the total number of Minnesota homes that have radon level of 4 picocuries per liter (pCi/L) or higher, with a 95% CI. The U.S. Environmental Protection Agency (2007) recommends fixing your home if the radon level is at least 4 pCi/L.

C. Working with Theory

21 Construct a small population and stratification for which $V(\hat{t}_{\text{str}})$ using proportional allocation is larger than the variance that would be obtained by taking an SRS with the same number of observations. HINT: Use (3.11).

22 A stratified sample is being designed to estimate the prevalence p of a rare characteristic, say the proportion of residents in Milwaukee, Wisconsin, who have Lyme disease. Stratum 1, with N_1 units, has a high prevalence of the characteristic; stratum 2, with N_2 units, has low prevalence. Assume that the cost to sample a unit (for example, the cost to select a person for the sample and determine whether he or she has Lyme disease) is the same for each stratum, and that at most 2000 units are to be sampled.

a Let p_1 and p_2 be the proportions in stratum 1 and stratum 2 with the rare characteristic. If $p_1 = 0.10$, $p_2 = 0.03$, and $N_1/N = 0.4$, what are n_1 and n_2 under optimal allocation?

b If $p_1 = 0.10$, $p_2 = 0.03$, and $N_1/N = 0.4$, what is $V(\hat{p}_{\text{str}})$ under proportional allocation? Under optimal allocation? What is the variance if you take an SRS of 2000 units from the population?

c (Use a spreadsheet for this part of the exercise.) Now fix $p = 0.05$. Let p_1 range from 0.05 to 0.50, and N_1/N range from 0.01 to 0.50 (these two values then determine the value of p_2). For each combination of p_1 and N_1/N, find the optimal allocation, and the variance under both proportional allocation and optimal allocation. Also find the variance from an SRS of 2000 units. When does the optimal allocation give a substantial increase in precision when compared to proportional allocation? When compared to an SRS?

23 (Requires probability.) We know from Section 2.3 that there are $\binom{N}{n} = \dfrac{N!}{n!(N-n)!}$ possible SRS's of size n from a population of size N. Suppose we stratify the population into H strata, where each stratum contains $N_h = N/H$ units. A stratified sample is to be selected using proportional allocation, so that $n_h = n/H$.

a How many possible stratified samples are there?

b Stirling's formula approximates $k!$, when k is large, by $k! \approx \sqrt{2\pi k}\,(k/e)^k$, where $e = \exp(1) \approx 2.718282$. Use Stirling's formula to approximate

$$\frac{\text{number of possible stratified samples of size } n}{\text{number of possible SRSs of size } n}.$$

24 (Requires calculus.) Show that the variance is minimized for a fixed cost with the cost function in (3.12) when $n_h \propto N_h S_h / \sqrt{c_h}$, as in (3.13). HINT: Use Lagrange multipliers.

25 Under Neyman allocation, discussed in Section 3.4.2, the optimal sample size in stratum h is

$$n_{h,\text{Neyman}} = \frac{N_h S_h}{\displaystyle\sum_{l=1}^{H} N_l S_l} n.$$

a Show that the variance of $\hat{t}_{\text{str}}$ if Neyman allocation is used is

$$V_{\text{Neyman}}(\hat{t}_{\text{str}}) = \frac{1}{n}\left(\sum_{h=1}^{H} N_h S_h\right)^2 - \sum_{h=1}^{H} N_h S_h^2.$$

b We showed in Section 3.4.1 that the variance of $\hat{t}_{\text{str}}$ if proportional allocation is used is

$$V_{\text{prop}}(\hat{t}_{\text{str}}) = \frac{N}{n}\sum_{h=1}^{H} N_h S_h^2 - \sum_{h=1}^{H} N_h S_h^2.$$

Prove that the theoretical variance from Neyman allocation is always less than or equal to the theoretical variance from proportional allocation by showing that

$$V_{\text{prop}}(\hat{t}_{\text{str}}) - V_{\text{Neyman}}(\hat{t}_{\text{str}}) = \frac{N^2}{n}\left[\sum_{h=1}^{H}\frac{N_h}{N} S_h^2 - \left(\sum_{h=1}^{H}\frac{N_h}{N} S_h\right)^2\right]$$

$$= \frac{N^2}{n}\sum_{h=1}^{H}\frac{N_h}{N}\left(S_h - \sum_{l=1}^{H}\frac{N_l}{N} S_l\right)^2.$$

c From (b), we see that the gain in precision from using Neyman allocation relative to using proportional allocation is higher if the stratum standard deviations S_h vary widely. When $H = 2$, show that

$$V_{\text{prop}}(\hat{t}_{\text{str}}) - V_{\text{Neyman}}(\hat{t}_{\text{str}}) = \frac{N_1 N_2}{n}(S_1 - S_2)^2.$$

26 (Requires calculus.) Suppose that there are K responses of interest, and response k has relative importance $a_k > 0$, where $\sum_{k=1}^{K} a_k = 1$. Let $\hat{t}_{yk}$ be the estimated population total for response k, and let S_{kh}^2 be the population variance for response k in stratum h. Then the optimal allocation problem is to minimize

$$\sum_{k=1}^{K} a_k V(\hat{t}_{yk})$$

subject to the constraint $C = c_0 + \sum_{h=1}^{H} n_h c_h$. Show that the optimal allocation for fixed total sample size n gives

$$n_h = n \frac{N_h \sqrt{\dfrac{\sum_{k=1}^{K} a_k S_{kh}^2}{c_h}}}{\displaystyle\sum_{j=1}^{H} N_j \sqrt{\dfrac{\sum_{k=1}^{K} a_k S_{kj}^2}{c_j}}}.$$

D. Projects and Activities

27 *Rectangles.* This activity continues Exercise 30 of Chapter 2. Divide the rectangles in the population of Figure 2.7 into two strata, based on your judgment of size. Now take a stratified sample of 10 rectangles. State how you decided on the sample size in each stratum. Estimate the total area of all the rectangles in the population, and give a 95% CI, based on your sample. How does your CI compare with that from the SRS in Chapter 2?

28 *Mutual funds.* In Exercise 31 of Chapter 2, you took an SRS of funds from a mutual fund company. Most companies have mutual funds in a number of different categories, for example, domestic stock funds, foreign stock funds, and bond funds, and the returns in these categories differ.

  a Divide the funds from the company into strata. You may use categories provided by the fund company, or other categories such as market capitalization. Create a table of the strata, with the number of mutual funds in each stratum.

  b Using proportional allocation, take a stratified random sample of size 25 from your population.

  c Find the mean and a 95% CI for the mean of the variable you studied in Exercise 31 of Chapter 2. How does your estimate from the stratified sample compare with the estimate from the SRS you found earlier?

29 The Consumer Bankruptcy Project of 2001 (Warren and Tyagi, 2003; data collection is described on pages 181–188) surveyed 2220 households who filed for Chapter 7 or Chapter 13 bankruptcy, with the goal of studying why families file for bankruptcy. Questionnaires were given to debtors attending the mandatory meeting with the bankruptcy trustee assigned to their case in the five districts selected by the investigators for the study (these districts included the cities of Nashville, Chicago, Dallas, Philadelphia, and Los Angeles) on specified target dates. Additional samples were taken from two rural districts in Tennessee and Iowa. Quota sampling was used in each district, with the goal of collecting 250 questionnaires from each district that had the same proportions of Chapter 7 and Chapter 13 bankruptcies as were filed in the district. Discuss the relative merits and disadvantages of using quota sampling for this study.

30 Read the article on estimating medical errors by Thomas et al. (2000). What is the purpose of this sample? How was stratification used in the survey design? Why do

TABLE 3.6

Table for Exercise 32.

	A	B	C	D	E	Other	Total
Apache	1	13	19	0	0	94	127
Cochise	12	5	0	637	40	0	694
Coconino	1	6	0	125	0	289	421
Gila	0	2	51	151	0	0	204
Graham	0	2	0	63	0	143	208
Greenlee	0	0	0	58	0	0	58
Maricopa	118	169	0	3,732	2,675	5,105	11,799
Mohave	4	6	0	44	0	476	530
Navajo	2	5	132	124	0	0	263
Pima	62	26	0	1,097	727	1,786	3,698
Pinal	5	10	13	22	360	478	888
Santa Cruz	0	5	0	118	150	0	273
Yavapai	7	8	0	173	0	198	386
Yuma	5	5	0	837	0	0	847
LaPaz	0	1	0	89	0	0	90
Total	217	263	215	7,270	3,952	8,569	20,486

you think the investigators chose the stratification variables they used? What are the possible sources of nonsampling error in this survey?

31 The U.S. Monthly Retail Trade and Food Services program, described at www.census. gov/mrts/www/mrts.html, provides estimates of sales at retail and food service stores. Read the documentation on Sample Design and Estimation Procedures. How does the survey use stratification in the design?

32 Suppose the Arizona Department of Health wishes to take a survey of 2-year-olds whose families receive medical assistance, to determine the proportion who have been immunized. The medical care is provided by several different health care organizations, and the state has 15 counties. Table 3.6 shows the population number of 2-year-olds for each county/organization combination.

The sample is to be stratified by county and organization. It is desired to select sample sizes for each combination so that

a the margin of error for estimating percentage immunized is 0.05 or less when the data are tabulated for each county (summing over all health care organizations)

b the margin of error for estimating percentage immunized is 0.05 or less when the data are tabulated for each health care organization (summing over all counties)

c at least two children (fewer, of course, if the cell does not have two children) are selected from every cell.

Note that for this problem, as for many survey designs, many different designs would be possible.

33   Example 3.7 discussed the use of stratified sampling in mutual funds.

    a   Locate an index fund or exchange traded fund that attempts to replicate an index. Summarize how they use stratified sampling.

    b   Suppose you were asked to devise a stratified sampling plan to represent the Wilshire 5000 Index using market capitalization classes as strata. Investigate how the index is constructed. Using a list of the stocks in the index, construct strata and develop a stratified sampling design.

34   *Trucks.* The Vehicle Inventory and Use Survey (VIUS) has been conducted by the U.S. government to provide information on the number of private and commercial trucks in each state. The stratified random sampling design is described in U.S. Census Bureau (2006b). For the 2002 survey, 255 strata were formed from the sampling frame of truck registrations using stratification variables *state* and *trucktype*. The 50 states plus the District of Columbia formed 51 geographic classes; in each, the truck registrations were partitioned into one of five classes:

1.   Pickups

2.   Minivans, other light vans, and sport utility vehicles

3.   Light single-unit trucks with gross vehicle weight less than 26,000 pounds

4.   Heavy single-unit trucks with gross vehicle weight greater than or equal to 26,000 pounds

5.   Truck-tractors

Consequently, the full data set has $51 \times 5 = 255$ strata. Selected variables from the data are in the data file vius.dat. For each question below, give a point estimate and a 95% CI.

    a   The sampling weights are found in variable *tabtrucks* and the stratification is given by variable *stratum*. Estimate the total number of trucks in the United States. (HINT: What should your response variable be?) Why is the standard deviation of your estimator essentially zero?

    b   Estimate the total number of truck miles driven in 2002 (variable *miles_annl*).

    c   Estimate the total number of truck miles driven in each of the five *trucktype* classes.

    d   Estimate the average miles per gallon (MPG) for the trucks in the population.

35   *Baseball data.* Exercise 32 of Chapter 2 described the population of baseball players in data file baseball.dat.

    a   Take a stratified random sample of 150 players from the file, using proportional allocation with the different teams as strata. Describe how you selected the sample.

    b   Find the mean of the variable *logsal*, using your stratified sample, and give a 95% CI.

    c   Estimate the proportion of players in the data set who are pitchers, and give a 95% CI.

    d   How do your estimates compare with those of Exercise 32 from Chapter 2?

    e   Examine the sample variances in each stratum. Do you think optimal allocation would be worthwhile for this problem?

f   Using the sample variances from (e) to estimate the population stratum variances, determine the optimal allocation for a sample in which the cost is the same in each stratum and the total sample size is 150. How much does the optimal allocation differ from the proportional allocation?

36 *Online bookstore.* In Exercise 33 from Chapter 2 you took an SRS of book titles from amazon.com. Use the same book genre for this problem.

   a   Stratify the population into two categories: hardcover and paperback. You can obtain the population counts in the paperback category by refining your search to include the word paperback.

   b   Take a stratified random sample of 40 books from your population using proportional allocation. Record the price and number of pages for each book.

   c   Give a point estimate and a 95% CI for the mean price of books and the mean number of pages for books in the population.

   d   Compare your CI's to those from Exercise 33 of Chapter 2. Does stratification appear to increase the precision of your estimate?

   e   Use your SRS from Chapter 2 to estimate the within-stratum variance of book price for each stratum. In this case, you are using the SRS as a pilot sample to help design a subsequent sample. Find the optimal allocation for a stratified random sample of 40 books. How does the optimal allocation differ from the proportional allocation?

37 *IPUMS exercises.* Exercise 37 of Chapter 2 described the IPUMS data.

   a   Using one or more of the following variables: *age*, *sex*, *race*, or *marstat*, divide the population into strata. Explain how you decided upon your stratification variable and how you chose the number of strata to use. (Note: It is NOT FAIR to use the values of *inctot* in the population to choose your strata! However, you may draw a pilot sample of size 200 using an SRS to aid you in constructing your strata.)

   b   Using the strata you constructed, draw a stratified random sample using proportional allocation. Use the same overall sample size you used for your SRS in Exercise 37 of Chapter 2. Explain how you calculated the sample size to be drawn from each stratum.

   c   Using the stratified sample you selected with proportional allocation, estimate the total income for the population, along with a 95% CI.

   d   Using the pilot sample of size 200 to estimate the within-stratum variances, use optimal allocation to determine sample stratum sizes. Use the same value of n as in part 37b, which is the same n from the SRS in Exercise 37 of Chapter 2. Draw a stratified random sample from the population along with a 95% CI.

   e   Under what conditions can optimal allocation be expected to perform much better than proportional allocation? Do these conditions exist for this population? Comment on the relative performance you observed between these two allocations.

   f   Overall, do you think your stratification was worthwhile for sampling from this population? How did your stratified estimates compare with the estimate from the SRS you took in Chapter 2? If you were to start over on the stratification, what would you do differently?

4

Ratio and Regression Estimation

The registers of births, which are kept with care in order to assure the condition of the citizens, can serve to determine the population of a great empire without resorting to a census of its inhabitants, an operation which is laborious and difficult to do with exactness. But for this it is necessary to know the ratio of the population to the annual births. The most precise means for this consists of, first, choosing subdivisions in the empire that are distributed in a nearly equal manner on its whole surface so as to render the general result independent of local circumstances; second, carefully enumerating the inhabitants of several communes in each of the subdivisions, for a specified time period; third, determining the corresponding mean number of annual births, by using the accounts of births during several years preceding and following this time period. This number, divided by that of the inhabitants, will give the ratio of the annual births to the population, in a manner that will be more reliable as the enumeration becomes larger.

—Pierre-Simon Laplace, *Essai Philosophique sur les Probabilités* (trans. S. Lohr)

France had no population census in 1802, and Laplace wanted to estimate the number of persons living there (Laplace, 1814; Cochran, 1978). He obtained a sample of 30 communes spread throughout the country. These communes had a total of 2,037,615 inhabitants on September 23, 1802. In the 3 years preceding September 23, 1802, a total of 215,599 births were registered in the 30 communes. Laplace determined the annual number of registered births in the 30 communes to be 215,599/3 = 71,866.33. Dividing 2,037,615 by 71,866.33, Laplace estimated that each year there was one registered birth for every 28.352845 persons. Reasoning that communes with large populations are also likely to have large numbers of registered births, and judging that the ratio of population to annual births in his sample would likely be similar to that throughout France, he concluded that one could estimate the total population of France by multiplying the total number of annual births in all of France by 28.352845. (For some reason, Laplace decided not to use the actual number of registered births in France in the year prior to September 22, 1802 in his calculation but instead multiplied the ratio by 1 million.)

Laplace was not interested in the total number of registered births for its own sake but used it as auxiliary information for estimating the total population of France. We often have auxiliary information in surveys. In Chapter 3, we used such auxiliary information in designing a survey. In this chapter, we use auxiliary information in the estimators. Ratio and regression estimation use variables that are correlated with the variable of interest to improve the precision of estimators of the mean and total of a population.

4.1
Ratio Estimation in a Simple Random Sample

For ratio estimation to apply, two quantities y_i and x_i must be measured on each sample unit; x_i is often called an **auxiliary variable** or **subsidiary variable**. In the population of size N

$$t_y = \sum_{i=1}^{N} y_i, \quad t_x = \sum_{i=1}^{N} x_i$$

and their ratio[1] is

$$B = \frac{t_y}{t_x} = \frac{\bar{y}_U}{\bar{x}_U}.$$

In the simplest use of ratio estimation, a simple random sample (SRS) of size n is taken, and the information in both x and y is used to estimate B, t_y, or $\bar{y}_U$.

Ratio and regression estimation both take advantage of the correlation of x and y in the population; the higher the correlation, the better they work. Define the **population correlation coefficient** of x and y to be

$$R = \frac{\sum_{i=1}^{N} (x_i - \bar{x}_U)(y_i - \bar{y}_U)}{(N-1)S_x S_y}. \tag{4.1}$$

Here, S_x is the population standard deviation of the x_i's, S_y is the population standard deviation of the y_i's, and R is simply the Pearson correlation coefficient of x and y for the N units in the population.

EXAMPLE **4.1**   Suppose the population consists of agricultural fields of different sizes. Let

$$y_i = \text{bushels of grain harvested in field } i$$
$$x_i = \text{acreage of field } i$$

Then

$$B = \text{average yield in bushels per acre}$$
$$\bar{y}_U = \text{average yield in bushels per field}$$
$$t_y = \text{total yield in bushels.} \quad \blacksquare$$

---

[1]Why use the letter B to represent the ratio? As we shall see in Section 4.6, ratio estimation is motivated by a regression model: $Y_i = \beta x_i + \varepsilon_i$, with $E[\varepsilon_i] = 0$ and $V[\varepsilon_i] = \sigma^2 x_i$. Thus the ratio of t_y and t_x is actually a regression coefficient.

If an SRS is taken, natural estimators for $B, t_y,$ and $\bar{y}_U$ are:

$$\hat{B} = \frac{\bar{y}}{\bar{x}} = \frac{\hat{t}_y}{\hat{t}_x}$$

$$\hat{t}_{yr} = \hat{B} t_x \qquad (4.2)$$

$$\hat{\bar{y}}_r = \hat{B} \bar{x}_U,$$

where t_x and $\bar{x}_U$ are assumed known.

4.1.1 Why Use Ratio Estimation?

1 Sometimes we simply want to estimate a ratio. In Example 4.1, B—the average yield per acre—is of interest and is estimated by the ratio of the sample means $\hat{B} = \bar{y}/\bar{x}$. If the fields differ in size, both numerator and denominator are random quantities; if a different sample is selected, both $\bar{y}$ and $\bar{x}$ are likely to change. In other survey situations, ratios of interest might be the ratio of liabilities to assets, the ratio of the number of fish caught to the number of hours spent fishing, or the per capita income of household members in Australia.

Some ratio estimates appear disguised because the denominator looks like it is just a regular sample size. To determine whether you need to use ratio estimation for a quantity, ask yourself "If I took a different sample, would the denominator be a different number?" If yes, then you are using ratio estimation. Suppose you are interested in the percentage of pages in *Good Housekeeping* magazine that contain at least one advertisement. You might take an SRS of 10 issues from the most recent 60 issues of the magazine and for each issue measure the following:

$$x_i = \text{total number of pages in issue } i$$
$$y_i = \text{total number of pages in issue } i$$
$$\text{that contain at least one advertisement.}$$

The proportion of interest can be estimated as

$$\hat{B} = \frac{\hat{t}_y}{\hat{t}_x}.$$

The denominator is the estimated total number of pages in the 60 issues. If a different sample of 10 issues is selected, the denominator will likely be different. In this example, we have an SRS of magazine issues; we have a **cluster sample** (we briefly discussed cluster samples in Section 2.1) of pages from the most recent 60 issues of *Good Housekeeping*. In Chapter 5, we shall see that ratio estimation is commonly used to estimate means in cluster sampling.

Technically, we are using ratio estimation every time we take an SRS and estimate a mean or proportion for a subpopulation, as will be discussed in Section 4.2.

2 Sometimes we want to estimate a population total, but the population size N is unknown. Then we cannot use the estimator $\hat{t}_y = N\bar{y}$ from Chapter 2. But we know that $N = t_x/\bar{x}_U$ and can estimate N by $t_x/\bar{x}$. We thus use another measure of size, t_x, instead of the population count N.

To estimate the total number of fish in a haul that are longer than 12 cm, you could take a random sample of fish, estimate the proportion that are larger than 12 cm, and multiply that proportion by the total number of fish, N. Such a procedure cannot be used if N is unknown. You can, however, weigh the total haul of fish, and use the fact that having a length of more than 12 cm (y) is related to weight (x), so

$$\hat{t}_{yr} = \bar{y}\,\frac{t_x}{\bar{x}}.$$

The total weight of the haul, t_x, is easily measured, and $t_x/\bar{x}$ estimates the total number of fish in the haul.

3 Ratio estimation is often used to increase the precision of estimated means and totals. Laplace used ratio estimation for this purpose in the example at the beginning of this chapter, and increasing precision will be the main use discussed in the chapter.

In Laplace's use of ratio estimation,

$$y_i = \text{number of persons in commune } i$$

$$x_i = \text{number of registered births in commune } i.$$

Laplace could have estimated the total population of France by multiplying the average number of persons in the 30 communes ($\bar{y}$) by the total number of communes in France (N). He reasoned that the ratio estimator would attain more precision: on average, the larger the population of a commune, the higher the number of registered births. Thus the population correlation coefficient R, defined in (4.1), is likely to be positive. Since $\bar{y}$ and $\bar{x}$ are then also positively correlated [see (A.11) in Appendix A], the sampling distribution of $\bar{y}/\bar{x}$ will have less variability than the sampling distribution of $\bar{y}/\bar{x}_U$. So if

$$t_x = \text{total number of registered births}$$

is known, the mean squared error (MSE) of $\hat{t}_{yr} = \hat{B}t_x$ is likely to be smaller than the MSE of $N\bar{y}$, an estimator that does not use the auxiliary information of registered births.

4 Ratio estimation is used to adjust estimates from the sample so that they reflect demographic totals. An SRS of 400 students taken at a university with 4000 students may contain 240 women and 160 men, with 84 of the sampled women and 40 of the sampled men planning to follow careers in teaching. Using only the information from the SRS, you would estimate that

$$\frac{4000}{400} \times 124 = 1240$$

students plan to be teachers. Knowing that the college has 2700 women and 1300 men, a better estimate of the number of students planning teaching careers might be

$$\frac{84}{240} \times 2700 + \frac{40}{160} \times 1300 = 1270.$$

Ratio estimation is used within each gender: In the sample, 60% are women, but 67.5% of the population are women, so we adjust the estimate of the total number of students planning a career in teaching accordingly. To estimate the total number of women who plan to follow a career in teaching, let

$$y_i = \begin{cases} 1 \text{ if woman and plans career in teaching} \\ 0 \text{ otherwise.} \end{cases}$$

$$x_i = \begin{cases} 1 \text{ if woman} \\ 0 \text{ otherwise.} \end{cases}$$

Then $(84/240) \times 2700 = (\bar{y}/\bar{x})t_x$ is a ratio estimate of the total number of women planning a career in teaching. Similarly, $(40/160) \times 1300$ is a ratio estimate of the total number of men planning a teaching career.

This use of ratio estimation, called **poststratification**, will be discussed in Section 4.4 and Chapters 7 and 8.

5 Ratio estimation may be used to adjust for nonresponse, as will be discussed in Chapter 8. Suppose a sample of businesses is taken; let y_i be the amount spent on health insurance by business i and x_i be the number of employees in business i. Assume that t_x, the total number of employees in all businesses in the population, is known. We expect that the amount a business spends on health insurance will be related to the number of employees. Some businesses may not respond to the survey, however. One method of adjusting for the nonresponse when estimating total insurance expenditures is to multiply the ratio $\bar{y}/\bar{x}$ (using data only from the respondents) by the population total t_x. If companies with few employees are less likely to respond to the survey, and if y_i is proportional to x_i, then we would expect the estimate $N\bar{y}$ to overestimate the population total t_y. In the ratio estimate $t_x\bar{y}/\bar{x}$, $t_x/\bar{x}$ is likely to be smaller than N because companies with many employees are more likely to respond to the survey. Thus a ratio estimate of total health care insurance expenditures may help to compensate for the nonresponse of companies with few employees.

EXAMPLE 4.2   Let's return to the SRS from the U.S. Census of Agriculture, described in Example 2.5. The file agsrs.dat contains data from an SRS of 300 of the 3078 counties.

For this example, suppose we know the population totals for 1987, but only have 1992 information on the SRS of 300 counties. When the same quantity is measured at different times, the response of interest at an earlier time often makes an excellent auxiliary variable. Let

$$y_i = \text{ total acreage of farms in county } i \text{ in 1992}$$
$$x_i = \text{ total acreage of farms in county } i \text{ in 1987.}$$

In 1987 a total of $t_x = 964{,}470{,}625$ acres were devoted to farms in the United States. The average acreage per county for the population is then $\bar{x}_U = 964{,}470{,}625/3078 = 313{,}343.3$ acres of farms per county. The data, and the line through the origin with slope $\hat{B}$, are plotted in Figure 4.1.

A portion of a spreadsheet with the 300 values of x_i and y_i is given in Table 4.1. Cells C305 and D305 contain the sample averages of y and x, respectively, so

$$\hat{B} = \frac{\bar{y}}{\bar{x}} = \frac{C305}{D305} = 0.986565,$$

FIGURE   4.1

The plot of acreage, 1992 vs. 1987, for an SRS of 300 counties. The line in the plot goes through the origin and has slope $\hat{B} = 0.9866$. Note that the variability about the line increases with x.

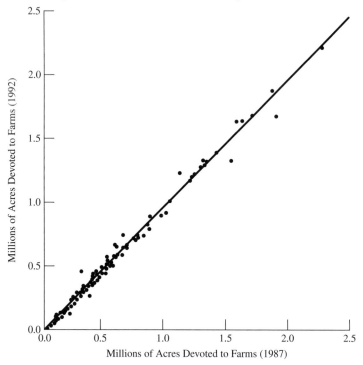

Millions of Acres Devoted to Farms (1992)

Millions of Acres Devoted to Farms (1987)

$$\hat{\bar{y}}_r = \hat{B}\bar{x}_U = (\hat{B})(313,343.283) = 309,133.6,$$

and

$$\hat{t}_{yr} = \hat{B}t_x = (\hat{B})(964,470,625) = 951,513,191.$$

Note that $\bar{y}$ for these data is 297,897.0, so $\hat{t}_{ySRS} = (3078)(\bar{y}) = 916,927,110$. In this example, $\bar{x}_S = 301,953.7$ is smaller than $\bar{x}_U = 313,343.3$. This means that our SRS of size 300 slightly underestimates the true population mean of the x's. Since the x's and y's are positively correlated, we have reason to believe that $\bar{y}_S$ may also underestimate the population value $\bar{y}_U$. Ratio estimation gives a more precise estimate of $\bar{y}_U$ by expanding $\bar{y}_S$ by the factor $\bar{x}_U/\bar{x}_S$. Figure 4.2 shows the ratio and SRS estimates of $\bar{y}_U$ on a graph of the center part of the data. ∎

4.1.2   Bias and Mean Squared Error of Ratio Estimators

Unlike the estimators $\bar{y}$ and $N\bar{y}$ in an SRS, ratio estimators are usually *biased* for estimating $\bar{y}_U$ and t_y. We start with the unbiased estimator $\bar{y}$—if we calculate $\bar{y}_S$ for each possible SRS $\mathcal{S}$, then the average of all of the sample means from the possible samples is the population mean $\bar{y}_U$. The estimation bias in ratio estimation arises

TABLE 4.1

Part of the Spreadsheet for the Census of Agriculture Data

	A	B	C	D	E
	County	State	*acres92* (y)	*acres87* (x)	Residual ($y - \hat{B}x$)
1	County	State	*acres92* (y)	*acres87* (x)	Residual ($y - \hat{B}x$)
2					
3	Coffee County	AL	175209	179311	−1693.00
4	Colbert County	AL	138135	145104	−5019.56
5	Lamar County	AL	56102	59861	−2954.78
6	Marengo County	AL	199117	220526	−18446.29
⋮	⋮	⋮	⋮	⋮	⋮
299	Rock County	WI	343115	357751	−9829.70
300	Kanawha County	WV	19956	21369	−1125.91
301	Pleasants County	WV	15650	15716	145.14
302	Putnam County	WV	55827	55635	939.44
303					
304	Column sum		89369114	90586117	3.96176E-09
305	Column average		297897.0467	301953.7233	
306	Column standard deviation		344551.8948	344829.5964	31657.21817
307	$\hat{B}$ = C305/D305=		0.986565237		

FIGURE 4.2

Detail of the center portion of Figure 4.1. Here, $\bar{x}_U$ is larger than $\bar{x}_S$, so $\hat{\bar{y}}_r$ is larger than $\bar{y}_S$.

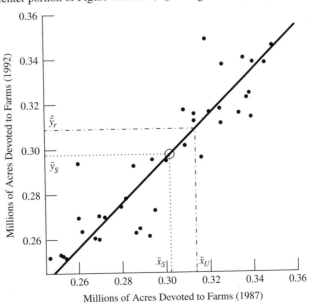

because $\bar{y}$ is multiplied by $\bar{x}_U/\bar{x}$; if we calculate $\hat{\bar{y}}_r$ for all possible SRSs $\mathcal{S}$, then the average of all the values of $\hat{\bar{y}}_r$ from the different samples will be close to $\bar{y}_U$, but will usually not equal $\bar{y}_U$ exactly.

The reduced variance of the ratio estimator usually compensates for the presence of bias—although $E[\hat{\bar{y}}_r] \neq \bar{y}_U$, the value of $\hat{\bar{y}}_r$ for any individual sample is likely to be closer to $\bar{y}_U$ than is the sample mean $\bar{y}_{\mathcal{S}}$. After all, we take only one sample in practice; most people would prefer to be able to say that their particular estimate from the sample is likely to be close to the true value, rather than that their particular value of $\bar{y}_{\mathcal{S}}$ may be quite far from $\bar{y}_U$, but that the average deviation $\bar{y}_{\mathcal{S}} - \bar{y}_U$, averaged over all possible samples $\mathcal{S}$ that could be obtained, is zero. For large samples, the sampling distributions of both $\bar{y}$ and $\hat{\bar{y}}_r$ will be approximately normal; if x and y are highly positively correlated, the following pictures illustrate the relative bias and variance of the two estimators:

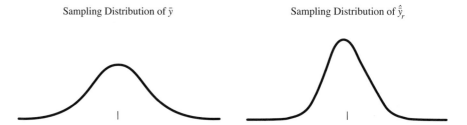

Sampling Distribution of $\bar{y}$ Sampling Distribution of $\hat{\bar{y}}_r$

$\bar{y}_U$ $\bar{y}_U$

To calculate the bias of the ratio estimator of $\bar{y}_U$, note that

$$\hat{\bar{y}}_r - \bar{y}_U = \frac{\bar{y}}{\bar{x}}\bar{x}_U - \bar{y}_U = \bar{y}\left(1 - \frac{\bar{x} - \bar{x}_U}{\bar{x}}\right) - \bar{y}_U.$$

Since $E[\bar{y}] = \bar{y}_U$,

$$\text{Bias}\,(\hat{\bar{y}}_r) = E[\hat{\bar{y}}_r - \bar{y}_U]$$
$$= E[\bar{y}] - \bar{y}_U - E\left[\frac{\bar{y}}{\bar{x}}(\bar{x} - \bar{x}_U)\right]$$
$$= -E\left[\hat{B}(\bar{x} - \bar{x}_U)\right]$$
$$= -\text{Cov}\,(\hat{B}, \bar{x}). \quad (4.3)$$

Consequently, as shown by Hartley and Ross (1954),

$$\frac{|\text{Bias}(\hat{\bar{y}}_r)|}{\sqrt{V(\hat{\bar{y}}_r)}} = \frac{|\text{Cov}\,(\hat{B}, \bar{x})|}{\bar{x}_U\sqrt{V(\hat{B})}} = \frac{|\text{Corr}\,(\hat{B}, \bar{x})|\sqrt{V(\hat{B})V(\bar{x})}}{\bar{x}_U\sqrt{V(\hat{B})}} \leq \frac{\sqrt{V(\bar{x})}}{\bar{x}_U} = \text{CV}\,(\bar{x}). \quad (4.4)$$

The absolute value of the bias of the ratio estimator is small relative to the standard deviation of the estimator if the coefficient of variation (CV) of $\bar{x}$ is small. For an SRS, CV $(\bar{x}) \leq S_x^2/(n\bar{x}_U)$, so that CV$(\bar{x})$ decreases as the sample size increases.

The result in Equation (4.3) is exact, but not necessarily easy to use with data. We now find approximations for the bias and variance of the ratio estimator that rely on the variances and covariance of $\bar{x}$ and $\bar{y}$. These approximations are an example of the linearization approach to approximating variances, to be discussed in Section 9.1. We write

$$\hat{\bar{y}}_r - \bar{y}_U = \frac{\bar{x}_U(\bar{y} - B\bar{x})}{\bar{x}} = (\bar{y} - B\bar{x})\left(1 - \frac{\bar{x} - \bar{x}_U}{\bar{x}}\right). \tag{4.5}$$

We can then show that (see Exercise 22)

$$\text{Bias}\,[\hat{\bar{y}}_r] = E[\hat{\bar{y}}_r - \bar{y}_U] \approx \frac{1}{\bar{x}_U}\left[B\,V(\bar{x}) - \text{Cov}(\bar{x}, \bar{y})\right]$$

$$= \left(1 - \frac{n}{N}\right)\frac{1}{n\bar{x}_U}(BS_x^2 - RS_xS_y), \tag{4.6}$$

with R the correlation between x and y. The bias of $\hat{\bar{y}}_r$ is thus small if:

- the sample size n is large
- the sampling fraction n/N is large
- $\bar{x}_U$ is large
- S_x is small
- the correlation R is close to 1.

Note that if all x's are the same value ($S_x = 0$), then the ratio estimator is the same as the SRS estimator $\bar{y}$ and the bias is zero.

For estimating the MSE of $\hat{\bar{y}}_r$, (4.5) gives:

$$E[(\hat{\bar{y}}_r - \bar{y}_U)^2] = E\left[\left\{(\bar{y} - B\bar{x})\left(1 - \frac{\bar{x} - \bar{x}_U}{\bar{x}}\right)\right\}^2\right]$$

$$= E\left[(\bar{y} - B\bar{x})^2 + (\bar{y} - B\bar{x})^2\left\{\left(\frac{\bar{x} - \bar{x}_U}{\bar{x}}\right)^2 - 2\frac{\bar{x} - \bar{x}_U}{\bar{x}}\right\}\right].$$

It can be shown (David and Sukhatme, 1974) that the second term is generally small compared with the first term, so the variance and MSE are approximated by

$$\text{MSE}\,(\hat{\bar{y}}_r) = E[(\hat{\bar{y}}_r - \bar{y}_U)^2] \approx E\left[(\bar{y} - B\bar{x})^2\right]. \tag{4.7}$$

The term $E\left[(\bar{y} - B\bar{x})^2\right]$ can also be written as

$$E\left[(\bar{y} - B\bar{x})^2\right] = V\left[\frac{1}{n}\sum_{i \in \mathcal{S}}(y_i - Bx_i)\right] = \left(1 - \frac{n}{N}\right)\frac{S_y^2 - 2BRS_xS_y + B^2S_x^2}{n}. \tag{4.8}$$

(See Exercise 18.)

From (4.7) and (4.8), the approximated MSE of $\hat{\bar{y}}_r$ will be small when

- the sample size n is large
- the sampling fraction n/N is large
- the deviations $y_i - Bx_i$ are small
- the correlation R is close to $+1$.

In large samples, the bias of $\hat{\bar{y}}_r$ is typically small relative to $V(\hat{\bar{y}}_r)$, so that MSE $(\hat{\bar{y}}_r) \approx V(\hat{\bar{y}}_r)$ (see Exercise 21). Thus, in the following, we use $\hat{V}(\hat{\bar{y}}_r)$ to estimate both the variance and the MSE.

Note from (4.8) that $E\left[(\bar{y} - B\bar{x})^2\right] = V(\bar{d})$, where $d_i = y_i - Bx_i$ and $\bar{d}_U = 0$. Since the deviations d_i depend on the unknown value B, define the new variable

$$e_i = y_i - \hat{B}x_i,$$

which is the ith residual from fitting the line $y = \hat{B}x$. Estimate $V(\hat{\bar{y}}_r)$ by

$$\hat{V}(\hat{\bar{y}}_r) = \left(1 - \frac{n}{N}\right)\left(\frac{\bar{x}_U}{\bar{x}}\right)^2 \frac{s_e^2}{n}, \tag{4.9}$$

where s_e^2 is the sample variance of the residuals e_i:

$$s_e^2 = \frac{1}{n-1}\sum_{i \in \mathcal{S}} e_i^2.$$

[Exercise 19 explains why we include the factor $\bar{x}_U/\bar{x}$ in (4.9). In large samples, we expect $\bar{x}_U/\bar{x}$ to be approximately equal to 1.] Similarly,

$$\hat{V}(\hat{B}) = \left(1 - \frac{n}{N}\right)\frac{s_e^2}{n\bar{x}^2} \tag{4.10}$$

and

$$\hat{V}(\hat{t}_{yr}) = \hat{V}(t_x\,\hat{B}) = \left(1 - \frac{n}{N}\right)\left(\frac{t_x}{\bar{x}}\right)^2 \frac{s_e^2}{n}. \tag{4.11}$$

If the sample sizes are sufficiently large, approximate 95% confidence intervals (CIs) can be constructed using the standard errors (SEs) as

$$\hat{B} \pm 1.96\,\mathrm{SE}(\hat{B}), \quad \hat{\bar{y}}_r \pm 1.96\,\mathrm{SE}(\hat{\bar{y}}_r), \quad \text{or} \quad \hat{t}_{yr} \pm 1.96\,\mathrm{SE}(\hat{t}_{yr}).$$

Some software packages, including SAS software, substitute a t percentile with $n-1$ degrees of freedom for the normal percentile 1.96.

EXAMPLE 4.3   Let's return to the sample taken from the Census of Agriculture in Example 4.2. In the spreadsheet in Table 4.1, we created Column E, containing the residuals $e_i = y_i - \hat{B}x_i$. The sample standard deviation of Column E was calculated in Cell E306 to be $s_e = 31{,}657.218$. Thus, using (4.11),

$$\mathrm{SE}(\hat{t}_{yr}) = 3078\sqrt{1 - \frac{300}{3078}}\left(\frac{313{,}343.283}{301{,}953.723}\right)\frac{31{,}657.218}{\sqrt{300}} = 5{,}546{,}162.$$

An approximate 95% CI for the total farm acreage, using the ratio estimator, is

$$951{,}513{,}191 \pm 1.96(5{,}546{,}162) = [940{,}642{,}713,\ 962{,}383{,}669].$$

The website gives SAS code for calculating the ratio $\hat{B} = \bar{y}/\bar{x}$ and its standard error, with the output:

```
           Ratio Analysis: acres92/acres87

Numerator Denominator    Ratio    Std Err      95% CL for Ratio
---------------------------------------------------------------
acres92    acres87      0.986565  0.005750 0.97524871  0.99788176
---------------------------------------------------------------
```

We then multiply each quantity in the output by $t_x = 964{,}470{,}625$ (we do the calculations on the computer to avoid roundoff error) to obtain $\hat{t}_{yr} = (964{,}470{,}625)$ $(0.986565237) = 951{,}513{,}191$ and 95% CI for t_y of $[940{,}598{,}734, 962{,}427{,}648]$. SAS PROC SURVEYMEANS uses the percentile from a t_{299} distribution, 1.96793, instead of the value 1.96 from the normal distribution, so the CI from SAS software is slightly larger than the one we obtained when doing calculations by hand.

Did using a ratio estimator for the population total improve the precision in this example? The standard error of $\hat{t}_y = N\bar{y}$ is more than 10 times as large:

$$SE(N\bar{y}) = 3078 \sqrt{\left(1 - \frac{150}{3078}\right)} \frac{s_y}{\sqrt{150}} = 58{,}169{,}381.$$

The estimated CV for the ratio estimator is $5{,}546{,}162/951{,}513{,}191 = 0.0058$, as compared with an estimated CV of 0.0634 for the SRS estimator $N\bar{y}$ which does not use the auxiliary information. Including the 1987 information through the ratio estimator has greatly increased the precision. If all quantities to be estimated were highly correlated with the 1987 acreage, we could dramatically reduce the sample size and still obtain high precision by using ratio estimators rather than $N\bar{y}$. ∎

EXAMPLE **4.4**   Let's take another look at the hypothetical population used in Example 2.2 and in Exercise 2 of Chapter 3. Now, though, instead of using x as a stratification variable in stratified sampling, we use it as auxiliary information for ratio estimation. The population values are the following:

Unit Number	x	y
1	4	1
2	5	2
3	5	4
4	6	4
5	8	7
6	7	7
7	7	7
8	5	8

TABLE   4.2

Sampling Distribution for $\hat{t}_{yr}$.

Sample Number	Sample, $\mathcal{S}$	$\bar{x}_{\mathcal{S}}$	$\bar{y}_{\mathcal{S}}$	$\hat{B}$	$\hat{t}_{SRS}$	$\hat{t}_{yr}$
1	{1,2,3,4}	5.00	2.75	0.55	22.00	25.85
2	{1,2,3,5}	5.50	3.50	0.64	28.00	29.91
3	{1,2,3,6}	5.25	3.50	0.67	28.00	31.33
4	{1,2,3,7}	5.25	3.50	0.67	28.00	31.33
⋮	⋮	⋮	⋮	⋮	⋮	⋮
67	{4,5,6,8}	6.50	6.50	1.00	52.00	47.00
68	{4,5,7,8}	6.50	6.50	1.00	52.00	47.00
69	{4,6,7,8}	6.25	6.50	1.04	52.00	48.88
70	{5,6,7,8}	6.75	7.25	1.07	58.00	50.48

Note that x and y are positively correlated. We can calculate population quantities since we know the entire population and sampling distribution:

$$t_x = 47 \qquad t_y = 40$$
$$S_x = 1.3562027 \qquad S_y = 2.618615$$
$$R = 0.6838403 \qquad B = 0.8510638$$

Part of the sampling distribution for $\hat{t}_{yr}$ for a sample of size $n = 4$ is given in Table 4.2; the full file for the possible samples is in file artifratio.dat. Figure 4.3 gives histograms for the sampling distributions of two estimators of t_y: $\hat{t}_{SRS} = N\bar{y}$, the estimator used in Chapter 2, and $\hat{t}_{yr}$. The sampling distribution for the ratio estimator is not spread out as much as the sampling distribution for $N\bar{y}$; it is also skewed rather than symmetric. The skewness leads to the slight estimation bias of the ratio estimator. The population total is $t_y = 40$; the mean value of the sampling distribution of $\hat{t}_{yr}$ is 39.85063.

The mean value of the sampling distribution of $\hat{B}$ is 0.8478857, so Bias$(\hat{B}) = -0.003178$. The approximate bias of $\hat{B}$, calculated by substituting the population quantities into (4.6) and noting from (4.2) that $\hat{B} = \hat{\bar{y}}_r / \bar{x}_U$, is

$$\left(1 - \frac{n}{N}\right) \frac{1}{n\bar{x}_U^2} (BS_x^2 - RS_x S_y) = -0.003126.$$

The variance of the sampling distribution of $\hat{B}$, calculated using the definition of variance in (2.5), is 0.015186446; the approximation using (4.8) is

$$\frac{4}{8} \frac{1}{(4)(5.875)^2} (S_y^2 - 2BRS_x S_y + B^2 S_x^2) = 0.01468762. \qquad \blacksquare$$

Example 4.4 demonstrates that the approximation to the MSE in (4.8) is in fact only an approximation; it happens to be a good approximation in that example even though the population and sample are both small.

FIGURE **4.3**

Sampling distributions for (a) $\hat{t}_{SRS}$ and (b) $\hat{t}_{yr}$.

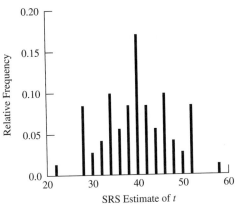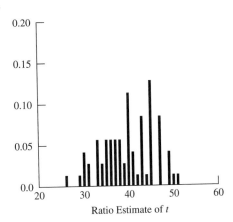

For (4.7) to be a good approximation to the MSE, the bias should be small and the terms discarded in the approximation of the variance should be small. If the CV of $\bar{x}$ is small—that is, if $\bar{x}_U$ is estimated with high relative precision, the bias is small relative to the square root of the variance. If we form a CI using $\hat{t}_{yr} \pm 1.96 \, \text{SE}[\hat{t}_{yr}]$, using (4.11) as the standard error, then the bias will not have a great effect on the coverage probability of the CI. A small CV $(\bar{x})$ also means that $\bar{x}$ is stable from sample to sample. In more complex sampling designs, though, the bias may be a matter of concern—we return to this issue in Section 4.5 and Chapter 9. For the approximation in (4.7) to work well, we want the sample size to be large, and CV $(\bar{x}) \le .1$, and CV $(\bar{y}) \le .1$. If these conditions are not met, then (4.7) may severely underestimate the true MSE.

4.1.3 Ratio Estimation with Proportions

Ratio estimation works exactly the same way when the quantity of interest is a proportion.

EXAMPLE 4.5 Peart (1994) collected the data shown in Table 4.3 as part of a study evaluating the effects of feral pig activity and drought on the native vegetation on Santa Cruz Island, California. She counted the number of woody seedlings in pig-protected areas under each of ten sampled oak trees in March 1992, following the drought-ending rains of 1991. She put a flag by each seedling, then determined how many were still alive in February 1994. The data (courtesy of Diann Peart) are plotted in Figure 4.4.

When most people who have had one introductory statistics course see data like these, they want to find the sample proportion of the 1992 seedlings that are still alive in 1994, and then calculate the standard error as though they had an SRS of 206 seedlings, obtaining a value of $\sqrt{(0.2961)(0.7039)/206} = 0.0318$. This calculation is *incorrect* for these data since plots, not individual seedlings, are the sampling units. Seedling survival depends on many factors such as local rainfall, amount of light, and predation. Such factors are likely to affect seedlings in the same plot to a similar

TABLE 4.3
Santa Cruz Island Seedling Data

Tree	$x =$ Number of Seedlings, 3/92	$y =$ Seedlings Alive, 2/94
1	1	0
2	0	0
3	8	1
4	2	2
5	76	10
6	60	15
7	25	3
8	2	2
9	1	1
10	31	27
Total	206	61
Average	20.6	6.1
Standard deviation	27.4720	8.8248

FIGURE 4.4
Plot of number of seedlings that survived (February 1994) vs. seedlings alive (March 1992), for ten oak trees.

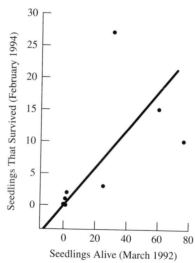

degree, leading different plots to have, in general, different survival rates. The sample size in this example is 10, not 206.

The design is actually a **cluster sample**; the clusters are the plots associated with each tree, and the observation units are individual seedlings in those plots. To look at this example from the framework of ratio estimation, let

$$y_i = \text{number of seedlings near tree } i \text{ that are alive in 1994}$$

$$x_i = \text{number of seedlings near tree } i \text{ that are alive in 1992}.$$

Then the ratio estimate of the proportion of seedlings still alive in 1994 is

$$\hat{B} = \hat{p} = \frac{\bar{y}}{\bar{x}} = \frac{6.1}{20.6} = 0.2961.$$

Using (4.10) and ignoring the finite population correction (fpc),

$$SE(\hat{B}) = \sqrt{\frac{1}{(10)(20.6)^2} \frac{\sum_{i \in S}(y_i - 0.2961165x_i)^2}{9}}$$

$$= \sqrt{\frac{56.3778}{(10)(20.6)^2}}$$

$$= 0.115.$$

The approximation to the variance of $\hat{B}$ in this example may be a little biased because the sample size is small. Note, however, the difference between the correctly calculated standard error of 0.115, and the incorrect value 0.0318 that would be obtained if one erroneously pretended that the seedlings were selected in an SRS.

SAS code for calculating these estimates is given on the website. The relevant output from PROC SURVEYMEANS is below:

```
                           Statistics

                                   Std Error
Variable            Mean           of Mean        95% CL for Mean
------------------------------------------------------------------
seed94           6.100000          2.790659    -0.2129093 12.4129093
seed92          20.600000          8.687411     0.9477108 40.2522892
------------------------------------------------------------------


               Ratio Analysis: seed94/seed92

Numerator Denominator    Ratio    Std Err    95% CL for Ratio
------------------------------------------------------------------
seed94      seed92      0.296117  0.115262  0.03537532  0.55685769
------------------------------------------------------------------                    ■
```

4.1.4 Ratio Estimation Using Weight Adjustments

In Section 2.4, we defined the sampling weight to be $w_i = 1/\pi_i$, and wrote the estimated population total as a function of the observations y_i and weights w_i:

$$\hat{t}_y = \sum_{i \in S} w_i y_i.$$

Note that

$$\hat{t}_{yr} = \frac{t_x}{\hat{t}_x}\hat{t}_y = \frac{t_x}{\hat{t}_x}\sum_{i\in S} w_i y_i.$$

We can think of the modification used in ratio estimation as an adjustment to each weight. Define

$$g_i = \frac{t_x}{\hat{t}_x}.$$

Then

$$\hat{t}_{yr} = \sum_{i\in S} w_i g_i y_i. \tag{4.12}$$

The estimator $\hat{t}_{yr}$ is a weighted sum of the observations, with weights $w_i^* = w_i g_i$. Unlike the original weights w_i, however, the adjusted weights w_i^* depend upon values from the sample: If a different sample is taken, the weight adjustment $g_i = t_x/\hat{t}_x$ will be different.

The weight adjustments g_i **calibrate** the estimates on the x variable. Since

$$\sum_{i\in S} w_i g_i x_i = t_x,$$

the adjusted weights force the estimated total for the x variable to equal the known population total t_x. The factors g_i are called the calibration factors.

The variance estimators in (4.9) and (4.11) can be calculated by forming the new variable $u_i = g_i e_i$. Then, for an SRS,

$$\hat{V}(\bar{u}) = \left(1 - \frac{n}{N}\right)\frac{1}{n(n-1)}\sum_{i\in S}(u_i - \bar{u})^2 = \left(1 - \frac{n}{N}\right)\frac{s_e^2}{n}\left(\frac{t_x}{\hat{t}_x}\right)^2 = \hat{V}(\hat{y}_r)$$

and, similarly, $\hat{V}(\hat{t}_u) = \hat{V}(\hat{t}_{yr})$.

EXAMPLE 4.6   For the Census of Agriculture data used in Examples 4.2 and 4.3, $g_i = t_x/\hat{t}_x = 964,470,625/929,413,560 = 1.037719554$ for each observation. Since the sample has $\hat{t}_x < t_x$, each observation's sampling weight is increased by a small amount. The sampling weight for the SRS design is $w_i = 3078/300 = 10.26$, so the ratio adjusted weight for each observation is

$$w_i^* = w_i g_i = (10.26)(1.037719554) = 10.64700262.$$

Then

$$\sum_{i\in S} w_i g_i x_i = \sum_{i\in S} 10.64700262\, x_i = 964,470,625 = t_x$$

and

$$\sum_{i\in S} w_i g_i y_i = \sum_{i\in S} 10.64700262\, y_i = 951,513,191 = \hat{t}_{yr}.$$

The adjusted weights, however, no longer sum to $N = 3078$:

$$\sum_{i\in S} w_i g_i = (300)(10.64700262) = 3194.$$

Thus, the ratio estimator is calibrated to the population total t_x of the x variable, but is no longer calibrated to the population size N. ∎

4.1.5 Advantages of Ratio Estimation

Ratio estimation is motivated by the desire to use information about a known auxiliary quantity x to obtain a more accurate estimator of t_y or $\bar{y}_U$. If x and y are perfectly correlated, that is, $y_i = Bx_i$ for all $i = 1, \ldots, N$, then $\hat{t}_{yr} = t_y$ and there is no estimation error. In general, if y_i is roughly proportional to x_i, the MSE will be small.

When does ratio estimation help? If the deviations of y_i from $\hat{B}x_i$ are smaller than the deviations of y_i from $\bar{y}$, then $\hat{V}(\hat{\bar{y}}_r) \le \hat{V}(\bar{y})$. Recall from Chapter 2 that

$$\text{MSE}(\bar{y}) = V(\bar{y}) = \left(1 - \frac{n}{N}\right)\frac{S_y^2}{n}.$$

Using the approximation in (4.7) and (4.8),

$$\text{MSE}(\hat{\bar{y}}_r) \approx \left(1 - \frac{n}{N}\right)\frac{1}{n}(S_y^2 - 2BRS_xS_y + B^2S_x^2).$$

Thus,

$$\text{MSE}(\hat{\bar{y}}_r) - \text{MSE}(\bar{y}) \approx \left(1 - \frac{n}{N}\right)\frac{1}{n}(S_y^2 - 2BRS_xS_y + B^2S_x^2 - S_y^2)$$

$$= \left(1 - \frac{n}{N}\right)\frac{1}{n}S_xB(-2RS_y + BS_x)$$

so to the accuracy of the approximation,

$$\text{MSE}(\hat{\bar{y}}_r) \le \text{MSE}(\bar{y}) \text{ if and only if } R \ge \frac{BS_x}{2S_y} = \frac{\text{CV}(x)}{2\text{CV}(y)}.$$

If the CVs are approximately equal, then it pays to use ratio estimation when the correlation between x and y is larger than 1/2.

Ratio estimation is most appropriate if a straight line through the origin summarizes the relationship between x_i and y_i and if the variance of y_i about the line is proportional to x_i. Under these conditions, $\hat{B}$ is the weighted least squares regression slope for the line through the origin with weights proportional to $1/x_i$—the slope $\hat{B}$ minimizes the sum of squares

$$\sum_{i\in S}\frac{1}{x_i}(y_i - \hat{B}x_i)^2.$$

4.2
Estimation in Domains

Often we want separate estimates for subpopulations; the subpopulations are called **domains** or **subdomains**. We may want to take an SRS of 1000 people from a population of 50,000 people and estimate the average salary for men and the average salary for women. There are two domains: men and women. We do not know which persons in the population belong to which domain until they are sampled, though. Thus, the number of persons in an SRS who fall into each domain is a random variable, with value unknown at the time the survey is designed.

Estimating domain means is a special case of ratio estimation. Suppose there are D domains. Let $\mathcal{U}_d$ be the index set of the units in the population that are in domain d, and let $\mathcal{S}_d$ be the index set of the units in the sample that are in domain d, for $d = 1, 2, \ldots, D$. Let N_d be the number of population units in $\mathcal{U}_d$, and n_d be the number of sample units in $\mathcal{S}_d$. Suppose we want to estimate the mean salary for the domain of women,

$$\bar{y}_{U_d} = \frac{\sum\limits_{i \in \mathcal{U}_d} y_i}{N_d} = \frac{\text{total salary for all women in population}}{\text{number of women in population}}.$$

A natural estimator of $\bar{y}_{U_d}$ is

$$\bar{y}_d = \frac{\sum\limits_{i \in \mathcal{S}_d} y_i}{n_d} = \frac{\text{total salary for women in sample}}{\text{number of women in sample}},$$

which looks at first just like the sample means studied in Chapter 2.

The quantity n_d is a random variable, however: If a different SRS is taken, we will very likely have a different value for n_d, the number of women in the sample. To see that $\bar{y}_d$ uses ratio estimation, let

$$x_i = \begin{cases} 1 & \text{if } i \in \mathcal{U}_d \\ 0 & \text{if } i \notin \mathcal{U}_d, \end{cases}$$

$$u_i = y_i x_i = \begin{cases} y_i & \text{if } i \in \mathcal{U}_d \\ 0 & \text{if } i \notin \mathcal{U}_d. \end{cases}$$

Then $t_x = \sum_{i=1}^{N} x_i = N_d$, $\bar{x}_U = N_d/N$, $t_u = \sum_{i=1}^{N} u_i$, $\bar{y}_{U_d} = t_u/t_x = B$, $\bar{x} = n_d/n$, and

$$\bar{y}_d = \hat{B} = \frac{\bar{u}}{\bar{x}} = \frac{\hat{t}_u}{\hat{t}_x}.$$

Because we are estimating a ratio, we use (4.10) to calculate the standard error:

$$\text{SE}(\bar{y}_d) = \sqrt{\left(1 - \frac{n}{N}\right) \frac{1}{n\bar{x}^2} \frac{\sum\limits_{i \in \mathcal{S}} (u_i - \hat{B}x_i)^2}{n - 1}}$$

$$= \sqrt{\left(1 - \frac{n}{N}\right) \frac{1}{n\bar{x}^2} \frac{\sum\limits_{i \in \mathcal{S}_d} (y_i - \hat{B})^2}{n - 1}}$$

$$= \sqrt{\left(1 - \frac{n}{N}\right) \frac{n}{n_d^2} \frac{(n_d - 1)s_{yd}^2}{n - 1}}, \tag{4.13}$$

where

$$s_{yd}^2 = \frac{\sum\limits_{i \in \mathcal{S}_d} (y_i - \bar{y}_d)^2}{n_d - 1}$$

is the sample variance of the sample observations in domain d. If $E(n_d)$ is large, then $(n_d - 1)/n_d \approx 1$ and

$$SE(\bar{y}_d) \approx \sqrt{\left(1 - \frac{n}{N}\right) \frac{s_{yd}^2}{n_d}}.$$

In a sufficiently large sample, the standard error of $\bar{y}_d$ is approximately the same as if we used formula (2.12).

The situation is a little more complicated when estimating a domain total. If N_d is known, we can estimate t_u by $N_d \bar{y}_d$. If N_d is unknown, though, we need to estimate t_u by

$$\hat{t}_{yd} = \hat{t}_u = N\bar{u}.$$

The standard error is

$$SE(\hat{t}_{yd}) = N \, SE(\bar{u}) = N\sqrt{\left(1 - \frac{n}{N}\right) \frac{s_u^2}{n}}.$$

EXAMPLE 4.7 In the SRS of size 300 from the Census of Agriculture (see Example 2.5), 39 counties are in western states.[2] What is the estimated total number of acres devoted to farming in the West?

The sample mean of the 39 counties is $\bar{y}_d = 598{,}681$, with sample standard deviation $s_{yd} = 516{,}157.7$. Using (4.13),

$$SE(\bar{y}_d) = \sqrt{\left(1 - \frac{300}{3078}\right) \frac{300}{39} \frac{38}{299} \frac{516{,}157.7}{\sqrt{39}}} = 77{,}637.$$

Thus, $\widehat{CV}[\bar{y}_d] = 0.1297$, and an approximate 95% CI for the mean farm acreage for counties in the western United States is $[445{,}897, \, 751{,}463]$.

For estimating the total number of acres devoted to farming in the West, suppose we do not know how many counties in the population are in the western United States. Define

$$x_i = \begin{cases} 1, & \text{if county } i \text{ is in the western United States} \\ 0, & \text{otherwise} \end{cases}$$

and define $u_i = y_i x_i$. Then

$$\hat{t}_{yd} = \hat{t}_u = \sum_{i \in \mathcal{S}} \frac{3078}{300} u_i = 239{,}556{,}051. \tag{4.14}$$

The standard error is

$$SE(\hat{t}_{yd}) = 3078\sqrt{\left(1 - \frac{300}{3078}\right) \frac{273005.4}{\sqrt{300}}} = 46{,}090{,}460.$$

The estimated CV for $\hat{t}_{yd}$ is $\widehat{CV}[\hat{t}_{yd}] = 46{,}090{,}460/239{,}556{,}051 = 0.1924$; had we known the number of counties in the western United States and been able to use that

---

[2] Alaska (AK), Arizona (AZ), California (CA), Colorado (CO), Hawaii (HI), Idaho (ID), Montana (MT), Nevada (NV), New Mexico (NM), Oregon (OR), Utah (UT), Washington (WA), and Wyoming (WY).

value in the estimate, the CV for the estimated total would have been 0.1297, the CV for the mean.

The SAS program on the website also contains the code for finding domain estimates. We define the domain indicator *west* to be 1 if the county is in the West and 0 otherwise. The relevant output is

```
                        Domain Analysis: west

                                   Std Error
     west     Variable      Mean     of Mean      95% CL for Mean
     ---------------------------------------------------------------
       0      acres92      252952      16834    219825.176 286079.583
       1      acres92      598681      77637    445897.252 751463.927
     ---------------------------------------------------------------


                        Domain Analysis: west

     west     Variable       Sum     Std Dev       95% CL for Sum
     ---------------------------------------------------------------
       0      acres92    677371058   47317687    584253179   770488938
       1      acres92    239556051   46090457    148853274   330258829
     ---------------------------------------------------------------
```

The output gives the estimates and CIs for both domains. ∎

EXAMPLE 4.8 An SRS of 1500 licensed boat owners in a state was sampled from a list of 400,000 names with currently licensed boats; 472 of the respondents said they owned an open motorboat longer than 16 feet. The 472 respondents with large motorboats reported having the following numbers of children:

Number of Children	Number of Respondents
0	76
1	139
2	166
3	63
4	19
5	5
6	3
8	1
Total	472

If we are interested in characteristics of persons who own large motorboats, there are two domains: persons who own large motorboats (domain 1) and persons who do not own large motorboats (domain 2). To estimate the percentage of large-motorboat owners who have children, we can use $\hat{p}_1 = 396/472 = 0.839$. This is a ratio estimator, but in this case, as shown in (4.13), the standard error is approximately what you would think it would be. Ignoring the fpc,

$$\text{SE}(\hat{p}_1) = \sqrt{\frac{0.839(1 - 0.839)}{472}} = 0.017.$$

To look at the average number of children per household among registered boat owners who register a motorboat more than 16 feet long, note that the average number of children for the 472 respondents in the domain is 1.667373, with variance 1.398678. Thus an approximate 95% CI for the average number of children in large-motorboat households is

$$1.667 \pm 1.96\sqrt{\frac{1.398678}{472}} = [1.56, 1.77].$$

To estimate the total number of children in the state whose parents register a large motorboat, we create a new variable u for the respondents that equals the number of children if the respondent has a motorboat, and 0 otherwise. The frequency distribution for the variable u is then

Number of Children	Number of Respondents
0	1104
1	139
2	166
3	63
4	19
5	5
6	3
8	1
Total	1500

Now $\bar{u} = 0.52466$ and $s_u^2 = 1.0394178$, so $\hat{t}_{yd} = 400,000(0.524666) = 209,867$ and

$$\text{SE}(\hat{t}_{yd}) = \text{SE}(\hat{t}_u) = \sqrt{\left(1 - \frac{1500}{400,000}\right)(400,000)^2 \frac{1.0394178}{1500}} = 10,510.$$

The variable u_i counts the number of children in household i that belong to a household with a large open motorboat. SAS code to find estimates for this example is given on the website. ∎

In this section, we have shown that estimating domain means is a special case of ratio estimation because the sample size in the domain varies from sample to sample. If the sample size for the domain in an SRS is sufficiently large, we can use SRS formulas for inference about the domain mean.

Inference about totals depends on whether the population size of the domain, N_d, is known. If N_d is known, then the estimated total is $N_d \bar{y}_d$. If N_d is unknown, then define a new variable u_i that equals y_i for observations in the domain and 0 for observations not in the domain; then use $\hat{t}_u$ to estimate the domain total.

The results of this section are only for SRSs, and the approximations depend on having a sufficiently large sample so that $E(n_d)$ is large. In Section 14.2, we discuss estimating domain means and totals if the data are collected using other sampling designs, or when the domain sample sizes are small.

4.3
Regression Estimation in Simple Random Sampling

4.3.1 Using a Straight-Line Model

Ratio estimation works best if the data are well fit by a straight line through the origin. Sometimes, data appear to be evenly scattered about a straight line that does not go through the origin—that is, the data look as though the usual straight-line regression model

$$y = B_0 + B_1 x$$

would provide a good fit.

Let $\hat{B}_1$ and $\hat{B}_0$ be the ordinary least squares regression coefficients of the slope and intercept. For the straight line regression model,

$$\hat{B}_1 = \frac{\displaystyle\sum_{i \in S} (x_i - \bar{x})(y_i - \bar{y})}{\displaystyle\sum_{i \in S} (x_i - \bar{x})^2} = \frac{rs_y}{s_x},$$

$$\hat{B}_0 = \bar{y} - \hat{B}_1\bar{x},$$

and r is the sample correlation coefficient of x and y.

In regression estimation, like ratio estimation, we use the correlation between x and y to obtain an estimator for $\bar{y}_U$ with (we hope) increased precision. Suppose we know $\bar{x}_U$, the population mean for the x's. Then the regression estimator of $\bar{y}_U$ is the predicted value of y from the fitted regression equation when $x = \bar{x}_U$:

$$\hat{\bar{y}}_{\text{reg}} = \hat{B}_0 + \hat{B}_1\bar{x}_U = \bar{y} + \hat{B}_1(\bar{x}_U - \bar{x}). \tag{4.15}$$

If $\bar{x}$ from the sample is smaller than the population mean $\bar{x}_U$ and x and y are positively correlated, then we would expect $\bar{y}$ to also be smaller than $\bar{y}_U$. The regression estimator adjusts $\bar{y}$ by the quantity $\hat{B}_1(\bar{x}_U - \bar{x})$.

Like the ratio estimator, the regression estimator is biased. Let B_1 be the least squares regression slope calculated from all the data in the population,

$$B_1 = \frac{\displaystyle\sum_{i=1}^{N} (x_i - \bar{x}_U)(y_i - \bar{y}_U)}{\displaystyle\sum_{i=1}^{N} (x_i - \bar{x}_U)^2} = \frac{RS_y}{S_x}.$$

Then, using (4.15), the bias of $\hat{\bar{y}}_{\text{reg}}$ is given by

$$E[\hat{\bar{y}}_{\text{reg}} - \bar{y}_U] = E[\bar{y} - \bar{y}_U] + E[\hat{B}_1(\bar{x}_U - \bar{x})] = -\text{Cov}(\hat{B}_1, \bar{x}). \tag{4.16}$$

If the regression line goes through all of the points (x_i, y_i) in the population, then the bias is zero: in that situation, $\hat{B}_1 = B_1$ for every sample, so $\text{Cov}(\hat{B}_1, \bar{x}) = 0$. As with ratio estimation, for large SRSs the MSE for regression estimation is approximately equal to the variance (see Exercise 29); the bias is often negligible in large samples.

The method used in approximating the MSE in ratio estimation can also be applied to regression estimation. Let $d_i = y_i - [\bar{y}_U + B_1(x_i - \bar{x}_U)]$. Then,

$$
\begin{aligned}
\text{MSE}(\hat{\bar{y}}_{\text{reg}}) &= E[\{\bar{y} + \hat{B}_1(\bar{x}_U - \bar{x}) - \bar{y}_U\}^2] \\
&\approx V(\bar{d}) \\
&= \left(1 - \frac{n}{N}\right) \frac{S_d^2}{n}.
\end{aligned}
\tag{4.17}
$$

Using the relation $B_1 = RS_y/S_x$, it may be shown that

$$
\begin{aligned}
\left(1 - \frac{n}{N}\right) \frac{S_d^2}{n} &= \left(1 - \frac{n}{N}\right) \frac{1}{n} \sum_{i=1}^{N} \frac{(y_i - \bar{y}_U - B_1[x_i - \bar{x}_U])^2}{N-1} \\
&= \left(1 - \frac{n}{N}\right) \frac{1}{n} S_y^2 (1 - R^2).
\end{aligned}
\tag{4.18}
$$

(See Exercise 28.) Thus, the approximate MSE is small when

- n is large
- n/N is large
- S_y is small
- the correlation R is close to -1 or $+1$.

The standard error may be calculated by substituting estimates for the population quantities in (4.17) or (4.18). We can estimate S_d^2 in (4.17) by using the residuals $e_i = y_i - (\hat{B}_0 + \hat{B}_1 x_i)$; then $s_e^2 = \sum_{i \in S} e_i^2/(n-1)$ estimates S_d^2 and

$$\text{SE}(\hat{\bar{y}}_{\text{reg}}) = \sqrt{\left(1 - \frac{n}{N}\right) \frac{s_e^2}{n}}. \tag{4.19}$$

In small samples, we may alternatively calculate s_e^2 using the MSE from a regression analysis: $s_e^2 = \sum_{i \in S} e_i^2/(n-2)$. This adjusts the estimator for the degrees of freedom in the regression. To estimate the variance using the formulation in (4.18), substitute the sample variance s_y^2 and the sample correlation r for the population quantities S_y^2 and R, obtaining

$$\text{SE}(\hat{\bar{y}}_{\text{reg}}) = \sqrt{\left(1 - \frac{n}{N}\right) \frac{1}{n} s_y^2 (1 - r^2)}. \tag{4.20}$$

EXAMPLE 4.9 To estimate the number of dead trees in an area, we divide the area into 100 square plots and count the number of dead trees on a photograph of each plot. Photo counts can be made quickly, but sometimes a tree is misclassified or not detected. So we select an SRS of 25 of the plots for field counts of dead trees. We know that the

FIGURE   **4.5**

The plot of photo and field tree-count data, along with the regression line. Note that $\hat{y}_{\text{reg}}$ is the predicted value from the regression equation when $x = \bar{x}_U$. The point $(\bar{x}, \bar{y})$ is marked by "+" on the graph.

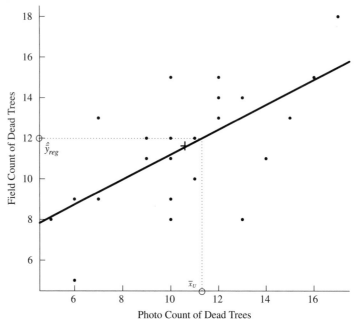

population mean number of dead trees per plot from the photo count is 11.3. The data—plotted in Figure 4.5—are given below.

Photo (x)	10	12	7	13	13	6	17	16	15	10	14	12	10
Field (y)	15	14	9	14	8	5	18	15	13	15	11	15	12

Photo (x)	5	12	10	10	9	6	11	7	9	11	10	10
Field (y)	8	13	9	11	12	9	12	13	11	10	9	8

For these data, $\bar{x} = 10.6$, $\bar{y} = 11.56$, $s_y^2 = 9.09$, and the sample correlation between x and y is $r = 0.62420$. Fitting a straight line regression model gives

$$\hat{y} = 5.059292 + 0.613274x$$

with $\hat{B}_0 = 5.059292$ and $\hat{B}_1 = 0.613274$. In this example, x and y are positively correlated so that $\bar{x}$ and $\bar{y}$ are also positively correlated. Since $\bar{x} < \bar{x}_U$, we expect that the sample mean $\bar{y}$ is also too small; the regression estimate adds the quantity $\hat{B}_1(\bar{x}_U - \bar{x}) = 0.613(11.3 - 10.6) = 0.43$ to $\bar{y}$ to compensate.

Using (4.15), the regression estimate of the mean is

$$\hat{\bar{y}}_{\text{reg}} = 5.059292 + 0.613274(11.3) = 11.99.$$

From (4.20), the standard error is

$$SE(\hat{\bar{y}}_{reg}) = \sqrt{\left(1 - \frac{25}{100}\right)(9.09)(1 - 0.62420^2)} = 0.408.$$

The standard error of $\hat{\bar{y}}_{reg}$ is less than that for $\bar{y}$:

$$SE[\bar{y}] = \sqrt{\left(1 - \frac{25}{100}\right)\frac{s_y^2}{25}} = 0.522.$$

We expect regression estimation to increase the precision in this example because the variables photo and field are positively correlated. To estimate the total number of dead trees, use

$$\hat{t}_{yreg} = (100)(11.99) = 1199;$$

$$SE(\hat{t}_{yreg}) = (100)(0.408) = 40.8.$$

In SAS software, PROC SURVEYREG calculates regression estimates. Code for this example is given on the website; partial output is given below.

```
                    Analysis of Estimable Functions

                            Standard                            95% Confidence
                                                                    Interval
Parameter      Estimate       Error    t Value  Pr > |t|
```

Parameter	Estimate	Standard Error	t Value	Pr > |t|	95% Confidence Interval	
Total field trees	1198.92920	42.7013825	28.08	<.0001	1110.79788	1287.06053
Mean field trees	11.98929	0.4270138	28.08	<.0001	11.10798	12.87061

The standard errors given by SAS software are slightly larger than those obtained by hand calculation because SAS uses a slightly different estimator (see Section 11.7). In practice, we recommend using survey regression software for regression estimation to avoid roundoff errors. ∎

4.3.2 Difference Estimation

Difference estimation is a special case of regression estimation, used when the investigator "knows" that the slope B_1 is 1. Difference estimation is often recommended in accounting when an SRS is taken. A list of accounts receivable consists of the book value for each account—the company's listing of how much is owed on each account. In the simplest sampling scheme, the auditor scrutinizes a random sample of the accounts to determine the audited value—the actual amount owed—in order to estimate the error in the total accounts receivable. The quantities considered are:

$$y_i = \text{audited value for company } i$$
$$x_i = \text{book value for company } i.$$

Then, $\bar{y} - \bar{x}$ is the mean difference for the audited accounts.

The estimated total difference is $\hat{t}_y - \hat{t}_x = N(\bar{y} - \bar{x})$; the estimated audited value for accounts receivable is

$$\hat{t}_{y\text{diff}} = t_x + (\hat{t}_y - \hat{t}_x).$$

The residuals from this model are $e_i = y_i - x_i$. The variance of $\hat{t}_{y\text{diff}}$ is

$$V(\hat{t}_{y\text{diff}}) = V[t_x + (\hat{t}_y - \hat{t}_x)] = V(\hat{t}_e),$$

where $\hat{t}_e = (N/n)\sum_{i\in S} e_i$. If the variability in the residuals e_i is smaller than the variability among the y_i, then difference estimation will increase precision.

Difference estimation works best if the population and sample have a large fraction of nonzero differences that are roughly equally divided between overstatements and understatements, and if the sample is large enough so that the sampling distribution of $(\bar{y} - \bar{x})$ is approximately normal. In auditing, it is possible that most of the audited values in the sample are all the same as the corresponding book value. In such a situation, the design-based variance estimator is unstable and a model-based approach may be preferred.

4.4
Poststratification

Suppose a sampling frame lists all households in an area, and you would like to estimate the average amount spent on food in a month. One desirable stratification variable might be household size because large households might be expected to have higher food bills than smaller households. From U.S. census data, the distribution of household size in the region is known:

Number of Persons in Household	Percentage of Households
1	25.75
2	31.17
3	17.50
4	15.58
5 or more	10.00

The sampling frame, however, does not include information on household size—it only lists the households. Thus, although you know the population size in each subgroup, you cannot take a stratified sample because you do not know the stratum membership of the units in your sampling frame. You can, however, take an SRS and record the amount spent on food as well as the household size for each household in your sample. If n, the size of the SRS, is large enough, then the sample is likely to resemble a stratified sample with proportional allocation: We would expect about 26% of the sample to be one-person households, about 31% to be two-person households, and so on.

Considering the different household-size groups to be different domains, we can use the methods from Section 4.2 to estimate the average amount spent on groceries for each domain. Take an SRS of size n. Let $n_1, n_2, \ldots, n_H$ be the numbers of units in the various household-size groups (domains) and let $\bar{y}_1, \ldots, \bar{y}_H$ be the sample means for the groups. In this case, since the poststrata are formed *after* the sample is taken, the sample domain sizes $n_1, n_2, \ldots, n_H$ are random quantities. If we selected another SRS from the population, the poststratum sizes in the sample would change. Since the poststratum sizes in the population are known, however, we can use the known values of N_h in the estimation.

To see how poststratification fits in the framework of ratio estimation, define $x_{ih} = 1$ if observation i is in poststratum h and 0 otherwise. Let $u_{ih} = y_i x_{ih}$. Then $t_{xh} = \sum_{i=1}^{N} x_{ih} = N_h$ and

$$t_{uh} = \sum_{i=1}^{N} u_{ih} = \text{population total of variable } y \text{ in poststratum } h.$$

For each poststratum h, we can estimate the total in the poststratum by

$$\hat{t}_{uh} = \sum_{i \in \mathcal{S}} \frac{N}{n} u_{ih}$$

[$\hat{t}_{uh}$ is the domain total estimator in (4.14)]. We can then use ratio estimation to obtain:

$$\hat{t}_{uhr} = \frac{t_{xh}}{\hat{t}_{xh}} \hat{t}_{uh} = \frac{N_h}{\hat{N}_h} \hat{t}_{uh} = N_h \bar{y}_h,$$

where $\bar{y}_h$ is the sample mean of the observations in poststratum h.

The poststratified estimator of the population total is

$$\hat{t}_{y\text{post}} = \sum_{h=1}^{H} \hat{t}_{uhr} = \sum_{h=1}^{H} \frac{N_h}{\hat{N}_h} \hat{t}_{uh} = \sum_{h=1}^{H} N_h \bar{y}_h;$$

ratio estimation is used within each poststratum to estimate the population total in that poststratum.

The poststratified estimator of $\bar{y}_U$ is

$$\bar{y}_{\text{post}} = \sum_{h=1}^{H} \frac{N_h}{N} \bar{y}_h. \tag{4.21}$$

If N_h/N is known, n_h is reasonably large (≥ 30 or so), and n is large, then we can use the variance for proportional allocation as an approximation to the poststratified variance:

$$\hat{V}(\bar{y}_{\text{post}}) \approx \left(1 - \frac{n}{N}\right) \sum_{h=1}^{H} \frac{N_h}{N} \frac{s_h^2}{n}. \tag{4.22}$$

This approximation is valid only when the expected sample sizes in each poststratum are large, however (see Exercise 37).

Many large surveys use poststratification to improve efficiency of the estimators or to correct for the effects of differential nonresponse in the poststrata (see Chapter 8). We discuss poststratification for general survey designs in Section 11.7.

4.5
Ratio Estimation with Stratified Samples*

The previous sections proposed ratio and regression estimators for use with SRSs. The concept of ratio estimation, however, is completely general and is easily extended to other sampling designs. In stratified sampling, for example, we can use estimators of the population totals for x and y from a stratified sample to give the **combined ratio estimator**

$$\hat{t}_{yrc} = \hat{B}t_x,$$

where

$$\hat{B} = \frac{\hat{t}_{y,\text{str}}}{\hat{t}_{x,\text{str}}}.$$

As in Sections 3.2 and 3.3,

$$\hat{t}_{y,\text{str}} = \sum_{h=1}^{H} N_h \bar{y}_h = \sum_{h=1}^{H} \sum_{j \in \mathcal{S}_h} w_{hj} y_{hj},$$

where the sampling weight is $w_{hj} = (N_h/n_h)$, and

$$\hat{t}_{x,\text{str}} = \sum_{h=1}^{H} N_h \bar{x}_h = \sum_{h=1}^{H} \sum_{j \in \mathcal{S}_h} w_{hj} x_{hj}.$$

Then, using the arguments in Section 4.1.2,

$$\text{MSE}(\hat{t}_{yrc}) \approx V(\hat{t}_{y,\text{str}} - B\hat{t}_{x,\text{str}}) = V\left[\sum_{h=1}^{H} \sum_{j \in \mathcal{S}_h} w_{hj}(y_{hj} - Bx_{hj})\right];$$

we estimate the MSE by

$$\hat{V}(\hat{t}_{yrc}) = \left(\frac{t_{x,\text{str}}}{\hat{t}_{x,\text{str}}}\right)^2 \hat{V}\left(\sum_{h=1}^{H} \sum_{j \in \mathcal{S}_h} w_{hj} e_{hj}\right)$$

$$= \left(\frac{t_{x,\text{str}}}{\hat{t}_{x,\text{str}}}\right)^2 \hat{V}(\hat{t}_{e,\text{str}})$$

$$= \left(\frac{t_{x,\text{str}}}{\hat{t}_{x,\text{str}}}\right)^2 \left[\hat{V}(\hat{t}_{y,\text{str}}) + \hat{B}^2 \hat{V}(\hat{t}_{x,\text{str}}) - 2\hat{B}\widehat{\text{Cov}}(\hat{t}_{y,\text{str}}, \hat{t}_{x,\text{str}})\right],$$

where $e_{hj} = y_{hj} - \hat{B}x_{hj}$. In the combined ratio estimator, first the strata are combined to estimate t_x and t_y, then ratio estimation is applied.

For the **separate ratio estimator**, ratio estimation is applied first, then the strata are combined. The estimator

$$\hat{t}_{yrs} = \sum_{h=1}^{H} \hat{t}_{yhr} = \sum_{h=1}^{H} t_{xh} \frac{\hat{t}_{yh}}{\hat{t}_{xh}},$$

uses ratio estimation separately in each stratum, with

$$\hat{V}(\hat{t}_{yrs}) = \sum_{h=1}^{H} \hat{V}(\hat{t}_{yhr}).$$

It can improve efficiency if the $\hat{t}_{yh}/\hat{t}_{xh}$ vary from stratum to stratum, but should not be used when strata sample sizes are small because each ratio is biased, and the bias can propagate through the strata. Note that poststratification (Section 4.4) is a special case of the separate ratio estimator.

The combined estimator has less bias when the sample sizes in some of the strata are small. When the ratios vary greatly from stratum to stratum, however, the combined estimator does not take advantage of the extra efficiency afforded by stratification as does the separate ratio estimator. Many survey software packages, including SAS, calculate the combined ratio estimator by default.

EXAMPLE 4.10   Steffey et al. (2006) describe the use of combined ratio estimation in the legal case *Labor Ready v. Gates McDonald*. The plaintiff alleged that the defendant had not thoroughly investigated claims for worker's compensation, resulting in overpayments for these claims by the plaintiff. A total of $N = 940$ claims were considered in 1997. For each of these, the incurred cost of the claim (x_i) was known, and consequently the total amount of incurred costs was known to be $t_x = \$9.407$ million. But the plaintiff contended that the incurred value amounts were unjustified, and that the assessed value (y_i) of some claims after a thorough review would differ from the incurred value.

A sampling plan was devised for estimating the total assessed value of all 940 claims. Since it was expected that the assessed value would be highly correlated with the incurred costs, ratio estimation is desirable here. Two strata were sampled: Stratum 1 consisted of the claims in which the incurred cost exceeded $25,000, and stratum 2 consisted of the smaller claims (incurred cost less than $25,000). Summary statistics for the strata are given in the following table, with r_h the sample correlation in stratum h:

Stratum	N_h	n_h	$\bar{x}_h$	s_{xh}	$\bar{y}_h$	s_{yh}	r_h
1	102	70	$59,549.55	$64,047.95	$38,247.80	$32,470.78	0.62
2	838	101	$5,718.84	$5,982.34	$3,833.16	$5,169.72	0.77

The sampling fraction was set much higher in stratum 1 than in stratum 2 because the variability is much higher in stratum 1 (the investigators used a modified form of the optimal allocation described in Section 3.4.2). We estimate

$$\hat{t}_{x,\text{str}} = \sum_{h=1}^{2} \hat{t}_{xh} = (102)(59,549.55) + (838)(5,718.84) = 10,866,442.02$$

$$\hat{t}_{y,\text{str}} = \sum_{h=1}^{2} \hat{t}_{yh} = (102)(38,247.80) + (838)(3,833.16) = 7,113,463.68$$

and

$$\hat{B} = \frac{\hat{t}_{y,\text{str}}}{\hat{t}_{x,\text{str}}} = \frac{7,113,463.68}{10,866,442.02} = 0.654626755.$$

Using formulas for variances of stratified samples,

$$\hat{V}(\hat{t}_{x,\text{str}}) = \left(1 - \frac{70}{102}\right)(102)^2\frac{(64,047.95)^2}{70} + \left(1 - \frac{101}{838}\right)(838)^2\frac{(5982.34)^2}{101}$$
$$= 410,119,750,555,$$

$$\hat{V}(\hat{t}_{y,\text{str}}) = \left(1 - \frac{70}{102}\right)(102)^2\frac{(32,470.78)^2}{70} + \left(1 - \frac{101}{838}\right)(838)^2\frac{(5169.72)^2}{101}$$
$$= 212,590,045,044,$$

and

$$\widehat{\text{Cov}}(\hat{t}_{x,\text{str}}, \hat{t}_{y,\text{str}}) = \left(1 - \frac{70}{102}\right)(102)^2\frac{(32,470.78)(64,047.95)(0.62)}{70}$$
$$+ \left(1 - \frac{101}{838}\right)(838)^2\frac{(5169.72)(5982.34)(0.77)}{101}$$
$$= 205,742,464,829.$$

Using the combined ratio estimator, the total assessed value of the claims is estimated by

$$\hat{t}_{yrc} = (9.407 \times 10^6)(0.654626755) = \$6.158 \text{ million}$$

with standard error

$$\text{SE}(\hat{t}_{yrc}) = \frac{10.866}{9.407}\sqrt{[2.126 + (0.6546)^2(4.101) - 2(0.6546)(2.057)] \times 10^{11}}$$
$$= \$0.371 \text{ million}.$$

We use $169 = $ (number of observations) $-$ (number of strata) degrees of freedom for the CI. An approximate 95% CI for the total assessed value of the claims is $6.158 \pm 1.97(0.371)$, or between $5.43 and $6.89 million. Note that the CI for t_y does not contain the total incurred value (t_x) of $9.407 million. This supported the plaintiff's case that the total incurred value was too high. ∎

4.6
Model-Based Theory for Ratio and Regression Estimation*

In the design-based theory presented in Sections 4.1 and 4.3, the form of the estimators $\hat{\bar{y}}_r$ and $\hat{\bar{y}}_{\text{reg}}$ were motivated by regression models. Properties of the estimators, however, depend only on the sampling design. Thus, we found in (4.8) that

$$V(\hat{\bar{y}}_r) \approx \frac{1}{n}\left(1 - \frac{n}{N}\right)\sum_{i=1}^{N}\frac{(y_i - Bx_i)^2}{N - 1} \text{ for an SRS. This variance approximation is}$$

derived from the simple random sampling formulas in Chapter 2 and does not rely on any assumptions about the model. If the model does not fit the data well, ratio or regression estimation might not increase precision for estimated means and totals, but CIs for the means or totals will be correct in the sense that a 95% CI will have coverage probability close to 0.95. Inferences about finite population quantities using ratio or regression estimation are correct even if the model does not fit the data well. For that reason, the ratio and regression estimators presented in Sections 4.1 and 4.3 are examples of **model-assisted estimators**—a model motivates the form of the estimator, but inference depends on the sampling design. Särndal et al. (1992) present the theory of model-assisted estimation, in which inference is based on randomization theory.

If you have studied regression analysis, you learned a different approach to model-fitting in which you make assumptions about the regression model, find the least squares estimators of the regression parameters under the model, and plot residuals and explore regression diagnostics to check how well the model fits the data. Such a model-based approach, pioneered by Brewer (1963) and Royall (1970), can also be followed with survey data. As in the model-based approach outlined in Section 2.9, the model is used to predict population values that are not in the sample. In this section we discuss models that give the point estimators in (4.2) and (4.15) for ratio and regression estimation. The variances under a model-based approach, however, are different, as we will see.

4.6.1 A Model for Ratio Estimation

We stated earlier that ratio estimation works well in an SRS when a straight line through the origin fits well and when the variance of the observations about the line is proportional to x. We can state these conditions as a linear regression model: Assume that $x_1, x_2, \ldots, x_N$ are known (and all are greater than zero) and that $Y_1, Y_2, \ldots, Y_N$ are independent and follow the model

$$Y_i = \beta x_i + \varepsilon_i, \tag{4.23}$$

where $E_M(\varepsilon_i) = 0$ and $V_M(\varepsilon_i) = \sigma^2 x_i$. The independence of observations in the model is an explicit statement that the sampling design gives no information that can be used in estimating quantities of interest; the sampling procedure has no effect on the validity of the model. Under the model, $T_y = \sum_{i=1}^{N} Y_i$ is a random variable and the population total of interest, t_y, is one realization of the random variable T_y (this is in contrast to the randomization approach, in which t_y is considered to be a fixed but unknown quantity and the only random variables are the sample indicators Z_i). If $\mathcal{S}$ represents the set of units in our sample, then

$$t_y = \sum_{i \in \mathcal{S}} y_i + \sum_{i \notin \mathcal{S}} y_i.$$

We observe the values of y_i for units in the sample, and predict those for units not in the sample as $\hat{\beta} x_i$, where $\hat{\beta} = \bar{y}/\bar{x}$ is the weighted least squares estimate of β under the model in (4.23) (see Exercise 32). Then a natural estimate of t_y is

$$\hat{t}_y = \sum_{i \in \mathcal{S}} y_i + \hat{\beta} \sum_{i \notin \mathcal{S}} x_i = n\bar{y} + \frac{\bar{y}}{\bar{x}} \sum_{i \notin \mathcal{S}} x_i = \frac{\bar{y}}{\bar{x}} \sum_{i=1}^{N} x_i = \frac{\bar{y}}{\bar{x}} t_x.$$

This model results in the ratio estimator of t_y given in Section 4.1. Indeed, the finite population ratio $B = t_y/t_x$ is the weighted least squares estimator of β, applied to the entire population.

In many common sampling schemes, we find that if we adopt a model consistent with the reasons we would adopt a certain sampling scheme or method of estimation, the point estimators obtained using the model are very close to the design-based estimators. The model-based variance, though, usually differs from the variance from the randomization theory. In **randomization theory**, or **design-based sampling**, the *sampling design* determines how sampling variability is estimated. In **model-based sampling**, the *model* determines how variability is estimated, and the sampling design is irrelevant—as long as the model holds, you could choose any n units you want to from the population.

The model-based estimator

$$\hat{T}_y = \sum_{i \in S} Y_i + \hat{\beta} \sum_{i \notin S} x_i$$

is unbiased under the assumed model in (4.23) since

$$E_M[\hat{T}_y - T] = E_M\left[\sum_{i \notin S} x_i - \sum_{i \notin S} Y_i\right] = 0.$$

The model-based variance is

$$V_M[\hat{T}_y - T] = V_M\left[\sum_{i \notin S} x_i - \sum_{i \notin S} Y_i\right]$$

$$= V_M\left[\beta \sum_{i \notin S} x_i\right] + V_M\left[\sum_{i \notin S} Y_i\right]$$

because $\hat{\beta}$ and $\sum_{i \notin S} Y_i$ are independent under the model assumptions. The model in (4.23) does not depend on which population units are selected to be the sample S, so S can be treated as though it is fixed. Consequently, using (4.23),

$$V_M\left[\sum_{i \notin S} Y_i\right] = V_M\left[\sum_{i \notin S} (\beta x_i + \varepsilon_i)\right] = V_M\left[\sum_{i \notin S} \varepsilon_i\right] = \sigma^2 \sum_{i \notin S} x_i,$$

and, similarly,

$$V_M\left[\hat{\beta} \sum_{i \notin S} x_i\right] = \left(\sum_{i \notin S} x_i\right)^2 V_M\left[\frac{\sum_{i \in S} Y_i}{\sum_{i \in S} x_i}\right] = \left(\sum_{i \notin S} x_i\right)^2 \frac{\sigma^2}{\sum_{i \in S} x_i}.$$

Combining the two terms,

$$V_M[\hat{T}_y - T] = \frac{\sigma^2 \sum_{i \notin S} x_i}{\sum_{i \in S} x_i} \left(\sum_{i \notin S} x_i + \sum_{i \in S} x_i \right)$$

$$= \frac{\sigma^2 \sum_{i \notin S} x_i}{\sum_{i \in S} x_i} t_x$$

$$= \left(1 - \frac{\sum_{i \in S} x_i}{t_x} \right) \frac{\sigma^2 t_x^2}{\sum_{i \in S} x_i}. \qquad (4.24)$$

Note that if the sample size is small relative to the population size, then

$$V_M[\hat{T}_y - T] \approx \frac{\sigma^2 t_x^2}{\sum_{i \in S} x_i};$$

the quantity $(1 - \sum_{i \in S} x_i/t_x)$ serves as an fpc in the model-based approach to ratio estimation.

EXAMPLE 4.11 Let's perform a model-based analysis of the data from the Census of Agriculture, used in Examples 4.2 and 4.3. We already plotted the data in Figure 4.1, and it looked as though a straight line through the origin would fit well, and that the variability about the line was greater for observations with larger values of x. For the data points with x positive, we can run a regression analysis with no intercept and with weight variable $1/x$. SAS PROC REG code used for this analysis is provided on the website. Only 299 observations are used in this analysis since observation 179, Hudson County, New Jersey, has $x_{179} = 0$.

```
Dependent Variable: acres92
NOTE: No intercept in model.  R-Square is redefined.
Weight: recacr87
```

Analysis of Variance

Source	DF	Sum of Squares	Mean Square	F Value	Pr > F
Model	1	88168461	88168461	41487.3	<.0001
Error	298	633307	2125.19126		
Uncorrected Total	299	88801768			

Root MSE	46.09980	R-Square	0.9929
Dependent Mean	38097	Adj R-Sq	0.9928
Coeff Var	0.12101		

Parameter Estimates

Variable	DF	Parameter Estimate	Standard Error	t Value	Pr > |t|
acres87	1	0.98657	0.00484	203.68	<.0001

The slope, 0.986565, and the model-based estimate of the total, 9.5151×10^8, are the same as the design-based estimates obtained in Example 4.2. The model-based standard error of the estimated total, using (4.24), is

$$\sqrt{\hat{\sigma}^2 \frac{t_x - \sum_{i \in \mathcal{S}} x_i}{\sum_{i \in \mathcal{S}} x_i} t_x}.$$

We can use the weighted residuals (for nonzero x_i)

$$r_i = \frac{y_i - \hat{\beta} x_i}{\sqrt{x_i}}$$

to estimate σ^2: If the model assumptions hold, $\hat{\sigma}^2 = \sum r_i^2 / (n-1)$ (given as the MSE in the ANOVA table) estimates σ^2. Thus

$$SE_M[\hat{T}_y] = \sqrt{(2125.19126) \left(\frac{964{,}470{,}625 - 90{,}586{,}117}{90{,}586{,}117} \right) (964{,}470{,}625)}$$
$$= 4{,}446{,}719.$$

Note that for this example, the model-based standard error is smaller than the standard error we calculated using randomization inference, which was 5,344,568. The model-based analysis assumes that $V_M(\varepsilon_i) = \sigma^2 x_i$; the design-based analysis does not require such an assumption. ∎

When adopting a model for a set of data, we need to check the assumptions of the model. The assumptions for the model used in this section are:

1 The model is correct, that is, $E_M(Y_i) = x_i \beta$.

2 The variance structure is correct, that is, $V_M(Y_i) = \sigma^2 x_i$.

3 The observations are independent.

Typically, assumptions 1 and 2 are checked by plotting the data and examining residuals from the model. Assumption 3, however, is difficult to check in practice, and requires knowledge of how the data were collected. Generally, if you take a random sample, then you may assume the observations are independent.

We can perform some checks on the appropriateness of a model with straight line through the origin for these data: If the variance of y_i about the line is proportional to

F I G U R E   4.6

The plot of weighted residuals vs. x, for the random sample from the agricultural census. A few counties may be outliers; overall, though, scatter appears to be fairly random.

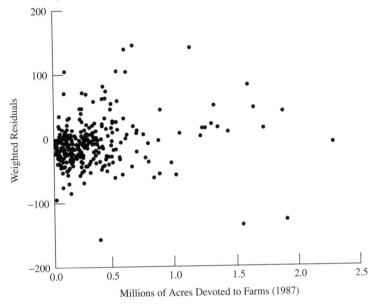

x_i, then a plot of the weighted residuals

$$\frac{y_i - \hat{\beta} x_i}{\sqrt{x_i}}$$

against x_i or $\log x_i$ should not exhibit any patterns. This plot is given for the agriculture census data in Figure 4.6; nothing appears in the plot to make us doubt the adequacy of this model for the observations in our sample.

4.6.2 A Model for Regression Estimation

A similar result occurs for regression estimation; for that, the model is

$$Y_i = \beta_0 + \beta_1 x_i + \varepsilon_i,$$

where the ε_i are independent and identically distributed with mean 0 and constant variance σ^2. The least squares estimators of β_0 and β_1 in this model are

$$\hat{\beta}_1 = \frac{\displaystyle\sum_{i \in \mathcal{S}} (x_i - \bar{x}_\mathcal{S})(Y_i - \bar{Y}_\mathcal{S})}{\displaystyle\sum_{i \in \mathcal{S}} (x_i - \bar{x}_\mathcal{S})^2}$$

and

$$\hat{\beta}_0 = \bar{Y}_\mathcal{S} - \hat{\beta}_1 \bar{x}_\mathcal{S}.$$

Then, using the predicted values in place of the units not sampled,

$$
\begin{aligned}
\hat{T}_y &= \sum_{i \in \mathcal{S}} Y_i + \sum_{i \notin \mathcal{S}} (\hat{\beta}_0 + \hat{\beta}_1 x_i) \\
&= n \bar{Y}_{\mathcal{S}} + \sum_{i \notin \mathcal{S}} (\hat{\beta}_0 + \hat{\beta}_1 x_i) \\
&= n(\hat{\beta}_0 + \hat{\beta}_1 \bar{x}_{\mathcal{S}}) + \sum_{i \notin \mathcal{S}} (\hat{\beta}_0 + \hat{\beta}_1 x_i) \\
&= \sum_{i=1}^{N} (\hat{\beta}_0 + \hat{\beta}_1 x_i) \\
&= N(\hat{\beta}_0 + \hat{\beta}_1 \bar{x}_U).
\end{aligned}
$$

The regression estimator of T_y is thus $N \times$ (predicted value under the model at $\bar{x}_U$).

In practice, if the sample size is small relative to the population size and we have an SRS, we can simply ignore the fpc and use the standard error for estimating the mean value of a response. From regression theory (see one of the regression books listed in the references for Chapter 11), the variance of $(\hat{\beta}_0 + \hat{\beta}_1 \bar{x}_U)$ is

$$
\sigma^2 \left[\frac{1}{n} + \frac{(\bar{x}_U - \bar{x})^2}{\sum_{i \in \mathcal{S}} (x_i - \bar{x})^2} \right].
$$

Thus if n/N is small,

$$
V_M[\hat{T}_y - T] \approx N^2 \sigma^2 \left[\frac{1}{n} + \frac{(\bar{x}_U - \bar{x}_{\mathcal{S}})^2}{\sum_{i \in \mathcal{S}} (x_i - \bar{x}_{\mathcal{S}})^2} \right]. \tag{4.25}
$$

EXAMPLE 4.12   In Example 4.9, the predicted value from the model when $x = 11.3$ is the regression estimator for $\bar{y}_U$. The predicted value is $(5.05929 + 0.61327 \times 11.3) = 11.9893$. The model-based standard error is obtained from (4.25):

$$
\text{SE}_M(\hat{\bar{Y}}_{\text{reg}}) = \sqrt{\hat{\sigma}^2 \left[\frac{1}{n} + \frac{(\bar{x}_U - \bar{x}_{\mathcal{S}})^2}{\sum_{i \in \mathcal{S}} (x_i - \bar{x}_{\mathcal{S}})^2} \right]} = \sqrt{5.79 \left[\frac{1}{25} + \frac{(11.3 - 10.6)^2}{226.006} \right]} = 0.494.
$$

These values can be calculated directly using the SAS PROC REG code on the website. The standard error from (4.25) does not incorporate the fpc: Exercise 34 examines the fpc in model-based regression. ∎

4.6.3 Differences Between Model-Based and Design-Based Estimators

Under the ratio model, the point estimator for the population total is the same as in the design-based approach, but the variance differs from that for the design-based estimator. Why aren't the standard errors the same as in randomization theory? That is, how can we have two different variances for the same estimator? The discrepancy is due to the different definitions of *variance*: In design-based sampling, the variance is the average squared deviation of the estimate from its expected value, averaged over all samples that could be obtained using a given design. If we are using a model, the variance is again the average squared deviation of the estimate from its expected value, but here the average is over all possible samples that could be generated from the population model.

The model-based estimator uses a prediction approach, in which the values of y_i not in the sample are predicted using the model. We have

$$\hat{T}_y = \sum_{i \in S} Y_i + \sum_{i \notin S} \hat{Y}_i = \sum_{i=1}^{N} \hat{Y}_i + \sum_{i \in S} (Y_i - \hat{Y}_i).$$

If you were absolutely certain that your model was correct, you could minimize the model-based variance of the regression estimator by including only the members of the population with the largest and smallest values of x to be in the sample, and excluding units with values of x between those extremes. No one would recommend such a design in practice, of course, because one never has that much assurance in a model. However, nothing in the model says that you should take an SRS or any other type of probability sample, or that the sample needs to be representative of the population—*as long as the model is correct.*

What if the model is wrong? The model-based estimates are only model-unbiased—that is, they are unbiased only within the structure of that particular model. If the model is wrong, the model-based estimators will be biased, but, from within the model, we will not necessarily be able to tell how big the bias is. Thus, if the model is wrong, the model-based estimator of the variance generally underestimates the MSE. When using model-based inference in sampling, then, you need to be very careful to check the assumptions of the model by examining residuals and using other diagnostic tools. The assumption of independence is typically the most difficult to check. You can (and should!) perform diagnostics to check some of the assumptions of the model for the sampled data, but need to realize that you are making a strong, untestable assumption that the model applies to population units you did not observe.

The randomization-based estimator of the MSE may be used whether any given model fits the data or not, because randomization inference depends only on how the sample was selected. But even the most die-hard randomization theorist relies on models for nonresponse, and for designing the survey. Hansen et al. (1983) point out that generally randomization theory samplers have a model in mind when designing the survey and take that model into account to improve efficiency.

We will return to this issue in Chapter 11.

4.7
Chapter Summary

Ratio and regression estimation use an auxiliary variable that is highly correlated with the variable of interest to reduce the MSE of estimated population means or totals. We "know" that y is correlated with x, and we know how far $\bar{x}$ is from $\bar{x}_U$, so we use this information to adjust $\bar{y}$ and (we hope) increase the precision of our estimate. The estimators in ratio and regression estimation come from models that we hope describe the data, but the randomization-theory properties of the estimators do not depend on these models.

As will be seen in Chapter 11, the ratio and regression estimators discussed in this chapter are special cases of a generalized regression estimator. All three estimators of the population total discussed so far—$\hat{t}_y$, $\hat{t}_{yr}$, and $\hat{t}_{yreg}$—can be expressed in terms of regression coefficients. For an SRS of size n, the estimators are given in the following table. For each, the estimated variance depends on s_e^2, the sample variance of the e_i.

	Estimator	e_i
SRS	$\hat{t}_y$	$y_i - \bar{y}$
Ratio	$\hat{t}_y \left(\dfrac{t_x}{\hat{t}_x} \right)$	$y_i - \hat{B}x_i$
Regression	$N[\bar{y} + \hat{B}_1(\bar{x}_U - \bar{x})]$	$y_i - \hat{B}_0 - \hat{B}_1 x_i$

In an SRS, ratio or regression estimators give greater precision than $\hat{t}_y$ when $\sum_{i \in S} e_i^2$ for the method is smaller than $\sum_{i \in S} (y_i - \bar{y})^2$. Ratio estimation is especially useful in cluster sampling, as we shall see in chapters 5 and 6.

We often want to find estimates for subpopulations of interest, for example, different age groups. If the sampling frame contains information on the age group for units in the population, a stratified sample can be designed as discussed in Chapter 3. If the sampling frame does not contain this information but we know the population sizes of the subpopulations, then the subpopulations are poststrata and we can estimate the population total in group h by $N_h \bar{y}_h$, where $\bar{y}_h$ is the mean of the sampled observations in the subpopulation. If we do not know the population sizes of the subpopulations, then they are domains and we estimate the population total in group h by $N(n_h/n)\bar{y}_h$.

In this chapter, we discussed ratio and regression estimation using just one auxiliary variable x. In practice, you may want to use several auxiliary variables. The principles for using multiple regression models will be the same; we shall present the theory for general surveys in Section 11.7.

Key Terms

Calibration: A procedure in which weights are adjusted so that estimated population totals of auxiliary variables coincide with the actual population totals of those variables.

Domain: A subpopulation for which estimates are desired. The domain sample sizes are generally random variables.

Poststratification: A form of ratio estimation in which sampled units are divided into subgroups based on characteristics measured in the sample; the population size of each subgroup is assumed known.

Ratio estimator: An estimator of the population mean or total based on a ratio with an auxiliary quantity for which the population mean or total is known.

Regression estimator: An estimator of the population mean or total based on a regression model using an auxiliary quantity for which the population mean or total is known.

For Further Reading

Raj (1968) and Cochran (1977) have good treatments of ratio and regression estimation in SRSs. For regression models in a general framework, discussed in this book in Chapter 11, see Särndal et al. (1992). The overview paper by Särndal (2007) summarizes the use of ratio and regression estimation for calibrating survey estimates to known population totals.

The books by Thompson (1997), Brewer (2002) and Valliant et al. (2000) describe differences between design-based and model-based approaches to survey inference. The articles by Hansen et al. (1983), Rao (1997), Lohr (2001), and Little (2004) discuss the relative merits of design- and model-based approaches to inference.

4.8
Exercises

A. Introductory Exercises

1   For each of the following situations, indicate how you might use ratio or regression estimation.

   a   Estimate the proportion of time in television news broadcasts in your city that is devoted to sports.

   b   Estimate the average number of fish caught per hour for anglers visiting a lake in August.

   c   Estimate the average amount that undergraduate students at your university spent on textbooks in fall semester.

   d   Estimate the total weight of usable meat (discarding bones, fat, and skin) in a shipment of chickens.

2   Consider the hypothetical population below, with population values:

Unit Number	x	y
1	13	10
2	7	7
3	11	13
4	12	17
5	4	8
6	3	1
7	11	15
8	3	7
9	5	4

a Find the values of the population quantities t_x, t_y, S_x, S_y, R, and B.

b Construct a table like that in Table 4.2, giving the sampling distribution of $N\bar{y}$ and of $\hat{t}_{yr}$, for a sample of size $n = 3$.

c Draw a histogram of the sampling distribution of $\hat{t}_{yr}$. Compare this histogram to a histogram of the sampling distribution of $N\bar{y}$.

d Find the mean and variance of the sampling distribution of $\hat{t}_{yr}$. How do these compare to the mean and variance of $N\bar{y}$? What is the bias of $\hat{t}_{yr}$?

e Use Equation (4.6), together with the population quantities you calculated in (a) to find an approximation to Bias $(\hat{t}_{yr}) = N$Bias $(\hat{\bar{y}}_r)$. How close is the approximation to the true bias in (c)?

3 Foresters want to estimate the average age of trees in a stand. Determining age is cumbersome, because one needs to count the tree rings on a core taken from the tree. In general, though, the older the tree, the larger the diameter, and diameter is easy to measure. The foresters measure the diameter of all 1132 trees and find that the population mean equals 10.3. They then randomly select 20 trees for age measurement.

Tree No.	Diameter, x	Age, y	Tree No.	Diameter, x	Age, y
1	12.0	125	11	5.7	61
2	11.4	119	12	8.0	80
3	7.9	83	13	10.3	114
4	9.0	85	14	12.0	147
5	10.5	99	15	9.2	122
6	7.9	117	16	8.5	106
7	7.3	69	17	7.0	82
8	10.2	133	18	10.7	88
9	11.7	154	19	9.3	97
10	11.3	168	20	8.2	99

a Draw a scatterplot of y vs. x.

b Estimate the population mean age of trees in the stand using ratio estimation and give an approximate standard error for your estimate.

c Repeat (b) using regression estimation.

d Label your estimates on your graph. How do they compare?

B. Working with Survey Data

4 Use the data in ssc.dat, described in Exercise 14 of Chapter 2, for this problem.

a Estimate the proportion of female members who are in academia. Note that this is a domain mean, with $x_i = 1$ if person i is female and 0 otherwise, and $y_i = 1$ if person i is female and in academia and 0 otherwise. Give a 95% CI.

b Estimate the total number of female members in academia, along with a 95% CI.

5 Use the data in file golfsrs.dat for this problem. Using the 18-hole courses only, estimate the average greens fee on a weekend to play 18 holes, along with its standard error.

6   For the 18-hole courses in file golfsrs.dat, plot the weekend 18-hole greens fee vs. the backtee yardage. Estimate the regression parameters for predicting weekend greens fees from backtee yardage. Is there a strong relationship between the two variables? Use regression estimation to estimate the weekend 18-hole greens fee with its standard error.

7   Use the data in file golfsrs.dat for this problem.

    a   Estimate the mean weekday greens fee to play 9 holes, for courses with a golf professional available.

    b   Now estimate the mean weekday greens fee to play 9 holes, for courses without a golf professional.

8   The data set agsrs.dat also contains information on the number of farms in 1987 for the SRS of $n = 300$ counties from the population of the $N = 3078$ counties in the United States (see Example 2.5). In 1987, the United States had a total of 2,087,759 farms.

    a   Plot the data.

    b   Use ratio estimation to estimate the total number of acres devoted to farming in 1992, using the number of farms in 1987 as the auxiliary variable.

    c   Repeat (b), using regression estimation.

    d   Which method gives the most precision: ratio estimation with auxiliary variable *acres87*, ratio estimation with auxiliary variable *farms87*, or regression estimation with auxiliary variable *farms87*? Why?

9   Using the data set agsrs.dat, estimate the total number of acres devoted to farming for each of two domains: (a) counties with fewer than 600 farms, and (b) counties with 600 or more farms. Give standard errors for your estimates.

10   The data set cherry.dat, from Hand et al. (1994), contains measurements of diameter (inches), height (feet), and timber volume (cubic feet) for a sample of 31 black cherry trees. Diameter and height of trees are easily measured, but volume is more difficult to measure.

    a   Plot volume vs. diameter for the 31 trees.

    b   Suppose that these trees are an SRS from a forest of $N = 2967$ trees and that the sum of the diameters for all trees in the forest is $t_x = 41{,}835$ inches. Use ratio estimation to estimate the total volume for all trees in the forest. Give a 95% CI.

    c   Use regression estimation to estimate the total volume for all trees in the forest. Give a 95% CI.

11   The data file counties.dat contains information on land area, population, number of physicians, unemployment, and a number of other quantities for an SRS of 100 of the 3141 counties in the United States (U.S. Census Bureau, 1994). The total land area for the United States is 3,536,278 square miles; 1993 population was estimated to be 255,077,536.

    a   Draw a histogram of the number of physicians for the 100 counties.

    b   Estimate the total number of physicians in the United States, along with its standard error, using $N\bar{y}$.

c Plot the number of physicians vs. population for each county. Which method do you think is more appropriate for these data: ratio estimation or regression estimation?

d Using the method you chose in (c), use the auxiliary variable population to estimate the total number of physicians in the United States, along with the standard error.

e The "true" value for total number of physicians in the population is 532,638. Which method of estimation came closer?

12 Repeat parts (a)–(d) of Exercise 11 with y = farm population and x = land area.

13 Repeat parts (a)–(d) of Exercise 11 with y = number of veterans and x = population.

14 (Model-based analysis; requires material in Section 4.6.) Refer to the situation in Exercise 11. Use a model-based analysis to estimate the total number of physicians in the United States. Which model did you choose, and why? What are the assumptions for the model? Do you think they are met? Be sure to examine the residual plots for evidence of the inadequacy of the model. How do your results differ from those you obtained in Exercise 11?

15 Jackson et al. (1987) compared the precision of systematic and stratified sampling for estimating the average concentration of lead and copper in the soil. The 1-km^2 area was divided into 100-m squares, and a soil sample was collected at each of the resulting 121 grid intersections. Summary statistics from this systematic sample are given below.

Element	n	Average (mg kg^{-1})	Range (mg kg^{-1})	Standard Deviation (mg kg^{-1})
Lead	121	127	22–942	146
Copper	121	35	15–90	16

The investigators also poststratified the same region. Stratum A consisted of farmland away from roads, villages, and woodlands. Stratum B contained areas within 50m of roads, and was expected to have larger concentrations of lead. Stratum C contained the woodlands, which were also expected to have larger concentrations of lead because the foliage would capture airborne particles. The data on concentration of lead and copper were not used in determining the strata. The data from the grid points falling in each stratum are in the following table:

Element	Stratum	n_h	Average (mg kg^{-1})	Range (mg kg^{-1})	Standard Deviation (mg kg^{-1})
Lead	A	82	71	22–201	28
Lead	B	31	259	36–942	232
Lead	C	8	189	88–308	79
Copper	A	82	28	15–68	9
Copper	B	31	50	22–90	18
Copper	C	8	45	31–69	15

a Calculate a 95% CI for the average concentration of lead in the area, using the systematic sample. (You may assume that this sample behaves like an SRS.) Repeat for the average concentration of copper.

b Now use the poststratified sample, and find 95% CIs for the average concentration of lead and copper. How do these compare with the CIs in (a)? Do you think that using stratification in future surveys would increase precision?

16 Poststratify the sample in data file agsrs.dat into the four census regions given in Example 3.2. Estimate the population mean $\bar{y}_U$ using (4.21) and approximate the variance using (4.22). How does the approximate 95% CI using poststratification compare with that from Example 2.10?

17 Using the data in Example 4.10, calculate the separate ratio estimate for the population total t_y, along with a 95% CI.

C. Working with Theory

18 (Requires probability.) Use covariances derived in Appendix A to show the result in (4.8).

19 (Requires computing.) In Equation (4.9), we used

$$\hat{V}_1[\hat{\bar{y}}_r] = \left(1 - \frac{n}{N}\right)\left(\frac{\bar{x}_U}{\bar{x}}\right)^2 \frac{s_e^2}{n}$$

to estimate $\left(1 - \frac{n}{N}\right)\frac{S_d^2}{n}$. An alternative estimator that has been proposed is

$$\hat{V}_2[\hat{\bar{y}}_r] = \left(1 - \frac{n}{N}\right)\frac{s_e^2}{n}.$$

Generate a population of size 1000 from the model $y_i = \beta x_i + \varepsilon_i$, where $\varepsilon_i \sim N(0, \sigma^2 x_i)$. Now take 100 different samples, each with $n = 50$. Compare $\hat{V}_1$ and $\hat{V}_2$.
    If the variability about the line $y = \beta x$ increases as x increases, as is the case for the data generated above, then, if $\bar{x} < \bar{x}_U$ we would expect $\frac{1}{n-1}\sum_{i \in S}(y_i - \hat{B}x_i)^2$

to be smaller than $\frac{1}{N-1}\sum_{i=1}^{N}(y_i - Bx_i)^2 = S_d^2$. Using $\hat{V}_1$ instead of $\hat{V}_2$ partially compensates for this. See Valliant (2002) for a discussion of why $\hat{V}_1$ is preferred to $\hat{V}_2$ from a conditional inference perspective.

20 Some books use the formula

$$\hat{V}[\hat{B}] = \left(1 - \frac{n}{N}\right)\frac{1}{n\bar{x}_U^2}(s_y^2 - 2\hat{B}rs_x s_y + \hat{B}^2 s_x^2),$$

where r is the sample correlation coefficient of x and y for the values in the sample, to estimate the variance of a ratio.

a Show that this formula is algebraically equivalent to (4.10).

b   It often does not work as well as (4.10) in practice, however: If s_x and s_y are large, many computer packages will truncate some of the significant digits so that the subtraction will be inaccurate. For the data in Example 4.2, calculate the values of s_y^2, s_x^2, r, and $\hat{B}$. Use the formula above to calculate the estimated variance of $\hat{B}$. Is it exactly the same as the value from (4.10)?

21 (Requires probability.) Recall from Section 2.2 that MSE = variance + (Bias)^2. Using (4.6) and other approximations in that section, show that $[E(\hat{\bar{y}}_r - \bar{y}_U)]^2$ is small compared to $V(\hat{\bar{y}}_r)$, when n is large.

22 (Requires probability.) Prove (4.6). HINT: Use (4.5) and the derivation of the covariance of $\bar{x}$ and $\bar{y}$ in (A.10) of Appendix A.

23 Use Equation (4.6) to find the approximate bias of $\hat{t}_{yr}$ and of $\hat{B}$.

24 *Comparing two domain means in an SRS.* Suppose there are two domains, defined by indicator variable

$$x_i = \begin{cases} 1 & \text{if unit } i \text{ is in domain 1} \\ 0 & \text{if unit } i \text{ is in domain 2} \end{cases}.$$

Then, letting $u_i = x_i y_i$, the population values of the two domain means are

$$\bar{y}_{U1} = \frac{\displaystyle\sum_{i=1}^{N} x_i y_i}{\displaystyle\sum_{i=1}^{N} x_i} = \frac{t_u}{t_x} = \frac{\bar{u}_U}{\bar{x}_U}$$

and

$$\bar{y}_{U2} = \frac{\displaystyle\sum_{i=1}^{N} (1 - x_i) y_i}{\displaystyle\sum_{i=1}^{N} (1 - x_i)} = \frac{t_y - t_u}{N - t_x} = \frac{\bar{y}_U - \bar{u}_U}{1 - \bar{x}_U}.$$

If an SRS of size n is taken from a population of size N, the population domain means may be estimated by

$$\bar{y}_1 = \frac{\hat{t}_u}{\hat{t}_x} = \frac{\bar{u}}{\bar{x}}, \qquad \bar{y}_2 = \frac{\hat{t}_y - \hat{t}_u}{N - \hat{t}_x} = \frac{\bar{y} - \bar{u}}{1 - \bar{x}}.$$

a   Use an argument similar to that in the discussion following (4.5) to show that

$$\text{Cov}\,(\bar{y}_1, \bar{y}_2) \approx \frac{1}{\bar{x}_U(1 - \bar{x}_U)} \text{Cov}\left[\left(\bar{u} - \frac{t_u}{t_x}\bar{x}\right), \left\{\bar{y} - \bar{u} - \frac{t_y - t_u}{N - t_x}(1 - \bar{x})\right\}\right].$$

$$(4.26)$$

b   For an SRS, show using (A.10) that

$$\text{Cov}\left[\left(\bar{u} - \frac{t_u}{t_x}\bar{x}\right), \left\{\bar{y} - \bar{u} - \frac{t_y - t_u}{N - t_x}(1 - \bar{x})\right\}\right] = 0.$$

[Consequently, since Property 7 of Expected Value in Section A.2 implies that $V(\bar{y}_1 - \bar{y}_2) = V(\bar{y}_1) + V(\bar{y}_2) - 2\,\mathrm{Cov}\,(\bar{y}_1, \bar{y}_2)$, in an SRS $V(\bar{y}_1 - \bar{y}_2) \approx V(\bar{y}_1) + V(\bar{y}_2)$ and an approximate 95% CI for $\bar{y}_{U1} - \bar{y}_{U2}$ is given by

$$\bar{y}_1 - \bar{y}_2 \pm 1.96\sqrt{\hat{V}(\bar{y}_1) + \hat{V}(\bar{y}_2)}.$$

Thus, for an SRS, the large-sample CI for the difference of two domain means is the same (if we ignore the fpc) as you learned in your introductory statistics class. Note, though, that this result holds only for an SRS. For more complex sampling designs the covariance of the estimated domain means may be nonzero, (see Exercise 21 of Chapter 6) so more general methods discussed in Section 11.3 must be used.]

25 (Requires mathematical statistics.) *Showing (4.8)*. Suppose that $n/N \to 0$ as $n \to \infty$, so that the fpc can be ignored. The central limit theorem tells us that under regularity conditions,

$$\sqrt{n}\begin{bmatrix} \bar{x} - \bar{x}_U \\ \bar{y} - \bar{y}_U \end{bmatrix} \xrightarrow{\mathcal{L}} N\left(\mathbf{0}, \begin{bmatrix} S_x^2 & RS_x S_y \\ RS_x S_y & S_y^2 \end{bmatrix} \right),$$

where $\mathcal{L}$ denotes convergence in distribution. Show that the limiting distribution of $\sqrt{n}(\hat{\bar{y}}_r - \bar{y}_U)$ has mean 0 and variance $S_y^2 - 2BRS_x S_y + B^2 S_x^2$.

26 Show that if we consider approximations to the MSE in (4.8) and (4.17) to be accurate, then the variance of $\hat{\bar{y}}_r$ from ratio estimation is at least as large as the variance of $\hat{\bar{y}}_{\text{reg}}$ from regression estimation. HINT: Look at $V(\hat{\bar{y}}_r) - V(\hat{\bar{y}}_{\text{reg}})$ using the formulas in (4.8) and (4.17), and show that the difference is non-negative.

27 Prove (4.16).

28 (Requires probability.) Prove (4.18).

29 (Requires probability.) Let $d_i = y_i - [\bar{y}_U + B_1(x_i - \bar{x}_U)]$. Show that for regression estimation,

$$E[\hat{\bar{y}}_{\text{reg}} - \bar{y}_U] \approx -\frac{1 - n/N}{n\,S_x^2} \sum_{i=1}^{N} \frac{d_i(x_i - \bar{x}_U)^2}{N - 1}.$$

As in Exercise 21, show that $(E[\hat{\bar{y}}_{\text{reg}} - \bar{y}_U])^2$ is small compared to $\mathrm{MSE}[\hat{\bar{y}}_{\text{reg}}]$, when n is large.

30 (Requires probability.) Consider the combined ratio estimator of the population total, $\hat{t}_{yrc}$, from Section 4.5.

　a Show that

$$\frac{\mathrm{Bias}|\hat{t}_{yrc}|}{\sqrt{V(\hat{t}_{yrc})}} \leq \mathrm{CV}\,(\hat{t}_x).$$

　HINT: See (4.4).

　b In a stratified random sample, find the approximate bias and MSE of $\hat{t}_{yrc}$.

31 (Requires probability.) Consider the separate ratio estimator of the population total, $\hat{t}_{yrs}$, from Section 4.5. Find the bias and an approximation to the MSE of $\hat{t}_{yrs}$ in a stratified random sample. Allow different ratios, B_h, in each stratum. When will the bias be small?

32 (Requires linear model theory.) Suppose we have a stochastic model

$$Y_i = \beta x_i + \varepsilon_i$$

where the ε_i's are independent with mean 0 and variance $\sigma^2 x_i$. Show that the weighted least squares estimator of β is $\hat{\beta} = \bar{Y}/\bar{x}$. Is the standard error for $\hat{\beta}$ that comes from weighted least squares the same as that for $\hat{B}$ in (4.10)?

33 (Requires linear model theory.) Suppose that the model in (4.23) misspecifies the variance structure and that a better model has $V_M[\varepsilon_i] = \sigma^2$.

  a What is the weighted least squares estimator of β if $V_M[\varepsilon_i] = \sigma^2$? What is the corresponding estimator of the population total for y?

  b Derive $V_M[\hat{T}_y - T_y]$.

  c Apply your estimators to the data in agsrs.dat. How do these estimates compare with those in Examples 4.2 and 4.11?

34 Equation (4.25) gave the model-based variance for a population total when it is assumed that the sample size is small relative to the population size. Derive the variance incorporating the finite population correction.

35 The quantity $\hat{B}$ used in ratio estimation is sometimes called the *ratio-of-means estimator*. An alternative that has been proposed is the *mean-of-ratios estimator*: Let $b_i = y_i/x_i$ for unit i; then the mean-of-ratios estimator is

$$\bar{b} = \frac{1}{n} \sum_{i \in \mathcal{S}} b_i.$$

  a Do you think the mean-of-ratios estimator is appropriate for the data in Example 4.5? Why, or why not?

  b Show that, for the ratio-of-means estimator $\hat{B}$, $t_x \hat{B} = t_y$ when the entire population is sampled (i.e., $\mathcal{S} = \mathcal{U}$).

  c Give an example to show that it is possible to have $t_x \bar{b} \neq t_y$ when the entire population is sampled.

  d Define

$$S_{bx} = \frac{1}{N-1} \sum_{i=1}^{N} (b_i - \bar{b}_U)(x_i - \bar{x}_U).$$

  Show that for an SRS of size n, the bias of $\bar{b}$ as an estimator of B is

$$E[\bar{b} - B] = -\frac{(N-1)S_{bx}}{t_x}.$$

  As a consequence, if $S_{bx} \neq 0$ the bias does not decrease as n increases.

  e (Requires linear model theory.) Show that $\bar{b}$ is the weighted least squares estimator of β under the model

$$Y_i = \beta x_i + \varepsilon_i$$

  when ε_i's are independent with mean 0 and variance $\sigma^2 x_i^2$.

36 (Requires computing.)

    a Generate 500 data sets, each with 30 pairs of observations (x_i, y_i). Use a bivariate normal distribution with means 0, standard deviations 1, and correlation 0.5 to generate each pair (x_i, y_i). For each data set, calculate $\bar{y}$ and $\hat{\bar{y}}_{reg}$, using $\bar{x}_U = 0$. Graph a histogram of the 500 values of $\bar{y}$ and another histogram of the 500 values of $\hat{\bar{y}}_{reg}$. What do you see?

    b Repeat part (a) for 500 data sets, each with 60 pairs of observations.

37 (Requires computing.) Use the population in agpop.dat for this exercise.

    a Take 500 SRSs, each of size $n = 40$, from the data set. For each SRS, calculate the sample mean $\bar{y}$, the poststratified mean $\bar{y}_{post}$ from (4.21), the estimated variance of $\bar{y}$, and the estimated variance of $\bar{y}_{post}$ using (4.22).

    b Calculate the sample variance of the 500 values of $\bar{y}$. This gives an estimate of the true value of $V(\bar{y})$. Compare your value with the average of the 500 values of $\hat{V}(\bar{y})$. Since $\hat{V}(\bar{y})$ is an unbiased estimator of $V(\bar{y})$ for any sample size, these values should be close.

    c Calculate the sample variance of the 500 values of $\bar{y}_{post}$. This gives an estimate of the true value of $V(\bar{y}_{post})$. Compare your value with the average of the 500 values of $\hat{V}(\bar{y}_{post})$.

D. Projects and Activities

38 Find a dictionary of a language you have studied. Choose 30 pages at random from the dictionary. For each, record

$$x = \text{number of words on the page}$$
$$y = \text{number of words that you know on the page (be honest!)}$$

How many words do you estimate are in the dictionary? How many do you estimate that you know? What percentage of the words do you know? Give standard errors for all your estimates.

39 The 2000 U.S. Presidential election generated controversy for many reasons; one part of the controversy was that television networks declared that candidate Gore would be the winner based on exit polls, which are surveys of voters as they leave polling places. Read the article by Mitofsky and Edelman (2002) on estimation in the 2000 election (the article is available at www.jos.nu). Describe how ratio estimation was used in the polls. What were the likely sources of bias in the 2000 exit polls?

40 *Online bookstore.* Use your sample from Exercise 33 in Chapter 2 for this exercise.

    a Estimate the ratio (average price/average number of pages) and give the standard error.

    b Consider two domains: hardcover books and paperback books. Estimate the mean price (with standard error) for books in each domain.

41 *Forest data.* Use your sample from Exercise 36 of Chapter 2 for this exercise.

    a Estimate the ratio of hillshade index at 9 am to hillshade index at noon. Include a 95% CI.

    b Estimate the average elevation for each of the 7 forest cover types, along with a 95% CI.

42 *Trucks.* Use the data described in Exercise 34 of Chapter 3 for this exercise.

   a The variable *business* describes the primary business in which the vehicle was used in 2002. Estimate the total miles driven for each type of business in 2002, along with a 95% CI. How is this a special case of estimating a domain total?

   b Estimate the average miles per gallon (MPG) for each of the transmission types (*transmssn*), along with a 95% CI.

   c Estimate the ratio of miles driven in 2002 (*miles_annl*) to lifetime miles driven (*miles_life*), along with a 95% CI.

43 *Baseball data.*

   a Using your SRS from Exercise 32 of Chapter 2, estimate the mean log salary for players in each position along with the standard errors.

   b Estimate the ratio (total number of home runs)/(number of runs scored) for the population and give a 95% CI.

44 *IPUMS exercises.*

   a Using your SRS from Exercise 37 of Chapter 2, try estimating total income *(inctot)* using ratio estimation with *age* as an auxiliary variable. Does it decrease the standard error? Why, or why not? (Include a plot as part of your answer.)

   b Using one of the following variables: *age, sex, race,* or *marstat,* use regression estimation to calibrate your estimate of total income to the category totals for the variable you chose.

5

Cluster Sampling with Equal Probabilities

"But averages aren't real," objected Milo; "they're just imaginary."

"That may be so," he agreed, "but they're also very useful at times. For instance, if you didn't have any money at all, but you happened to be with four other people who had ten dollars apiece, then you'd each have an average of eight dollars. Isn't that right?"

"I guess so," said Milo weakly.

"Well, think how much better off you'd be, just because of averages," he explained convincingly. "And think of the poor farmer when it doesn't rain all year: if there wasn't an average yearly rainfall of 37 inches in this part of the country, all his crops would wither and die."

It all sounded terribly confusing to Milo, for he had always had trouble in school with just this subject.

"There are still other advantages," continued the child. "For instance, if one rat were cornered by nine cats, then, on the average, each cat would be 10 per cent rat and the rat would be 90 per cent cat. If you happened to be a rat, you can see how much nicer it would make things."

—Norton Juster, *The Phantom Tollbooth*

In all the sampling procedures discussed so far, we have assumed that the population is given and all we must do is reach in and take a suitable sample of units. But units are not necessarily nicely defined, even when the population is. There may be several ways of listing the units, and the unit size we choose may very well contain smaller subunits.

Suppose we want to find out how many bicycles are owned by residents in a community of 10,000 households. We could take a simple random sample (SRS) of 400 households, or we could divide the community into blocks of about 20 households each and sample every household (or subsample some of the households) in each of 20 blocks selected at random from the 500 blocks in the community. The latter plan is an example of cluster sampling. The blocks are the **primary sampling units** (psus), or **clusters**. (In this chapter, we use the terms cluster and psu interchangeably.) The households are the **secondary sampling units** (ssus); often the ssus are the elements in the population.

The cluster sample of 400 households is likely to give less precision than an SRS of 400 households; some blocks of the community are composed mainly of families (with more bicycles), while the residents of other blocks are mainly retirees (with fewer bicycles). Twenty households in the same block are not as likely to mirror the diversity of the community as well as 20 households chosen at random. Thus, cluster sampling in this situation will probably result in less information per observation than an SRS of the same size. However, if you conduct the survey in person, it is much cheaper and easier to interview all 20 households in a block than 20 households selected at random from the community, so cluster sampling may well result in more information per dollar spent.

In cluster sampling, individual elements of the population are allowed in the sample only if they belong to a cluster (psu) that is included in the sample. The sampling unit (psu) is not the same as the observation unit (ssu), and the two sizes of experimental units must be considered when calculating standard errors from cluster samples.

Why use cluster samples?

1   Constructing a sampling frame list of observation units may be difficult, expensive, or impossible. We cannot list all honeybees in a region or all customers of a store; we may be able to construct a list of all trees in a stand of northern hardwood forest or a list of individuals in a city for which we only have a list of housing units, but constructing the list will be time consuming and expensive.

2   The population may be widely distributed geographically or may occur in natural clusters such as households or schools, and it is less expensive to take a sample of clusters rather than an SRS of individuals. If the target population is residents of nursing homes in the United States, it is much cheaper to sample nursing homes and interview every resident in the selected homes than to interview an SRS of nursing home residents: With an SRS of residents, you might have to travel to a nursing home just to interview one resident. If taking an archaeological survey, you would examine all of the artifacts found in a region—you would not just choose points at random and examine only artifacts occurring at those isolated points.

Clusters bear a superficial resemblance to strata: A cluster, like a stratum, is a grouping of the members of the population. The selection process, though, is quite different in the two methods. Similarities and differences between cluster samples and stratified samples are illustrated in Figure 5.1.

Whereas stratification generally increases precision when compared with simple random sampling, cluster sampling generally decreases it. Members of the same cluster tend to be more similar than elements selected at random from the whole population—members of the same household tend to have similar political views; fish in the same lake tend to have similar concentrations of mercury; residents of the same nursing home tend to have similar opinions of the quality of care. These similarities usually arise because of some underlying factors that may or may not be measurable— residents of the same nursing home may have similar opinions because the care is poor, or the concentration of mercury in the fish may reflect the concentration of mercury in the lake. Thus, we do not obtain as much information about all nursing home residents in the United States by sampling two residents in the same home as by sampling two residents in different homes, because the two residents in the same home are likely to have more similar opinions. By sampling everyone in the cluster,

FIGURE 5.1

Similarities and differences between stratified sampling and one-stage cluster sampling

Stratified Sampling	Cluster Sampling
Each element of the population is in exactly one stratum.	Each element of the population is in exactly one cluster.
Population of H strata; stratum h has n_h elements: 	One-stage cluster sampling; population of N clusters: 
Take an SRS from *every* stratum: 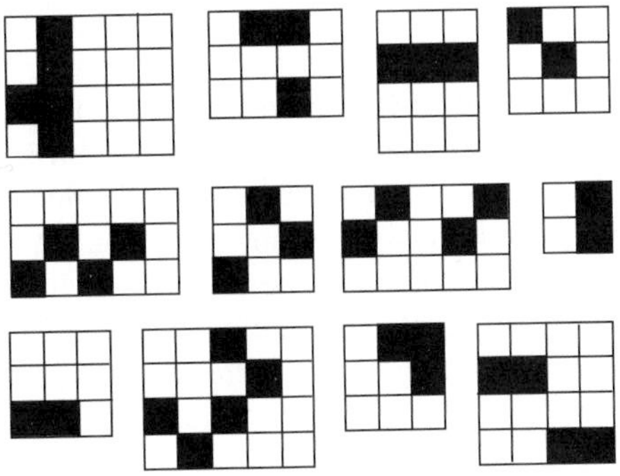	Take an SRS of clusters; observe all elements within the clusters in the sample: 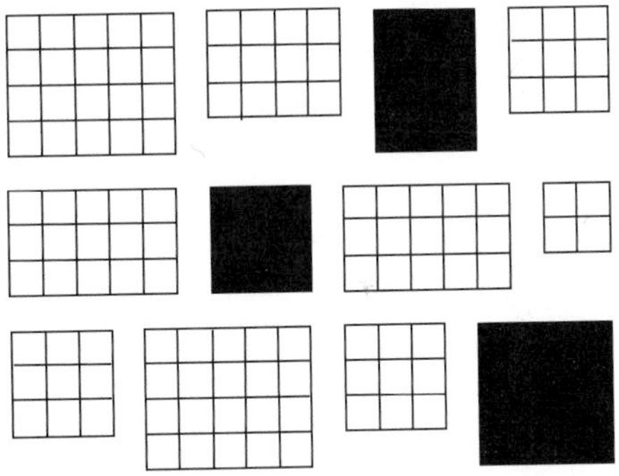
Variance of the estimate of $\bar{y}_U$ depends on the variability of values *within* strata.	The cluster is the sampling unit; the more clusters we sample, the smaller the variance. The variance of the estimate of $\bar{y}_U$ depends primarily on the variability *between* cluster means.
For greatest precision, individual elements within each stratum should have similar values, but stratum means should differ from each other as much as possible.	For greatest precision, individual elements within each cluster should be heterogeneous, and cluster means should be similar to one another.

we partially repeat the same information instead of obtaining new information, and that gives us less precision for estimates of population quantities. Cluster sampling is used in practice because it is usually much cheaper and more convenient to sample in clusters than randomly in the population. Most large household surveys carried out by the U.S. government, or by commercial or academic institutions, use cluster sampling because of the cost savings.

One of the biggest mistakes made by researchers using survey data is to analyze a cluster sample as if it were an SRS. Such confusion usually results in the researchers reporting standard errors that are much smaller than they should be; this gives the impression that the survey results are much more precise than they really are. Exercise 33 presents an activity for exploring what happens to properties of confidence intervals (CIs) when clustered data are analyzed incorrectly.

EXAMPLE 5.1   Basow and Silberg (1987) report results of their research on whether students evaluate female college professors differently than they evaluate male college professors. The authors matched 16 female professors with 16 male professors by subject taught, years of teaching experience, and tenure status, and gave evaluation questionnaires to students in those professors' classes. The sample size for analyzing this study is $n = 32$, the number of faculty studied; it is not 1029, the number of students who returned questionnaires. Students' evaluations of faculty reflect the different styles of faculty teaching; students within the same class are likely to have some agreement in their rating of the professor and should not be treated as independent observations because their ratings will probably be positively correlated. If this positive correlation is ignored and the student ratings treated as independent observations, differences will be declared statistically significant far more often than they should be.   ∎

After a brief journey into "notation land" in Section 5.1, we begin by discussing **one-stage cluster sampling**, in which every element within a sampled cluster is included in the sample. We then generalize the results to **two-stage cluster sampling**, in which we subsample only some of the elements of selected clusters, in Section 5.3. In Section 5.4, we discuss design issues for cluster sampling, including selection of subsample and sample sizes. In Section 5.5, we return to systematic sampling, which we previously discussed in Section 2.7, and show that it is a special case of cluster sampling. The chapter concludes with theory of cluster sampling from the model-based perspective; we shall derive the design-based theory in the more general setting of Section 6.6.

5.1
Notation for Cluster Sampling

In simple random sampling, the units sampled are also the elements observed. In cluster sampling, the sampling units are the clusters (psus) and the elements observed are the ssus within the clusters. The universe $\mathcal{U}$ is the population of N psus; $\mathcal{S}$ designates the sample of psus chosen from the population of psus, and $\mathcal{S}_i$ is the sample of ssus chosen from the ith psu. The measured quantities are

$$y_{ij} = \text{measurement for } j\text{th element in } i\text{th psu,}$$

but in cluster sampling, it is easiest to think at the psu level in terms of cluster totals. No matter how you define it, the notation for cluster sampling is messy because you need notation for both the psu and the ssu levels. The notation used in this chapter and Chapter 6 is presented in this section for easy reference. Note that in Chapters 5 and 6, N is the number of psus, not the number of observation units.

psu Level—Population Quantities

$$N = \text{number of psus in the population}$$

$$M_i = \text{number of ssus in psu } i$$

$$M_0 = \sum_{i=1}^{N} M_i = \text{total number of ssus in the population}$$

$$t_i = \sum_{j=1}^{M_i} y_{ij} = \text{total in psu } i$$

$$t = \sum_{i=1}^{N} t_i = \sum_{i=1}^{N} \sum_{j=1}^{M_i} y_{ij} = \text{population total}$$

$$S_t^2 = \frac{1}{N-1} \sum_{i=1}^{N} \left(t_i - \frac{t}{N} \right)^2 = \text{population variance of the psu totals}$$

ssu Level—Population Quantities

$$\bar{y}_U = \sum_{i=1}^{N} \sum_{j=1}^{M_i} \frac{y_{ij}}{M_0} = \text{population mean}$$

$$\bar{y}_{iU} = \sum_{j=1}^{M_i} \frac{y_{ij}}{M_i} = \frac{t_i}{M_i} = \text{population mean in psu } i$$

$$S^2 = \sum_{i=1}^{N} \sum_{j=1}^{M_i} \frac{(y_{ij} - \bar{y}_U)^2}{M_0 - 1} = \text{population variance (per ssu)}$$

$$S_i^2 = \sum_{j=1}^{M_i} \frac{(y_{ij} - \bar{y}_{iU})^2}{M_i - 1} = \text{population variance within psu } i$$

Sample Quantities

$$n = \text{number of psus in the sample}$$

$$m_i = \text{number of ssus in the sample from psu } i$$

$$\bar{y}_i = \sum_{j \in \mathcal{S}_i} \frac{y_{ij}}{m_i} = \text{sample mean (per ssu) for psu } i$$

$$\hat{t}_i = \sum_{j \in \mathcal{S}_i} \frac{M_i}{m_i} y_{ij} = \text{estimated total for psu } i$$

$$\hat{t}_{\text{unb}} = \sum_{i \in \mathcal{S}} \frac{N}{n} \hat{t}_i = \text{unbiased estimator of population total}$$

$$s_t^2 = \frac{1}{n-1} \sum_{i \in \mathcal{S}} \left(\hat{t}_i - \frac{\hat{t}_{\text{unb}}}{N} \right)^2$$

$$s_i^2 = \sum_{j \in S_i} \frac{(y_{ij} - \bar{y}_i)^2}{m_i - 1} = \text{sample variance within psu } i$$

$$w_{ij} = \text{sampling weight for ssu } j \text{ in psu } i$$

5.2
One-Stage Cluster Sampling

In one-stage cluster sampling, either all or none of the elements that compose a cluster (= psu) are in the sample. One-stage cluster sampling is used in many surveys in which the cost of sampling ssus is negligible compared with the cost of sampling psus. For education surveys, a natural psu is the classroom; all students in a selected classroom are often included as the ssus since little extra cost is added by handing out a questionnaire to all students in the classroom rather than some.

In the population of N psus, the ith psu contains M_i ssus (elements). In the simplest design, we take an SRS of n psus from the population and measure our variable of interest on *every* element in the sampled psus. Thus, for one-stage cluster sampling, $M_i = m_i$.

5.2.1   Clusters of Equal Sizes: Estimation

Let's consider the simplest case in which each psu has the same number of elements, with $M_i = m_i = M$. Most naturally occurring clusters of people do not fit into this framework, but it can occur in agricultural and industrial sampling. Estimating population means or totals is simple: We treat the psu means or totals as the observations and simply ignore the individual elements.

Thus, we have an SRS of n data points $\{t_i, i \in S\}$; t_i is the total for all the elements in psu i. Then $\bar{t}_S = \sum_{i \in S} t_i/n$ estimates the average of the cluster totals. In a household survey to estimate income in two-person households, the individual observations y_{ij} are the incomes of individual persons within the household, t_i is the total income for household i (t_i is *known* for sampled households because both persons are interviewed), $\bar{t}_U$ is the average income per household, and $\bar{y}_U$ is the average income per person. To estimate the total income t, we can use the estimator

$$\hat{t} = \frac{N}{n} \sum_{i \in S} t_i. \tag{5.1}$$

The results in sections 2.3 and 2.8 apply to $\hat{t}$ because we have an SRS of n units from a population of N units. As a result, $\hat{t}$ is an unbiased estimator of t, with variance given by

$$V(\hat{t}) = N^2 \left(1 - \frac{n}{N}\right) \frac{S_t^2}{n} \tag{5.2}$$

and with standard error

$$\text{SE}(\hat{t}) = N \sqrt{\left(1 - \frac{n}{N}\right) \frac{s_t^2}{n}}, \tag{5.3}$$

where S_t^2 and s_t^2 are the population and sample variance, respectively, of the psu totals:

$$S_t^2 = \frac{1}{N-1} \sum_{i=1}^{N} \left(t_i - \frac{t}{N} \right)^2$$

and

$$s_t^2 = \frac{1}{n-1} \sum_{i \in \mathcal{S}} \left(t_i - \frac{\hat{t}}{N} \right)^2 .$$

To estimate $\bar{y}_U$, divide the estimated total by the number of persons, obtaining

$$\hat{\bar{y}} = \frac{\hat{t}}{NM}, \tag{5.4}$$

with

$$V(\hat{\bar{y}}) = \left(1 - \frac{n}{N} \right) \frac{S_t^2}{nM^2} \tag{5.5}$$

and

$$SE(\hat{\bar{y}}) = \frac{1}{M} \sqrt{\left(1 - \frac{n}{N} \right) \frac{s_t^2}{n}} . \tag{5.6}$$

No new ideas are introduced to carry out one-stage cluster sampling; we simply use the results for simple random sampling with the psu totals as the observations.

EXAMPLE 5.2 A student wants to estimate the average grade point average (GPA) in his dormitory. Instead of obtaining a listing of all students in the dorm and conducting an SRS, he notices that the dorm consists of 100 suites, each with four students; he chooses 5 of those suites at random, and asks every person in the 5 suites what her or his GPA is. The results are as follows:

Person Number	Suite (psu)				
	1	2	3	4	5
1	3.08	2.36	2.00	3.00	2.68
2	2.60	3.04	2.56	2.88	1.92
3	3.44	3.28	2.52	3.44	3.28
4	3.04	2.68	1.88	3.64	3.20
Total	12.16	11.36	8.96	12.96	11.08

The psus are the suites, so $N = 100, n = 5$, and $M = 4$. The estimate of the population total (the estimated sum of all the GPAs for everyone in the dorm—a meaningless quantity for this example but useful for demonstrating the procedure) is

$$\hat{t} = \frac{100}{5} (12.16 + 11.36 + 8.96 + 12.96 + 11.08) = 1130.4.$$

The average of the suite totals is estimated by $\bar{t} = 1130.4/100 = 11.304$, and

$$s_t^2 = \frac{1}{5-1} \left[(12.16 - 11.304)^2 + \cdots + (11.08 - 11.304)^2 \right] = 2.256.$$

Note that s_t^2 is simply the usual sample variance of the 5 suite totals. Thus, using (5.4) and (5.6), $\hat{\bar{y}} = 1130.4/400 = 2.826$, and

$$\mathrm{SE}(\hat{\bar{y}}) = \sqrt{\left(1 - \frac{5}{100}\right) \frac{2.256}{(5)(4)^2}} = 0.164.$$

Note that in these calculations, only the "total" row of the data table is used—the individual GPAs are only used for their contribution to the suite total. ∎

One-stage cluster sampling with an SRS of psus produces a self-weighting sample. The weight for each observation unit is

$$w_{ij} = \frac{1}{P\{\text{ssu } j \text{ of psu } i \text{ is in sample}\}} = \frac{N}{n}.$$

For the data in Example 5.2, then,

$$\hat{t} = \sum_{i \in S} \sum_{j \in S_i} w_{ij} y_{ij}$$

$$= \frac{N}{n}(3.08 + 2.60 + \cdots + 3.28 + 3.20)$$

$$= \frac{100}{5}(56.52) = 1130.4.$$

Thus, as in stratified sampling, we can estimate a population total by summing the product of the observed values and the sampling weights. The population mean is estimated by

$$\hat{\bar{y}} = \frac{\displaystyle\sum_{i \in S} \sum_{j \in S_i} w_{ij} y_{ij}}{\displaystyle\sum_{i \in S} \sum_{j \in S_i} w_{ij}}$$

$$= \frac{1130.4}{NM}$$

$$= 2.826.$$

SAS code for analyzing these data is given on the website. The output follows. The only indication from the output that the analysis uses the clustering is in the data summary line giving the number of clusters. The sum of the weights can sometimes be used to diagnose problems in your weight calculations; if the sum of the weights is far from the number of observations, you may have calculated the weights incorrectly.

```
          Data Summary

Number of Clusters                5
Number of Observations           20
Sum of Weights                  400
```

```
                         Statistics

                                  Std Error
Variable          N       Mean    of Mean       95% CL for Mean
--------------------------------------------------------------------
gpa              20    2.826000   0.163665   2.37159339 3.28040661
--------------------------------------------------------------------
```

If we had taken an SRS of nM elements, each element in the sample would have been assigned weight $(NM)/(nM) = N/n$—the same weights we obtain for cluster sampling. The precision obtained for the two types of sampling, however, can differ greatly; the difference in precision is explored in the next section.

5.2.2 Clusters of Equal Sizes: Theory

In this section we compare cluster sampling with simple random sampling: Cluster sampling almost always provides less precision for the estimators than one would obtain by taking an SRS with the same number of elements.

As in stratified sampling, let's look at the ANOVA table (Table 5.1) for the whole population. In stratified sampling, the variance of the estimator of t depended on the variability *within* the strata; Equation (3.3) and Table 3.3 imply that the variance in stratified sampling is small if SSW is small relative to SSTO, or equivalently, if the within mean square (MSW) is small relative to S^2. In stratified sampling, you have some information about *every* stratum, so you need not worry about variability due to unsampled strata. If MSB/MSW is large—that is, the variability among the strata means is large when compared with the variability within strata—then stratified sampling increases precision.

The opposite situation occurs in cluster sampling. In one-stage cluster sampling when each psu has M ssus, the variability of the unbiased estimator of t depends entirely on the *between*-psu part of the variability, because

$$S_t^2 = \sum_{i=1}^{N} \frac{(t_i - \bar{t}_U)^2}{N - 1} = \sum_{i=1}^{N} \frac{M^2(\bar{y}_{iU} - \bar{y}_U)^2}{N - 1} = M\,(\text{MSB}).$$

Thus, for cluster sampling,

$$V(\hat{t}_{\text{cluster}}) = N^2 \left(1 - \frac{n}{N}\right) \frac{M(\text{MSB})}{n}. \tag{5.7}$$

If MSB/MSW is large in cluster sampling, then cluster sampling decreases precision. In that situation, MSB is relatively large because it measures the cluster-to-cluster variability: Elements in different clusters often vary more than elements in the same cluster because different clusters have different means. If we took a cluster sample of classes and sampled all students within the selected classes, we would likely find

TABLE 5.1

Population ANOVA Table—Cluster Sampling

Source	df	Sum of Squares	Mean Square
Between psus	$N - 1$	$\text{SSB} = \sum_{i=1}^{N}\sum_{j=1}^{M}(\bar{y}_{iU} - \bar{y}_{U})^2$	MSB
Within psus	$N(M - 1)$	$\text{SSW} = \sum_{i=1}^{N}\sum_{j=1}^{M}(y_{ij} - \bar{y}_{iU})^2$	MSW
Total, about $\bar{y}_U$	$NM - 1$	$\text{SSTO} = \sum_{i=1}^{N}\sum_{j=1}^{M}(y_{ij} - \bar{y}_{U})^2$	S^2

that average reading scores varied from class to class. An excellent reading teacher might raise the reading scores for the entire class; a class of students from an area with much poverty might tend to be undernourished and not score as highly at reading. Unmeasured factors, such as teaching skill or poverty, can affect the overall mean for a cluster, and thus cause MSB to be large.

Within a class, too, students' reading scores vary. The MSW is the pooled value of the within-cluster variances: the variance from element to element, present for all elements of the population. If the clusters are relatively homogeneous—if, for example, students in the same class have similar scores—the MSW will be small.

Now let's compare cluster sampling to simple random sampling. If, instead of taking a cluster sample of M elements in each of n clusters, we had taken an SRS with nM observations, the variance of the estimated total would have been

$$V(\hat{t}_{\text{SRS}}) = (NM)^2 \left(1 - \frac{nM}{NM}\right)\frac{S^2}{nM} = N^2\left(1 - \frac{n}{N}\right)\frac{M S^2}{n}.$$

Comparing this with (5.7), we see that if $\text{MSB} > S^2$, then cluster sampling is less efficient than simple random sampling.

The **intraclass** (sometimes called **intracluster**) **correlation coefficient** (ICC) tells us how similar elements in the same cluster are. It provides a **measure of homogeneity** within the clusters. ICC is defined to be the Pearson correlation coefficient for the $NM(M - 1)$ pairs (y_{ij}, y_{ik}) for i between 1 and N and $j \neq k$ (see Exercise 22) and can be written in terms of the population ANOVA table quantities as

$$\text{ICC} = 1 - \frac{M}{M - 1}\frac{\text{SSW}}{\text{SSTO}}. \tag{5.8}$$

Because $0 \leq \text{SSW/SSTO} \leq 1$, it follows from (5.8) that

$$-\frac{1}{M - 1} \leq \text{ICC} \leq 1.$$

If the clusters are perfectly homogeneous and hence SSW = 0, then ICC = 1. Equation (5.8) also implies that

$$\text{MSB} = \frac{NM - 1}{M(N - 1)} S^2 [1 + (M - 1)\text{ICC}].\qquad(5.9)$$

How much precision do we lose by taking a cluster sample? From (5.7) and (5.9),

$$\frac{V(\hat{t}_{\text{cluster}})}{V(\hat{t}_{\text{SRS}})} = \frac{\text{MSB}}{S^2} = \frac{NM - 1}{M(N - 1)}[1 + (M - 1)\text{ICC}].\qquad(5.10)$$

If N, the number of psus in the population, is large so that $NM - 1 \approx M(N - 1)$, then the ratio of the variances in (5.10) is approximately $1 + (M - 1)\text{ICC}$. So $1 + (M - 1)\text{ICC}$ ssus, taken in a one-stage cluster sample, give us approximately the same amount of information as one ssu from an SRS. If ICC = 1/2 and M=5, then $1 + (M - 1)\text{ICC} = 3$, and we would need to measure 300 elements using a cluster sample to obtain the same precision as an SRS of 100 elements. We hope, though, that because it is often much cheaper and easier to collect data in a cluster sample, that we will have more precision per dollar spent in cluster sampling.

The ICC provides a measure of homogeneity for the clusters. The ICC is positive if elements within a psu tend to be similar; then, SSW will be small relative to SSTO, and the ICC relatively large. When the ICC is positive, cluster sampling is less efficient than simple random sampling of elements.

If the clusters occur naturally in the population, the ICC is usually positive. Elements within the same cluster tend to be more similar than elements selected at random from the population. This may occur because the elements in a cluster share a similar environment—we would expect wells in the same geographic cluster to have similar levels of pesticides, or we would expect one area of a city to have a different incidence of measles than another area of a city. In human populations, personal choice as well as interactions among household members or neighbors may cause the ICC to be positive—wealthy households tend to live in similar neighborhoods, and persons in the same neighborhood may share similar opinions.

The ICC is negative if elements within a cluster are dispersed *more* than a randomly chosen group would be. This forces the cluster means to be very nearly equal—because SSTO = SSW + SSB, if SSTO is held fixed and SSW is large, then SSB must be small. If ICC < 0, cluster sampling is more efficient than simple random sampling of elements. The ICC is rarely negative in naturally occurring clusters, but negative values can occur in some systematic samples or artificial clusters, as discussed in Section 5.5.

The ICC is only defined for clusters of equal sizes. An alternative measure of homogeneity in general populations is the adjusted R^2, called R_a^2 and defined as

$$R_a^2 = 1 - \frac{\text{MSW}}{S^2}.\qquad(5.11)$$

If all psus are of the same size, then the increase in variance due to cluster sampling is

$$\frac{V(\hat{t}_{\text{cluster}})}{V(\hat{t}_{\text{SRS}})} = \frac{\text{MSB}}{S^2} = 1 + \frac{N(M - 1)}{N - 1} R_a^2;$$

by comparing with (5.10), you can see that for many populations, R_a^2 is close to the ICC. The quantity R_a^2 is a reasonable measure of homogeneity because of its interpretation in linear regression: It is the relative amount of variability in the population explained by the psu means, adjusted for the number of degrees of freedom. If the psus are homogeneous, then the psu means are highly variable relative to the variation within psus, and R_a^2 will be high.

EXAMPLE 5.3   Consider two artificial populations, each having three psus with three elements per psu.

	Population A			Population B		
psu 1	10	20	30	9	10	11
psu 2	11	20	32	17	20	20
psu 3	9	17	31	31	32	30

The elements are the same in the two populations, so the populations share the values $\bar{y}_U = 20$ and $S^2 = 84.5$. In population A, the psu means are similar and most of the variability occurs within psus; in population B, most of the variability occurs between psus.

	Population A		Population B	
	$\bar{y}_{iU}$	S_i^2	$\bar{y}_{iU}$	S_i^2
psu 1	20	100	10	1
psu 2	21	111	19	3
psu 3	19	124	31	1

ANOVA Table for Population A:

Source	df	SS	MS
Between psus	2	6	3
Within psus	6	670	111.67
Total, about mean	8	676	84.5

ANOVA Table for Population B:

Source	df	SS	MS
Between psus	2	666	333
Within psus	6	10	1.67
Total, about mean	8	676	84.5

For population A:

$$R_a^2 = 1 - \frac{111.67}{84.5} = -0.3215$$

$$\text{ICC} = 1 - \left(\frac{3}{2}\right)\frac{670}{676} = -0.4867$$

For population B:

$$R_a^2 = 1 - \frac{1.67}{84.5} = 0.9803$$

$$\text{ICC} = 1 - \left(\frac{3}{2}\right)\frac{10}{676} = 0.9778$$

Population A has much variation among elements within the psus, but little variation among the psu means. This is reflected in the large negative values of the ICC and R_a^2: Elements in the same cluster are actually less similar than randomly selected elements from the whole population. For this situation, cluster sampling is more efficient than simple random sampling.

The opposite situation occurs in population B: Most of the variability occurs between psus, and the psus themselves are relatively homogeneous. The ICC and R_a^2 are very close to 1, indicating that little new information would be gleaned by sampling more than one element per psu. Here, one-stage cluster sampling is much less efficient than simple random sampling. ∎

Most real-life populations fall somewhere between these two extremes. The ICC is usually positive, but not overly close to 1. Thus, there is a penalty in efficiency for using cluster sampling, and that decreased efficiency should be offset by cost savings.

EXAMPLE 5.4   When all psus are the same size, we can estimate the variance of $\hat{t}$ as well as the ICC from the sample ANOVA table. Here is the sample ANOVA table for the GPA data from Example 5.2:

Source	df	SS	MS
Between suites	4	2.2557	0.56392
Within suites	15	2.7756	0.18504
Total	19	5.0313	0.26480

In one-stage cluster sampling with equal psu sizes, the mean squares for within suites and between suites are unbiased estimators of the corresponding quantities in the population ANOVA table (see Exercise 25). Thus,

$$E\left[\widehat{\mathrm{MSB}}\right] = \mathrm{MSB} = \frac{S_t^2}{M}$$

and, using (5.7),

$$\mathrm{SE}\,(\bar{\hat{y}}) = \sqrt{\left(1 - \frac{n}{N}\right)\frac{\widehat{\mathrm{MSB}}}{nM}} = \sqrt{\left(1 - \frac{5}{100}\right)\frac{0.56392}{(5)(4)}} = 0.164,$$

as calculated in Example 5.2.

The sample mean square total is biased for estimating S^2, though (see Exercise 26). Note that we can estimate the sums of squares from the population ANOVA table by $\widehat{\mathrm{SSB}} = (N-1)\widehat{\mathrm{MSB}}$ and $\widehat{\mathrm{SSW}} = N(M-1)\widehat{\mathrm{MSW}}$, so an unbiased estimator of S^2 is

$$\hat{S}^2 = \frac{(N-1)\widehat{\mathrm{MSB}} + N(M-1)\widehat{\mathrm{MSW}}}{NM-1}.$$

For the GPA data, $\widehat{\mathrm{SSB}} = (99)(0.56392) = 55.828$ and $\widehat{\mathrm{SSW}} = (300)(0.18504) = 55.512$. Consequently, $\widehat{\mathrm{SSTO}} = 55.828 + 55.512 = 111.340$. The estimates of the population sums of squares are given in the following table:

	df	$\widehat{\mathrm{SS}}$ (estimated)	$\widehat{\mathrm{MS}}$
Between suites	99	55.828	0.56392
Within suites	300	55.512	0.18504
Total	399	111.340	0.279

Using these estimates, $\hat{S}^2 = 111.340/399 = 0.279$ (note the difference between this estimate and the one from the sample ANOVA table, 0.265). In addition,

$$\widehat{ICC} = 1 - \frac{M}{M-1}\frac{\widehat{SSW}}{\widehat{SSB} + \widehat{SSW}} = 1 - \left(\frac{4}{3}\right)\frac{55.512}{111.34} = 0.335$$

and

$$\hat{R}_a^2 = 1 - \frac{\widehat{MSW}}{\hat{S}^2} = 1 - \frac{0.18504}{0.279} = 0.337.$$

The increase in variance for using cluster sampling is estimated to be

$$\frac{\widehat{MSB}}{\hat{S}^2} = \frac{0.56392}{0.279} = 2.02.$$

This says that we need to sample about $2.02\,n$ elements in a cluster sample to get the same precision as an SRS of size n. There are 4 persons in each psu, so in terms of precision, one psu is worth about $4/2.02 = 1.98$ SRS persons. ∎

EXAMPLE 5.5   When is a cluster not a cluster? When it's the whole population.

Consider the situation of sampling oak trees on Santa Cruz Island, described in Example 4.5. There, the sampling unit was one tree, and an observation unit was a seedling by the tree. The population of interest was seedlings of oak trees on Santa Cruz Island. An SRS of trees was used to estimate quantities of interest about the population of oak trees on the island.

But suppose the investigator had been interested in seedling survival in all of California, had divided the regions with oak trees into equal-sized areas, and had randomly selected five of those areas to be in the study. Then the primary sampling unit is the area, and trees are subsampled in each area. If Santa Cruz Island had been selected as one of the five areas, we could no longer treat the ten trees on Santa Cruz Island as though they were part of a random sample of trees from the population; instead, those trees are part of the Santa Cruz Island cluster. We would expect all ten trees on Santa Cruz Island to experience, as a group, different environmental factors (such as weather conditions and numbers of seedling eaters) than the ten trees selected in the Santa Ynez Valley on the mainland. Thus the ICC within each cluster (area) would likely be positive.

However, suppose we were only interested in the seedlings from Tree #10 on Santa Cruz Island. Then the population is all seedlings from Tree #10, and the primary sampling unit is the seedling. In this situation, then, the tree is not a cluster but is the entire population. ∎

5.2.3 Clusters of Unequal Sizes

Clusters are rarely of equal sizes in social surveys. In one of the early probability samples (Converse, 1987), the Enumerative Check Census of 1937, a 2% sample of postal routes was chosen, and questionnaires were distributed to all households on each chosen postal route with the goal of checking unemployment figures. Since postal routes had different numbers of households, the cluster sizes could vary greatly.

In a one-stage cluster sample of n of the N psus, we know how to estimate population totals and means in two ways: using unbiased estimation and using ratio estimation.

Unbiased Estimation. An **unbiased** estimator of t is calculated exactly as in (5.1):

$$\hat{t}_{\text{unb}} = \frac{N}{n} \sum_{i \in S} t_i, \tag{5.12}$$

and, by (5.3),

$$\text{SE}(\hat{t}_{\text{unb}}) = N \sqrt{\left(1 - \frac{n}{N}\right) \frac{s_t^2}{n}}. \tag{5.13}$$

The difference between unequal- and equal-sized clusters is that the variation among the individual cluster totals t_i is likely to be large when the clusters have different sizes. The investigators conducting the Enumerative Check Census of 1937 were interested in the total number of unemployed persons, and t_i would be the number of unemployed persons in postal route i. One would expect to find more persons, and hence more unemployed persons, on a postal route with a large number of households than on a postal route with a small number of households. So we would expect that t_i would be large when the psu size M_i is large, and small when M_i is small. Often, then, s_t^2 is larger in a cluster sample when the psus have unequal sizes than when the psus all have the same number of ssus.

The probability that a psu is in the sample is n/N, as an SRS of n of the N psus is taken. Since one-stage cluster sampling is used, an ssu is included in the sample whenever its psu is included in the sample. Thus, as in Section 5.2.1,

$$w_{ij} = \frac{1}{P\{\text{ssu } j \text{ of psu } i \text{ is in sample}\}} = \frac{N}{n}.$$

One-stage cluster sampling produces a self-weighting sample when the psus are selected with equal probabilities. Using the weights, (5.12) may be written as

$$\hat{t}_{\text{unb}} = \sum_{i \in S} \sum_{j \in S_i} w_{ij} y_{ij}. \tag{5.14}$$

We can use (5.12) and (5.13) to derive an unbiased estimator for $\bar{y}_U$ and to find its standard error. Define

$$M_0 = \sum_{i=1}^{N} M_i$$

as the total number of ssus in the population; then $\hat{\bar{y}}_{\text{unb}} = \hat{t}_{\text{unb}}/M_0$ and $\text{SE}(\hat{\bar{y}}_{\text{unb}}) = \text{SE}(\hat{t}_{\text{unb}})/M_0$. The unbiased estimator of the mean $\hat{\bar{y}}_{\text{unb}}$ can be inefficient when the values of M_i are unequal since it, like $\hat{t}_{\text{unb}}$, depends on the variability of the cluster totals t_i. It also requires that M_0 be known; however, we often know M_i only for the sampled clusters. In the Enumerative Check Census, for example, the number of households on a postal route would only be ascertained for the postal routes actually chosen to be in the sample. We now examine another estimator for $\bar{y}_U$ that is usually more efficient when the population psu sizes are unequal.

Ratio Estimation. We usually expect t_i to be positively correlated with M_i. If psus are counties, we would expect the total number of households living in poverty in county i (t_i) to be roughly proportional to the total number of households in county i (M_i). The population mean $\bar{y}_U$ is a ratio

$$\bar{y}_U = \frac{\sum_{i=1}^{N} t_i}{\sum_{i=1}^{N} M_i} = \frac{t}{M_0},$$

where t_i and M_i are usually positively correlated. Thus, $\bar{y}_U = B$ as in Section 4.1 (substituting t_i for y_i and using M_i as the auxiliary variable x_i). Define

$$\hat{\bar{y}}_r = \frac{\hat{t}_{\text{unb}}}{\hat{M}_0} = \frac{\sum_{i \in S} t_i}{\sum_{i \in S} M_i} = \frac{\sum_{i \in S} M_i \bar{y}_i}{\sum_{i \in S} M_i}. \tag{5.15}$$

Note that $\hat{\bar{y}}_r$ from (5.15) may also be calculated using the weights w_{ij}, as

$$\hat{\bar{y}}_r = \frac{\hat{t}_{\text{unb}}}{\hat{M}_0} = \frac{\sum_{i \in S} \sum_{j \in S_i} w_{ij} y_{ij}}{\sum_{i \in S} \sum_{j \in S_i} w_{ij}}. \tag{5.16}$$

Since an SRS of clusters is selected, all the weights are the same with $w_{ij} = N/n$.

The estimator $\hat{\bar{y}}_r$ in (5.15) is the quantity $\hat{B}$ in (4.2): the denominator is a random quantity that depends on which particular psus are included in the sample. If the M_i's are unequal and a different cluster sample of size n is taken, the denominator will likely be different. From (4.10),

$$\text{SE}(\hat{\bar{y}}_r) = \sqrt{\left(1 - \frac{n}{N}\right) \frac{1}{n\overline{M}^2} \frac{\sum_{i \in S} (t_i - \hat{\bar{y}}_r M_i)^2}{n - 1}}$$

$$= \sqrt{\left(1 - \frac{n}{N}\right) \frac{1}{n\overline{M}^2} \frac{\sum_{i \in S} M_i^2 (\bar{y}_i - \hat{\bar{y}}_r)^2}{n - 1}}. \tag{5.17}$$

The variance of the ratio estimator depends on the variability of the means per element in the clusters, and can be much smaller than that of the unbiased estimator $\hat{\bar{y}}_{\text{unb}}$.

If the total number of elements in the population, $M_0 = \sum_{i=1}^{N} M_i$, is known, we can also use ratio estimation to estimate the population total: the ratio estimator is $\hat{t}_r = M_0 \hat{\bar{y}}_r$ with $\text{SE}(\hat{t}_r) = M_0 \text{SE}(\hat{\bar{y}}_r)$. Note, though, that $\hat{t}_r$ requires that we know the total number of elements in the population, M_0; the unbiased estimator in (5.12) makes no such requirement.

EXAMPLE 5.6   One-stage cluster samples are often used in educational studies, since students are naturally clustered into classrooms or schools. Consider a population of 187 high school algebra classes in a city. An investigator takes an SRS of 12 of those classes and gives each student in the sampled classes a test about function knowledge. The (hypothetical) data are given in the file algebra.dat, with the following summary statistics.

Class Number	M_i	$\bar{y}_i$	t_i	$M_i^2(\bar{y}_i - \hat{\bar{y}}_r)^2$
23	20	61.5	1,230	456.7298
37	26	64.2	1,670	1,867.7428
38	24	58.4	1,402	9,929.2225
39	34	58.0	1,972	24,127.7518
41	26	58.0	1,508	14,109.3082
44	28	64.9	1,816	4,106.2808
46	19	55.2	1,048	19,825.3937
51	32	72.1	2,308	93,517.3218
58	17	58.2	989	5,574.9446
62	21	66.6	1,398	7,066.1174
106	26	62.3	1,621	33.4386
108	26	67.2	1,746	14212.7867
Total	299		18,708	194,827.0387

We can use either (5.15) or (5.16) to estimate the mean score in the population: Using (5.15),

$$\hat{\bar{y}}_r = \frac{\sum\limits_{i \in S} M_i \bar{y}_i}{\sum\limits_{i \in S} M_i} = \frac{18,708}{299} = 62.57.$$

The standard error, from (5.17), is

$$SE(\hat{\bar{y}}_r) = \sqrt{\left(1 - \frac{12}{187}\right) \frac{1}{(12)(24.92^2)} \frac{194,827}{11}} = 1.49.$$

The weight for each observation is $w_{ij} = 187/12 = 15.5833$; we can alternatively calculate $\hat{\bar{y}}_r$ using (5.16) as

$$\hat{\bar{y}}_r = \frac{\sum\limits_{i \in S} \sum\limits_{j=1}^{M_i} w_{ij} y_{ij}}{\sum\limits_{i \in S} \sum\limits_{j=1}^{M_i} w_{ij}} = \frac{291,533}{4659.41667} = 62.57.$$

SAS software uses (5.16) to estimate $\bar{y}_U$. SAS code for calculating these estimates and producing the following output is given on the website.

```
                    Data Summary

Number of Clusters                    12
Number of Observations               299
Sum of Weights                 4659.41667


                    Statistics


                                                          Std Error
Variable              N           DF           Mean        of Mean
-----------------------------------------------------------------------
score                299          11        62.568562      1.491578
-----------------------------------------------------------------------
```

The sum of the weights for the sample, 4659.41667, estimates the total number of students in the 187 high school algebra classes.   ∎

5.3
Two-Stage Cluster Sampling

In one-stage cluster sampling, we observe all the ssus within the selected psus. In many situations, though, the elements in a cluster may be so similar that measuring all subunits within a psu wastes resources; alternatively, it may be expensive to measure ssus relative to the cost of sampling psus. In these situations, it may be much cheaper to take a subsample within each psu selected. The stages within a two-stage cluster sample, when we sample the psus and subsample the ssus with equal probabilities, are:

1   Select an SRS S of n psus from the population of N psus.
2   Select an SRS of ssus from each selected psu. The SRS of m_i elements from the ith psu is denoted S_i.

The difference between one-stage and two-stage cluster sampling is illustrated in Figure 5.2. The extra stage complicates the notation and estimators, as we need to consider variability arising from both stages of data collection. The point estimators of t and $\bar{y}_U$ are analogous to those in one-stage cluster sampling, but the variance formulas become messier.

In one-stage cluster sampling, we could estimate the population total by $\hat{t}_{\text{unb}} = (N/n)\sum_{i \in S} t_i$; the psu totals t_i were known because we sampled every ssu in the selected psus. In two-stage cluster sampling, however, since we do not observe every ssu in the sampled psus, we need to estimate the individual psu totals by

$$\hat{t}_i = \sum_{j \in S_i} \frac{M_i}{m_i} y_{ij} = M_i \bar{y}_i$$

FIGURE 5.2

The difference between one-stage and two-stage cluster sampling.

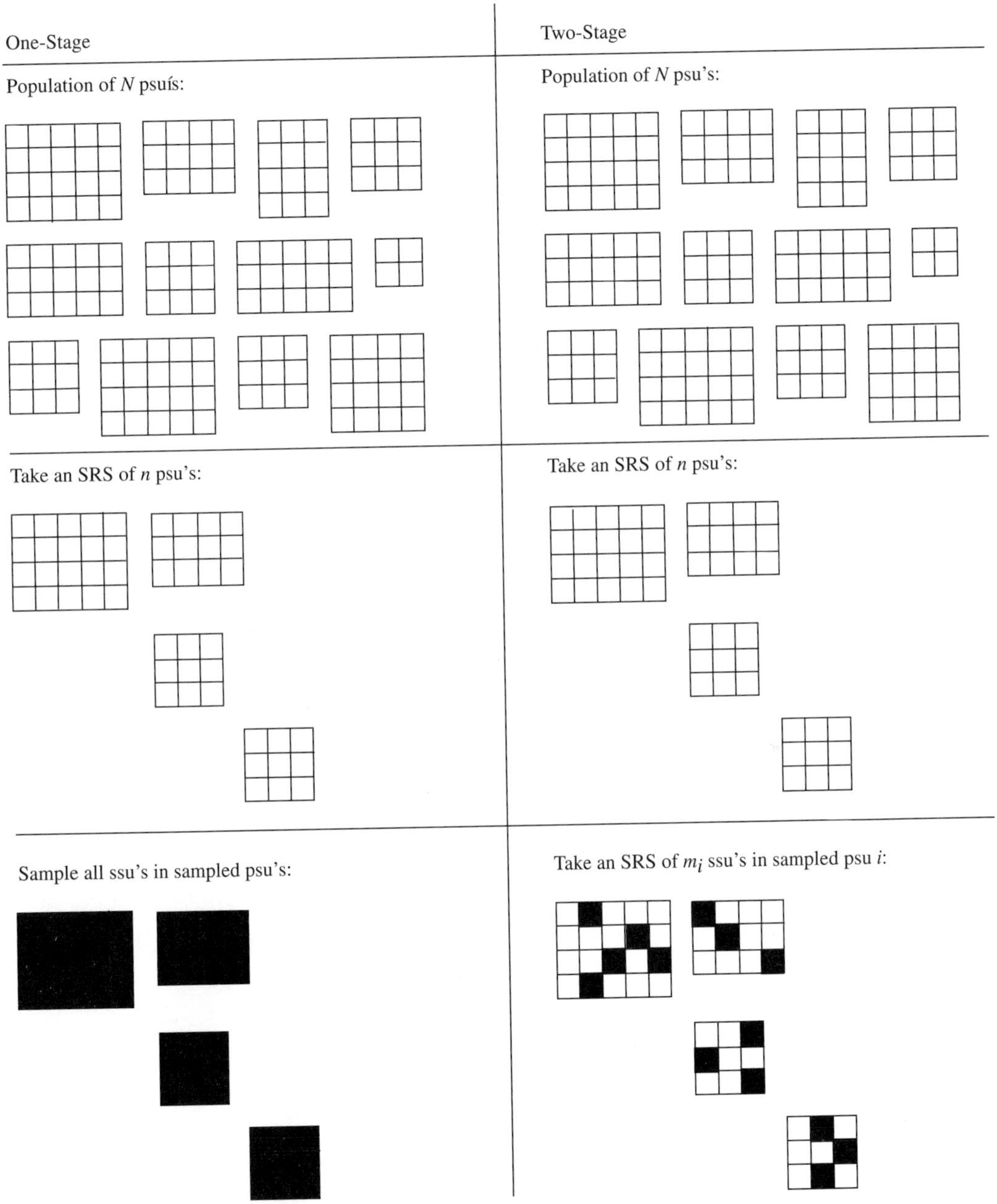

One-Stage	Two-Stage
Population of N psuís:	Population of N psu's:

| Take an SRS of n psu's: | Take an SRS of n psu's: |

| Sample all ssu's in sampled psu's: | Take an SRS of m_i ssu's in sampled psu i: |

and an unbiased estimator of the population total is

$$\hat{t}_{\text{unb}} = \frac{N}{n} \sum_{i \in \mathcal{S}} \hat{t}_i = \frac{N}{n} \sum_{i \in \mathcal{S}} M_i \bar{y}_i = \sum_{i \in \mathcal{S}} \sum_{j \in \mathcal{S}_i} \frac{N}{n} \frac{M_i}{m_i} y_{ij}. \qquad (5.18)$$

For estimating means and totals in cluster samples, most survey statisticians use sampling weights. Equation (5.18) suggests that the sampling weight for ssu j of psu i is $\dfrac{N}{n} \dfrac{M_i}{m_i}$, and we can see that this is so by calculating the inclusion probability. For cluster sampling,

$P(\,j$th ssu in ith psu is selected)

$$= P(i\text{th psu selected}) \times P(j\text{th ssu selected} \mid i\text{th psu selected})$$

$$= \frac{n}{N} \frac{m_i}{M_i}.$$

Recall from Section 2.4 that the weight of an element is the reciprocal of the probability of its selection. Thus,

$$w_{ij} = \frac{NM_i}{nm_i}. \tag{5.19}$$

If psus are blocks, for example, and ssus are households, then household j in psu i represents $(NM_i)/(nm_i)$ households in the population: itself, and $(NM_i)/(nm_i) - 1$ households that are not sampled. Then,

$$\hat{t}_{\text{unb}} = \sum_{i \in \mathcal{S}} \sum_{j \in \mathcal{S}_i} w_{ij} y_{ij}. \tag{5.20}$$

In two-stage cluster sampling, a self-weighting design has each ssu representing the same number of ssus in the population. To take a cluster sample of persons in Illinois, we could take an SRS of counties in Illinois and then take an SRS of m_i of the M_i persons from county i in the sample. To have every person in the sample represent the same number of persons in the population, m_i needs to be proportional to M_i, so that m_i/M_i is approximately constant. Thus, we would subsample more persons in the large counties than in the small counties to have a self-weighting sample.

The sampling weights provide a convenient way of calculating point estimates; they do not avoid associated shortcomings such as large variances. Also, the sampling weights give no information on how to find standard errors: We need to derive the formula for the variance using the sampling design.

In two-stage sampling, the $\hat{t}_i$'s are random variables. Consequently, the variance of $\hat{t}_{\text{unb}}$ has two components: (1) the variability between psus and (2) the variability of ssus within psus. We do not have to worry about component (2) in one-stage cluster sampling.

The variance of $\hat{t}_{\text{unb}}$ in (5.18) equals the variance of $\hat{t}_{\text{unb}}$ from one-stage cluster sampling plus an extra term to account for the extra variance due to estimating the $\hat{t}_i$'s rather than measuring them directly. For two-stage cluster sampling,

$$V(\hat{t}_{\text{unb}}) = N^2 \left(1 - \frac{n}{N}\right) \frac{S_t^2}{n} + \frac{N}{n} \sum_{i=1}^{N} \left(1 - \frac{m_i}{M_i}\right) M_i^2 \frac{S_i^2}{m_i}, \tag{5.21}$$

where S_t^2 is the population variance of the cluster totals, and S_i^2 is the population variance among the elements within cluster i. The first term in (5.21) is the variance from one-stage cluster sampling, and the second term is the additional variance due to subsampling within the psus. If $m_i = M_i$ for each psu i, as occurs in one-stage cluster sampling, then the second term in (5.21) is 0. To prove (5.21), we need to condition on the units included in the sample. This is more easily done in the general setting of unequal probability sampling; to avoid proving the same result twice, we shall prove the general result in Section 6.6.[1]

To estimate $V(\hat{t}_{\text{unb}})$, let

$$s_t^2 = \frac{1}{n-1} \sum_{i \in \mathcal{S}} \left(\hat{t}_i - \frac{\hat{t}_{\text{unb}}}{N} \right)^2 \tag{5.22}$$

be the sample variance among the estimated psu totals and let

$$s_i^2 = \frac{1}{m_i - 1} \sum_{j \in \mathcal{S}_i} (y_{ij} - \bar{y}_i)^2 \tag{5.23}$$

be the sample variance of the ssus sampled in psu i. As will be shown in Section 6.6, an unbiased estimator of the variance in (5.21) is given by

$$\hat{V}(\hat{t}_{\text{unb}}) = N^2 \left(1 - \frac{n}{N} \right) \frac{s_t^2}{n} + \frac{N}{n} \sum_{i \in \mathcal{S}} \left(1 - \frac{m_i}{M_i} \right) M_i^2 \frac{s_i^2}{m_i}. \tag{5.24}$$

The standard error, $\text{SE}(\hat{t}_{\text{unb}})$, is of course the square root of (5.24).

Remark. In many situations when N is large, the contribution of the second term in (5.24) to the variance estimator is negligible compared with that of the first term. We show in Section 6.6 that

$$E[s_t^2] = S_t^2 + \frac{1}{N} \sum_{i=1}^{N} \left(1 - \frac{m_i}{M_i} \right) M_i^2 \frac{S_i^2}{m_i}.$$

We expect the sample variance of the estimated psu totals $\hat{t}_i$ to be larger than the sample variance of the true psu totals t_i because $\hat{t}_i$ will be different if we take a different subsample in psu i. Thus, if N is large, the first term in (5.24) is approximately unbiased for the theoretical variance in (5.21). To simplify calculations, most software packages for analyzing survey data (including SAS software) estimate the variance using only the first term of (5.24), often omitting the finite population correction (fpc), $(1 - n/N)$. The estimator

$$\hat{V}_{\text{WR}}(\hat{t}_{\text{unb}}) = N^2 \frac{s_t^2}{n} \tag{5.25}$$

estimates the with-replacement variance for a cluster sample, as will be seen in Section 6.3. If the first-stage sampling fraction n/N is small, there is little difference between the variance from a with-replacement sample and that from a without-replacement sample. Alternatively, a replication method of variance estimation from Chapter 9 can be used.

If we know the total number of elements in the population, M_0, we can estimate the population mean by $\hat{\bar{y}}_{\text{unb}} = \hat{t}_{\text{unb}}/M_0$ with standard error $\text{SE}(\hat{\bar{y}}_{\text{unb}}) = \text{SE}(\hat{t}_{\text{unb}})/M_0$.

---

[1]Working with the additional level of abstraction will allow us to see the structure of the variance more clearly, without floundering in the notation of the special case of equal probabilities discussed in this chapter. If you prefer to see the proof before you use the variance results, read Section 6.6 now.

As in one-stage cluster sampling with unequal cluster sizes, s_t^2 can be very large since it is affected both by variations in the unit sizes (the M_i) and by variations in the $\bar{y}_i$. If the cluster sizes are disparate, this component is large, even if the cluster means are fairly constant.

Ratio Estimation.   As in one-stage cluster sampling, we use a ratio estimator for the population mean. Again, the y's of Chapter 4 are the psu totals (now estimated by $\hat{t}_i$) and the x's are the psu sizes M_i. As in (5.15),

$$\hat{\bar{y}}_r = \frac{\sum\limits_{i \in \mathcal{S}} \hat{t}_i}{\sum\limits_{i \in \mathcal{S}} M_i} = \frac{\sum\limits_{i \in \mathcal{S}} M_i \bar{y}_i}{\sum\limits_{i \in \mathcal{S}} M_i}. \tag{5.26}$$

Using the sampling weights in (5.19) with $w_{ij} = (NM_i)/(nm_i)$, we can rewrite $\hat{\bar{y}}_r$ as

$$\hat{\bar{y}}_r = \frac{\hat{t}_{\text{unb}}}{\hat{M}_0} = \frac{\sum\limits_{i \in \mathcal{S}} \sum\limits_{j \in \mathcal{S}_i} w_{ij} y_{ij}}{\sum\limits_{i \in \mathcal{S}} \sum\limits_{j \in \mathcal{S}_i} w_{ij}}. \tag{5.27}$$

The weights are different, but the form of the estimator is the same as in (5.16). The variance estimator is again based on the approximation in (4.10):

$$\hat{V}(\hat{\bar{y}}_r) = \frac{1}{\bar{M}^2}\left(1 - \frac{n}{N}\right)\frac{s_r^2}{n} + \frac{1}{nN\bar{M}^2}\sum\limits_{i \in \mathcal{S}} M_i^2\left(1 - \frac{m_i}{M_i}\right)\frac{s_i^2}{m_i}, \tag{5.28}$$

where s_i^2 is defined in (5.23),

$$s_r^2 = \frac{1}{n-1}\sum\limits_{i \in \mathcal{S}}(M_i\bar{y}_i - M_i\hat{\bar{y}}_r)^2, \tag{5.29}$$

and $\bar{M}$ is the average psu size. As with $\hat{t}_{\text{unb}}$, the second term in (5.28) is usually negligible compared with the first term, and most survey software packages calculate the variance using only the first term.

EXAMPLE **5.7**   The data in the file coots.dat come from Arnold's (1991) work on egg size and volume of American Coot eggs in Minnedosa, Manitoba. In this data set, we look at volumes of a subsample of eggs in clutches (nests of eggs) with at least two eggs available for measurement.

The data are plotted in Figures 5.3 and 5.4. Data from a cluster sample can be plotted in many ways, and you often need to construct more than one type of plot to see features of the data. Because we have only two observations per clutch, we can plot the individual data points. If we had many observations per clutch, we could instead construct side-by-side boxplots, with one boxplot for each psu (we did a similar plot in Figure 3.1 for a stratified sample, constructing a boxplot for each stratum). We shall return to the issue of plotting data from complex surveys in Section 7.4.

Next, we use a spreadsheet (partly shown in Table 5.2; the full spreadsheet is on the website) to calculate summary statistics for each clutch. The summary statistics may then be used to estimate the average egg volume and its variance. The numbers

FIGURE **5.3**

Plot of egg volume data. Note the wide variation in the means from clutch to clutch. This indicates that eggs within the same clutch tend to be more similar than two randomly selected eggs from different clutches, and that clustering does not provide as much information per egg as would an SRS of eggs.

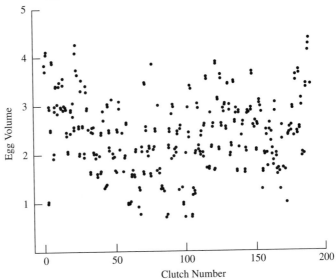

FIGURE **5.4**

Another plot of egg volume data. Here, we ordered the clutches from smallest mean to largest mean, and drew the line connecting the two measurements of volume from the eggs in the clutch. Clutch number 88, represented by the long line in the middle of the graph, has an unusually large difference between the two eggs: One egg has volume 1.85, and the other has volume 2.84.

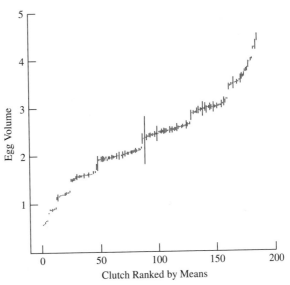

TABLE 5.2

Part of Spreadsheet Used for Calculations in Example 5.7

Clutch	M_i	$\bar{y}_i$	s_i^2	$\hat{t}_i$	$(1 - \frac{2}{M_i})M_i^2 \frac{s_i^2}{m_i}$	$(\hat{t}_i - M_i\hat{\bar{y}}_r)^2$
1	13	3.86	0.0094	50.23594	0.671901	318.9232
2	13	4.19	0.0009	54.52438	0.065615	490.4832
3	6	0.92	0.0005	5.49750	0.005777	89.22633
4	11	3.00	0.0008	32.98168	0.039354	31.19576
$\vdots$	$\vdots$	$\vdots$	$\vdots$	$\vdots$	$\vdots$	$\vdots$
182	13	4.22	0.00003	54.85854	0.002625	505.3962
183	13	4.41	0.0088	57.39262	0.630563	625.7549
184	12	3.48	0.000006	41.81168	0.000400	142.1994
sum	1757			4375.94652	42.174452	11,439.5794
$\hat{\bar{y}}_r =$		2.490579				

have been rounded so that they fit on the page; in practice, of course, you should carry out all calculations to machine precision.

We use the ratio estimator to estimate the mean egg volume. From (5.26),

$$\hat{\bar{y}}_r = \frac{\sum\limits_{i \in S} \hat{t}_i}{\sum\limits_{i \in S} M_i} = \frac{4375.947}{1757} = 2.49.$$

From the spreadsheet (Table 5.2),

$$s_r^2 = \frac{1}{n-1} \sum\limits_{i \in S} (\hat{t}_i - M_i\hat{\bar{y}}_r)^2 = \frac{11,439.58}{183} = 62.51$$

and $\bar{M}_S = 1757/184 = 9.549$. Using (5.28), then,

$$\hat{V}(\hat{\bar{y}}_r) = \frac{1}{9.549^2} \left[\left(1 - \frac{184}{N}\right) \frac{62.511}{184} + \frac{1}{N} \frac{42.17}{184} \right].$$

Now N, the total number of clutches in the population, is unknown but presumed to be large (and known to be larger than 184). Thus, we may take the psu-level fpc to be 1, and note that the second term in the estimated variance will be very small relative to the first term. We then use

$$SE(\hat{\bar{y}}_r) = \frac{1}{9.549} \sqrt{\frac{62.511}{184}} = 0.061.$$

The estimated coefficient of variation for $\hat{\bar{y}}_r$ is

$$\frac{SE(\hat{\bar{y}}_r)}{\hat{\bar{y}}_r} = \frac{0.061}{2.49} = 0.0245.$$

TABLE   5.3

Part of Spreadsheet for Egg Volume Calculations Using Relative Weights

clutch	*csize*	*volume*	*relweight*	*volume×relweight*
1	13	3.795757	6.5	24.67242
1	13	3.93285	6.5	25.56352
2	13	4.215604	6.5	27.40142
2	13	4.172762	6.5	27.12295
3	6	0.931765	3.0	2.795294
3	6	0.900736	3.0	2.702209
⋮	⋮	⋮	⋮	⋮
183	13	4.481221	6.5	29.12794
183	13	4.348412	6.5	28.26468
184	12	3.486132	6.0	20.91679
184	12	3.482482	6.0	20.89489
sum	3514		1757	4375.947

We used a spreadsheet to illustrate calculating the variance estimate with the formulas. Survey software packages will calculate $\hat{\bar{y}}_r$ and its standard error for you. For this example, SAS code provided on the website results in the values:

```
                             Std Error
Variable    N      Mean      of Mean       95% CL for Mean
-----------------------------------------------------------------
volume     368   2.490579   0.061040    2.37014533 2.61101179
-----------------------------------------------------------------
```

SAS software calculates the mean using (5.27) with the weights. The weight for egg j in clutch i is:

$$w_{ij} = \frac{N}{n} \frac{M_i}{m_i} = \frac{N}{184} \frac{M_i}{2}.$$

Because N is unknown, we display the relative weights $M_i/2$ in a spreadsheet (Table 5.3). Column 5 is set equal to y_i times the relative weight; using (5.27), $\hat{\bar{y}}_r = 4378.3/1758 = 2.49$. The weights do not allow us to calculate the standard error, however; we still need to use (5.28) for that.   ∎

In Example 5.7, we could only use the ratio estimator because we know neither N nor M_0. The M_i's, however, did not vary widely, so the unbiased estimator would probably have had similar coefficient of variation. If all the M_i's are equal, in fact, the unbiased estimator is the same as the ratio estimator (see Exercise 25); if the M_i's vary, the unbiased estimator often performs poorly. The next example illustrates that the unbiased estimator of t may have large variance when the cluster sizes are highly variable.

EXAMPLE 5.8   **The case of the six-legged puppy.** Suppose we want to estimate the average number of legs on the healthy puppies in Sample City puppy homes. Sample City has two puppy homes: Puppy Palace with 30 puppies, and Dog's Life with 10 puppies. Let's select one puppy home with probability 1/2. After the home is selected, then select 2 puppies at random from the home, and use $\hat{\bar{y}}_{\text{unb}}$ to estimate the average number of legs per puppy.

Suppose we select Puppy Palace. Not surprisingly, each of the two puppies sampled has four legs, so $\hat{t}_{\text{PP}} = 30 \times 4 = 120$. Then, using (5.18), an unbiased estimate for the total number of puppy legs in both homes is

$$\hat{t}_{\text{unb}} = \frac{2}{1}\hat{t}_{\text{PP}} = 240.$$

We divide the estimated total number of legs by the total number of puppies to estimate the mean number of legs per puppy as 240/40 = 6.

If we select Dog's Life instead, $\hat{t}_{\text{DL}} = 10 \times 4 = 40$, and

$$\hat{t}_{\text{unb}} = \frac{2}{1}\hat{t}_{\text{DL}} = 80.$$

If Dog's Life is selected, the unbiased estimate of the mean number of legs per puppy is 80/40 = 2.

These are not good estimates of the number of legs per puppy. But the estimator is mathematically unbiased: (6 + 2)/2 = 4, so averaging over all possible samples results in the right number. The poor quality of the estimator is reflected in the very large variance of the estimator, calculated using (5.21):

$$V(\hat{t}_{\text{unb}}) = \left(1 - \frac{1}{2}\right) 2^2 \frac{S_t^2}{1} + \frac{2}{1} \sum_{i=1}^{2} \left(1 - \frac{m_i}{M_i}\right) M_i^2 \frac{S_i^2}{m_i}$$

$$= \frac{1}{2}(4)(3200) = 6400.$$

The ratio estimator, on the other hand, is right on target: If Puppy Palace is selected, $\hat{\bar{y}}_r = 120/30 = 4$; if Dog's Life is selected, $\hat{\bar{y}}_r = 40/10 = 4$. Because the estimate is the same for all possible samples, $V(\hat{\bar{y}}_r) = 0$.   ∎

In general, the unbiased estimator of the population total is inefficient if the cluster sizes are unequal and t_i is roughly proportional to M_i. The variance of $\hat{t}_{\text{unb}}$ depends on the variance of the t_i's, and that variance may be large if the M_i's are unequal.

The ratio estimator, however, generally performs well when t_i is roughly proportional to M_i. Recall from (4.7) that the approximate mean squared error (MSE) of the estimator $\hat{B}$ is proportional to the variance of the residuals from the model: Using the notation of this chapter, the approximate MSE of $\hat{\bar{y}}_r (= \hat{B})$ is proportional to $\sum_{i=1}^{N} (t_i - \bar{y}_U M_i)^2$. When t_i (the response variable) is highly positively correlated with M_i (the auxiliary variable), the residuals are small. In Example 5.8, the total number of puppy legs in a puppy home (t_i) is exactly four times the total number of puppies in the home (M_i), so the variance of the ratio estimator is zero.

This is an important issue, since many naturally occurring clusters are of unequal sizes, and we expect that the cluster totals will often be proportional to the number of ssus. In a cluster sample of nursing homes, we expect that a larger number of residents will be satisfied with the level of care in a home with 500 residents than in a home with 20 residents, even though the proportions of residents who are satisfied may be the same. The total of the math scores for all students in a class will be much greater for large classes than for small classes. In general, we expect to see more honeybees in a large area than a small area. For all of these situations, then, while the estimator $\hat{\bar{y}}_r$ works well, the estimator $\hat{t}_{unb}$ tends to have large variability. In Chapter 6, we discuss an alternative design and estimator for cluster sampling that result in a much lower variance for the estimated population total when t_i is proportional to M_i.

5.4
Designing a Cluster Sample

Persons and organizations taking an expensive, large-scale survey need to devote a great deal of time to designing the survey; typically, large surveys administered by the U.S. Census Bureau take several years to design and test. Even then, the Fundamental Principle of Survey Design often holds true: You can best design the survey you should have taken after you have finished the survey. After the survey is completed, you can assess the effect of the clustering on the estimates, and know where you could have allocated more resources to obtain better information.

The more you know about a population, the better you can design an efficient sampling scheme to study it. If you know the value of y_{ij} for every person in your population, then you can design a flawless (but unnecessary because you already know everything!) survey for studying the population. If you know very little about the population, chances are that you will gain information about it after collecting the survey, but you may not have the most efficient design possible for that survey. You may, however, be able to use your newly gained knowledge to make your next survey more efficient.

When designing a cluster sample, you need to decide four major issues:

1 What overall precision is needed?

2 What size should the psus be?

3 How many ssus should be sampled in each psu selected for the sample?

4 How many psus should be sampled?

Question 1 must be faced in any survey design. To answer questions 2 through 4, you need to know the cost of sampling a psu for possible psu sizes, the cost of sampling a ssu, and a measure of homogeneity (R_a^2 or ICC) for the possible sizes of psu.

5.4.1   Choosing the psu Size

The psu size is often a natural unit. In Example 5.7, a clutch of eggs was an obvious cluster unit. A survey to estimate calf mortality might use farms as the psus; a survey of sixth-grade students might use classes or schools as the psus.

In other surveys, however, the investigator may have a wide choice for psu size. In a survey to estimate the sex and age ratios of mule deer in a region of Colorado [see Bowden et al. (1984) for more discussion of the problem], psus might be designated

TABLE   5.4

Relative Net Precision in the Potato Beetle Study

Number of Stems Sampled per Site	$\hat{\bar{y}}$	SE($\hat{\bar{y}}$)	Cost to Sample One Field	Relative Net Precision
1	1.12	0.15	31.67	0.24
2	1.01	0.10	33.33	0.30
3	0.96	0.08	35.00	0.34
4	0.91	0.07	36.67	0.35
5	0.91	0.06	38.33	0.40

areas and ssus might be individual deer or groups of deer in those areas. But should the size of the psus be 1 km^2, 2 km^2, or 100 m^2?

A general principle in area surveys is that the larger the psu size, the more variability you expect to see within a psu. Hence you expect R_a^2 and ICC to be smaller with a large psu than with a small psu. If the psu size is too large, however, you may lose the cost savings of cluster sampling.

Bellhouse (1984) reviews optimal designs for sampling, and the theory provides useful guidance for designing your own survey. There are many ways to "try out" different psu sizes before taking your survey. One way is to postulate a model for the relationship between R_a^2 or MSW and M, and to fit the model using preliminary data or information from other studies. Then use different combinations of R_a^2 and M, and compare the costs. Another way is to perform an experiment and collect data on relative costs and variances with different psu sizes.

EXAMPLE 5.9   The Colorado potato beetle has long been considered a major pest of potatoes. Zehnder et al. (1990) studied different sizes of sampling units that could be used to estimate potato beetle counts. Ten randomly selected sites were sampled from each of ten fields. The investigators visually inspected each site for small larvae, large larvae, and adults on all foliage from a single stem on each of five adjacent plants.

They then considered different sizes of psu, ranging from one stem per site to five stems per site. To study the efficiency of a one-stem-per-site design, they examined data from stem 1 of each site. Similarly, the data from stems 1 and 2 of each site gave a cluster sample with two ssus per psu, and so on. It takes about 30 minutes to walk among the sites in each field; sampling one stem requires about 10 seconds during the early part of the season. Thus the total cost to sample all ten sites with the one-stem-per-site design is estimated to be 30 + 100/60 = 31.67 minutes. Data for estimating the number of small larvae are given in Table 5.4.

The relative net precision is calculated as 1000/[(cost)CV($\hat{\bar{y}}$)]. For this example, since the cost to sample additional stems at a site is small compared with the time to traverse the field, the five-stem-per-site design is most efficient among those studied. ∎

5.4.2 Choosing Subsampling Sizes

The goal in designing a survey is generally to get the most information possible for the least cost and inconvenience. In this section, we concentrate on designing a two-stage cluster survey when all psus have the same number, M, of ssus; designing cluster samples will be treated more generally in Chapters 6 and 7. One approach for equal-sized clusters, discussed in Cochran (1977), is to minimize the variance in (5.21) for a fixed cost. If $M_i = M$ and $m_i = m$ for all psus, then $V(\hat{\bar{y}}_{unb})$ may be rewritten (see Exercise 24) as:

$$V(\hat{\bar{y}}_{unb}) = \left(1 - \frac{n}{N}\right)\frac{MSB}{nM} + \left(1 - \frac{m}{M}\right)\frac{MSW}{nm} \tag{5.30}$$

where MSB and MSW are the between and within mean squares, respectively, in Table 5.1, the population ANOVA table.

If MSW $= 0$ and hence $R_a^2 = 1$, for R_a^2 defined in (5.11), then each element within a psu equals the psu mean. In that case you may as well take $m = 1$; examining more than one element per psu just costs extra time and money without increasing precision. For other values of R_a^2, the optimal allocation depends on the relative costs of sampling psus and ssus.

Consider the simple cost function

$$\text{total cost} = C = c_1 n + c_2 nm, \tag{5.31}$$

where c_1 is the cost per psu (not including the cost of measuring ssus) and c_2 is the cost of measuring each ssu. One can easily determine, using calculus, that the values

$$n_{opt} = \frac{C}{c_1 + c_2 m_{opt}}$$

and

$$m_{opt} = \sqrt{\frac{c_1 M(N-1)(1-R_a^2)}{c_2(NM-1)R_a^2}} \tag{5.32}$$

minimize the variance for fixed total cost C under this cost function (see Exercise 27); often, though, a number of different values will work about equally well, and graphing the projected variance of the estimator will give more information than merely computing one fixed solution. A graphical approach also allows you to perform what-if analyses on the designs: What if the costs or the cost function are slightly different? Or the value of R_a^2 is changed slightly? You can also explore different cost functions with this approach.

In (5.32), the value R_a^2 is from the population ANOVA table. In practice, we can estimate it from pilot survey data by $\hat{R}_a^2 = \widehat{MSW}/\hat{S}^2$. In large populations, the ratio $M(N-1)/(NM-1)$ will be close to 1, so we can use $\hat{m}_{opt} = \sqrt{c_1(1-\hat{R}_a^2)/(c_2\hat{R}_a^2)}$.

EXAMPLE **5.10** Would subsampling have been more efficient for Example 5.2 than the one-stage cluster sample that was used? We do not know the population quantities, but have

FIGURE   5.5

Estimated variance that would be obtained for the GPA example, for different values of c_1 and c_2 and different values of m. The sample estimate of 0.337 was used for R_a^2. The total cost used for this graph was $C = 300$. If it takes 40 minutes per suite and 5 minutes per person, then one-stage cluster sampling should be used; if it takes 10 minutes per suite and 20 minutes per person, then only one person should be sampled per suite; if it takes 20 minutes per suite and 10 minutes per person, the minimum is reached at $m \approx 2$, although the flatness of the curve indicates that any subsampling size would be acceptable.

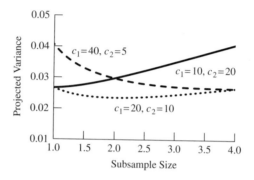

FIGURE   5.6

Estimated variance that would be obtained for the GPA example, for different values of R_a^2 and different values of m. The costs used in constructing this graph were $C = 300$, $c_1 = 20$, and $c_2 = 10$. The higher the value of R_a^2, the smaller the subsample size m should be.

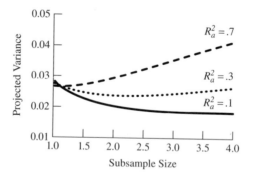

information from the sample that can be used for planning future studies. Recall that $\hat{S}^2 = 0.279$, and we estimated R_a^2 by 0.337. Figures 5.5 and 5.6 show the estimated variance that would be achieved for different subsample sizes for different values of c_1 and c_2, and for different values of R_a^2.   ∎

For design purposes, we only need a rough estimate of R_a^2 or of MSW and MSB. The adjusted R^2 from the ANOVA table from sample data usually provides a good starting point, even though the sample value of the mean square total often underestimates S^2 when the number of psus in the sample is small (see Exercise 26).

EXAMPLE 5.11   We obtain the following ANOVA table for the coots data in Example 5.7.

Source	DF	Sum of Squares	Mean Square	F Value
Model	183	257.4175336	1.4066532	237.44
Error	184	1.0900782	0.0059243	
Corrected Total	367	258.5076118		

R-Square	C.V.	Root MSE	VOLUME Mean
0.995783	3.298616	0.076970	2.333394

If a future survey were planned to estimate average egg volume, one might explore subsample sizes using R_a^2's around $1 - 0.0059243/(258.5/367) = 0.99$. These data indicate a high degree of homogeneity within clutches for egg volume. For this survey, however, locating and accessing a clutch is much more time consuming than measuring the eggs in a clutch. Thus, it might be best to take $m_i = M_i$ despite the high degree of homogeneity, because the additional information can be used to answer other research questions concerning variability from clutch to clutch or possible effects of egg-laying sequence. ∎

Although we discussed only designs where all M_i's are equal, we can use these methods with unequal M_i's as well: just substitute $\bar{M}$ for M in the above work, and decide the average subsample size $\bar{m}$ to take. Then either take $\bar{m}$ observations in every cluster, or allocate observations so that

$$\frac{m_i}{M_i} = \text{constant}.$$

As long as the M_i's do not vary too much, this should produce a reasonable design. If the M_i's are widely variable, and the t_i's are correlated with the M_i's, a cluster sample with equal probabilities is not necessarily very efficient; an alternative design is presented in Chapter 6. The file clusterselect.sas tells how to select a two-stage cluster sample using SAS software.

5.4.3 Choosing the Sample Size (Number of psus)

After the psu size is determined and the subsampling fraction set, we then look at the number of psus to sample, n. Like any survey design, design of a cluster sample is an iterative process: (1) Determine a desired precision, (2) choose the psu and subsample sizes, (3) conjecture the variance that will be achieved with that design, (4) set n to achieve the precision, and (5) iterate (adding stratification and auxiliary variables to use in ratio estimation) until the cost of the survey is within your budget.

If clusters are of equal size and we ignore the psu-level fpc, (5.30) implies that

$$V(\hat{\bar{y}}_{\text{unb}}) \le \frac{1}{n}\left[\frac{\text{MSB}}{M} + \left(1 - \frac{m}{M}\right)\frac{\text{MSW}}{m}\right] = \frac{1}{n}v.$$

An approximate $100(1 - \alpha)\%$ CI will be

$$\hat{\bar{y}}_{\text{unb}} \pm z_{\alpha/2}\sqrt{\frac{1}{n}v}.$$

Thus, to achieve a desired CI half-width e, set $n = z_{\alpha/2}^2 v /e^2$. Of course, this approach presupposes that you have some knowledge of v, perhaps from a prior survey. In Section 7.5, we examine how to determine sample sizes for any situation in which you know the efficiency of the specified design relative to an SRS design.

5.5
Systematic Sampling

Systematic sampling, discussed briefly in Chapter 2, is really a special case of cluster sampling. Suppose we want to take a sample of size 3 from a population that has 12 elements:

$$1 \quad 2 \quad 3 \quad 4 \quad 5 \quad 6 \quad 7 \quad 8 \quad 9 \quad 10 \quad 11 \quad 12.$$

To take a systematic sample, choose a number randomly between 1 and 4. Draw that element and every fourth element thereafter. Thus, the population contains four psus (they are clusters even though the elements are not contiguous):

$$\{1, 5, 9\} \quad \{2, 6, 10\} \quad \{3, 7, 11\} \quad \{4, 8, 12\}.$$

Now we take an SRS of one psu.

In a population of NM elements, there are N possible choices for the systematic sample, each of size M. We observe only the mean of the one psu that comprises our systematic sample,

$$\bar{y}_i = \bar{y}_{iU} = \hat{\bar{y}}_{\text{sys}}.$$

From the results in Section 5.2.1, $E[\hat{\bar{y}}_{\text{sys}}] = \bar{y}_U$. For a simple systematic sample, we select $n = 1$ of the N psus, so by (5.5) and (5.10), the theoretical variance is

$$V(\hat{\bar{y}}_{\text{sys}}) = \left(1 - \frac{1}{N}\right)\frac{S_t^2}{M^2} = \left(1 - \frac{1}{N}\right)\frac{\text{MSB}}{M} \approx \frac{S^2}{M}[1 + (M - 1)\text{ICC}]. \quad (5.33)$$

In the notation for cluster sampling, M is the size of the systematic sample. Ignoring the fpc, we see that systematic sampling is more precise than an SRS of size M if the ICC is negative. Systematic sampling is more precise than simple random sampling when the variance within the possible systematic samples (psus) is *larger* than the overall population variance—then the psu means will be more similar. If there is little variation within the systematic samples relative to that in the population (that is, ICC is large), then the elements in the sample all give similar information, and systematic sampling would be expected to be have higher variance than an SRS.

Since $n = 1$, however, we cannot calculate $V(\hat{\bar{y}}_{\text{sys}})$ using (5.6); we need to know something about the structure of the population to estimate the variance. Let's look at three different population structures.

1 *The list is in random order.* Systematic sampling is likely to produce a sample that behaves like an SRS. In many situations, the ordering of the population is unrelated

to the characteristics of interest, as when the list of persons in the sampling frame is ordered by the last four digits of their telephone numbers. There is no reason to believe that the persons in a systematic sample will be more or less similar than a random sample of persons: We expect that ICC ≈ 0. In this situation, simple random and systematic sampling will give similar results. We can use SRS results and formulas to estimate $V(\hat{\bar{y}}_{sys})$.

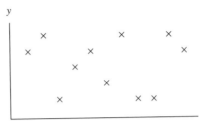

Position in Sampling Frame

2 *The sampling frame is in increasing or decreasing order.* Systematic sampling is likely to be more precise than simple random sampling. Financial records may be listed with the largest amounts first and the smallest amounts last. Such a population is said to have **positive autocorrelation**: adjacent elements tend to be more similar than elements that are farther apart. In this case, $V(\hat{\bar{y}}_{sys})$ is less than the variance of the sample mean in an SRS of the same size since ICC < 0. A systematic sample forces the sample values to be spread out; it is possible that an SRS would consist of all low values or all high values. When the frame is in increasing or decreasing order, you may use the SRS formula for standard error, but it will likely be an overestimate and CIs constructed using the SRS standard error will be too wide.

Position in Sampling Frame

Stratified sampling may work better than systematic sampling for positively autocorrelated populations: If the random start is close to either end of the sampling interval, it will tend to give an estimate that is too low or too high.

3 *The sampling frame has a periodic pattern.* If we sample at the same interval as the periodicity, systematic sampling will be less precise than simple random sampling. Systematic sampling is most dangerous when the population is in a cyclical or periodic order, and the sampling interval coincides with a multiple of the period.

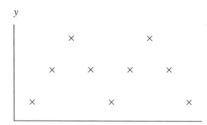

Position in Sampling Frame

Suppose the population values (in order) are

$$1 \quad 2 \quad 3 \quad 1 \quad 2 \quad 3 \quad 1 \quad 2 \quad 3 \quad 1 \quad 2 \quad 3$$

and the sampling interval is 3. Then all elements in the systematic sample will be the same; if we use the SRS formula to estimate the variance, we will have $\hat{V}(\hat{\bar{y}}_{sys}) = 0$. But the true value of $V(\hat{\bar{y}}_{sys})$ for this population is $2/3 = S^2$; this sample is no more precise than a single observation chosen randomly from the population.

Systematic sampling is often used when a researcher wants a representative sample of the population, but does not have the resources to construct a sampling frame in advance. It is commonly used to select elements at the bottom stage of a cluster sample. In many situations in which systematic sampling is used, the systematic sample can be treated as if it were an SRS.

EXAMPLE **5.12**   **Sampling for Hazardous Waste Sites.** Many dumps and landfills in the United States contain toxic materials. These materials may have been sealed in containers when deposited, but may now be suspected of leaking. But we no longer know where the materials were deposited—containers of hazardous waste may be randomly distributed throughout the landfill, or they may be concentrated in one area, or there may be none at all.

A common practice is to take a systematic sample of grid points and to take soil samples from each to look for evidence of contamination. Choose a point at random in the area, then construct a grid containing that point so that grid points are an equal distance apart. One such grid is shown in Figure 5.7. The advantages of taking a systematic sample rather than an SRS are that the systematic sample forces an even coverage of the region and is easier to implement in the field. If you are not worried about periodic patterns in the distribution of toxic materials, and you have little prior knowledge where the toxic materials might be, a systematic sample is a good design.

With any grid in systematic sampling, you need to worry if the toxic materials are regularly placed so that the grid may miss all of them, as shown in Figure 5.8. If this is a concern, you would be better off taking a stratified sample. Lay out the grid, but select a point at random in each square at which to take the soil sample.   ∎

If periodicity is a concern in a population, one solution is to use **interpenetrating systematic samples** (Mahalanobis, 1946). Instead of taking one systematic sample,

FIGURE 5.7
A grid used for detecting hazardous wastes

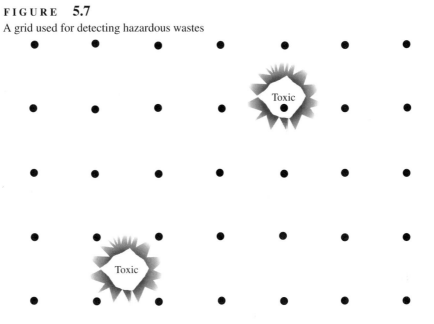

FIGURE 5.8
A grid used for detecting hazardous wastes: the worst-case scenario. Since the waste occurs in a similar pattern to the grid, the systematic sample misses every deposit of toxic waste.

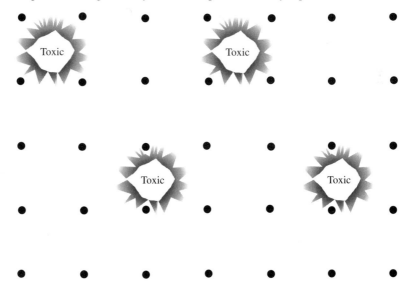

take several systematic samples from the population. Then you can use the formulas for cluster samples to estimate variances; each systematic sample acts as one psu. This approach is explored in Exercise 21.

5.6
Model-Based Inference in Cluster Sampling*

The one-way ANOVA model with fixed effects provides a theoretical framework for stratified sampling; one possible analogous model for cluster sampling is the one-way ANOVA model with random effects (Scott and Smith, 1969). Let's look at a simple version of this model:

$$M1 : Y_{ij} = \mu + A_i + \varepsilon_{ij} \tag{5.34}$$

with A_i generated by a distribution with mean 0 and variance σ_A^2, ε_{ij} generated by a distribution with mean 0 and variance σ^2, and all A_i's and ε_{ij}'s independent.

Let $T_i = \sum_{j=1}^{M_i} Y_{ij}$. Model M1 implies that the expected total for a cluster increases linearly with the number of elements in the cluster, because $E_{M1}[Y_{ij}] = \mu$ and

$$E_{M1}[T_i] = E_{M1}\left[\sum_{j=1}^{M_i} Y_{ij} \right] = M_i\mu.$$

This assumption is often appropriate for cluster samples taken in practice. Suppose we are taking a two-stage cluster sample to estimate total hospital charges for delivering babies; hospitals are selected at the first stage, and birth records are selected at the second stage (twins and triplets count as one record). We expect total costs billed by a hospital to be larger if the hospital delivers more babies.

The average cost per birth, however, varies from hospital to hospital—some hospitals may have higher personnel costs, and others may serve a higher-risk population or have more expensive equipment. That variation is reflected in the model by the random effects A_i: A_i is the random variable representing the average cost per birth in the ith hospital minus μ, and σ_A^2 is the population variance among the hospital means. In addition, costs vary from birth to birth within the hospitals; that variation is incorporated into the model by the term ε_{ij} with variance σ^2. These ideas are illustrated in Figure 5.9, presuming that the A_i's and ε_{ij}'s are normally distributed.

Figure 5.9 illustrates that, according to the model in (5.34), costs for births in the same hospital tend to be more similar than costs for births selected randomly across the entire population of hospital births, because the cost for a birth in a given hospital incorporates the hospital characteristics such as personnel costs or nurse/patient ratios. The model-based intraclass correlation coefficient for Model M1 is defined to be

$$\rho = \frac{\sigma_A^2}{\sigma_A^2 + \sigma^2}. \tag{5.35}$$

Note that ρ in Model M1 is always nonnegative, in contrast to ICC which can take on negative values.[2] Thus, if Model M1 describes the data, cluster sampling *must* be

---

[2]Model M1, with $\rho \geq 0$, would not be appropriate if there is competition within clusters so that one member of a cluster profits at the expense of another. For example, if other environmental factors can be discounted, competition within the uterus might cause some fraternal twins to be more variable than non-twin full siblings.

FIGURE 5.9

Illustration of random effects for hospitals and births.

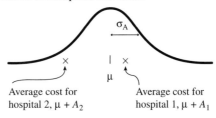

Average cost for
hospital 2, $\mu + A_2$

Average cost for
hospital 1, $\mu + A_1$

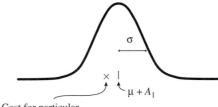

Cost for particular
birth at hospital 1

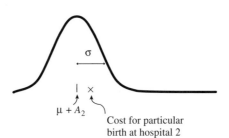

Cost for particular
birth at hospital 2

less efficient than an SRS of equal size. With Model M1,

$$\text{Cov}_{\text{M1}}[Y_{ij}, Y_{kl}] = \begin{cases} \sigma^2 + \sigma_A^2 & \text{if } i = k \text{ and } j = l \\ \sigma_A^2 & \text{if } i = k \text{ and } j \neq l \\ 0 & \text{if } i \neq k \end{cases}.$$

5.6.1 Estimation Using Models

Now let's find properties of various estimators under Model M1. To save some work later, we look at a general linear estimator of the form

$$\hat{T} = \sum_{i \in \mathcal{S}} \sum_{j \in \mathcal{S}_i} b_{ij} Y_{ij}$$

for b_{ij} any constants. The random variable representing the finite population total is

$$T = \sum_{i=1}^{N} \sum_{j=1}^{M_i} Y_{ij}.$$

Inference in a model-based approach is conditional on the units selected to be in the sample; that is, the inference treats $\mathcal{S}$ and $\mathcal{S}_i$ as fixed, and treats Y_{ij} as a random variable. Then, the bias is

$$E_{M1}[\hat{T} - T] = E_{M1}\left[\sum_{i \in \mathcal{S}} \sum_{j \in \mathcal{S}_i} b_{ij} Y_{ij} - \sum_{i=1}^{N} \sum_{j=1}^{M_i} Y_{ij} \right]$$

$$= \mu \left(\sum_{i \in \mathcal{S}} \sum_{j \in \mathcal{S}_i} b_{ij} - M_0 \right).$$

Thus, $\hat{T}$ is model-unbiased when $\sum_{i \in \mathcal{S}} \sum_{j \in \mathcal{S}_i} b_{ij} = M_0$. The model-based (for model M1) variance of $\hat{T} - T$ is

$$V_{M1}[\hat{T} - T] = \sigma_A^2 \left[\sum_{i \in \mathcal{S}} \left(\sum_{j \in \mathcal{S}_i} b_{ij} - M_i \right)^2 + \sum_{i \notin \mathcal{S}} M_i^2 \right] + \sigma^2 \left[\sum_{i \in \mathcal{S}} \sum_{j \in \mathcal{S}_i} (b_{ij}^2 - 2b_{ij}) + M_0 \right].$$

$$(5.36)$$

(See Exercise 30.)

Now let's look at what happens with design-based estimators under Model M1. The random variable for the design-unbiased estimator is

$$\hat{T}_{\text{unb}} = \sum_{i \in \mathcal{S}} \sum_{j \in \mathcal{S}_i} \frac{NM_i}{nm_i} Y_{ij};$$

the coefficients b_{ij} are the sampling weights $(NM_i)/(nm_i)$. But

$$\sum_{i \in \mathcal{S}} \sum_{j \in \mathcal{S}_i} b_{ij} = \frac{N}{n} \sum_{i \in \mathcal{S}} \sum_{j \in \mathcal{S}_i} \frac{M_i}{m_i} = \frac{N}{n} \sum_{i \in \mathcal{S}} M_i,$$

so the bias under model M1 in (5.34) is

$$\mu \left(\frac{N}{n} \sum_{i \in \mathcal{S}} M_i - M_0 \right).$$

Note that the bias depends on which sample is taken, and the estimator is model-unbiased under (5.34) only when the average of the M_i's in the sample equals the average of the M_i's in the population, such as will occur when all M_i's are the same.

For the ratio estimator, the coefficients are $b_{ij} = M_0(M_i/m_i)/\sum_{k \in \mathcal{S}} M_k$, and

$$\hat{T}_r = \frac{M_0 \sum_{i \in \mathcal{S}} \sum_{j \in \mathcal{S}_i} \frac{M_i}{m_i} Y_{ij}}{\sum_{k \in \mathcal{S}} M_k}.$$

For these b_{ij}'s,

$$\sum_{i \in \mathcal{S}} \sum_{j \in \mathcal{S}_i} b_{ij} = \sum_{i \in \mathcal{S}} \sum_{j \in \mathcal{S}_i} \frac{M_0 M_i}{m_i \sum_{k \in \mathcal{S}} M_k} = M_0,$$

so the ratio estimator is model-unbiased under Model M1. If Model M1 describes the population, then the ratio estimator adjusts for the sizes of the particular psus chosen for the sample; it uses M_i, a quantity that is correlated with the ith psu total, to compensate for the possibility that the sample may have a different proportion of large psus than does the population.

The variance expression in (5.36) is complicated; if $M_i = M$ and $m_i = m$ for all i, then $\hat{T}_{\text{unb}} = \hat{T}_r$, $b_{ij} = (NM)(nm)$, and the variance in (5.36) simplifies to

$$V_{\text{M1}}[\hat{T}_{\text{unb}} - T] = M_0 M(N - n)\frac{\sigma_A^2}{n} + M_0(MN - mn)\frac{\sigma^2}{mn}. \qquad (5.37)$$

EXAMPLE 5.13 Let's return to the puppy homes discussed in Example 5.8. They certainly follow Model M1: All puppies have four legs, so $Y_{ij} = \mu = 4$ for all i and j. Consequently, $\sigma_A^2 = \sigma^2 = 0$. The model-based variance of the estimate $\hat{T}_{\text{unb}}$ is therefore 0, no matter which puppy home and puppies are chosen. If Puppy Palace is selected for the sample, the bias under the model in (5.34) is $4(2 \times 30 - 40) = 80$; if Dog's Life is selected, the bias is $4(2 \times 10 - 40) = -80$. The large variance in the design-based approach thus becomes a bias when a model-based approach is adopted. It is not surprising that $\hat{T}_{\text{unb}}$ performs poorly for the puppy homes; it is a poor estimator for a model that describes the situation well. The model-based bias and variance for $\hat{T}_r$, though, are both zero. ∎

The above results are only for Model M1. Suppose that a better model for the population is

$$\text{M2}: Y_{ij} = B_i + \varepsilon_{ij}, \qquad (5.38)$$

with $E[B_i] = \mu/M_i$, $V[M_i B_i] = \sigma_B^2$, $E[\varepsilon_{ij}] = 0$, $V[\varepsilon_{ij}] = \sigma^2$, and all B_i and ε_{ij} independent. Under Model M2, then, the cluster totals all have expected value μ, regardless of cluster size. Examples that are described by this model are harder to come by in practice, but let's construct one based on the principle that tasks expand to fill up the allotted time. All students at Idyllic College have 100 hours available for writing term papers, but an individual student may have from one to five papers assigned. It would never occur to an Idyllic student to finish a paper quickly and relax in the extra time, so a student with one paper spends all 100 hours on the paper, a student with two papers spends 50 on each, and so on. Thus, the expected total amount of time spent writing term papers, $E[T_i]$, is 100 for each student, although the numbers of papers assigned (M_i) vary.

The estimator $\hat{T}_{\text{unb}}$ is unbiased under Model M2:

$$E_{\text{M2}}[\hat{T}_{\text{unb}} - T] = E_{\text{M2}}\left[\sum_{i \in \mathcal{S}} \sum_{j \in \mathcal{S}_i} \frac{NM_i}{nm_i} Y_{ij} - \sum_{i=1}^{N} \sum_{j=1}^{M_i} Y_{ij} \right]$$

$$= \sum_{i \in \mathcal{S}} \sum_{j \in \mathcal{S}_i} \frac{NM_i}{nm_i} \frac{\mu}{M_i} - \sum_{i=1}^{N} \sum_{j=1}^{M_i} \frac{\mu}{M_i} = 0.$$

Thus, $\hat{T}_{\text{unb}}$ performs poorly if model (5.34) is appropriate, but often quite well if model (5.38) is appropriate. Of course, these are not the only two possible models: Royall (1976) derives results for a general class of possible models that includes both (5.34) and (5.38), and allows unequal variances for different clusters. Chapter 8

FIGURE 5.10

Plot of $\hat{t}_i$ vs. M_i, for the coots data.

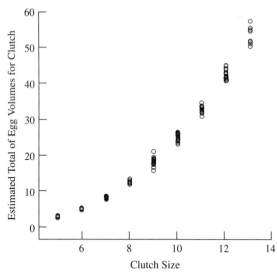

of Valliant et al. (2000) describes other correlation structures for clustered populations.

If you decide to use a model-based approach to analyze cluster sample data, you need to be very careful that the model chosen is appropriate. We saw in the puppy example that the Model M1 variance for $\hat{T}_{\text{unb}}$ is 0 but the bias is large; we could only evaluate the bias because we knew the results for the whole population. A person who sampled only Puppy Palace and did not know the results for Dog's Life would not be able to evaluate the bias, and might conclude that puppies average six legs each! Thus, assessing the adequacy of the model is crucial in any model-based analysis. You must check the assumption that $V[\varepsilon_{ij}] = \sigma^2$ by plotting the variances of each cluster, just as you assess the equal variance assumption in ANOVA. A plot of $\hat{t}_i$ versus M_i is often useful in assessing the appropriateness of a model for the data in the sample. As always in model-based inference, we must assume that the model also holds for population elements not in the sample.

EXAMPLE 5.14 Let's fit Model M1, a one-way random effects model, to the coots data. Looking at Figure 5.4, it seems plausible (except for one clutch) that the within-clutch variance is the same for each clutch. Figure 5.10 shows the plot of $\hat{t}_i$ vs. M_i for the coots data.

For these data, $\widehat{\text{Corr}}(\hat{t}_i, M_i) = 0.97$. If Model M1 is appropriate for the data, we expect that $\hat{t}_i$ will increase with M_i; if Model M2 were appropriate, we would expect a horizontal line to fit the plotted points. For these data, $\hat{t}_i$ and M_i are clearly related, although the relationship does not appear to be a straight line.

Using SAS PROC MIXED, the estimated variance components are $\hat{\sigma}_A^2 = 0.70036$ and $\hat{\sigma}^2 = 0.00592$. Using $b_{ij} = M_i/(m_i \sum_{k \in \mathcal{S}} M_k)$, the estimated mean egg volume is

2.492; adapting (5.36) to ignore the fpc (see Exercise 30), the estimated model-based variance is

$$\sum_{i \in \mathcal{S}} \left(\frac{M_i}{\sum_{k \in \mathcal{S}} M_k} \right)^2 \hat{\sigma}_A^2 + \sum_{i \in \mathcal{S}} \frac{1}{m_i} \left(\frac{M_i}{\sum_{k \in \mathcal{S}} M_k} \right)^2 \hat{\sigma}^2 = 0.003944 + 0.000017 = 0.00396.$$

If a different model were adopted, the estimated variance would be different. ∎

Most statisticians who use model-based analyses with cluster samples adopt a model such as M1 in (5.34) to estimate the population mean μ or regression parameters. Binder and Roberts (2003) describe the use of models for this situation. We shall return to models for cluster samples in Section 11.5.

5.6.2 Design Using Models

Models are extremely useful for designing a cluster sample. Using a model for design does not mean you have to use a model for analysis of your survey data when it is collected; rather, the model provides a useful way of summarizing information you can use to make the survey more efficient. Much research has been done on using models for design: see Rao (1979b), Bellhouse (1984), and Royall (1992b) for literature reviews.

Suppose that Model M1 seems reasonable for your population, and that all psu sizes in the population are equal. Then you would like to design the survey to minimize the variance in (5.37), subject to cost constraints. Then, using the cost function in (5.31), the model-based variance is minimized when

$$m = \sqrt{\frac{c_1 \sigma^2}{c_2 \sigma_A^2}}.$$

Suppose that the M_i's are unequal and that Model M1 holds. We can use the variance in (5.36) to determine the optimal subsampling size m_i for each cluster. This approach was used by Royall (1976) for more general models than considered in this section. For $\hat{T}_r$, $b_{ij} = M_i/(m_i \sum_{k \in \mathcal{S}} M_k)$, and the variance is minimized when m_i is proportional to M_i (see Exercise 32).

5.7
Chapter Summary

Cluster sampling is commonly used in large surveys, but estimates obtained from cluster samples usually have greater variance than if we were able to measure the same number of observation units using an SRS. If it is much less expensive to sample clusters than individual elements, though, cluster sampling can provide more precision per dollar spent.

All of the formulas in this chapter for cluster sampling with equal probabilities are special cases of the general results for two-stage cluster sampling with unequal psu sizes, to be derived in Chapter 6. They can be applied to any two-stage cluster sample in which the psus are selected with equal probability. These formulas were given in (5.18), (5.24), (5.26), and (5.28) and are repeated here:

$$\hat{t}_{\text{unb}} = \frac{N}{n} \sum_{i \in S} \hat{t}_i = \frac{N}{n} \sum_{i \in S} M_i \bar{y}_i, \tag{5.39}$$

$$\hat{V}(\hat{t}_{\text{unb}}) = N^2 \left(1 - \frac{n}{N}\right) \frac{s_t^2}{n} + \frac{N}{n} \sum_{i \in S} \left(1 - \frac{m_i}{M_i}\right) M_i^2 \frac{s_i^2}{m_i}, \tag{5.40}$$

$$\hat{\bar{y}}_r = \frac{\displaystyle\sum_{i \in S} M_i \bar{y}_i}{\displaystyle\sum_{i \in S} M_i}, \tag{5.41}$$

$$\hat{V}(\hat{\bar{y}}_r) = \frac{1}{\bar{M}^2} \left[\left(1 - \frac{n}{N}\right) \frac{s_r^2}{n} + \frac{1}{nN} \sum_{i \in S} \left(1 - \frac{m_i}{M_i}\right) M_i^2 \frac{s_i^2}{m_i} \right], \tag{5.42}$$

with

$$s_t^2 = \frac{1}{n-1} \sum_{i \in S} \left(\hat{t}_i - \frac{\hat{t}_{\text{unb}}}{N}\right)^2$$

and

$$s_r^2 = \frac{1}{n-1} \sum_{i \in S} (M_i \bar{y}_i - M_i \hat{\bar{y}}_r)^2.$$

In one-stage cluster sampling, the second term in (5.40) and (5.42) is zero since $m_i = M_i$. The variance estimators depend mostly on the variability between psus.

Point estimates of the population mean and total are usually calculated using weights. If an SRS of n of the N population psus is chosen, and an SRS of m_i of the M_i ssus in psu i is taken, then the sampling weight for observation j of psu i is

$$w_{ij} = \frac{NM_i}{nm_i}.$$

Then

$$\hat{t}_{\text{unb}} = \sum_{i \in S} \sum_{j \in S_i} w_{ij} y_{ij}$$

and

$$\hat{\bar{y}}_r = \frac{\displaystyle\sum_{i \in S} \sum_{j \in S_i} w_{ij} y_{ij}}{\displaystyle\sum_{i \in S} \sum_{j \in S_i} w_{ij}}.$$

While weights can be used to find estimated means and totals, they do not provide sufficient information to estimate the variance in a cluster sample. You need to use the formulas in (5.40) and (5.42), or a method such as jackknife from Chapter 9, to calculate standard errors.

Key Terms

Cluster: See Primary sampling unit.

Cluster sampling: A probability sampling design in which observations are grouped into clusters (psu). A probability sample of psus is selected from the population of psus.

Intraclass correlation coefficient (ICC): The Pearson correlation coefficient of all pairs of units within the same cluster.

One-stage cluster sampling: A cluster sampling design in which all ssus in selected psus are observed.

Primary sampling unit (psu): The unit that is sampled from the population.

Secondary sampling unit (ssu): A subunit that is subsampled from the selected psus.

Two-stage cluster sampling: A cluster sampling design in which the ssus in selected psus are subsampled.

For Further Reading

Stuart (1984) gives a great deal of intuition into cluster sampling with clear illustrations and examples. Cochran's (1977) classic book thoroughly covers the theory of unbiased estimation in cluster samples; Cochran (1939) used ANOVA tables in sample surveys. Skinner et al. (1989b) and Binder and Roberts (2003) delineate the issues involved in different approaches to inference in cluster samples. Royall (1976) applies best linear unbiased estimation to finite population sampling problems with naturally occurring clusters. The book by Valliant et al. (2000) describes model-based methods for cluster samples.

   The classic paper by Mahalanobis (1946) gives insight into many issues in survey sampling. Among other concepts, Mahalanobis developed the technique of interpenetrating subsampling, in which the sample is drawn as two smaller, independent subsamples. We mentioned this technique briefly for estimating the variance of systematic samples. Ultimately, Mahalanobis' idea led to the replication methods (discussed in Sections 9.2 and 9.3) now commonly used for variance estimation in complex surveys.

5.8
Exercises

A. Introductory Exercises

1   A city council of a small city wants to know the proportion of eligible voters that oppose having a incinerator of Phoenix garbage opened just outside of the city limits. They randomly select 100 residential numbers from the city's telephone book that contains 3,000 such numbers. Each selected residence is then called and asked for (a) the total number of eligible voters and (b) the number of voters opposed to the incinerator. A total of 157 voters were surveyed; of these, 23 refused to answer the

question. Of the remaining 134 voters, 112 opposed the incinerator, so the council estimates the proportion by

$$\hat{p} = 112/134 = .83582$$

with

$$\hat{V}(\hat{p}) = .83582(1 - .83582)/134 = 0.00102.$$

Are these estimates valid? Why, or why not?

2   Senturia et al. (1994) describe a survey taken to study how many children have access to guns in their households. Questionnaires were distributed to all parents who attended selected clinics in the Chicago area during a one-week period for well or sick child visits.

a   Suppose that the quantity of interest is percentage of the households with guns. Describe why this is a cluster sample. What is the psu? The ssu? Is it a one-stage or two-stage cluster sample? How would you estimate the percentage of households with guns, and the standard error of your estimate?

b   What is the sampling population for this study? Do you think this sampling procedure results in a representative sample of households with children? Why, or why not?

3   Kleppel et al. (2004) report on a study of wetlands in upstate New York. Four wetlands were selected for the study: Two of the wetlands drain watersheds from small towns and the other two drain suburban watersheds. Quantities such as pH were measured at two to four randomly selected sites within each of the four wetlands.

a   Describe why this is a cluster sample. What are the psus? The ssus? How would you estimate the average pH in the suburban wetlands?

b   The authors used Student's two-sample t test to compare the average pH from the sites in the suburban wetlands with the average pH from the sites in the small town wetlands, treating all sites as independent. Is this analysis appropriate? Why, or why not?

4   Survey evidence is often introduced in court cases involving trademark violation and employment discrimination. There has been controversy, however, about whether nonprobability samples are acceptable as evidence in litigation. Jacoby and Handlin (1991) selected 26 from a list of 1285 scholarly journals in the social and behavioral sciences. They examined all articles published during 1988 for the selected journals, and recorded (1) the number of articles in the journal that described empirical research from a survey (they excluded articles in which the authors analyzed survey data which had been collected by someone else) and (2) the total number of articles for each journal which used probability sampling, nonprobability sampling, or for which the sampling method could not be determined. The data are in file journal.dat.

a   Explain why this is a cluster sample.

b   Estimate the proportion of articles in the 1285 journals that use nonprobability sampling, and give the standard error of your estimate.

c   The authors conclude that, because "an overwhelming proportion of ... recognized scholarly and practitioner experts rely on non-probability sampling designs," courts "should have no problem admitting otherwise well-conducted non-probability surveys and according them due weight" (p. 175). Comment on this statement.

5 A language school owner takes an SRS of 10 of the 72 Introductory Spanish classes offered by the school. Each student in each of the sampled classes is given a vocabulary test and is also asked whether he or she is planning a trip to a Spanish-speaking country in the next year. The data are in file spanish.dat.

   a Estimate the total number of students planning a trip to a Spanish-speaking country in the next year, and give a 95% CI.

   b Estimate the mean vocabulary test score for Introductory Spanish students in the language school, and give a 95% CI.

6 An inspector samples cans from a truckload of canned creamed corn to estimate the total number of worm fragments in the truckload. The truck has 580 cases; each case contains 24 cans. The inspector samples 12 cases at random, and subsamples 3 cans randomly from each selected case.

	Case											
	1	2	3	4	5	6	7	8	9	10	11	12
Can 1	1	4	0	3	4	0	5	3	7	3	4	0
Can 2	5	2	1	6	9	7	5	0	3	1	7	0
Can 3	7	4	2	6	8	3	1	2	5	4	9	0

Using (5.20) and (5.24), estimate the total number of worm fragments, along with a 95% CI. Compare the estimated value of the variance from (5.24) with the approximation that is used by SAS software, given in (5.25).

7 The new candy Green Globules is being test-marketed in an area of upstate New York. The market research firm decided to sample 6 cities from the 45 cities in the area and then to sample supermarkets within cities, wanting to know the number of cases of Green Globules sold.

City	Number of Supermarkets	Number of Cases Sold
1	52	146, 180, 251, 152, 72, 181, 171, 361, 73, 186
2	19	99, 101, 52, 121
3	37	199, 179, 98, 63, 126, 87, 62
4	39	226, 129, 57, 46, 86, 43, 85, 165
5	8	12, 23
6	14	87, 43, 59

Obtain summary statistics for each cluster. Plot the data, and estimate the total number of cases sold, and the average number sold per supermarket, along with the standard errors of your estimates.

8 A homeowner with a large library needs to estimate the purchase cost and replacement value of the book collection for insurance purposes. She has 44 shelves containing books, and selects 12 shelves at random. To prepare for the second stage of sampling,

she counts the number of books M_i, on the selected shelves. She generates five random numbers between 1 and M_i for each selected shelf, to determine which specific books, numbered from left to right, to examine more closely. She then looks up the replacement value for the sampled books in *Books in Print*. The data are given in the file books.dat.

   a Draw side-by-side boxplots for the replacement costs of books on each shelf. Does it appear that the means are about the same? The variances?

   b Estimate the total replacement cost for the library, and find the standard error of your estimate. What is the estimated coefficient of variation?

   c Estimate the average replacement cost per book, along with the standard error. What is the estimated coefficient of variation?

 9 Repeat Exercise 8 for the purchase cost for each book. Plot the data, and estimate the total and average amount she has spent for books, along with the standard errors.

10 Construct a sample ANOVA table for the replacement cost data in Exercise 8. What is your estimate for R_a^2? Do books on the same shelf tend to have more similar replacement costs? Suppose that $c_1 = 10$ and $c_2 = 4$. If all shelves had 30 books, how many books should be sampled per shelf?

B. Working with Survey Data

11 An accounting firm is interested in estimating the error rate in a compliance audit it is conducting. The population contains 828 claims, and the firm audits an SRS of 85 of those claims. In each of the 85 sampled claims, 215 fields are checked for errors. One claim has errors in 4 of the 215 fields, 1 claim has 3 errors, 4 claims have 2 errors, 22 claims have 1 error, and the remaining 57 claims have no errors. (Data courtesy of Fritz Scheuren.)

   a Treating the claims as psus and the observations for each field as ssus, estimate the error rate, defined to be the average number of errors per field, along with the standard error for your estimate.

   b Estimate (with standard error) the total number of errors in the 828 claims.

   c Suppose that instead of taking a cluster sample, the firm had taken an SRS of $85 \times 215 = 18{,}275$ fields from the 178,020 fields in the population. If the estimated error rate from the SRS had been the same as in (a), what would the estimated variance $\hat{V}(\hat{p}_{SRS})$ be? How does this compare with the estimated variance from (a)?

12 Use the data in coots.dat to estimate the average egg length, along with its standard error. Be sure to plot the data appropriately.

13 The Arizona Health Care Cost Containment System (AHCCCS) provides medical assistance to low-income households in Arizona. Each county determines whether households are eligible for assistance. Sometimes, however, households are certified to be eligible when they are really not eligible. The Arizona Statutes, section 36-2905.01, mandate the collection of a "statistically valid quality control sample of the eligibility certifications made by each county." The certification error rate for each county is to be determined "by dividing the number of members in the sample who were erroneously certified by the total number of members in the sample." Quality control audits are done by sampling household records, however; once a household record is selected and audited, it costs the same amount to evaluate one person in the household as to evaluate all persons in the household.

  a Explain how to use cluster sampling to estimate the certification error rate for a county.

  b Suppose that a county certified 1572 households to be eligible for medical assistance in 1995. In past years, the certification error rate per household has been about 10%. How many households should be included in your sample so that the half-width of a 95% CI for estimating the per-person certification error rate is less than 0.03? What assumptions did you need to make to arrive at your sample size? Calculate the sample size for different values of M and homogeneity.

14 A researcher took an SRS of 4 high schools from a region with 29 high schools for a study on the prevalence of smoking among female high school students in the region. The results were as follows:

School	Number of Students	Number of Female Students in School	Number of Female Students Interviewed	Number of Smokers
1	1471	792	25	10
2	890	447	15	3
3	1021	511	20	6
4	1587	800	40	27

  a Estimate the percentage of female high school students in the region who smoke, along with a 95% CI.

  b Estimate the total number of female high school students in the region who smoke, along with a 95% CI.

  c The researcher now wants to study the prevalence of smoking and other risk behaviors among female high school students in a different region with 35 high schools. She intends to drive to n of the schools and then interview some or all of the female students in the selected schools. Assuming that MSB and MSW are similar in the two regions, use information from the study of 4 schools to estimate R_a^2 and design a cluster sample for the new study. Suppose it takes about 50 hours per school to contact school officials, obtain permission, obtain a list of female students, and travel back and forth. Although interviews themselves are only about 10 minutes, it takes about 30 minutes per interview obtained to allow for additional scheduling of no-shows, obtaining parental permission, and other administrative tasks. The investigator would like to spend 300 hours or less on the data collection.

15 Gnap (1995) conducted a survey to estimate the teacher workload in Maricopa County, Arizona, public school districts. Her target population was all first through sixth grade full-time public school teachers with at least one year of experience. In 1994, Maricopa County had 46 school districts with 311 elementary schools and 15,086 teachers. Gnap stratified the schools by size of school district; the large stratum, consisting of schools in districts with more than 5000 students, is considered in this exercise. The stratum contained 245 schools; 23 participated in the survey. All teachers in the selected schools were asked to fill out the questionnaire. Due to nonresponse, however, some questionnaires were not returned. (We shall examine possible effects of nonresponse in Exercise 12 of Chapter 8.) The data are in file teachers.dat, with psu information in teachmi.dat.

a   Why would a cluster sample be a better design than an SRS for this study? Consider issues such as cost, ease of collecting data, and confidentiality of respondent. What are some disadvantages of using a cluster sample?

b   Calculate the mean and standard deviation of *hrwork* for each school in the "large" stratum. Construct a graph of the means for each school and a separate graph of the standard deviations. Does there seem to be more variation within a school, or does more of the variability occur between different schools? How did you deal with missing values (coded as -9)?

c   Construct a scatterplot of the standard deviations versus the means for the schools, for the variable *hrwork*. Is there more variability in schools with higher workloads? Less? No apparent relation?

d   Estimate the average of *hrwork* in the large stratum in Maricopa County, along with its standard error. Use *popteach* in teachmi.dat for the M_i's.

16   The file measles.dat contains data consistent with that obtained in a survey of parents whose children had not been immunized for measles during a recent campaign to immunize all children between the ages of 11 and 15. During the campaign, 7633 children from the 46 schools in the area were immunized; 9962 children whose records showed no previous immunization were not immunized. In a follow-up survey to explore why the children had not been immunized during the campaign, Roberts et al. (1995) sent questionnaires to the parents of a cluster sample of the 9962 children. Ten schools were randomly selected, then a sample of the M_i nonimmunized children from each school was selected and the parents of those children were sent a questionnaire. Not all parents responded to the questionnaire; you will examine the effects of nonresponse in Exercise 13 of Chapter 8.

a   Estimate, separately for each school, the percentage of parents who returned a consent form (variable *returnf*). For this exercise, treat the "no answer" responses (value 9) as not returned.

b   Using the number of respondents in school *i* as m_i, construct the sampling weight for each observation.

c   Estimate the overall percentage of parents who received a consent form along with a 95% CI.

d   How do your estimate and interval in part (c) compare with the results you would have obtained if you had ignored the clustering and analyzed the data as an SRS? Find the ratio:

$$\frac{\text{estimated variance from (c)}}{\text{estimated variance if the data were analyzed as an SRS}}.$$

What is the effect of clustering?

17 Repeat Exercise 16, for estimating the percentage of children who had previously had measles.

18 Refer to Example 5.9. Later in the potato growing season, it takes more time to inspect stems. Suppose that it takes two minutes to inspect each stem. Which psu size is most efficient?

19 **a** For the SRS from the Census of Agriculture data in the file agsrs.dat (discussed in Example 2.5), find the sample ANOVA table of *acres92*, using *state* as the cluster variable. Estimate R_a^2 from the sample. Is there a clustering effect?

b Suppose that $c_1 = 15c_2$, where c_1 is the cost to sample a state, and c_2 is the cost to sample a county within a state. What should $\bar{m}$ be, if it is desired to sample a total of 300 counties? How many states would be sampled (that is, what is n)?

20 Using the value of n determined in Exercise 19, draw a self-weighting cluster sample of 300 counties from agpop.dat, using *state* as the cluster variable. Plot the data using side-by-side boxplots. Estimate the total number of acres devoted to farms in the United States, along with the standard error, using both the unbiased estimate and the ratio estimate. How do these values compare with each other and with values from the SRS and stratified sample from Examples 2.5 and 3.2?

21 The file ozone.dat contains hourly ozone readings from Eskdalemuir, Scotland, for 1994 and 1995.

a Construct a histogram of the population values. Find the mean, standard deviation, and median of the population.

b Take a systematic sample with period 24. To do this, select a random integer k between 1 and 24, and select the column containing the observations with GMT k. Construct a histogram of the sample values.

c Now suppose you treated your systematic sample as though it was an SRS. Find the sample mean, standard deviation, and median. Construct an interval estimate of the population mean using the procedure in Section 2.5. Does your interval contain the true value of the population mean from (a)?

d Take four independent systematic samples, each with period 96. Now use formulas from cluster sampling to estimate the population mean, and construct a 95% CI for the mean.

C. Working with Theory

22 The ICC is defined as the Pearson correlation coefficient for the $NM(M-1)$ pairs (y_{ij}, y_{ik}) for i between 1 and N and $j \neq k$:

$$\text{ICC} = \frac{\sum_{i=1}^{N}\sum_{j=1}^{M}\sum_{k \neq j}^{M}(y_{ij} - \bar{y}_U)(y_{ik} - \bar{y}_U)}{(NM-1)(M-1)S^2}. \tag{5.43}$$

Show that the above definition is equivalent to (5.8). HINT: First show that

$$\sum_{i=1}^{N}\sum_{j=1}^{M}\sum_{k\neq j}^{M}(y_{ij}-\bar{y}_U)(y_{ik}-\bar{y}_U) + \sum_{i=1}^{N}\sum_{j=1}^{M}(y_{ij}-\bar{y}_U)^2 = M(\text{SSB}).$$

23 For the quantities in the population ANOVA table (Table 5.1), show that

$$\text{MSW} = \frac{NM-1}{NM}S^2(1-\text{ICC})$$

and

$$\text{MSB} = \frac{NM-1}{M(N-1)}S^2[1+(M-1)\text{ICC}].$$

24 Suppose in a two-stage cluster sample that all population cluster sizes are equal ($M_i = M$ for all i), and that all sample sizes for the clusters are equal ($m_i = m$ for all i).

a Show (5.30).

b Show that $\text{MSW} = S^2(1-R_a^2)$ and that

$$\text{MSB} = S^2\left[\frac{N(M-1)R_a^2}{N-1}+1\right].$$

c Using (a) and (b), express $V(\hat{\bar{y}})$ as a function of n, m, N, M, and R_a^2.

d Show that if S^2 and the sample and population sizes are fixed, and if $(m-1)/m > n/N$, then $V(\hat{\bar{y}})$ is an increasing function of R_a^2.

25 Suppose in a two-stage cluster sample that all population cluster sizes are equal ($M_i = M$ for all i), and that all sample sizes for the clusters are equal ($m_i = m$ for all i).

a Show that $\hat{t}_{\text{unb}} = \hat{t}_r$, and, hence, that $\hat{\bar{y}}_{\text{unb}} = \hat{\bar{y}}_r$.

b Fill in the formulas for the sums of squares in the ANOVA table below, for the sample data.

Source	df	Sum of Squares	Mean Square
Between psus	$n-1$		msb
Within psus	$n(m-1)$		msw
Total	$nm-1$		msto

c Show that $E[\text{msw}] = \text{MSW}$ and

$$E[\text{msb}] = \frac{m}{M}\text{MSB} + \left(1-\frac{m}{M}\right)\text{MSW},$$

where MSB and MSW are the between and within mean squares, respectively, from the *population* ANOVA table given in Table 5.1.

d Show that

$$\widehat{\text{MSB}} = \frac{M}{m}\text{msb} - \left(\frac{M}{m}-1\right)\text{msw}$$

is an unbiased estimator of MSB.

e   Show, using (5.24) or (5.28), that

$$\hat{V}(\bar{y}_{unb}) = \left(1 - \frac{n}{N}\right)\frac{msb}{nm} + \frac{1}{N}\left(1 - \frac{m}{M}\right)\frac{msw}{m}.$$

26   For the situation in Exercise 25, let msto represent the mean square total from the sample ANOVA table.

a   Write msto as a function of msb and msw, and use the results of Exercise 25(c) to find $E[msto]$.

b   Show that $E[msto] \approx S^2$ if n and N are large.

c   Show that

$$\hat{S}^2 = \frac{M(N-1)}{m(NM-1)}msb + \frac{(m-1)NM + M - m}{m(NM-1)}msw$$

is an unbiased estimator of S^2.

27   (Requires calculus.) Show that if $M_i = M$ and $m_i = m$ for all i, and if the cost function is $C = c_1 n + c_2 nm$, then

$$m_{opt} = \sqrt{\frac{c_1 M(MSW)}{c_2(MSB - MSW)}} = \sqrt{\frac{c_1 M(N-1)(1-R_a^2)}{c_2(NM-1)R_a^2}}$$

minimizes the variance for fixed total cost C. HINT: Show the result with MSW and MSB first, then use Exercise 24(b).

28   (Requires trigonometry.) In Example 5.12, a systematic sampling scheme was proposed for detecting hazardous wastes in landfills. How far apart should sampling points be placed? Suppose that if there is leakage, it will spread to a circular region with radius R. Let D be the distance between adjacent sampling points in the same row or column.

a   Calculate the probability with which a contaminant will be detected. HINT: Consider three cases, with $R < D$, $D \le R \le \sqrt{2}D$, and $R > \sqrt{2}D$.

b   Propose a sampling design that gives a higher probability that a contaminant will be detected than the square grid, but does not increase the number of sampling points.

29   (Requires knowledge of random effects models.) Under Model M1 in (5.34), a one-way random effects model, the intraclass correlation coefficient ρ may be estimated by

$$\hat{\rho} = \frac{\hat{\sigma}_A^2}{\hat{\sigma}_A^2 + \hat{\sigma}^2},$$

where $\hat{\sigma}_A^2$ and $\hat{\sigma}^2$ estimate the variance components σ_A^2 and σ^2. The methods of moments estimators for one-stage cluster sampling when all clusters are of the same size are $\hat{\sigma}^2 = msw$ and $\hat{\sigma}_A^2 = (msb - msw)/M$, where msw and msb are the within and between mean squares from the sample ANOVA table.

a   What is $\hat{\rho}$ in Example 5.4? How does it compare with $\widehat{ICC}$?

b   Calculate $\hat{\rho}$ for Populations A and B in Example 5.3. Why do these differ from the ICC?

30   (Requires knowledge of random effects models.)

a   Suppose we ignore the fpc of a model-based estimator. Find

$$V_{M1}\left(\sum_{i\in S}\sum_{j\in S_i} b_{ij} Y_{ij}\right).$$

b   Prove (5.36). HINT: let

$$c_{ij} = \begin{cases} b_{ij} - 1 & \text{if } i \in S \text{ and } j \in S_i \\ -1 & \text{otherwise.} \end{cases}$$

Then, $\hat{T} - T = \sum_{i=1}^{N} \sum_{j=1}^{M_i} c_{ij} Y_{ij}$.

31   (Requires linear algebra and calculus.) Although $\hat{T}_r$ is unbiased for Model M1, it is possible to construct an estimator with smaller variance: let

$$c_k = \frac{m_k}{1 + \rho(m_k - 1)}$$

and

$$\hat{T}_{\text{opt}} = \sum_{i\in S}\sum_{j\in S_i} \frac{c_i}{m_i}\left[\rho M_i + \frac{M_0 - \rho \sum_{k\in S} c_k M_k}{\sum_{k\in S} c_k}\right] Y_{ij}.$$

Show that $\hat{T}_{\text{opt}}$ is unbiased and minimizes the variance in (5.36) among all unbiased estimators for Model (5.34).

32   (Requires calculus.) Suppose that the M_i's are unequal and Model M1 holds. The budget allows you to take a total of L measurements on subunits. Show that the variance in (5.36) is minimized for $\hat{T}_r$ when m_i is proportional to M_i. HINT: Use Lagrange multipliers, with the constraint $\sum_{i\in S} m_i = L$.

D. Projects and Activities

33   (Requires the R statistical software package.) Alf and Lohr (2007) present an R program, *intervals*, that explores the effects of ignoring clustering on CIs. The program and its usage are described on the website.

a   After copying the program into R, type `intervals(0)` to generate 100 samples, and their associated confidence intervals, from a population with ICC= 0. What is the effect of ignoring clustering when ICC= 0?

b   Now type `intervals(0.5)` to see what happens with a population with ICC = 1/2. What percentage of the interval estimates, calculated ignoring the clustering, include the true mean? What about the intervals calculated using the formulae for cluster samples? Compare the widths of the two interval estimates.

c   What do you think will happen if you try the intervals program with ICC = 1? What will happen to the widths of the correctly calculated confidence intervals? Do you expect the percentage of interval estimates that include the true mean to increase or decrease for the two methods? Type `intervals(1)` to test your predictions.

d Explore how the estimated coverage probability depends on the ICC and M. Run the program with all 9 combinations of $M \in \{2, 10, 25\}$ and ICC $\in \{0, 0.2, 0.7, 1\}$. Plot the coverage probability versus ICC, drawing a curve for each value of M.

34 The January 1994 issue of *The Nation* ranked 22 columnists by how much they used the words "I," "me," and "myself." Select your favorite newspaper columnist or blogger. Randomly select 5 of the columnist's or blogger's columns that appeared in the past year, and use one-stage cluster sampling to estimate the proportion of total words taken up by "I," "me," and "myself." What is your psu? Your ssu?

35 *Online bookstore.* You may have noticed in Exercise 33 of Chapter 2 that it took quite a bit of time to locate the records chosen for the SRS. It may be faster to take an SRS of pages from the website, then look at some or all of the books listed on that page. Use the following procedure to take a cluster sample of books from the genre you studied in Chapter 2, recording the amount of time you spend selecting the sample and collecting the data. Take an SRS of 10 pages, then sample 5 books per page. For each sampled book, record the price, number of pages, and whether the book is paperback or hardback. Estimate the mean of each variable, and give a 95% CI. Do you think clustering decreased precision relative to an SRS? Compare the precision per unit time for the SRS and the cluster sample by calculating 1/[(estimated variance) × time] for each method.

36 *Baseball data.*

    a Use the population in the file baseball.dat to take a one-stage cluster sample with the teams as the psus. Your sample should have approximately 150 players altogether, as in the SRS from Exercise 32 of Chapter 2. Describe how you selected your sample. The SAS code in the file clusterselect.sas may be helpful.

    b Draw side-by-side boxplots of *logsal* for the teams in your sample.

    c Use your sample to estimate the mean of the variable *logsal*, and give a 95% CI.

    d Estimate the proportion of players in the data set who are pitchers, and give a 95% CI.

    e Compare your estimates with those from Exercise 32 of Chapter 2. Which estimates have smaller CIs? Why do you think that happened?

37 *Baseball data: Two-stage sample.*

    a Use your SRS from Exercise 32 of Chapter 2 to estimate the population value of R_a^2. If we treat teams as the psus, and if all teams had the same size, what would the optimal subsampling size be (assume that $c_1 = c_2$).

    b Use the population in the file baseball.dat to take a two-stage cluster sample with the teams as the psus, using the subsampling fraction from part (a). Your sample should have approximately 150 players altogether, as in the SRS from Exercise 32 of Chapter 2. Describe how you selected your sample.

    c Draw side-by-side boxplots of *logsal* for the teams in your sample.

    d Use your sample to estimate the mean of the variable *logsal*, and give a 95% CI.

    e Estimate the proportion of players in the data set who are pitchers, and give a 95% CI.

f Compare your estimates with those from Exercise 32 of Chapter 2. Which estimates have smaller CIs? Why do you think that happened?

g Compare your estimates with those from Exercise 36.

38 *IPUMS exercises.*

  a Generate a frequency table of the number of persons within each psu.

  b Suppose that it costs $50 per interview to collect data using an SRS. If a cluster sample is taken, it costs $100 per psu chosen, plus $20 for each interview taken. Select an SRS of 10 psus. In each of the selected psus, take a subsample of persons with sample size proportional to the population size within that psu. Your total cost for the sample should be about the same as for the SRS you took in Chapter 2.

  c Using the sample you selected, estimate the population mean of *inctot* and give the standard error of your estimate. Also estimate the population total of *inctot* and give its standard error. How do these estimates compare with those from the SRS you took in Chapter 2?

6

Sampling with Unequal Probabilities

'Personally I never care for fiction or storybooks. What I like to read about are facts and statistics of any kind. If they are only facts about the raising of radishes, they interest me. Just now, for instance, before you came in'—he pointed to an encyclopædia on the shelves—'I was reading an article about "Mathematics." Perfectly pure mathematics.

'My own knowledge of mathematics stops at "twelve times twelve," but I enjoyed that article immensely. I didn't understand a word of it; but facts, or what a man believes to be facts, are always delightful. That mathematical fellow believed in his facts. So do I. Get your facts first, and'—the voice dies away to an almost inaudible drone—'then you can distort 'em as much as you please.'

—Mark Twain, quoted in Rudyard Kipling, *From Sea to Sea*

Up to now, we have only discussed sampling schemes in which the probabilities of choosing sampling units are equal. Equal probabilities give schemes that are often easy to design and explain. Such schemes are not, however, always possible or, if practicable, as efficient as schemes using unequal probabilities. We saw in Example 5.8 that a cluster sample with equal probabilities may result in a large variance for the design-unbiased estimator of the population mean and total.

EXAMPLE **6.1** O'Brien et al. (1995) took a sample of nursing home residents in the Philadelphia area, with the objective of determining residents' preferences on life-sustaining treatments. Do they wish to have cardiopulmonary resuscitation (CPR) if the heart stops beating, or to be transferred to a hospital if a serious illness develops, or to be fed through an enteral tube if no longer able to eat? The target population was all residents of licensed nursing homes in the Philadelphia area. There were 294 such homes, with a total of 37,652 beds (before sampling, they only knew number of beds, not number of residents).

Because the survey was to be done in person, cluster sampling was essential for keeping survey costs manageable. Had the researchers chosen to use cluster sampling with equal probabilities of selection, they would have taken a simple random sample (SRS) of nursing homes, then another SRS of residents within each selected home.

In a cluster sample with equal probabilities, however, a nursing home with 20 beds is as likely to be chosen for the sample as a nursing home with 1000 beds. The sample is only self-weighting if the subsample size for each home is proportional to the number of beds in the home. Each bed sampled represents the same number of beds in the population if one-stage cluster sampling is used, or if 10% (or any other fixed percentage) of beds are sampled in each selected home.

Sampling homes with equal probabilities would result in a mathematically valid estimator, but it has three major shortcomings. First, you would expect that the total number of patients in a home who desire CPR (t_i) would be proportional to the number of beds in the home (M_i), so estimators from Chapter 5 may have large variance. Second, a self-weighting equal-probability sample may be cumbersome to administer. It may require driving out to a nursing home just to interview one or two residents, and equalizing workloads of interviewers may be difficult. Third, the cost of the sample is unknown in advance—a random sample of 40 homes may consist primarily of large nursing homes, which would lead to greater expense than anticipated.

Instead of taking a cluster sample of homes with equal probabilities, the investigators randomly drew a sample of 57 nursing homes with probabilities proportional to the number of beds. They then took an SRS of 30 beds (and their occupants) from a list of all beds within the nursing home. If the number of residents equals the number of beds, and if a home has the same number of beds when visited as are listed in the sampling frame, then the sampling design results in every resident having the same probability of being included in the sample. The cost is known before selecting the sample, the same number of interviews are taken at each home, and the estimator of a population total will likely have a smaller variance than estimators in Chapter 5.

Since this sample is self-weighting, you can easily obtain point estimates (but *not* standard errors) of desired quantities by usual methods. You can estimate the median age of the nursing home residents by finding the sample median of the residents in the sample, or the 70th percentile by finding the 70th percentile of the sample. If a sample is not self-weighting, point estimates are still easily calculated using weights. A warning, though: Always consider the cluster design when calculating the precision of your estimates.   ∎

In Chapter 3 we noted that sometimes stratified sampling is used to sample units with different probabilities. In a survey to estimate total business expenditures on advertising, we might want to stratify by company sales or income. The largest companies such as IBM would be in one stratum, medium-sized companies would be in a number of different strata, and very small companies such as Robin's Tailor Shop would be in yet other strata. An optimal allocation scheme would sample a very high fraction (perhaps 100%) in the stratum with the largest companies, and a small fraction of companies in the strata with the smallest companies; the variance from company to company will be much higher among IBM, General Motors, and Microsoft than among Robin's Tailor Shop, Pat's Shoe Repair, and Flowers by Leslie. The variance is larger in the large companies just because the amounts of money involved are so much larger. Thus, the sampling variance is decreased by assigning unequal probabilities to sampling units in different strata.

To estimate the total spent on advertising using this stratified sample, we assign higher weights to companies with lower inclusion probabilities. As discussed in Section 3.3, the probability that a company in stratum h will be included in the sample is

n_h/N_h; the sampling weight for that company is N_h/n_h. Each company sampled in stratum h represents N_h/n_h companies in the population, and $\hat{t}_{str} = \sum_{h=1}^{H} \sum_{j \in S_h} (N_h/n_h) y_{hj}$.

We can also use unequal inclusion probabilities to decrease variances without explicitly stratifying. When sampling with unequal probabilities, we deliberately vary the probabilities that we will select different primary sampling units (psus) for the sample, and compensate by providing suitable weights in the estimation. The key is that we *know* the probabilities[1] with which we will select a given unit:

$$P(\text{unit } i \text{ selected on first draw}) = \psi_i \qquad (6.1)$$

$$P(\text{unit } i \text{ in sample}) = \pi_i. \qquad (6.2)$$

The deliberate selection of psus with known but unequal probabilities differs greatly from the selection bias discussed in Chapter 1. Many surveys with selection bias do sample with unequal probabilities, but the probabilities of selection are unknown and unestimable, so the survey takers cannot compensate for the unequal probabilities in the weighting. If you take a survey of students by asking students who walk by the library to participate, you certainly are sampling with unequal probabilities—students who use the library frequently are more likely to be asked to participate in the survey, while other students never go by the library at all. But you have no idea how many students in the population are represented by a participant in your survey, and no way of correcting for the unequal probabilities of selection in the estimation. In addition, some students in your target population never walk by the library, so they cannot be included in your sample.

When first presented with the idea of unequal-probability sampling, some people think of it as "unnatural" or "contrived." On the contrary, for many populations with clustering, unequal-probability sampling at the psu level produces a sample that mirrors the population better than an equal-probability sample. Examples of unequal-probability samples are given in Section 6.5. To understand these examples and to design your own samples, it is essential that you have an understanding of probability. We consider with-replacement sampling first, starting with the simple design of selecting one psu. Many large sample surveys are analyzed as though the sampling was done with replacement, even if a without-replacement sample was collected, because the estimators of the variance for with-replacement samples have simple form and require less information. In Section 6.4, we consider unequal-probability sampling without replacement. Notation used in this chapter is defined in Section 5.1.

6.1
Sampling One Primary Sampling Unit

As a special case, suppose we select just one ($n = 1$) of the N psus to be in the sample. The total for psu i is denoted by t_i, and we want to estimate the population total, t. Sampling one psu will demonstrate the ideas of unequal-probability sampling without introducing the complications.

---

[1] We consider two different probabilities in this chapter, because when sampling with unequal probabilities without replacement, as considered in Section 6.4, selecting a unit on the first draw can affect the selection probabilities for other units.

Let's start out by looking at what happens for a situation in which we know the whole population. A town has four supermarkets, ranging in size from 100 square meters (m²) to 1000 m². We want to estimate the total amount of sales in the four stores for last month by sampling just one of the stores. (Of course, this is just an illustration—if we really had only four supermarkets we would probably take a census.) You might expect that a larger store would have more sales than a smaller store, and that the variability in total sales among several 1000-m² stores will be greater than the variability in total sales among several 100-m² stores.

Since we sample only one store, the probability that a store is selected on the first draw (ψ_i) is the same as the probability that the store is included in the sample (π_i). For this example, take

$$\pi_i = \psi_i = P(\text{Store } i \text{ selected})$$

proportional to the size of the store. Since Store A accounts for 1/16 of the total floor area of the four stores, it is sampled with probability 1/16. For illustrative purposes, we know the values of t_i for the whole population:

Store	Size (m²)	ψ_i	t_i (in Thousands)
A	100	$\dfrac{1}{16}$	11
B	200	$\dfrac{2}{16}$	20
C	300	$\dfrac{3}{16}$	24
D	1000	$\dfrac{10}{16}$	245
Total	1600	1	300

We could select a probability sample of size 1 with the probabilities given above by shuffling cards numbered 1 through 16 and choosing one card. If the card's number is 1, choose store A; if 2 or 3, choose B; if 4, 5, or 6, choose C; and if 7 through 16, choose D. Or we could spin once on a spinner like this:

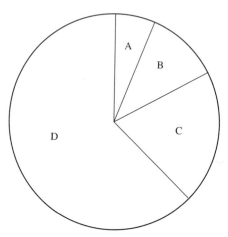

We compensate for the unequal probabilities of selection by also using ψ_i in the estimator. We have already seen such compensation for unequal probabilities in stratified sampling: If we select 10% of the units in stratum 1 and 20% of the units in stratum 2, the sampling weight is 10 for each unit in stratum 1 and 5 for each unit in stratum 2. Here, we select store A with probability 1/16, so store A's sampling weight is 16. If the size of the store is roughly proportional to the total sales for that store, we would expect that store A also has about 1/16 of the total sales and that multiplying store A's sales by 16 would estimate the total sales for all four stores. As always, the sampling weight of unit i is the reciprocal of the probability of selection:

$$w_i = \frac{1}{P(\text{unit } i \text{ in sample})} = \frac{1}{\psi_i}.$$

Thus, our estimator of the population total from an unequal-probability sample of size 1 is

$$\hat{t}_\psi = \sum_{i \in S} w_i t_i = \sum_{i \in S} \frac{t_i}{\psi_i}.$$

Four samples of size 1 are possible from this simple population:

Sample	ψ_i	t_i	$\hat{t}_\psi$	$(\hat{t}_\psi - t)^2$
{A}	$\frac{1}{16}$	11	176	15,376
{B}	$\frac{2}{16}$	20	160	19,600
{C}	$\frac{3}{16}$	24	128	29,584
{D}	$\frac{10}{16}$	245	392	8,464

Using the definition of expected value in (2.3),

$$E[\hat{t}_\psi] = \sum_{\substack{\text{possible} \\ \text{samples } S}} P(S)\hat{t}_{\psi S}$$

$$= \frac{1}{16}(176) + \frac{2}{16}(160) + \frac{3}{16}(128) + \frac{10}{16}(392) = 300.$$

Of course, $\hat{t}_\psi$ will always be unbiased because in general,

$$E[\hat{t}_\psi] = \sum_{i=1}^{N} \psi_i \frac{t_i}{\psi_i} = t. \tag{6.3}$$

The variance of $\hat{t}_\psi$ is

$$V[\hat{t}_\psi] = E[(\hat{t}_\psi - t)^2]$$

$$= \sum_{\substack{\text{possible} \\ \text{samples } \mathcal{S}}} P(\mathcal{S})(\hat{t}_{\psi\mathcal{S}} - t)^2$$

$$= \sum_{i=1}^{N} \psi_i \left(\frac{t_i}{\psi_i} - t \right)^2. \tag{6.4}$$

For this example,

$$V[\hat{t}_\psi] = \frac{1}{16}(15{,}376) + \frac{2}{16}(19{,}600) + \frac{3}{16}(29{,}584) + \frac{10}{16}(8{,}464) = 14{,}248.$$

Compare these results to those from an SRS of size 1, in which the probability of selecting each unit is $\psi_i = 1/4$, so $1/\psi_i = 4 = N$. Note that if all of the probabilities of selection are equal, as in simple random sampling, $1/\psi_i$ always equals N. For the SRS design:

Sample	ψ_i	t_i	$\hat{t}_\psi$	$(\hat{t}_\psi - t)^2$
{A}	$\frac{1}{4}$	11	44	65,536
{B}	$\frac{1}{4}$	20	80	48,400
{C}	$\frac{1}{4}$	24	96	41,616
{D}	$\frac{1}{4}$	245	980	462,400

As always, $\hat{t}_{\text{SRS}}$ is unbiased and thus has expectation 300, but for this example the SRS variance is much larger than the variance from the unequal-probability design:

$$V[\hat{t}_{\text{SRS}}] = \frac{1}{4}(65{,}536) + \frac{1}{4}(48{,}400) + \frac{1}{4}(41{,}616) + \frac{1}{4}(462{,}400) = 154{,}488.$$

The variance from the unequal-probability design, 14,248, is much smaller because the design uses auxiliary information: We expect the store size to be related to the sales, and use that information in the sample design.

We believe that t_i is correlated to the size of the store, which is known. Since Store D accounts for 10/16 of the total floor area of supermarkets, it is reasonable to believe that Store D will account for about 10/16 of the total sales as well. Thus, if store D is chosen and is believed to account for about 10/16 of the total sales, we would have a good estimate of total sales by multiplying Store D's sales by 16/10.

What if Store D only accounts for 4/16 of the total sales? Then the unequal-probability estimator $\hat{t}_\psi$ will still be unbiased over repeated sampling, but it will have a large variance (see Exercise 3). The method still works mathematically, but is not as efficient as if t_i is roughly proportional to ψ_i.

Sampling only one psu is not as unusual as you might think. Many large complex surveys are so highly stratified that each stratum contains only a few psus. A large number of strata is used to increase the precision of the survey estimates. In such a survey, it may be perfectly reasonable to want to select only one psu from each stratum. But, with only one psu per stratum in the sample, we do not have an estimate of the variability between psus within a stratum. When large survey organizations sample only one psu per stratum, they often divide the psus into pseudo-psus for variance estimation.

6.2
One-Stage Sampling with Replacement

Now suppose $n > 1$, and we sample *with* replacement. Sampling with replacement means that the selection probabilities do not change after we have drawn the first unit. Let

$$\psi_i = P(\text{select unit } i \text{ on first draw}).$$

If we sample with replacement, then ψ_i is also the probability that unit i is selected on the second draw, or the third draw, or any other given draw.

The idea behind unequal-probability sampling is simple. Draw n psus with replacement. Then estimate the population total, using the estimator from the previous section, separately for each psu drawn. Some psus may be drawn more than once—the estimated population total, calculated using a given psu, is included as many times as the psu is drawn. Since the psus are drawn with replacement, we have n independent estimates of the population total. Estimate the population total t by averaging those n independent estimates of t. The estimated variance is the sample variance of the n independent estimates of t, divided by n.

6.2.1 Selecting Primary Sampling Units

6.2.1.1 The Cumulative-Size Method

There are several ways to sample psus with unequal probabilities. All require that you have a measure of size for all psus in the population. The cumulative-size method extends the method used in the previous section, in which random numbers are generated and psus corresponding to those numbers are included in the sample. For the supermarkets, we drew cards from a deck with cards numbered 1 through 16. If the card's number is 1, choose store A; if 2 or 3, choose B; if 4, 5, or 6, choose C; and if 7 through 16, choose D. To sample with replacement, put the card back after selecting a psu and draw again.

EXAMPLE **6.2** Consider the population of introductory statistics classes at a college shown in Table 6.1. The college has 15 such classes; class i has M_i students, for a total of 647 students in introductory statistics courses. We decide to sample 5 classes with replacement, with probability proportional to M_i, and then collect a questionnaire from each student in the sampled classes. For this example, then, $\psi_i = M_i/647$.

To select the sample, generate five random integers with replacement between 1 and 647. Then the psus to be chosen for the sample are those whose range in the

TABLE 6.1

Population of Introductory Statistics Classes

Class Number	M_i	ψ_i	Cumulative M_i Range	
1	44	0.068006	1	44
2	33	0.051005	45	77
3	26	0.040185	78	103
4	22	0.034003	104	125
5	76	0.117465	126	201
6	63	0.097372	202	264
7	20	0.030912	265	284
8	44	0.068006	285	328
9	54	0.083462	329	382
10	34	0.052550	383	416
11	46	0.071097	417	462
12	24	0.037094	463	486
13	46	0.071097	487	532
14	100	0.154560	533	632
15	15	0.023184	633	647
Total	647	1		

cumulative M_i includes the randomly generated numbers. The set of five random numbers {487, 369, 221, 326, 282} results in the sample of units {13, 9, 6, 8, 7}. The cumulative-size method allows the same unit to appear more than once: the five random numbers {553, 082, 245, 594, 150} leads to the sample {14, 3, 6, 14, 5}—psu 14 is then included twice in the data. SAS code for selecting a sample from this population is given on the website. ∎

Of course, we can take an unequal-probability sample when the ψ_i's are not proportional to the M_i's: Simply form a cumulative ψ_i range instead, and sample uniform random numbers between 0 and 1. This variation of the method is discussed in Exercise 2.

Systematic sampling is often used to select psus in large complex samples, rather than generating random numbers with replacement. Systematic sampling usually gives a sample without replacement, but in large populations sampling without replacement and sampling with replacement are very similar, as the probability that a unit will be selected twice is small. To sample psus systematically, list the population elements for the first psu in the sample, followed by the elements for the second psu, and so on. Then take a systematic sample of the elements. The psus to be included in the sample are those in which at least one element is in the systematic sample of elements. The larger the psu, the higher the probability it will be in the sample.

The statistics classes have a total of 647 students. To take a (roughly, because 647 is not a multiple of 5) systematic sample, choose a random number k between 1 and 129 and select the psu containing student k, the psu containing student $129 + k$, the psu containing student $2(129) + k$, and so on. Suppose the random number we select as a start value is 112. Then the systematic sample of elements results in the following psus being chosen:

Number in Systematic Sample	psu Chosen
112	4
241	6
370	9
499	13
628	14

Larger classes (psus) have a higher chance of being in the sample because it is more likely that a multiple of the random number chosen will be one of the numbered elements in a large psu. Systematic sampling does not give us a true random sample with replacement, though, because it is impossible for classes with 129 or fewer students to occur in the sample more than once, and classes with more than 129 students are sampled with probability 1. In many populations, however, it is much easier to implement than methods that do give a random sample. If the psus are arranged geographically, taking a systematic sample may force the selected psus to be spread out over more of the region, and may give better results than a random sample with replacement.

6.2.1.2   Lahiri's Method

Lahiri's (1951) method may be more tractable than the cumulative-size method when the number of psus is large. It is an example of a *rejective* method, because you generate pairs of random numbers to select psus and then reject some of them if the psu size is too small. Let N = number of psus in population and max$\{M_i\}$ = maximum psu size. You will show that Lahiri's method produces a with-replacement sample with the desired probabilities in Exercise 15.

1   Draw a random number between 1 and N. This indicates which psu you are considering.

2   Draw a random number between 1 and max$\{M_i\}$. If this random number is less than or equal to M_i, then include psu i in the sample; otherwise go back to step 1.

3   Repeat until desired sample size is obtained.

EXAMPLE **6.3**   Let's use Lahiri's method for the classes in Example 6.2. The largest class has max$\{M_i\} = 100$ students, so we generate pairs of random integers, the first between 1 and 15, the second between 1 and 100, until the sample has five psus (Table 6.2). The psus to be sampled are $\{12, 14, 14, 5, 1\}$. ∎

TABLE 6.2

Lahiri's Method, for Example 6.3

First Random Number (psu i)	Second Random Number	M_i	Action
12	6	24	$6 < 24$; include psu 12 in sample
14	24	100	Include in sample
1	65	44	$65 > 44$; discard pair of numbers and try again
7	84	20	$84 > 20$; try again
10	49	34	Try again
14	47	100	Include
15	43	15	Try again
5	24	76	Include
11	87	46	Try again
1	36	44	Include

6.2.2 Theory of Estimation

Because we are sampling with replacement, the sample may contain the same psu more than once. Let $\mathcal{R}$ denote the set of n units in the sample, including the repeats. For Example 6.3, $\mathcal{R} = \{12, 14, 14, 5, 1\}$; unit 14 is included twice in $\mathcal{R}$. We saw in Section 6.1 that for a sample of size 1, $u_i = t_i/\psi_i$ is an unbiased estimator of the population total t. When we sample n psus with replacement, we have n independent estimators of t, so we average them:

$$\hat{t}_\psi = \frac{1}{n} \sum_{i \in \mathcal{R}} \frac{t_i}{\psi_i} = \frac{1}{n} \sum_{i \in \mathcal{R}} u_i = \bar{u}. \tag{6.5}$$

We estimate $V(\hat{t}_\psi)$ by

$$\hat{V}(\hat{t}_\psi) = \frac{s_u^2}{n} = \frac{1}{n}\frac{1}{n-1} \sum_{i \in \mathcal{R}} (u_i - \bar{u})^2 = \frac{1}{n}\frac{1}{n-1} \sum_{i \in \mathcal{R}} \left(\frac{t_i}{\psi_i} - \hat{t}_\psi\right)^2. \tag{6.6}$$

The estimator $\hat{t}_\psi$ in (6.5) is often referred to as the Hansen–Hurwitz (1943) estimator.

Equation (6.6) is the estimated variance of the average $\bar{u}$ from a simple random sample with replacement. Where are the unequal probabilities in the variance estimator? To prove that $\hat{t}_\psi$ and $\hat{V}(\hat{t}_\psi)$ are unbiased estimators of t and $V(\hat{t}_\psi)$, respectively, we need random variables to keep track of which psus occur multiple times in the sample. Define

$$Q_i = \text{number of times unit } i \text{ occurs in the sample};$$

Q_i is a with-replacement analogue of the random variable Z_i used to indicate sample inclusion for without-replacement sampling in (2.27). Then, $\hat{t}_\psi$ is the average of all

t_i/ψ_i for units chosen to be in the sample, including each unit as many times as it appears in the sample:

$$\hat{t}_\psi = \frac{1}{n}\sum_{i\in\mathcal{R}}\frac{t_i}{\psi_i} = \frac{1}{n}\sum_{i=1}^{N}Q_i\frac{t_i}{\psi_i}. \tag{6.7}$$

If a unit appears k times in the sample, it is counted k times in the estimator. Note that $\sum_{i=1}^{N}Q_i = n$ and $E[Q_i] = n\psi_i$ (see Exercise 16), so $\hat{t}_\psi$ is an unbiased estimator of t.

To calculate the variance, note that the estimator in (6.7) is the average of n independent observations, each with variance $\sum_{i=1}^{N}\psi_i(t_i/\psi_i - t)^2$ [from (6.4)], so

$$V(\hat{t}_\psi) = \frac{1}{n}\sum_{i=1}^{N}\psi_i\left(\frac{t_i}{\psi_i} - t\right)^2. \tag{6.8}$$

To show that the variance estimator in (6.6) is unbiased for $V(\hat{t}_\psi)$, we write it in terms of the random variables Q_i:

$$\hat{V}(\hat{t}_\psi) = \frac{1}{n}\frac{1}{n-1}\sum_{i\in\mathcal{R}}\left(\frac{t_i}{\psi_i} - \hat{t}_\psi\right)^2 = \frac{1}{n}\frac{1}{n-1}\sum_{i=1}^{N}Q_i\left(\frac{t_i}{\psi_i} - \hat{t}_\psi\right)^2. \tag{6.9}$$

Equation (6.8) involves a weighted average of the N values of $(t_i/\psi_i - t)^2$, weighted by the unequal selection probabilities ψ_i. In taking the sample, we have already used the unequal probabilities—they appear in the random variables Q_i in (6.7). The ith psu appears Q_i times in the with-replacement sample. Because the n units are selected independently, $E[Q_i] = n\psi_i$, so including the squared deviation $(t_i/\psi_i - \hat{t}_\psi)^2$ a total of Q_i times in the variance estimator causes (6.9) to be an unbiased estimator of the variance in (6.8):

$$E[\hat{V}(\hat{t}_\psi)] = \frac{1}{n(n-1)}\sum_{i=1}^{N}E\left[Q_i\left(\frac{t_i}{\psi_i} - \hat{t}_\psi\right)^2\right]$$

$$= \frac{1}{n(n-1)}E\left[\sum_{i=1}^{N}Q_i\left(\frac{t_i}{\psi_i} - t + t - \hat{t}_\psi\right)^2\right]$$

$$= \frac{1}{n(n-1)}E\left[\sum_{i=1}^{N}Q_i\left(\frac{t_i}{\psi_i} - t\right)^2 + \sum_{i=1}^{N}Q_i(\hat{t}_\psi - t)^2 - 2\sum_{i=1}^{N}Q_i\left(\frac{t_i}{\psi_i} - t\right)(\hat{t}_\psi - t)\right]$$

$$= \frac{1}{n(n-1)}E\left[\sum_{i=1}^{N}Q_i\left(\frac{t_i}{\psi_i} - t\right)^2 + n(\hat{t}_\psi - t)^2 - 2n(\hat{t}_\psi - t)^2\right]$$

$$= \frac{1}{n(n-1)}\left[\sum_{i=1}^{N}n\psi_i\left(\frac{t_i}{\psi_i} - t\right)^2 - nV(\hat{t}_\psi)\right]$$

$$= V(\hat{t}_\psi).$$

In line 4 of the argument, we use the facts that $\sum_{i=1}^{N} Q_i = n$ and $\sum_{i=1}^{N} Q_i t_i / \psi_i = n\hat{t}_\psi$. In Exercise 7, you will show that the variance estimator for simple random sampling with replacement is a special case of (6.9).

Warning: If N is small or some of the ψ_i's are unusually large, it is possible that the sample will consist of one psu sampled n times. In that case, the estimated variance is zero; it is better to use sampling without replacement (see Section 6.4) if this may occur.

We estimate the population mean $\bar{y}_U$ by

$$\hat{\bar{y}}_\psi = \frac{\hat{t}_\psi}{\hat{M}_{0\psi}}, \tag{6.10}$$

where

$$\hat{M}_{0\psi} = \frac{1}{n} \sum_{i \in \mathcal{R}} \frac{M_i}{\psi_i} \tag{6.11}$$

estimates the total number of elements in the population. In (6.10), $\hat{\bar{y}}_\psi$ is a ratio; using results in Chapter 4, we calculate the residuals $t_i/\psi_i - \hat{\bar{y}}_\psi M_i/\psi_i$ to estimate the variance:

$$\hat{V}(\hat{\bar{y}}_\psi) = \frac{1}{(\hat{M}_{0\psi})^2} \frac{1}{n} \frac{1}{n-1} \sum_{i \in \mathcal{R}} \left(\frac{t_i}{\psi_i} - \frac{\hat{\bar{y}}_\psi M_i}{\psi_i} \right)^2. \tag{6.12}$$

Note that (6.12) is an estimated variance of the same form as (6.6) with the values $(t_i - \hat{\bar{y}}_\psi M_i)/\hat{M}_{0\psi}$ substituted for t_i.

EXAMPLE 6.4   For the situation in Example 6.3, suppose we sample the psus selected by Lahiri's method, {12, 14, 14, 5, 1}. The response t_i is the total number of hours all students in class i spent studying statistics last week, with the following data:

Class	ψ_i	t_i	$\dfrac{t_i}{\psi_i}$
12	$\dfrac{24}{647}$	75	2021.875
14	$\dfrac{100}{647}$	203	1313.410
14	$\dfrac{100}{647}$	203	1313.410
5	$\dfrac{76}{647}$	191	1626.013
1	$\dfrac{44}{647}$	168	2470.364

The numbers in the last column of the table are the estimates of t that would be obtained if that psu were the only one selected in a sample of size 1. The population total is estimated by averaging the five values of t_i/ψ_i, using (6.5):

$$\hat{t}_\psi = \frac{2021.875 + 1313.410 + 1313.410 + 1626.013 + 2470.364}{5} = 1749.014.$$

The standard error of $\hat{t}_\psi$ is simply $s/\sqrt{n}$ [see Equation (6.6)], where s is the sample standard deviation of the five numbers in the rightmost column of the table:

$$SE(\hat{t}_\psi) = \frac{1}{\sqrt{5}}\sqrt{\frac{(2021.875 - 1749.014)^2 + \cdots + (2470.364 - 1749.014)^2}{4}}$$

$$= 222.42.$$

Since $\psi_i = M_i/M_0$ for this sample, we have $\hat{M}_0 = M_0 = 647$. The average amount of time a student spent studying statistics is estimated as

$$\hat{\bar{y}}_\psi = \frac{1749.014}{647} = 2.70$$

hours. For this example, with $\psi_i = M_i/M_0$, (6.12) simplifies to

$$\hat{V}(\hat{\bar{y}}_\psi) = \frac{1}{(M_0)^2}\frac{1}{n}\frac{1}{n-1}\sum_{i\in\mathcal{R}}\left(\frac{t_i}{\psi_i} - \frac{\hat{t}_\psi M_i}{M_0\psi_i}\right)^2 = \frac{\hat{V}(\hat{t}_\psi)}{M_0^2},$$

so the standard error of $\hat{\bar{y}}_\psi$ is 222.42/647 = 0.34 hours. [In other examples, if $\hat{M}_0$ can vary from sample to sample, this simplification of (6.12) does not occur.] SAS code for finding $\hat{t}_\psi$ and $\hat{\bar{y}}_\psi$ is provided on the website. ∎

6.2.3 Designing the Selection Probabilities

We would like to choose the ψ_i's so that the variance of $\hat{t}_\psi$ is as small as possible. Ideally, we would use $\psi_i = t_i/t$ (then $\hat{t}_\psi = t$ for all samples and $V[\hat{t}_\psi] = 0$), so if t_i is the annual income of the ith household, ψ_i would be the proportion of total income in the population that came from the ith household. But of course, the t_i's are unknown until sampled. Even if the income were known before the survey was taken, we are often interested in more than one quantity; using income for designing the probabilities of selection may not work well for estimating other quantities.

Because many totals in a psu are related to the number of elements in a psu, we often take ψ_i to be the proportion of elements in psu i or the relative size of psu i. Then, a large psu has a greater chance of being in the sample than a small psu. With M_i the number of elements in the ith psu and $M_0 = \sum_{i=1}^N M_i$ the number of elements in the population, we take $\psi_i = M_i/M_0$. With this choice of the probabilities ψ_i, we have **probability proportional to size** (pps) sampling. We used pps sampling in Example 6.2.

Then for one-stage pps sampling, $t_i/\psi_i = t_i M_0/M_i = M_0\bar{y}_i$, so $\hat{t}_\psi = \frac{1}{n}\sum_{i\in\mathcal{R}} M_0\bar{y}_i$

and $\hat{\bar{y}}_\psi = \frac{1}{n}\sum_{i\in\mathcal{R}}\bar{y}_i$. With $\psi_i = M_i/M_0$, $\hat{\bar{y}}_\psi$ is the average of the sampled psu means.

Also, for $\psi_i = M_i/M_0$, (6.11) implies that $\hat{M}_{0\psi} = M_0$ for every possible sample, so

FIGURE   6.1

Selected plots for pps sample estimating the total number of physicians in the United States. (a) Plot of t_i versus ψ_i; there is a strong linear relationship between the variables, which indicates that pps sampling increases efficiency. The unusual observation (marked by the '+') is New York County, New York. (b) Histogram of the 100 values of t_i/ψ_i. Each value estimates t.

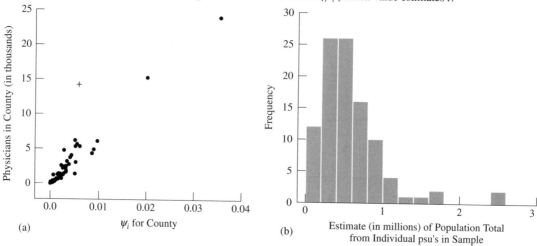

(a) ψ_i for County

(b)   Estimate (in millions) of Population Total from Individual psu's in Sample

from (6.12), $\hat{V}(\hat{\bar{y}}_\psi) = \dfrac{1}{n}\dfrac{1}{n-1}\displaystyle\sum_{i\in\mathcal{R}}(\bar{y}_i - \hat{\bar{y}}_\psi)^2$. Note that $\hat{V}(\hat{\bar{y}}_\psi)$ is of the form s^2/n, where s^2 is the sample variance of the psu means $\bar{y}_i$.

All of the work in pps sampling has been done in the sampling design itself. The pps estimates can be calculated simply by treating the $\bar{y}_i$'s as individual observations, and finding their mean and sample variance. In practice, however, there are usually some deviations from a strict pps scheme, so you should use (6.5) and (6.6) for estimating the population total and its estimated variance.

EXAMPLE   6.5   The file statepop.dat contains data from an unequal-probability sample of 100 counties in the United States. Counties were chosen using the cumulative-size method from the listings in the *County and City Data Book* (U.S. Census Bureau, 1994) with probabilities proportional to their populations. The total population for all counties is $M_0 = \sum_{i=1}^{N} M_i = 255{,}077{,}536$. Sampling was done with replacement, so very large counties occur multiple times in the sample: Los Angeles County, with the largest population in the United States, occurs four times.

One of the quantities recorded for each county was the number of physicians in the county. You would expect larger counties to have more physicians, so pps sampling should work well for estimating the total number of physicians in the United States.

You must be careful in plotting data from an unequal-probability sample, as you need to consider the unequal probabilities when interpreting the plots. A plot of t_i versus ψ_i (Figure 6.1a) tells the efficiency of the unequal-probability design: The design works well when the plot shows positive correlation. A histogram of t_i in a pps sample will not give a representative view of the population of psus, as psus with large ψ_i's are overrepresented in the sample. A histogram of t_i/ψ_i, however, may give

TABLE 6.3

Sampled Counties in Example 6.5

State	County	Population Size, M_i	ψ_i	Number of Physicians, t_i	$\dfrac{t_i}{\psi_i}$
AL	Wilcox	13,672	0.00005360	4	74,627.72
AZ	Maricopa	2,209,567	0.00866233	4320	498,710.81
AZ	Maricopa	2,209,567	0.00866233	4320	498,710.81
AZ	Pinal	120,786	0.00047353	61	128,820.64
AR	Garland	76,100	0.00029834	131	439,095.36
AR	Mississippi	55,060	0.00021586	48	222,370.54
CA	Contra Costa	840,585	0.00329541	1761	534,379.68
⋮	⋮	⋮	⋮	⋮	⋮
VA	Chesterfield	225,225	0.00088297	181	204,990.72
WA	King	1,557,537	0.00610613	5280	864,704.59
WI	Lincoln	27,822	0.00010907	28	256,709.47
WI	Waukesha	320,306	0.00125572	687	547,096.42
		average			570,304.30
		std. dev.			414,012.30

an idea of the spread involved in the population estimates, and may help you identify unusual psus (Figure 6.1b).

The sample was chosen using the cumulative-size method; Table 6.3 shows the sampled counties arranged alphabetically by state. The ψ_i's were calculated using $\psi_i = M_i/M_0$. The average of the t_i/ψ_i column is 570,304.3, the estimated total number of physicians in the United States. The standard error of the estimate is $414{,}012.3/\sqrt{100} = 41{,}401$. For comparison, the *County and City Data Book* lists a total of 532,638 physicians in the United States; a 95% CI using our estimate includes the true value.

These estimates can be found using the SAS code on the website. Partial output is given below:

```
        Data Summary

Number of Observations          100
Sum of Weights             2450.71956

                    Statistics

                                Std Error
Variable      N        Mean     of Mean      95% CL for Mean
------------------------------------------------------------
physicns     100   232.708918   48.859302   135.761463 329.656372
------------------------------------------------------------
```

```
                              Statistics

Variable              Sum           Std Dev       95% CL for Sum
- - - - - - - - - - - - - - - - - - - - - - - - - - - - - - - - - - - -
physicns           570304            41401      488155.273 652453.317
- - - - - - - - - - - - - - - - - - - - - - - - - - - - - - - - - - - -
                                                                      ■
```

What if you do not know the value of M_i for each psu in the population? In that case, you may know the value of a quantity that is related to M_i. If sampling fish, you may not know the number of fish in a haul but you may know the total weight of fish in a haul. You can then use x_i = (total weight of fish in haul i) to set the selection probability for haul i as $\psi_i = x_i/t_x$. Since x_i/t_x is not exactly the same as M_i/M_0, you then must use (6.5) and (6.6) for estimating the population total of y and its standard error.

6.2.4   Weights in Unequal-Probability Sampling with Replacement

As in other types of sampling, we can estimate the population total t using weights. In without-replacement sampling, we use the reciprocal of the inclusion probability ($= 1/E[Z_i]$) as the weight for a unit; $E[Z_i]$ is the expected number of times unit i appears in the sample (expected number of "hits"). In with-replacement sampling, we use the first-stage weight

$$w_i = \frac{1}{\text{expected number of hits}} = \frac{1}{E[Q_i]} = \frac{1}{n\psi_i}.$$

With this choice of weight, we have, for $\hat{t}_\psi$ in Equation (6.5),

$$\hat{t}_\psi = \sum_{i \in \mathcal{R}} w_i t_i.$$

In one-stage cluster sampling with replacement, we observe all of the M_i ssus every time psu i is selected, so we define

$$w_{ij} = w_i = \frac{1}{n\psi_i}.$$

Then, in terms of the elements,

$$\hat{t}_\psi = \sum_{i \in \mathcal{R}} \sum_{j=1}^{M_i} w_{ij} y_{ij}$$

and

$$\hat{\bar{y}}_\psi = \frac{\displaystyle\sum_{i \in \mathcal{R}} \sum_{j=1}^{M_i} w_{ij} y_{ij}}{\displaystyle\sum_{i \in \mathcal{R}} \sum_{j=1}^{M_i} w_{ij}}.$$

If the selection probabilities ψ_i are unequal, the sample is not self-weighting. In one-stage pps sampling, elements in large psus have smaller weights than elements in small psus.

6.3
Two-Stage Sampling with Replacement

The estimators for two-stage unequal-probability sampling with replacement are almost the same as those for one-stage sampling. Take a sample of psus with replacement, choosing the ith psu with known probability ψ_i. As in one-stage sampling with replacement, Q_i is the number of times psu i occurs in the sample. Then take a probability sample of m_i subunits in the ith psu. Simple random sampling without replacement or systematic sampling is often used to select the subsample, although any probability sampling method may be used.

The only difference between two-stage sampling with replacement and one-stage sampling with replacement is that in two-stage sampling, we must estimate t_i. If psu i is in the sample more than once, there are Q_i estimators of the total for psu i: $\hat{t}_{i1}, \hat{t}_{i2}, \ldots, \hat{t}_{iQ_i}$.

The subsampling procedure needs to meet two requirements:

1   Whenever psu i is selected to be in the sample, the same subsampling design is used to select secondary sampling units (ssus) from that psu. Different subsamples from the same psu, though, must be sampled independently. Thus, if you decide before sampling that you will take an SRS of size 5 from psu 42 if it is selected, every time psu 42 appears in the sample you must generate a different set of random numbers to select 5 of the ssus in psu 42. If you just take one subsample of size 5, and use it more than once for psu 42, you do not have independent subsamples and (6.14) will not be an unbiased estimator of the variance.

2   The jth subsample taken from psu i (for $j = 1, \ldots, Q_i$) is selected in such a way that $E[\hat{t}_{ij}] = t_i$. Because the same procedure is used each time psu i is selected for the sample, we can define $V[\hat{t}_{ij}] = V_i$ for all j.

The estimators from one-stage unequal sampling with replacement are modified slightly to allow for different subsamples in psus that are selected more than once:

$$\hat{t}_\psi = \frac{1}{n} \sum_{i=1}^{N} \sum_{j=1}^{Q_i} \frac{\hat{t}_{ij}}{\psi_i}. \tag{6.13}$$

$$\hat{V}(\hat{t}_\psi) = \frac{1}{n} \frac{1}{n-1} \sum_{i=1}^{N} \sum_{j=1}^{Q_i} \left(\frac{\hat{t}_{ij}}{\psi_i} - \hat{t}_\psi \right)^2. \tag{6.14}$$

In Exercise 16, you will show that (6.14) is an unbiased estimator of the variance $V(\hat{t}_\psi)$, given in (6.46). Because sampling is with replacement, and hence it is possible to have more than one subsample from a given psu, the variance estimator captures both parts of the variance: the part due to the variability among psus, and the part that arises because t_i is estimated from a subsample rather than observed. When the psus are sampled with replacement, and when an independent subsample is chosen each time a psu is selected, the variance estimator can be calculated in the same way as if the psu totals were measured rather than estimated.

The estimator of the population mean $\bar{y}_U$ has a form similar to (6.10):

$$\hat{\bar{y}}_\psi = \frac{\hat{t}_\psi}{\hat{M}_{0\psi}},$$

where

$$\hat{M}_{0\psi} = \frac{1}{n} \sum_{i \in \mathcal{R}} \frac{M_i}{\psi_i}$$

estimates the total number of elements in the population. The variance estimator again uses the ratio results in (6.12):

$$\hat{V}(\hat{\bar{y}}_\psi) = \frac{1}{(\hat{M}_{0\psi})^2} \frac{1}{n} \frac{1}{n-1} \sum_{i=1}^{N} \sum_{j=1}^{Q_i} \left(\frac{\hat{t}_{ij}}{\psi_i} - \frac{\hat{\bar{y}}_\psi M_i}{\psi_i} \right)^2. \tag{6.15}$$

The weights for the observation units include a factor to reflect the subsampling within each psu. If an SRS of size m_i is taken in psu i, the weight for ssu j in psu i is

$$w_{ij} = \frac{1}{n\psi_i} \frac{M_i}{m_i}. \tag{6.16}$$

In a pps sample, in which the ith psu is selected with probability $\psi_i = M_i/M_0$, the weight for ssu j of psu i is $w_{ij} = \dfrac{M_0}{nM_i} \dfrac{M_i}{m_i} = \dfrac{M_0}{nm_i}$; a pps sample is self-weighting if all m_i's are equal.

In summary, here are the steps for taking a two-stage unequal-probability sample with replacement:

1   Determine the probabilities of selection ψ_i, the number n of psus to be sampled, and the subsampling procedure to be used within each psu. With any method of selecting the psus, we take a probability sample of ssus within the psus: often in two-stage cluster sampling, we take an SRS (without replacement) of elements within the chosen psus.

2   Select n psus with probabilities ψ_i and with replacement. Either the cumulative-size method or Lahiri's method may be used to select the psus for the sample.

3   Use the procedure determined in step 1 to select subsamples from the psus chosen. If a psu occurs in the sample more than once, independent subsamples are used for each replicate.

4   Estimate the population total t from each psu in the sample as though it were the only one selected. The result is n estimates of the form $\hat{t}_{ij}/\psi_i$.

5   $\hat{t}_\psi$ is the average of the n estimates in step 4. Alternatively, calculate $\hat{t}_\psi = \sum_{i \in \mathcal{R}} \sum_{i \in \mathcal{S}_i} w_{ij} y_{ij}$.

6   $\text{SE}(\hat{t}_\psi) = (1/\sqrt{n})$ (sample standard deviation of the n estimates in step 4).

TABLE 6.4

Spreadsheet for Calculations in Example 6.6

Class	M_i	ψ_i	y_{ij}	$\bar{y}_i$	$\hat{t}_i$	$\hat{t}_i/\psi_i$
12	24	0.0371	2, 3, 2.5, 3, 1.5	2.4	57.6	1552.8
14	100	0.1546	2.5, 2, 3, 0, 0.5	1.6	160.0	1035.2
14	100	0.1546	3, 0.5, 1.5, 2, 3	2.0	200.0	1294.0
5	76	0.1175	1, 2.5, 3, 5, 2.5	2.8	212.8	1811.6
1	44	0.0680	4, 4.5, 3, 2, 5	3.7	162.8	2393.9
			average			1617.5
			std. dev.			521.628

EXAMPLE 6.6   Let's return to the situation in Example 6.4. Now suppose we subsample five students in each class rather than observing t_i. The estimation process is almost the same as in Example 6.4. The response y_{ij} is the total number of hours student j in class i spent studying statistics last week (Table 6.4). Note that class 14 appears twice in the sample; each time it appears, a different subsample is collected.

Thus, $\hat{t}_\psi = 1617.5$ and $SE(\hat{t}_\psi) = 521.628/\sqrt{5} = 233.28$. From this sample, the average amount of time a student spent studying statistics is

$$\hat{\bar{y}}_\psi = \frac{1617.5}{647} = 2.5$$

hours with standard error $233.28/647 = 0.36$ hour.

Here is SAS output finding these estimates (the SAS code is on the website). Note that the sum of the weights is 647, which is the number of students (the value is exact here, since pps sampling was used; in general, the sum of the weights will estimate the number of elements).

```
              Data Summary

Number of Clusters                    5
Number of Observations               25
Sum of Weights                      647

                  Statistics

                                   Std Error
Variable      N        Mean        of Mean       95% CL for Mean
-----------------------------------------------------------------------
hours        25     2.500000     0.360555    1.49893848 3.50106152
-----------------------------------------------------------------------


                  Statistics

Variable       Sum        Std Dev        95% CL for Sum
-----------------------------------------------------------------------
hours      1617.500000   233.279168   969.813197 2265.18680
-----------------------------------------------------------------------
```

Classes were selected with probability proportional to number of students in the class, so $\psi_i = M_i/M_0$. Subsampling the same number of students in each class results in a self-weighting sample, with each student having weight

$$w_{ij} = \frac{M_0}{nM_i}\frac{M_i}{5} = \frac{647}{(5)(5)} = 25.88.$$

The population total is equivalently estimated as

$$25.88(2 + 3 + 2.5 + \cdots + 3 + 2 + 5) = 1617.5. \quad \blacksquare$$

EXAMPLE 6.7   Let's see what happens if we use unequal-probability sampling on the puppy homes considered in Example 5.8. Take ψ_i proportional to the number of puppies in the home, so that Puppy Palace with 30 puppies is sampled with probability 3/4 and Dog's Life with 10 puppies is sampled with probability 1/4. As before, once a puppy home is chosen, take an SRS of two puppies in the home. Then if Puppy Palace is selected, $\hat{t}_\psi = \hat{t}_{PP}/(3/4) = (30)(4)/(3/4) = 160$. If Dog's Life is chosen, $\hat{t}_\psi = \hat{t}_{DL}/(1/4) = (10)(4)/(1/4) = 160$. Thus, either possible sample results in an estimated average of $\hat{\bar{y}}_\psi = 160/40 = 4$ legs per puppy, and the variance of the estimator is zero. $\quad \blacksquare$

Sampling with replacement has the advantage that it is very easy to select the sample and to obtain estimates of the population total and its variance. If N is small, however, as occurs in many highly stratified complex surveys with few clusters in each stratum, sampling with replacement may be less efficient than sampling without replacement. The next section discusses advantages and challenges of sampling without replacement.

6.4
Unequal-Probability Sampling Without Replacement

Generally, sampling with replacement is less efficient than sampling without replacement; with-replacement sampling is introduced first because of the ease in selecting and analyzing samples. Nevertheless, in large surveys with many small strata, the inefficiencies may wipe out the gains in convenience. Much research has been done on unequal-probability sampling without replacement; the theory is more complicated because the probability that a unit is selected is different for the first unit chosen than for the second, third, and subsequent units. When you understand the probabilistic arguments involved, however, you can find the properties of any sampling scheme.

EXAMPLE 6.8   The supermarket example from Section 6.1 can be used to illustrate some of the features of unequal-probability sampling with replacement. Here is the population again:

Store	Size (m^2)	ψ_i	t_i (in Thousands)
A	100	$\dfrac{1}{16}$	11
B	200	$\dfrac{2}{16}$	20
C	300	$\dfrac{3}{16}$	24
D	1000	$\dfrac{10}{16}$	245
Total	1600	1	300

Let's select two psus without replacement and with unequal probabilities. As in Sections 6.1 to 6.3, let

$$\psi_i = P(\text{Select unit } i \text{ on first draw}).$$

Since we are sampling without replacement, though, the probability that unit j is selected on the second draw depends on which unit was selected on the first draw.

One way to select the units with unequal probabilities is to use ψ_i as the probability of selecting unit i on the first draw, and then adjust the probabilities of selecting the other stores on the second draw. If store A was chosen on the first draw, then for selecting the second store we would spin the wheel on page 222 while blocking out the section for store A, or shuffle the deck and redeal without Card 1. Thus,

$$P(\text{store A chosen on first draw}) = \psi_A = \frac{1}{16}$$

and

$$P(\text{B chosen on second draw} \mid \text{A chosen on first draw}) = \frac{\dfrac{2}{16}}{1 - \dfrac{1}{16}} = \frac{\psi_B}{1 - \psi_A}.$$

The denominator is the sum of the ψ_i's for stores B, C, and D. In general,

$P(\text{unit } i \text{ chosen first, unit } k \text{ chosen second})$

$\quad = P(\text{unit } i \text{ chosen first}) \, P(\text{unit } k \text{ chosen second} \mid \text{unit } i \text{ chosen first})$

$\quad = \psi_i \dfrac{\psi_k}{1 - \psi_i}.$

Similarly,

$$P(\text{unit } k \text{ chosen first, unit } i \text{ chosen second}) = \psi_k \frac{\psi_i}{1 - \psi_k}.$$

Note that $P(\text{unit } i \text{ chosen first, unit } k \text{ chosen second})$ is not the same as $P(\text{unit } k \text{ chosen first, unit } i \text{ chosen second})$: The order of selection makes a difference! By adding the probabilities of the two choices, though, we can find the probability that a sample of

TABLE   6.5

Inclusion probabilities (π_i) and joint inclusion probabilities (π_{ik}) for samples of size 2 that could be selected using the method in Example 6.8. The entries of the table are the π_{ik}'s for each pair of stores (rounded to four decimal places); the margins give the π_i's for the four stores

		Store k			
		A	B	C	D	π_i
	A	—	0.0173	0.0269	0.1458	0.1900
Store i	B	0.0173	—	0.0556	0.2976	0.3705
	C	0.0269	0.0556	—	0.4567	0.5393
	D	0.1458	0.2976	0.4567	—	0.9002
	π_k	0.1900	0.3705	0.5393	0.9002	2.0000

size 2 consists of psus i and k:

$$\text{For } n = 2, \quad P(\text{units } i \text{ and } k \text{ in sample}) = \pi_{ik} = \psi_i \frac{\psi_k}{1 - \psi_i} + \psi_k \frac{\psi_i}{1 - \psi_k}.$$

The probability that psu i is in the sample is then

$$\pi_i = \sum_{S:i\in S} P(S).$$

Table 6.5 gives the π_i's and π_{ik}'s for the supermarkets.   ∎

6.4.1   The Horvitz–Thompson Estimator for One-Stage Sampling

Assume we have a without-replacement sample of n psus, and we know the **inclusion probability**

$$\pi_i = P(\text{unit } i \text{ in sample})$$

and the **joint inclusion probability**

$$\pi_{ik} = P(\text{units } i \text{ and } k \text{ are both in the sample}).$$

The inclusion probability π_i can be calculated as the sum of the probabilities of all samples containing the ith unit and has the property that

$$\sum_{i=1}^{N} \pi_i = n. \tag{6.17}$$

For the π_{ik}'s, as shown in Theorem 6.1 of Section 6.6,

$$\sum_{\substack{k=1 \\ k\neq i}}^{N} \pi_{ik} = (n-1)\pi_i. \tag{6.18}$$

Because the inclusion probabilities sum to n, we can think of π_i/n as the "average probability" that a unit will be selected on one of the draws. Recall that for one-stage sampling with replacement, $\hat{t}_\psi$ is the average of the values of t_i/ψ_i for psus in the sample. But when samples are drawn without replacement, the probabilities of selection depend on what was drawn before. Instead of dividing the total t_i from psu i by ψ_i, we divide by the *average* probability of selecting that unit in a draw, π_i/n. We then have the **Horvitz–Thompson (HT) estimator** of the population total (Horvitz and Thompson, 1952):

$$\hat{t}_{\text{HT}} = \sum_{i \in S} \frac{t_i}{\pi_i} = \sum_{i=1}^{N} Z_i \frac{t_i}{\pi_i}, \tag{6.19}$$

where $Z_i = 1$ if psu i is in the sample, and 0 otherwise.

The Horvitz–Thompson estimator is shown to be unbiased for t by using Theorem 6.2 in Section 6.6. Here, $P(Z_i = 1) = \pi_i$, so by (6.38),

$$E[\hat{t}_{\text{HT}}] = \sum_{i=1}^{N} \pi_i \frac{t_i}{\pi_i} = t.$$

We shall show in Section 6.6, using Equations (6.39) through (6.41), that the variance of the Horvitz–Thompson estimator in one-stage sampling is

$$V(\hat{t}_{\text{HT}}) = \sum_{i=1}^{N} \frac{1 - \pi_i}{\pi_i} t_i^2 + \sum_{i=1}^{N} \sum_{k \neq i}^{N} \frac{\pi_{ik} - \pi_i \pi_k}{\pi_i \pi_k} t_i t_k \tag{6.20}$$

$$= \frac{1}{2} \sum_{i=1}^{N} \sum_{\substack{k=1 \\ k \neq i}}^{N} (\pi_i \pi_k - \pi_{ik}) \left(\frac{t_i}{\pi_i} - \frac{t_k}{\pi_k} \right)^2. \tag{6.21}$$

The expression in (6.21) is the Sen-Yates-Grundy (SYG) form of the variance (Sen, 1953; Yates and Grundy, 1953). You can see from (6.21) that the variance of the Horvitz–Thompson estimator is 0 if t_i is proportional to π_i.

The expressions for the variance in (6.20) and (6.21) are algebraically identical (this is shown in Theorem 6.2 of Section 6.6). When the inclusion probabilities π_i or the joint inclusion probabilities π_{ik} are unequal, however, substituting sample quantities into (6.20) or (6.21) leads to different estimators of the variance.

The estimator of the variance starting from (6.20), suggested by Horvitz and Thompson (1952), is

$$\hat{V}_{\text{HT}}(\hat{t}_{\text{HT}}) = \sum_{i \in S} (1 - \pi_i) \frac{t_i^2}{\pi_i^2} + \sum_{i \in S} \sum_{\substack{k \in S \\ k \neq i}} \frac{\pi_{ik} - \pi_i \pi_k}{\pi_{ik}} \frac{t_i}{\pi_i} \frac{t_k}{\pi_k}. \tag{6.22}$$

The SYG estimator, working from (6.21), is

$$\hat{V}_{\text{SYG}}(\hat{t}_{\text{HT}}) = \frac{1}{2} \sum_{i \in S} \sum_{\substack{k \in S \\ k \neq i}} \frac{\pi_i \pi_k - \pi_{ik}}{\pi_{ik}} \left(\frac{t_i}{\pi_i} - \frac{t_k}{\pi_k} \right)^2. \tag{6.23}$$

Theorem 6.4 in Section 6.6 shows that (6.22) and (6.23) are both unbiased estimators of the variance in (6.21). Both require $\pi_{ik} > 0$ for all units in the sample. The SYG form in (6.23) is generally the more stable of the two variance estimators.

EXAMPLE 6.9   Let's look at the Horvitz–Thompson estimator for a sample of 2 supermarkets in Example 6.8 with joint inclusion probabilities given in Table 6.5. We use the draw-by-draw method to select the sample. To select the first psu, we generate a random integer from $\{1, \ldots, 16\}$: the random integer we generate is 12, which tells us that store D is selected on the first draw. We then remove the values $\{7, \ldots, 16\}$ corresponding to store D, and generate a second random integer from $\{1, \ldots, 6\}$; we generate 6, which tells us to select store C on the second draw. The Horvitz–Thompson estimate of the total sales for sample $\{C, D\}$ is then

$$\hat{t}_{HT} = \sum_{i \in S} \frac{t_i}{\pi_i} = \frac{245}{0.9002} + \frac{24}{0.5393} = 316.6639.$$

Since for this example we know the entire population, we can calculate the theoretical variance of $\hat{t}_{HT}$ using (6.21):

$$V(\hat{t}_{HT}) = \frac{1}{2} \sum_{i=1}^{N} \sum_{k \neq i}^{N} (\pi_i \pi_k - \pi_{ik}) \left(\frac{t_i}{\pi_i} - \frac{t_k}{\pi_k} \right)^2 = 4383.6.$$

[We obtain the same value, 4383.6, if we use the equivalent formulation in (6.20).] We have two estimates of the variance from sample $\{C, D\}$: from (6.22),

$$\hat{V}_{HT}(\hat{t}_{HT}) = \frac{(1 - 0.9002)(245)^2}{(0.9002)^2} + \frac{(1 - 0.5393)(24)^2}{(0.5393)^2}$$
$$+ 2 \frac{0.4567 - (0.9002)(0.5393)}{0.4567} \left(\frac{245}{0.9002} \right) \left(\frac{24}{0.5393} \right)$$
$$= 6782.8.$$

The SYG estimate, from (6.23), is

$$\hat{V}_{SYG}[\hat{t}_{HT}] = \frac{(0.9002)(0.5393) - 0.4567}{0.4567} \left(\frac{245}{0.9002} - \frac{24}{0.5393} \right)^2 = 3259.8.$$

Because all values in this population are known, we can examine the estimators for all possible samples selected according to the probabilities in Table 6.5. Results are given in Table 6.6. For three of the possible samples, $\hat{V}_{HT}(\hat{t}_{HT})$ is negative! This is true even though $\hat{V}_{HT}(\hat{t}_{HT})$ and $\hat{V}_{SYG}(\hat{t}_{HT})$ are unbiased estimators of $V_{HT}(\hat{t}_{HT})$; it is easy to check for this example that

$$\sum_{\text{possible samples } S} P(S)\hat{V}_{HT}(\hat{t}_{HT,S}) = \sum_{\text{possible samples } S} P(S)\hat{V}_{SYG}(\hat{t}_{HT,S}) = 4383.6.$$

∎

Example 6.9 demonstrates a problem that can arise in estimating the variance of $\hat{t}_{HT}$: The unbiased estimators in (6.22) or (6.23) can take on negative values in some unequal-probability designs! [See Exercise 24 for a situation in which (6.23) is negative.] In some designs, the estimates of the variance can be widely disparate for

TABLE 6.6

Variance estimates for all possible without-replacement samples of size 2, for the supermarket example

Sample, $\mathcal{S}$	$P(\mathcal{S})$	$\hat{t}_{HT}$	$\hat{V}_{HT}(\hat{t}_{HT})$	$\hat{V}_{SYG}(\hat{t}_{HT})$
{A, B}	0.01726	111.87	−14,691.5	47.1
{A, C}	0.02692	102.39	−10,832.1	502.8
{A, D}	0.14583	330.06	4,659.3	7,939.8
{B, C}	0.05563	98.48	−9,705.1	232.7
{B, D}	0.29762	326.15	5,682.8	5,744.1
{C, D}	0.45673	316.67	6,782.8	3,259.8

different samples. The stability can sometimes be improved by careful choice of the sampling design, but in general, the calculations are cumbersome.

In addition, the estimators in (6.22) and (6.23) can be difficult to use in practice because they require knowledge of the joint inclusion probabilities π_{ik} (see Särndal, 1996). Since π_{ik} appears in the denominator, the joint inclusion probability π_{ik} must be strictly positive for every pair of psus. Public use data sets from large-scale surveys commonly include a variable of weights that can be used to calculate the Horvitz–Thompson estimator. But it is generally impractical to provide the joint inclusion probabilities π_{ik}—this would require an additional $n(n-1)/2$ values to be included in the data set, where n is often large. In addition, for many surveys it is challenging to calculate the joint inclusion probabilities π_{ik}.

An alternative suggested by Durbin (1953), which avoids some of the potential instability and computational complexity, is to pretend the units were selected with replacement and use the with-replacement variance estimator in (6.9) rather than (6.22) or (6.23). The with-replacement variance estimator, setting $\psi_i = \pi_i/n$, is

$$\hat{V}_{WR}(\hat{t}_{HT}) = \frac{1}{n}\frac{1}{n-1}\sum_{i \in \mathcal{S}}\left(\frac{t_i}{\psi_i} - \hat{t}_{HT}\right)^2 = \frac{n}{n-1}\sum_{i \in \mathcal{S}}\left(\frac{t_i}{\pi_i} - \frac{\hat{t}_{HT}}{n}\right)^2. \quad (6.24)$$

The variance estimator in (6.24) is always nonnegative, so you can avoid the potential embarrassment of trying to explain a negative variance estimate. In addition, the with-replacement variance estimator does not require knowledge of the joint inclusion probabilities π_{ik}. If without-replacement sampling is more efficient than with-replacement sampling, the with-replacement variance estimator in (6.24) is expected to overestimate the variance and result in conservative confidence intervals (CIs), but the bias is expected to be small if the sampling fraction n/N is small. The commonly used computer-intensive methods described in Chapter 9 calculate the with-replacement variance.

In general, we recommend using the with-replacement variance estimator in (6.24). When the sampling fraction n/N is large, however, this can overestimate the variance. Some survey software packages will calculate the SYG variance estimate if the user provides the π_{ik}'s. Berger (2004) and Brewer and Donadio (2003) suggest alternatives for estimating $V(\hat{t}_{HT})$ when the joint inclusion probabilities are unknown. These methods are presented in Exercises 29 and 30.

EXAMPLE 6.10   Let's select an unequal-probability sample without replacement of size 15 from the file agpop.dat. In Example 4.2, we used the variable *acres87* as auxiliary information in ratio estimation. We now use it in the sample design, selecting counties with probability proportional to *acres87*. The SAS code used to select and analyze this sample is given on the website. The data for the sample, along with the joint inclusion probabilities, are in file agpps.dat.

The Horvitz–Thompson estimate of the total for *acres92* is

$$\hat{t}_{HT} = \sum_{i \in S} \frac{t_i}{\pi_i} = 992{,}665{,}083,$$

where t_i is the value of *acres92* for county i in the sample. The three variance estimates are: $\hat{V}_{HT}(\hat{t}_{HT}) = 5.31 \times 10^{15}$, $\hat{V}_{SYG}(\hat{t}_{HT}) = 1.22 \times 10^{14}$, and $\hat{V}_{WR}(\hat{t}_{HT}) = 1.33 \times 10^{14}$. Because of the instability of $\hat{V}_{HT}(\hat{t}_{HT})$, we prefer to use either $\hat{V}_{SYG}(\hat{t}_{HT})$ or $\hat{V}_{WR}(\hat{t}_{HT})$ to estimate the variance. For this sample, $\hat{V}_{WR}(\hat{t}_{HT})$ is quite close to the SYG estimate because the sampling fraction n/N is small. Using the SAS code on the website, we obtain $\mathrm{SE}\,(\hat{t}_{HT}) = 11{,}543{,}326 = \sqrt{1.33 \times 10^{14}}$, which is the square root of the with-replacement variance estimate.

Note the gain in efficiency from using unequal-probability sampling. From Example 2.6, an SRS of size 300 gave a standard error of 58,169,381 for the estimated total of *acres92*. The unequal-probability sample has a smaller standard error even though the sample size is only 15 because of the high correlation between *acres92* and *acres87*. Using the auxiliary information in the variable *acres87* in the design results in a large gain in efficiency. ∎

6.4.2   Selecting the psus

For the supermarkets in Example 6.8, the draw-by-draw selection probabilities ψ_i are proportional to the store sizes. The inclusion probabilities π_i's, however, are not proportional to the sizes of the stores—in fact, they cannot be proportional to the store sizes, because Store D accounts for more than half of the total floor area but cannot be sampled with a probability greater than one. The π_i's that result from this draw-by-draw method due to Yates and Grundy (1953) may or may not be the desired probabilities of inclusion in the sample; you may need to adjust the ψ_i's to obtain a pre-specified set of π_i's. Such adjustments become difficult for large populations and for sample sizes larger than two.

Many methods have been proposed for selecting psus without replacement so that desired inclusion probabilities are attained. Systematic sampling can be used to draw a sample without replacement and is relatively simple to implement (hence its widespread use), but many of the π_{ik}'s for the population are zero. If psus are selected using systematic sampling, you need to use the with-replacement estimator of variance in (6.24), since the without-replacement variance estimators in (6.22) and (6.23) contain π_{ik} in the denominator and hence are undefined. Brewer and Hanif

(1983) present more than 50 methods for selecting without-replacement unequal-probability samples. Most of these methods are for $n = 2$; three of the methods are described in Exercises 25, 27, and 28. Some methods are easier to compute, some are more suitable for specific applications, and some result in a more stable estimator of $V(\hat{t}_{HT})$. Tillé (2006) gives general algorithms for selecting without-replacement unequal-probability samples.

SAS software (PROC SURVEYSELECT) will select samples with unequal probabilities. The website has examples of SAS programs (example0602.sas and ppsselect.sas) that can be used to select without-replacement unequal-probability samples. In Example 6.10, we used a method developed by Hanurav (1967) and Vijayan (1968) to select the sample.

6.4.3 The Horvitz–Thompson Estimator for Two-Stage Sampling

The Horvitz–Thompson estimator for two-stage sampling is similar to the estimator for one-stage sampling in (6.19): We substitute an unbiased estimator $\hat{t}_i$ of the psu total for the unknown value of t_i, obtaining

$$\hat{t}_{HT} = \sum_{i \in S} \frac{\hat{t}_i}{\pi_i} = \sum_{i=1}^{N} Z_i \frac{\hat{t}_i}{\pi_i}, \tag{6.25}$$

where $Z_i = 1$ if psu i is in the sample, and 0 otherwise.

The two-stage Horvitz–Thompson estimator is an unbiased estimator of t as long as $E[\hat{t}_i] = t_i$ for each psu i (see Theorem 6.2 in Section 6.6). We shall show in Section 6.6, using Equations (6.39) through (6.41), that the variance of the Horvitz–Thompson estimator is

$$V(\hat{t}_{HT}) = \sum_{i=1}^{N} \frac{1 - \pi_i}{\pi_i} t_i^2 + \sum_{i=1}^{N} \sum_{k \neq i}^{N} \frac{\pi_{ik} - \pi_i \pi_k}{\pi_i \pi_k} t_i t_k + \sum_{i=1}^{N} \frac{V(\hat{t}_i)}{\pi_i} \tag{6.26}$$

$$= \frac{1}{2} \sum_{i=1}^{N} \sum_{\substack{k=1 \\ k \neq i}}^{N} (\pi_i \pi_k - \pi_{ik}) \left(\frac{t_i}{\pi_i} - \frac{t_k}{\pi_k} \right)^2 + \sum_{i=1}^{N} \frac{V(\hat{t}_i)}{\pi_i}. \tag{6.27}$$

The expression in (6.27) is again the SYG form. The first part of the variance is the same as for one-stage sampling [see (6.20) and (6.21)]. The last term is the additional variability due to estimating the t_i's rather than measuring them exactly.

The Horvitz–Thompson estimator of the variance in two-stage cluster sampling is

$$\hat{V}_{HT}(\hat{t}_{HT}) = \sum_{i \in S} (1 - \pi_i) \frac{\hat{t}_i^2}{\pi_i^2} + \sum_{i \in S} \sum_{\substack{k \in S \\ k \neq i}} \frac{\pi_{ik} - \pi_i \pi_k}{\pi_{ik}} \frac{\hat{t}_i}{\pi_i} \frac{\hat{t}_k}{\pi_k} + \sum_{i \in S} \frac{\hat{V}(\hat{t}_i)}{\pi_i}, \tag{6.28}$$

and the SYG estimator is

$$\hat{V}_{SYG}(\hat{t}_{HT}) = \frac{1}{2} \sum_{i \in S} \sum_{\substack{k \in S \\ k \neq i}} \frac{\pi_i \pi_k - \pi_{ik}}{\pi_{ik}} \left(\frac{\hat{t}_i}{\pi_i} - \frac{\hat{t}_k}{\pi_k} \right)^2 + \sum_{i \in S} \frac{\hat{V}(\hat{t}_i)}{\pi_i}. \tag{6.29}$$

Theorem 6.4 in Section 6.6 shows that both are unbiased estimators of the variance in (6.27); however, just as in one-stage sampling, either can be negative in practice.

For most situations, we recommend using the with-replacement sampling variance estimator:

$$\hat{V}_{WR}(\hat{t}_{HT}) = \frac{1}{n}\frac{1}{n-1}\sum_{i\in S}\left(\frac{n\hat{t}_i}{\pi_i} - \hat{t}_{HT}\right)^2 = \frac{n}{n-1}\sum_{i\in S}\left(\frac{\hat{t}_i}{\pi_i} - \frac{\hat{t}_{HT}}{n}\right)^2. \qquad (6.30)$$

The with-replacement variance estimator for two-stage sampling has exactly the same form as the estimator in (6.24) for one-stage sampling; the only difference is that we substitute the estimator $\hat{t}_i$ for the ith psu population total t_i. We saw in Section 6.3 that the with-replacement variance estimator captures the variability at both stages of sampling. This results in the tremendous practical advantage that the variance estimation method depends only on information at the first-stage level of the design. You do not have to use properties of the subsampling design at all for the variance estimation.

6.4.4 Weights in Unequal-Probability Samples

All without-replacement sampling schemes discussed so far in the book can be considered as special cases of two-stage cluster sampling with (possibly) unequal probabilities. The formulas for unbiased estimation of totals in without-replacement sampling in Chapters 2, 3, 5, and 6 are special cases of (6.25) through (6.29). In Example 6.15, we will derive the formulas in Chapter 5 from the general Horvitz–Thompson results. You will show that the formulas for stratified sampling are a special case of Horvitz–Thompson estimation in Exercise 18.

As in earlier chapters, we can write the Horvitz–Thompson estimator using sampling weights. The first-stage sampling weight for psu i is

$$w_i = \frac{1}{\pi_i}.$$

Thus, the Horvitz–Thompson estimator for the population total is

$$\hat{t}_{HT} = \sum_{i\in S} w_i\hat{t}_i.$$

For a without-replacement probability sample of ssus within psus, we define, using the notation of Särndal et al. (1992),

$$\pi_{j|i} = P(j\text{th ssu in }i\text{th psu included in sample} \mid i\text{th psu is in the sample}).$$

Then,

$$\hat{t}_i = \sum_{j\in S_i} \frac{y_{ij}}{\pi_{j|i}}.$$

The overall probability that ssu j of psu i is included in the sample is $\pi_{j|i}\pi_i$. Thus, we can define the sampling weight for the (i,j)th ssu as

$$w_{ij} = \frac{1}{\pi_{j|i}\pi_i} \qquad (6.31)$$

and the Horvitz–Thompson estimator of the population total as

$$\hat{t}_{HT} = \sum_{i \in S} \sum_{j \in S_i} w_{ij} y_{ij}. \tag{6.32}$$

The population mean is estimated by

$$\hat{\bar{y}}_{HT} = \frac{\displaystyle\sum_{i \in S} \sum_{j \in S_i} w_{ij} y_{ij}}{\displaystyle\sum_{i \in S} \sum_{j \in S_i} w_{ij}}. \tag{6.33}$$

The estimator $\hat{\bar{y}}_{HT}$ is a ratio, so, using the results from Chapter 4, we estimate its variance by forming the residuals from the estimated psu totals. Let

$$\hat{e}_i = \hat{t}_i - \hat{\bar{y}}_{HT} \hat{M}_i,$$

where $\hat{M}_i = \sum_{j \in S_i} (1/\pi_{j|i})$ estimates the number of ssus in psu i. Note that $\hat{e}_i/\pi_i = \sum_{j \in S_i} w_{ij}(y_{ij} - \hat{\bar{y}}_{HT})$ and $\sum_{i \in S} \hat{e}_i/\pi_i = 0$. We then use the with-replacement variance in (6.30), with $\hat{e}_i/\hat{M}_0$ substituted for $\hat{t}_i$, to obtain:

$$\hat{V}_{WR}(\hat{\bar{y}}_{HT}) = \frac{n}{n-1} \sum_{i \in S} \left(\frac{\hat{e}_i}{\hat{M}_0 \pi_i} \right)^2 = \frac{n}{n-1} \sum_{i \in S} \left(\frac{\displaystyle\sum_{j \in S_i} w_{ij}(y_{ij} - \hat{\bar{y}}_{HT})}{\displaystyle\sum_{k \in S} \sum_{j \in S_i} w_{kj}} \right)^2, \tag{6.34}$$

where $\hat{M}_0 = \sum_{i \in S} \hat{M}_i = \sum_{i \in S} \sum_{j \in S_i} w_{ij}$ estimates M_0, the number of ssus in the population. Survey software will calculate these quantities for you.

EXAMPLE 6.11 Let's take a two-stage unequal-probability sample without replacement from the population of statistics classes in Example 6.2. We want the psu inclusion probabilities to be proportional to the class sizes M_i given in Table 6.1. SAS code used to select and analyze the sample is given on the website; the data are in file classpps.dat and in Table 6.7.

We calculate the weight for each student in the sample as

$$w_{ij} = \frac{1}{\pi_i \pi_{j|i}} = \frac{1}{\pi_i (4/M_i)}.$$

Since the same number of students ($m_i = 4$) is selected from each class and since the psu inclusion probabilities π_i are proportional to the class sizes M_i, the sample of students is self-weighting.

The estimated total number of hours spent studying statistics is

$$\hat{t}_{HT} = \sum_{i \in S} \sum_{j \in S_i} w_{ij} y_{ij} = 2232.15.$$

This can also be calculated by $\hat{t}_{HT} = \sum_{i \in S} \hat{t}_i/\pi_i = 2232.15$. Using the with-replacement variance estimate in (6.30),

$$\hat{V}_{WR}(\hat{t}_{HT}) = \frac{n}{n-1} \sum_{i \in S} \left(\frac{\hat{t}_i}{\pi_i} - \frac{\hat{t}_{HT}}{n} \right)^2 = \frac{5}{4} 77{,}749.9 = 97{,}187.4,$$

TABLE   6.7

Data from Two-Stage Sample of Introductory Statistics Classes

Class	M_i	π_i	w_{ij}	y_{ij}	$w_{ij}y_{ij}$	$\hat{t}_i$	$\dfrac{\hat{t}_i}{\pi_i}$	$\left(\dfrac{\hat{t}_i}{\pi_i} - \dfrac{\hat{t}_{HT}}{5}\right)^2$	$\left(\dfrac{\hat{e}_i}{\hat{M}_0\pi_i}\right)^2$
4	22	0.17002	32.35	5	161.750	110.00	646.983	40,222.54	0.09609
4	22	0.17002	32.35	4.5	145.575				
4	22	0.17002	32.35	5.5	177.925				
4	22	0.17002	32.35	5	161.750				
10	34	0.26275	32.35	2	64.700	106.25	404.377	1,768.23	0.00423
10	34	0.26275	32.35	4	129.400				
10	34	0.26275	32.35	3	97.050				
10	34	0.26275	32.35	3.5	113.225				
1	44	0.34003	32.35	5	161.750	154.00	452.901	41.91	0.00010
1	44	0.34003	32.35	3	97.050				
1	44	0.34003	32.35	4	129.400				
1	44	0.34003	32.35	2	64.700				
9	54	0.41731	32.35	3.5	113.225	195.75	469.076	512.96	0.00123
9	54	0.41731	32.35	4	129.400				
9	54	0.41731	32.35	1	32.350				
9	54	0.41731	32.35	6	194.100				
14	100	0.77280	32.35	2	64.700	200.00	258.799	35,204.25	0.08410
14	100	0.77280	32.35	1.5	48.525				
14	100	0.77280	32.35	1.5	48.525				
14	100	0.77280	32.35	3	97.050				
Sum			647.00		2232.15	2232.15		77,749.90	0.18574

giving a standard error of $\sqrt{97{,}187.4} = 311.7$. For this example, since $n = 5$ is large relative to $N = 15$, this standard error is likely an overestimate; in Exercise 14 you will calculate the without-replacement variance estimates in (6.28) and (6.29), as well as an approximation to the without-replacement variance used by SAS software.

We estimate the mean number of hours spent studying statistics by

$$\hat{\bar{y}}_{HT} = \frac{\displaystyle\sum_{i\in S}\sum_{j\in S_i} w_{ij}y_{ij}}{\displaystyle\sum_{i\in S}\sum_{j\in S_i} w_{ij}} = \frac{2232.15}{647} = 3.45.$$

Using (6.34),

$$\hat{V}_{WR}(\hat{\bar{y}}_{HT}) = \frac{n}{n-1}\sum_{i\in S}\left(\frac{e_i}{\hat{M}_0\pi_i}\right)^2 = \frac{5}{4}(0.18574) = 0.23218,$$

so $\text{SE}(\hat{\bar{y}}_{HT}) = \sqrt{0.23218} = 0.482.$ ∎

6.5
Examples of Unequal-Probability Samples

Many sampling situations are well suited for unequal-probability samples. This section gives three examples of sampling designs in common use.

EXAMPLE **6.12** *Random Digit Dialing.*

In telephone surveys, it is important to have a well-defined and efficient procedure to select telephone numbers for the sample. In the early days of telephone surveys, many organizations simply took numbers from the telephone directory. That approach leads to selection bias, however, because telephone numbers that are unlisted or added since publication do not appear in the directory. Modifications of sampling from the directory have been suggested to allow inclusion of unlisted numbers, but most have some difficulties with undercoverage.

Random Digit Dialing Element Sampling. Generating telephone numbers at random from the frame of all possible telephone numbers avoids undercoverage of unlisted numbers. In the United States, telephone numbers consist of

area code + prefix (or exchange) + suffix
(3 digits)   (3 digits)   (4 digits)

Thus a random sample of telephone numbers in the United States can be chosen by randomly selecting a 10-digit number. If the random number chosen does not belong to a household, the number is discarded and a new 10-digit number tried. The procedure is repeated until the desired sample size is obtained.

This method is simple to understand and explain, and, assuming no nonresponse, produces an SRS of telephone numbers from the frame of all possible telephone numbers. In practice, the method can be expensive: Even with the frame of telephone numbers restricted to area codes and prefixes known to be in use, many telephone numbers generated by this method will not belong to a household. Multiple calls to a number may be needed to ascertain whether the number is residential or not.

The Mitofsky Waksberg Method. Mitofsky (1970) and Waksberg (1978) developed a cluster-sampling method for sampling residential telephone numbers. The following description is of the "sampler's utopia" procedure in which everyone answers the phone (see Brick and Tucker, 2007).

First, form the sampling frame of psus. Construct a list of all area codes and prefixes in the area of interest. Form a list of psus by appending each of the numbers 00 to 99 to each possible combination of area code and prefix. The resulting list of psus consists of the set of possible first eight digits for the 10-digit telephone numbers in the population. Each psu in the frame contains the numbers (abc)-def-gh00 to (abc)-def-gh99, and is called a 100-bank of numbers.

The Mitofsky–Waksberg method then uses a method similar to Lahiri's (1951) method to sample psus with probabilities proportional to the number of residential telephone numbers. Select a psu at random from the list of all psus, and also select a number randomly between 00 and 99 to serve as the last two digits. Dial that telephone number. If the selected number is residential, interview the household and choose its psu to be in the sample; the associated psu is the block of 100 telephone numbers that have the same first eight digits as the selected number. For example, if the randomly selected telephone number (202) 456-1414 is determined to be residential, then the psu of all numbers of the form (202) 456-14xx is included in the sample. Continue sampling in that psu until a total of k interviews are obtained. If the original number selected in the psu is not residential, reject that psu. Continue the procedure until the desired number of psus, n, is selected.

Lepkowski (1988) found that in the late 1980s, 60% of telephone numbers chosen with the Mitofsky–Waksberg method reached households, compared with 25% for random digit element sampling. The method worked well because the psus of 100 telephone numbers were clustered—some psus were unassigned, some tended to be assigned to commercial establishments, and some were largely residential. The procedure eliminates sampling unassigned psus at the second stage, and reduces the probability of selecting psus with few residential telephone numbers.

Under ideal conditions, the Mitofsky–Waksberg procedure samples psus with probabilities proportional to the number of residential telephone numbers in the psus. If the second stage prescribes selecting an additional $(k - 1)$ residential telephone numbers in each sampled psu, and if all psus in the sample have at least k residential telephone numbers, then the Mitofsky–Waksberg procedure gives each residential telephone number the same probability of being selected in the sample—the result is a self-weighting sample of residential telephone numbers.

To see this, let M_i be the number of residential telephone numbers in psu i, and let N be the total number of psus in the sampling frame. The probability that psu i is selected to be in the sample on the first iteration of the procedure is M_i/M_0, where $M_0 = \sum_{i=1}^{N} M_i$ (see Exercise 32), even though the values of M_i and M_0 are unknown. Then, if each psu in the population has either $M_i = 0$ or $M_i \geq k$,

$$P(\text{number selected}) = P(\text{psu } i \text{ selected}) \, P(\text{number selected} \mid \text{psu } i \text{ selected})$$

$$\propto \frac{M_i}{M_0} \frac{k}{M_i} = \frac{k}{M_0}.$$

The sampling weight for each number in the sample is M_0/k; to estimate a population total, you would need to know M_0, the total number of residential telephone numbers in the population. To estimate an average or proportion, the typical goal of telephone surveys, you do not need to know M_0. You only need to know a "relative weight" w_{ij} for each response y_{ij} in the sample, and can estimate the population mean as

$$\hat{\bar{y}} = \frac{\displaystyle\sum_{i \in S} \sum_{j \in S_i} w_{ij} y_{ij}}{\displaystyle\sum_{i \in S} \sum_{j \in S_i} w_{ij}}.$$

Here, with a self-weighting sample, you can use relative weights of $w_{ij} = 1$.

Note that although under ideal conditions the Mitofsky–Waksberg method leads to a self-weighting sample of residential telephone numbers, it does *not* give a self-weighting sample of households—some households may have more than one telephone number; others may not have a telephone. In practice, someone using the Mitofsky–Waksberg method would adjust the weights to compensate for multiple telephone lines and nonresponse, as will be discussed in Chapter 8.

Although in ideal situations the Mitofsky–Waksberg method produces a self-weighting sample of residential telephone numbers, those ideal situations are rarely encountered in practice. The inclusion of a psu in the sample depends on the determination of whether the first number dialed is residential or not. But a household belonging to that number may not respond, or may require many attempts to be reached, which delays the decision about whether to include that psu in the sample or results in an incorrect rejection of the psu. Many survey organizations currently use list-assisted random digit dialing (Casady and Lepkowski, 1993), in which telephone numbers are selected from 100-banks constructed from telephone directories. The telephone numbers in a 100-bank are included in the sampling frame if the directory contains at least one telephone number in that 100-bank. The 100-banks with no numbers in the directory are not included in the sampling frame. With list-assisted random digit dialing, there is undercoverage of households that are in a 100-bank where everyone has an unlisted number, but the undercoverage is thought to be small. Tucker et al. (2002) discuss these methods in view of changes in the assignment of residential telephone numbers. The increased prevalence of cell-only households has increased the coverage problems of random digit dialing surveys based on directories of landline numbers; Lavrakas et al. (2007) discuss the challenges involved in sampling cellular telephone households. ∎

EXAMPLE **6.13** *3-P Sampling*

Probability **P**roportional to **P**rediction (3-P) sampling, described by Schreuder et al. (1968), is commonly recommended as a sampling scheme in forestry. Suppose an investigator wants to estimate the total volume of timber in an area. Several options are available: (1) Estimate the volume for each tree in the area. There may be thousands of trees, however, and this can be very time consuming. (2) Use a cluster sample in which plots of equal areas are selected, and the volume of every tree in the selected plots measured. (3) Use an unequal-probability sampling scheme in which points in the area are selected at random, and the trees closest to the points are included in the sample. In this design, a tree is selected with probability proportional to the area of the region that is closer to that tree than to any other tree. (4) Estimate the volume of each tree by eye and then select trees with probability proportional to the estimated volume. When done in one pass, with trees selected as the volume is estimated, this is 3-P sampling—the prediction *P* stands for the predicted (estimated) volume used in determining the π_i's.

The largest trees tend to produce the most timber and contribute most to the variability of the estimate of total volume. Thus, unequal-probability sampling can be expected to lead to less sampling effort. Theoretically, you could estimate the volume of each of the *N* trees in the forest by eye, obtaining a value x_i for tree *i*. Then, you could revisit trees randomly selected with probabilities proportional to x_i, and carefully measure the volume t_i. Such a procedure, however, requires you to make two trips through the forest and adds much work to the sampling process. In 3-P sampling, only one trip is made through the forest, and trees are selected for the sample at the same time the x_i's are measured. The procedure is as follows:

1   Estimate or guess what the maximum value of x_i for the trees is likely to be. Define a value L that is larger than your estimated maximum value of x_i.

2   Proceed to a tree in the forest, and determine x_i for that tree. Generate a random number u_i in $[0, L]$. If $u_i \leq x_i$, then measure the volume y_i on that tree; otherwise, skip that tree and go on to the next tree.

3   Repeat step 2 on every tree in the forest.

The unequal-probability sampling in this case essentially gives every board-foot of timber an equal chance of being selected for the sample. Note that the size of the unequal-probability sample is unknown until sampling is completed. The probability that tree i is included in the sample is $\pi_i = x_i/L$. The Horvitz–Thompson estimator is

$$\hat{t}_{HT} = \sum_{i \in S} \frac{y_i}{\pi_i} = L \sum_{i \in S} \frac{y_i}{x_i} = \sum_{i=1}^{N} Z_i \frac{y_i}{\pi_i},$$

where $Z_i = 1$ if tree i is in the sample, and 0 otherwise. The Z_i's are independent Bernoulli random variables (Z_i has success probability π_i), so 3-P sampling is a special case of a method known as **Poisson sampling**. The sample size is the random variable $\sum_{i=1}^{N} Z_i$ with expected value $\sum_{i=1}^{N} x_i/L$.

Because the sample size is variable rather than fixed, Poisson sampling provides a different method of unequal-probability sampling than those discussed in Sections 6.1 through 6.4. Särndal et al. (1992) give additional theory and references for Poisson sampling. ∎

Unequal-probability methods are common in natural resource sampling. Overton and Stehman (1995) give a number of other examples.

EXAMPLE **6.14**   *Dollar Unit Sampling.*

An accountant auditing the accounts receivable amounts for a company often takes a sample to estimate the true total accounts receivable balance. The book value x_i is known for each account in the population; the audited value t_i will be known only for accounts in the sample. In Section 4.3 we saw how the auxiliary information x_i could be used in difference estimation to improve the precision from an SRS of accounts. Ratio or regression estimation could be used similarly.

Instead of being used in the analysis, the book values could be used in the design of the sample. You could stratify the accounts by the value of x_i, or you could take an unequal-probability sample with inclusion probabilities proportional to x_i. (Or you could do both: First stratify, then sample with unequal probabilities within each stratum.) If you sample accounts with probabilities proportional to x_i, then each individual dollar in the book values has the same probability of being selected in the sample (hence the name **dollar unit sampling**). With each dollar equally likely to be included in the sample, an account with book value $10,000 is ten times as likely to be in the sample as an account with book value $1000.

Consider a client with 87 accounts receivable, with a book balance of $612,824. The auditor has decided that a sample of size 25 will be sufficient for estimating

TABLE   **6.8**

Account Selection for Audit Sample

Account (Audit Unit)	Book Value	Cumulative Book Value	Random Number	
1	2,459	2,459		
2	2,343	4,802		
3	6,842	11,644	11,016	
4	4,179	15,823		
5	750	16,573		
6	2,708	19,281		
7	3,073	22,354		
8	4,742	27,096		
9	16,350	43,446	31,056	38,500
10	5,424	48,870		
11	9,539	58,409		
12	3,108	61,517		
13	3,935	65,452	63,047	
14	900	66,352		

the error in accounts receivable and takes a random sample with replacement of the 612,824 dollars in the book value population. As individual dollars can only be audited as part of the whole account, each dollar selected serves as a "hook" to snag the whole account for audit. The cumulative-size method is used to select psus (accounts) for this example; often, in practice, auditors take a systematic sample of dollars and their accompanying psus. A systematic sample guarantees that accounts with book values greater than the sampling interval will be included in the sample. Table 6.8 shows the first few lines of the account selection; the full table is in file auditselect.dat. Here, accounts 3 and 13 are included once, and account 9 is included twice (but only needs to be audited once since this is a one-stage cluster sample). This is thus an example of one-stage pps sampling with replacement, as discussed in Section 6.2.

The selected accounts are audited, and the audit values are recorded in file auditresult.dat. The overstatement in each sampled account is calculated as (book value − audit value). Table 6.9 gives part of a spreadsheet (the full spreadsheet is on the website) that may be used to estimate the total overstatement. Using the results from Section 6.2, the total overstatement is estimated from (6.5) to be $4334 with standard error $13,547/\sqrt{25} = \$2709$ from (6.6). In many auditing situations, however, most of the audited values agree with the book values, so most of the differences are zeros. A CI based on a normal approximation does not perform well in this situation, so auditors typically use a CI based on the Poisson or multinomial distribution (see Neter et al., 1978) rather than a CI of the form ($\hat{t} \pm 1.96$ SE).

Another way of looking at the unequal-probability estimate is to find the overstatement for each individual dollar in the sample. Account 24, for example, has a book value of $7090 and an error of $40. The error is prorated to every dollar in the book value, leading to an overstatement of $0.00564 for each of the 7090 dollars. The

TABLE   6.9

Results of the Audit on Accounts in the Sample

Account (Audit Unit)	Book Value (BV)	ψ_i	Audit Value (AV)	BV − AV Difference	Diff/ψ_i	Difference per Dollar
3	6,842	0.0111647	6,842	0	0	0.00000
9	16,350	0.0266798	16,350	0	0	0.00000
9	16,350	0.0266798	16,350	0	0	0.00000
13	3,935	0.0064211	3,935	0	0	0.00000
24	7,090	0.0115694	7,050	40	3,457	0.00564
29	5,533	0.0090287	5,533	0	0	0.00000
⋮	⋮	⋮	⋮	⋮	⋮	⋮
75	2,291	0.0037384	2,191	100	26,749	0.04365
79	4,667	0.0076156	4,667	0	0	0.00000
81	31,257	0.0510049	31,257	0	0	0.00000
		average			4,334	0.007071874
		std. dev.			13,547	0.02210527

average overstatement for the individual dollars in the sample is $0.007071874, so the total overstatement for the population is estimated as (0.007071874)(612824) = 4334. ∎

6.6
Randomization Theory Results and Proofs*

In two-stage cluster sampling, we select the psus first and then select subunits within the sampled psus. One approach to calculate a theoretical variance for any estimator in multistage sampling is to condition on which psus are included in the sample. To do this, we need to use Properties 4 (successive conditioning) and 5 (calculating variances conditionally) of conditional expectation, stated in Section A.4.

In this section, we state and prove Theorem 6.2, the Horvitz–Thompson Theorem (Horvitz and Thompson, 1952), which gives the properties of the estimator in (6.25). In Theorem 6.4, we find unbiased estimators of the variance. We then show that the variance for cluster sampling with equal probabilities in (5.21) follows as a special case of these theorems. First, however, we prove (6.17) and (6.18).

Throughout this section, let

$$Z_i = \begin{cases} 1 & \text{if psu } i \text{ is in the sample} \\ 0 & \text{if psu } i \text{ is not in the sample} \end{cases} \tag{6.35}$$

denote the random variable specifying whether psu i is included in the sample or not. The probability that psu i is included in the sample is

$$\pi_i = P(Z_i = 1) = E(Z_i); \tag{6.36}$$

the probability that both psu i and psu k $(i \neq k)$ are included in the sample is

$$\pi_{ik} = P(Z_i = 1 \text{ and } Z_k = 1) = E(Z_i Z_k). \tag{6.37}$$

THEOREM **6.1**

For a without-replacement probability sample of n units, let Z_i, π_i, and π_{ik} be as defined in (6.35)–(6.37). Then

$$\sum_{i=1}^{N} \pi_i = n$$

and

$$\sum_{\substack{k=1 \\ k \neq i}}^{N} \pi_{ik} = (n-1)\pi_i.$$

Proof Since the sample size is n, $\sum_{i=1}^{N} Z_i = n$ for every possible sample. Also,

$$E[Z_i] = E[Z_i^2] = \pi_i$$

because $P(Z_i = 1) = \pi_i$. Consequently,

$$n = E \left[\sum_{i=1}^{N} Z_i \right] = \sum_{i=1}^{N} \pi_i.$$

In addition,

$$\sum_{\substack{k=1 \\ k \neq i}}^{N} \pi_{ik} = \sum_{\substack{k=1 \\ k \neq i}}^{N} E[Z_i Z_k] = E[Z_i(n - Z_i)] = \pi_i(n-1),$$

which completes the proof. ∎

THEOREM **6.2**
Horvitz–Thompson

Let Z_i, π_i, and π_{ik} be as defined in (6.35)–(6.37). Suppose that sampling is done at the second stage so that sampling in any psu is independent of the sampling in any other psu, and that $\hat{t}_i$ is independent of $(Z_1, \ldots, Z_N)$ with $E[\hat{t}_i] = E[\hat{t}_i \mid Z_1, \ldots, Z_N] = t_i$. Then

$$E \left[\sum_{i=1}^{N} Z_i \frac{\hat{t}_i}{\pi_i} \right] = \sum_{i=1}^{N} \pi_i \frac{t_i}{\pi_i} = t \tag{6.38}$$

and

$$V \left[\sum_{i=1}^{N} Z_i \frac{\hat{t}_i}{\pi_i} \right] = V_{\text{psu}} + V_{\text{ssu}}, \tag{6.39}$$

where

$$V_{\text{psu}} = V\left[\sum_{i=1}^{N} Z_i \frac{t_i}{\pi_i}\right] = \sum_{i=1}^{N}(1-\pi_i)\frac{t_i^2}{\pi_i} + \sum_{i=1}^{N}\sum_{\substack{k=1 \\ k \neq i}}^{N}(\pi_{ik}-\pi_i\pi_k)\frac{t_i}{\pi_i}\frac{t_k}{\pi_k} \qquad (6.40)$$

and

$$V_{\text{ssu}} = \sum_{i=1}^{N}\frac{V(\hat{t}_i)}{\pi_i}. \qquad (6.41)$$

Proof   First note that

$$\text{Cov}(Z_i, Z_k) = \begin{cases} \pi_i(1-\pi_i) & \text{if } i = k \\ \pi_{ik} - \pi_i\pi_k & \text{if } i \neq k. \end{cases}$$

We use successive conditioning to show (6.38):

$$E\left[\sum_{i=1}^{N} Z_i \frac{\hat{t}_i}{\pi_i}\right] = E\left\{E\left[\sum_{i=1}^{N} Z_i \frac{\hat{t}_i}{\pi_i}\middle| Z_1, \dots, Z_N\right]\right\} = E\left[\sum_{i=1}^{N} Z_i \frac{t_i}{\pi_i}\right] = \sum_{i=1}^{N} \pi_i \frac{t_i}{\pi_i} = t.$$

The first step above simply applies successive conditioning; in the second step, we use the independence of $\hat{t}_i$ and $(Z_1, \dots, Z_N)$.

To find the variance, we use the expression for calculating the variance conditionally in Property 5 of Section A.4, and again use the independence of $\hat{t}_i$ and $(Z_1, \dots, Z_N)$:

$$V\left[\sum_{i=1}^{N} Z_i \frac{\hat{t}_i}{\pi_i}\right] = V\left[E\left(\sum_{i=1}^{N} Z_i \frac{\hat{t}_i}{\pi_i}\middle| Z_1, \dots, Z_N\right)\right] + E\left[V\left(\sum_{i=1}^{N} Z_i \frac{\hat{t}_i}{\pi_i}\middle| Z_1, \dots, Z_N\right)\right]$$

$$= V\left[\sum_{i=1}^{N} Z_i \frac{t_i}{\pi_i}\right] + E\left[\sum_{i=1}^{N} Z_i^2 \frac{V(\hat{t}_i)}{\pi_i^2}\right]$$

$$= \sum_{i=1}^{N}\sum_{k=1}^{N}\frac{t_i}{\pi_i}\frac{t_k}{\pi_k}\text{Cov}(Z_i, Z_k) + \sum_{i=1}^{N}\pi_i\frac{V(\hat{t}_i)}{\pi_i^2}$$

$$= \sum_{i=1}^{N}\pi_i(1-\pi_i)\frac{t_i^2}{\pi_i^2} + \sum_{i=1}^{N}\sum_{\substack{k=1 \\ k \neq i}}^{N}(\pi_{ik}-\pi_i\pi_k)\frac{t_i}{\pi_i}\frac{t_k}{\pi_k} + \sum_{i=1}^{N}\frac{V(\hat{t}_i)}{\pi_i}. \qquad \blacksquare$$

Equation (6.38) establishes that the Horvitz–Thompson estimator is unbiased, and (6.39) through (6.41) show that (6.26) is the variance of the Horvitz–Thompson estimator. In one-stage cluster sampling, $V(\hat{t}_i) = 0$ for $i \in S$, so $V_{\text{ssu}} = 0$ and $V(\hat{t}_{\text{HT}}) = V_{\text{psu}}$ as given in (6.20).

We now show that the Horvitz–Thompson form of the variance in (6.20) and the SYG form in (6.21) are equivalent.

THEOREM 6.3

Let V_{psu} be as defined in (6.40). Then

$$V_{psu} = \sum_{i=1}^{N}(1-\pi_i)\frac{t_i^2}{\pi_i} + \sum_{i=1}^{N}\sum_{\substack{k=1\\k\neq i}}^{N}(\pi_{ik}-\pi_i\pi_k)\frac{t_i}{\pi_i}\frac{t_k}{\pi_k}$$

$$= \frac{1}{2}\sum_{i=1}^{N}\sum_{\substack{k=1\\k\neq i}}^{N}(\pi_i\pi_k-\pi_{ik})\left(\frac{t_i}{\pi_i}-\frac{t_k}{\pi_k}\right)^2.$$

Proof Starting with the SYG form in (6.21),

$$\frac{1}{2}\sum_{i=1}^{N}\sum_{\substack{k=1\\k\neq i}}^{N}(\pi_i\pi_k-\pi_{ik})\left(\frac{t_i}{\pi_i}-\frac{t_k}{\pi_k}\right)^2 = \frac{1}{2}\sum_{i=1}^{N}\sum_{\substack{k=1\\k\neq i}}^{N}(\pi_i\pi_k-\pi_{ik})\left(\frac{t_i^2}{\pi_i^2}+\frac{t_k^2}{\pi_k^2}-2\frac{t_i}{\pi_i}\frac{t_k}{\pi_k}\right).$$

From results (6.17) and (6.18) proven in Theorem 6.1, noting that $\pi_{ik}=\pi_{ki}$,

$$\frac{1}{2}\sum_{i=1}^{N}\sum_{\substack{k=1\\k\neq i}}^{N}\pi_i\pi_k\left(\frac{t_i^2}{\pi_i^2}+\frac{t_k^2}{\pi_k^2}\right) = \frac{1}{2}\sum_{i=1}^{N}\sum_{k=1}^{N}\pi_i\pi_k\left(\frac{t_i^2}{\pi_i^2}+\frac{t_k^2}{\pi_k^2}\right)-\sum_{i=1}^{N}\pi_i^2\frac{t_i^2}{\pi_i^2} = \sum_{i=1}^{N}(n-\pi_i)\frac{t_i^2}{\pi_i},$$

$$\sum_{i=1}^{N}\sum_{\substack{k=1\\k\neq i}}^{N}\pi_{ik}\frac{t_i^2}{\pi_i^2} = \sum_{i=1}^{N}\sum_{\substack{k=1\\k\neq i}}^{N}\pi_{ik}\frac{t_k^2}{\pi_k^2} = (n-1)\sum_{i=1}^{N}\frac{t_i^2}{\pi_i}.$$

Thus,

$$\frac{1}{2}\sum_{i=1}^{N}\sum_{\substack{k=1\\k\neq i}}^{N}(\pi_i\pi_k-\pi_{ik})\left(\frac{t_i}{\pi_i}-\frac{t_k}{\pi_k}\right)^2 = \sum_{i=1}^{N}[n-\pi_i-(n-1)]\frac{t_i^2}{\pi_i^2}-\sum_{i=1}^{N}\sum_{\substack{k=1\\k\neq i}}^{N}(\pi_i\pi_k-\pi_{ik})\frac{t_i}{\pi_i}\frac{t_k}{\pi_k},$$

which shows the equality of the two expressions for the variance. ∎

Theorem 6.4 shows that (6.28) and (6.29) are unbiased estimators for the variance in (6.26) and (6.27); the one-stage variance estimators in (6.20) and (6.21) follow as a special case when $V(\hat{t}_i) = 0$.

THEOREM 6.4

Suppose the conditions of Theorem 6.2 hold, and that $\hat{V}(\hat{t}_i)$ is an unbiased estimator of $V(\hat{t}_i)$ that is independent of Z_i. Then,

$$E\left[\sum_{i=1}^{N}Z_i\frac{\hat{V}(\hat{t}_i)}{\pi_i^2}\right] = V_{ssu}, \tag{6.42}$$

$$E\left[\sum_{i=1}^{N} Z_i(1-\pi_i)\frac{\hat{t}_i^2}{\pi_i^2} + \sum_{i=1}^{N}\sum_{\substack{k=1 \\ k \ne i}}^{N} Z_i Z_k \frac{\pi_{ik}-\pi_i\pi_k}{\pi_{ik}}\frac{\hat{t}_i}{\pi_i}\frac{\hat{t}_k}{\pi_k}\right]$$

$$= E\left[\frac{1}{2}\sum_{i=1}^{N}\sum_{\substack{k=1 \\ k \ne i}}^{N} Z_i Z_k \frac{\pi_i\pi_k-\pi_{ik}}{\pi_{ik}}\left(\frac{\hat{t}_i}{\pi_i}-\frac{\hat{t}_k}{\pi_k}\right)^2\right]$$

$$= V_{\text{psu}} + \sum_{i=1}^{N}(1-\pi_i)\frac{V(\hat{t}_i)}{\pi_i}, \qquad (6.43)$$

and

$$E\left[\hat{V}_{\text{HT}}(\hat{t}_{\text{HT}})\right] = E\left[\hat{V}_{\text{SYG}}(\hat{t}_{\text{HT}})\right] = V_{\text{psu}} + V_{\text{ssu}}. \qquad (6.44)$$

Proof   We prove (6.42) by using successive conditioning:

$$E\left[Z_i\frac{\hat{V}(\hat{t}_i)}{\pi_i^2}\right] = E\left[E\left(Z_i\frac{\hat{V}(\hat{t}_i)}{\pi_i^2}\,\Big|\,Z_i\right)\right] = E\left[Z_i\frac{V(\hat{t}_i)}{\pi_i^2}\right] = \frac{V(\hat{t}_i)}{\pi_i}.$$

Result (6.42) follows by summation.

To prove (6.43), note that because $\hat{t}_i$ and $(Z_1,\dots,Z_N)$ are independent,

$$E[\hat{t}_i^2 \mid Z_1,\dots,Z_N] = E[\hat{t}_i^2] = t_i^2 + V(\hat{t}_i).$$

Thus,

$$E\left[\sum_{i=1}^{N} Z_i(1-\pi_i)\frac{\hat{t}_i^2}{\pi_i^2}\right] = E\left[E\left(\sum_{i=1}^{N} Z_i(1-\pi_i)\frac{\hat{t}_i^2}{\pi_i^2}\,\Big|\,Z_1,\dots,Z_N\right)\right]$$

$$= E\left[\sum_{i=1}^{N} Z_i\frac{1-\pi_i}{\pi_i^2}\left\{t_i^2 + V(\hat{t}_i)\right\}\right]$$

$$= \sum_{i=1}^{N} \frac{1-\pi_i}{\pi_i}[t_i^2 + V(\hat{t}_i)].$$

Because subsampling is done independently in different psus, $E[\hat{t}_i\,\hat{t}_k] = t_i\,t_k$ for $k \ne i$, so

$$E\left[\sum_{i=1}^{N}\sum_{\substack{k=1 \\ k \ne i}}^{N} Z_i Z_k \frac{\pi_{ik}-\pi_i\pi_k}{\pi_{ik}}\frac{\hat{t}_i}{\pi_i}\frac{\hat{t}_k}{\pi_k}\right]$$

$$= E\left[E\left(\sum_{i=1}^{N}\sum_{\substack{k=1 \\ k \ne i}}^{N} Z_i Z_k \frac{\pi_{ik}-\pi_i\pi_k}{\pi_{ik}}\frac{\hat{t}_i}{\pi_i}\frac{\hat{t}_k}{\pi_k}\,\Big|\,Z_1,\dots,Z_N\right)\right]$$

$$= E\left[\sum_{i=1}^{N}\sum_{\substack{k=1 \\ k \ne i}}^{N} Z_i Z_k \frac{\pi_{ik}-\pi_i\pi_k}{\pi_{ik}}\frac{t_i}{\pi_i}\frac{t_k}{\pi_k}\right]$$

$$= \sum_{i=1}^{N}\sum_{\substack{k=1 \\ k \ne i}}^{N} (\pi_{ik}-\pi_i\pi_k)\frac{t_i}{\pi_i}\frac{t_k}{\pi_k}.$$

Combining the two results, we see that

$$E\left[\sum_{i=1}^{N} Z_i(1-\pi_i)\frac{\hat{t}_i^2}{\pi_i^2} + \sum_{i=1}^{N}\sum_{\substack{k=1\\k\neq i}}^{N} Z_iZ_k\frac{\pi_{ik}-\pi_i\pi_k}{\pi_{ik}}\frac{\hat{t}_i}{\pi_i}\frac{\hat{t}_k}{\pi_k}\right] = V_{\text{psu}} + \sum_{i=1}^{N}\frac{1-\pi_i}{\pi_i}V(\hat{t}_i),$$

which proves the first part of (6.43). We show the second part of (6.43) similarly, using results from Theorem 6.1:

$$E\left[\frac{1}{2}\sum_{i=1}^{N}\sum_{\substack{k=1\\k\neq i}}^{N} Z_iZ_k\frac{\pi_i\pi_k-\pi_{ik}}{\pi_{ik}}\left(\frac{\hat{t}_i}{\pi_i}-\frac{\hat{t}_k}{\pi_k}\right)^2\right]$$

$$= E\left\{E\left[\sum_{i=1}^{N}\sum_{\substack{k=1\\k\neq i}}^{N} Z_iZ_k\frac{\pi_i\pi_k-\pi_{ik}}{\pi_{ik}}\left(\frac{\hat{t}_i^2}{\pi_i^2}-\frac{\hat{t}_i}{\pi_i}\frac{\hat{t}_k}{\pi_k}\right)\Big| Z_1,\dots,Z_N\right]\right\}$$

$$= E\left[\sum_{i=1}^{N}\sum_{\substack{k=1\\k\neq i}}^{N} Z_iZ_k\frac{\pi_i\pi_k-\pi_{ik}}{\pi_{ik}}\left(\frac{t_i^2+V(\hat{t}_i)}{\pi_i^2}-\frac{t_i}{\pi_i}\frac{t_k}{\pi_k}\right)\right]$$

$$= \sum_{i=1}^{N}\sum_{\substack{k=1\\k\neq i}}^{N}(\pi_i\pi_k-\pi_{ik})\left(\frac{t_i^2+V(\hat{t}_i)}{\pi_i^2}-\frac{t_i}{\pi_i}\frac{t_k}{\pi_k}\right)$$

$$= \sum_{i=1}^{N}[\pi_i(n-\pi_i)-(n-1)\pi_i]\left(\frac{t_i^2+V(\hat{t}_i)}{\pi_i^2}\right)+\sum_{i=1}^{N}\sum_{\substack{k=1\\k\neq i}}^{N}(\pi_{ik}-\pi_i\pi_k)\frac{t_i}{\pi_i}\frac{t_k}{\pi_k}$$

$$= \sum_{i=1}^{N}\frac{1-\pi_i}{\pi_i}t_i^2+\sum_{i=1}^{N}\sum_{\substack{k=1\\k\neq i}}^{N}(\pi_{ik}-\pi_i\pi_k)\frac{t_i}{\pi_i}\frac{t_k}{\pi_k}+\sum_{i=1}^{N}\frac{1-\pi_i}{\pi_i}V(\hat{t}_i).$$

Equation (6.44) follows because

$$E\left[\sum_{i\in\mathcal{S}}\frac{\hat{V}(\hat{t}_i)}{\pi_i}\right] = E\left[\sum_{i=1}^{N}Z_i\frac{\hat{V}(\hat{t}_i)}{\pi_i}\right] = \sum_{i=1}^{N}V(\hat{t}_i). \qquad \blacksquare$$

EXAMPLE **6.15** We now show that the results in Section 5.3 are special cases of Theorems 6.2 and 6.4. If psus are selected with equal probabilities,

$$P(Z_i=1)=\pi_i=\frac{n}{N},$$

$$P(Z_i=1 \text{ and } Z_k=1)=\pi_{ik}=\frac{n}{N}\frac{n-1}{N-1},$$

and

$$\hat{t}_{\text{unb}}=\sum_{i\in\mathcal{S}}\frac{N}{n}\hat{t}_i=\sum_{i=1}^{N}Z_i\frac{N}{n}\hat{t}_i=\sum_{i=1}^{N}Z_i\frac{\hat{t}_i}{\pi_i},$$

so we can apply Theorem 6.2 with $\pi_i = n/N$. Then,

$$E[\hat{t}_{\mathrm{unb}}] = \sum_{i=1}^{N} \frac{n}{N}\frac{N}{n} t_i = t,$$

and, from (6.40),

$$
\begin{aligned}
V_{\mathrm{psu}}[\hat{t}_{\mathrm{unb}}] &= \sum_{i=1}^{N} \frac{1-\pi_i}{\pi_i} t_i^2 + \sum_{i=1}^{N}\sum_{k\neq i}^{N} \frac{\pi_{ik}-\pi_i\pi_k}{\pi_i\pi_k} t_i t_k \\
&= \sum_{i=1}^{N}\left(1-\frac{n}{N}\right)\left(\frac{N}{n}\right) t_i^2 + \sum_{i=1}^{N}\sum_{\substack{k=1\\k\neq i}}^{N}\left[\frac{n}{N}\frac{n-1}{N-1}-\left(\frac{n}{N}\right)^2\right]\left(\frac{N}{n}\right)^2 t_i t_k \\
&= \frac{N}{n}\left(1-\frac{n}{N}\right)\left[\sum_{i=1}^{N} t_i^2 - \frac{1}{N-1}\sum_{i=1}^{N}\sum_{\substack{k=1\\k\neq i}}^{N} t_i t_k\right] \\
&= \frac{N}{n(N-1)}\left(1-\frac{n}{N}\right)\left[(N-1)\sum_{i=1}^{N} t_i^2 - \sum_{i=1}^{N}\sum_{k=1}^{N} t_i t_k + \sum_{i=1}^{N} t_i^2\right] \\
&= \frac{N}{n(N-1)}\left(1-\frac{n}{N}\right)\left[N\sum_{i=1}^{N} t_i^2 - t^2\right] \\
&= N^2\left(1-\frac{n}{N}\right)\frac{S_t^2}{n}.
\end{aligned}
$$

By result (2.9) from SRS theory,

$$V(\hat{t}_i) = M_i^2\left(1-\frac{m_i}{M_i}\right)\frac{S_i^2}{m_i},$$

so, using (6.41),

$$V_{\mathrm{ssu}} = \sum_{i=1}^{N} \frac{N}{n} M_i^2\left(1-\frac{m_i}{M_i}\right)\frac{S_i^2}{m_i}.$$

This completes the proof of (5.21). In the special case of an SRS, $t_i = y_i$ and S_t^2 is the variance among population elements, so V_{psu} reduces to the formula in (2.16).

For two-stage cluster sampling with equal probabilities, we defined

$$s_t^2 = \frac{1}{n-1}\sum_{i\in\mathcal{S}}\left(\hat{t}_i - \frac{\hat{t}_{\mathrm{unb}}}{N}\right)^2$$

to be the sample variance among the estimated psu totals in (5.22). We now show that, when $\pi_i = n/N$ and $\pi_{ik} = n(n-1)/[N(N-1)]$,

$$\sum_{i=1}^{N} Z_i(1-\pi_i)\frac{\hat{t}_i^2}{\pi_i^2} + \sum_{i=1}^{N}\sum_{\substack{k=1\\k\neq i}}^{N} Z_i Z_k \frac{\pi_{ik}-\pi_i\pi_k}{\pi_{ik}}\frac{\hat{t}_i}{\pi_i}\frac{\hat{t}_k}{\pi_k} = N^2\left(1-\frac{n}{N}\right)\frac{s_t^2}{n},$$

so that Theorem 6.4 can be applied. Substituting n/N for π_i and $n(n-1)/[N(N-1)]$ for π_{ik},

$$\sum_{i=1}^{N} Z_i(1-\pi_i)\frac{\hat{t}_i^2}{\pi_i^2} + \sum_{i=1}^{N}\sum_{\substack{k=1\\k\neq i}}^{N} Z_i Z_k \frac{\pi_{ik}-\pi_i\pi_k}{\pi_{ik}}\frac{\hat{t}_i}{\pi_i}\frac{\hat{t}_k}{\pi_k}$$

$$= \left(\frac{N}{n}\right)^2 \left(1-\frac{n}{N}\right)\sum_{i=1}^{N} Z_i\hat{t}_i^2 + \left(\frac{N}{n}\right)^2\left[1-\frac{n(N-1)}{N(n-1)}\right]\sum_{i=1}^{N}\sum_{\substack{k=1\\k\neq i}}^{N} Z_i Z_k \hat{t}_i\hat{t}_k$$

$$= \left(\frac{N}{n}\right)^2 \left(1-\frac{n}{N}\right)\sum_{i=1}^{N} Z_i\hat{t}_i^2 - \left(\frac{N}{n}\right)^2\frac{1}{n-1}\left(1-\frac{n}{N}\right)\sum_{i=1}^{N}\sum_{\substack{k=1\\k\neq i}}^{N} Z_i Z_k \hat{t}_i\hat{t}_k$$

$$= \left(\frac{N}{n}\right)^2 \left(1-\frac{n}{N}\right)\left(\sum_{i=1}^{N} Z_i\hat{t}_i^2 - \frac{1}{n-1}\sum_{i=1}^{N}\sum_{k=1}^{N} Z_i Z_k \hat{t}_i\hat{t}_k + \frac{1}{n-1}\sum_{i=1}^{N} Z_i\hat{t}_i^2\right)$$

$$= \left(\frac{N}{n}\right)^2 \left(1-\frac{n}{N}\right)\frac{1}{n-1}\left[n\sum_{i=1}^{N} Z_i\hat{t}_i^2 - \left(\sum_{k=1}^{N} Z_k\hat{t}_k\right)^2\right]$$

$$= \left(\frac{N}{n}\right)^2 \left(1-\frac{n}{N}\right)\frac{1}{n-1}\left[n\sum_{i=1}^{N} Z_i\hat{t}_i^2 - n^2\left(\frac{\hat{t}_{\text{unb}}}{N}\right)^2\right]$$

$$= \left(\frac{N^2}{n}\right)\left(1-\frac{n}{N}\right)\frac{1}{n-1}\sum_{i=1}^{N} Z_i\left(\hat{t}_i - \frac{\hat{t}_{\text{unb}}}{N}\right)^2$$

$$= N^2\left(1-\frac{n}{N}\right)\frac{s_t^2}{n}.$$

Thus, by (6.43),

$$E\left[N^2\left(1-\frac{n}{N}\right)\frac{s_t^2}{n}\right] = N^2\left(1-\frac{n}{N}\right)\frac{S_t^2}{n} + \frac{N}{n}\left(1-\frac{n}{N}\right)\sum_{i=1}^{N} V(\hat{t}_i). \qquad (6.45)$$

Note that the expected value of s_t^2 is larger than S_t^2: s_t^2 includes the variation from psu total to psu total, plus variation from not knowing the psu totals.

Because

$$\hat{V}(\hat{t}_i) = \left(1 - \frac{m_i}{M_i}\right)M_i^2\frac{s_i^2}{m_i}$$

is an unbiased estimator of $V(\hat{t}_i)$, Theorem 6.4 implies that

$$E\left[\sum_{i=1}^{N} Z_i \left(\frac{N}{n}\right)^2 \hat{V}(\hat{t}_i)\right] = E\left[\sum_{i \in \mathcal{S}} \left(\frac{N}{n}\right)^2 \hat{V}(\hat{t}_i)\right] = V_{\text{ssu}}.$$

Using (6.45), then,

$$E\left[N^2 \left(1 - \frac{n}{N}\right) \frac{s_t^2}{n} + \frac{N}{n} \sum_{i \in \mathcal{S}} \hat{V}(\hat{t}_i)\right]$$

$$= N^2 \left(1 - \frac{n}{N}\right) \frac{S_t^2}{n} + \frac{N}{n} \left(1 - \frac{n}{N}\right) \sum_{i=1}^{N} V(\hat{t}_i) + \sum_{i=1}^{N} V(\hat{t}_i)$$

$$= N^2 \left(1 - \frac{n}{N}\right) \frac{S_t^2}{n} + \frac{N}{n} \sum_{i=1}^{N} V(\hat{t}_i),$$

so (5.24) is an unbiased estimator of (5.21). ∎

The methods used in these proofs can be applied to any number of levels of clustering. You may want to sample schools, then classes within schools, then students within classes. Exercise 36 asks you to find an expression for the variance in three-stage cluster sampling. Rao (1979a) presents an alternative and elegant approach, relying on properties of nonnegative definite matrices, for deriving mean squared errors and variance estimators for linear estimators of population totals.

6.7
Models and Unequal-Probability Sampling*

In general, data from a good sampling design should produce reasonable inferences from either a model-based or randomization approach. Let's see how the Horvitz–Thompson estimator performs for Model M1 from (5.34). The model is

$$\text{M1} : Y_{ij} = \mu + A_i + \varepsilon_{ij}$$

with the A_i's generated by a distribution with mean 0 and variance σ_A^2, the ε_{ij}'s generated by a distribution with mean 0 and variance σ^2, and all A_i's and ε_{ij}'s independent.

As we did for the estimators in Chapter 5, we can write the estimator as a linear combination of the random variables Y_{ij}. For a pps design, $\psi_i = M_i/M_0$ and $\pi_i = n\psi_i$, so

$$\hat{T}_P = \sum_{i \in \mathcal{S}} \frac{M_0}{nM_i} \hat{T}_i = \sum_{i \in \mathcal{S}} M_0 \frac{\overline{Y}_{\mathcal{S}_i}}{n} = \sum_{i \in \mathcal{S}} \sum_{j \in \mathcal{S}_i} \frac{M_0}{nm_i} Y_{ij}.$$

Note that $\sum_{i \in S} \sum_{j \in S_i} M_0/(nm_i) = M_0$, so $\hat{T}_P$ is unbiased under Model M1 in (5.34). In addition, from (5.36),

$$
\begin{aligned}
V_{\text{M1}}[\hat{T}_P - T] &= \sigma_A^2 \left[\sum_{i \in S} \left(\sum_{j \in S_i} \frac{M_0}{nm_i} - M_i \right)^2 + \sum_{i \notin S} M_i^2 \right] \\
&\quad + \sigma^2 \left[\sum_{i \in S} \sum_{j \in S_i} \left\{ \left(\frac{M_0}{nm_i} \right)^2 - 2 \frac{M_0}{nm_i} \right\} + M_0 \right] \\
&= \sigma_A^2 \left[\frac{M_0^2}{n} - 2 \frac{M_0}{n} \sum_{i \in S} M_i + \sum_{i=1}^N M_i^2 \right] + \sigma^2 \left[\sum_{i \in S} \frac{M_0^2}{n^2 m_i} - M_0 \right].
\end{aligned}
$$

The model-based variance for $\hat{T}_P$ has implications for design. Suppose a sample is desired that will minimize $V_{\text{M1}}[\hat{T}_P - T]$. The psu sizes M_i for the sample units appear only in the term $-2\sigma_A^2(M_0/n)\sum_{i \in S} M_i$, so for fixed n the variance is smallest when the n units with largest M_i's are included in the sample. If, in addition, a constraint is placed on the number of subunits that can be examined, $\sum_{i \in S} (1/m_i)$ is smallest when all m_i's are equal.

Inference in the model-based approach does not depend on the sampling design. As long as model M1 holds for the population, $\hat{T}_P$ is model-unbiased with variance given above. In a model-based approach, an investigator with complete faith in the model can simply select the psus with the largest values of M_i to be the sample. In practice, however, this would not be done—no one has complete faith in a model, especially before data collection. Royall and Eberhardt (1975) suggested using balanced sampling, in which the sample is selected in such a way that inferences are robust to certain forms of model misspecification.

As described in Section 6.2, pps sampling can be thought of as a way of introducing randomness into the optimal design for model M1 and estimator $\hat{T}_P$. The self-weighting design of taking all m_i's to be equal also minimizes the variance in the model-based approach. Thus, if model M1 is thought to describe the data, pps sampling and estimation should perform well in practice.

We conclude our discussion with a widely quoted example from Basu (1971, pp. 212–213), often used to argue that Horvitz–Thompson estimates can be as silly as any other statistical procedures improperly applied.

> The circus owner is planning to ship his 50 adult elephants and so he needs a rough estimate of the total weight of the elephants. As weighing an elephant is a cumbersome process, the owner wants to estimate the total weight by weighing just one elephant. Which elephant should he weigh? So the owner looks back on his records and discovers a list of the elephants' weights taken 3 years ago. He finds that 3 years ago Sambo the middle-sized elephant was the average (in weight) elephant in his herd. He checks with the elephant trainer who reassures him (the owner) that Sambo may still be considered to be the average elephant in the herd. Therefore, the owner plans to weigh Sambo and take 50 y (where y is the present weight of Sambo) as an estimate of the total weight $Y = Y_1 + Y_2 + \cdots + Y_{50}$ of the 50 elephants. But the circus statistician is horrified when he learns of the owner's purposive sampling plan. "How can you get an unbiased estimate of Y this way?" protests the statistician. So, together they work

out a compromise sampling plan. With the help of a table of random numbers they devise a plan that allots a selection probability of 99/100 to Sambo and equal selection probabilities of 1/4900 to each of the other 49 elephants. Naturally, Sambo is selected and the owner is happy. "How are you going to estimate Y?", asks the statistician. "Why? The estimate ought to be $50y$ of course," says the owner. "Oh! No! That cannot possibly be right," says the statistician, "I recently read an article in the Annals of Mathematical Statistics where it is proved that the Horvitz–Thompson estimator is the unique hyperadmissible estimator in the class of all generalized polynomial unbiased estimators." "What is the Horvitz–Thompson estimate in this case?" asks the owner, duly impressed. "Since the selection probability for Sambo in our plan was 99/100," says the statistician, "the proper estimate of Y is $100y/99$ and not $50y$." "And how would you have estimated Y," inquires the incredulous owner, "if our sampling plan made us select, say, the big elephant Jumbo?" "According to what I understand of the Horvitz–Thompson estimation method," says the unhappy statistician, "the proper estimate of Y would then have been $4900y$, where y is Jumbo's weight." That is how the statistician lost his circus job (and perhaps became a teacher of statistics!)

Should the circus statistician have been fired? A statistician desiring to use a model in analyzing survey data would say yes: The circus statistician is using the model $y_i \propto 99/100$ for Sambo, and $y_i \propto 1/4900$ for all other elephants in the herd—certainly not a model that fits the data well. A randomization-inference statistician would also say yes: Even though models are not used explicitly in the Horvitz–Thompson theory, the estimator is most efficient (has the smallest variance) when the psu total (here, y_i) is proportional to the probability of selection. The silly design used by the circus statistician leads to a huge variance for the Horvitz–Thompson estimator. If that were not reason enough, the statistician proposes a sample of size 1—he can neither check the validity of the model in a model-based approach nor estimate the variance of the Horvitz–Thompson estimator!

Had the circus statistician used a ratio estimator in the design-based setting, he might have saved his job even though he did not use a good design. He wants to estimate the population total, t (called Y in Basu's paper). The ratio estimator is

$$\hat{t}_{yr} = \frac{\hat{t}_y}{\hat{t}_x} t_x.$$

Thus, if Sambo is selected,

$$\hat{t}_{yr} = \frac{y_{\text{Sambo}}/\pi_{\text{Sambo}}}{x_{\text{Sambo}}/\pi_{\text{Sambo}}} t_x = \frac{y_{\text{Sambo}}}{x_{\text{Sambo}}} t_x.$$

Similarly, if Jumbo is selected,

$$\hat{t}_{yr} = \frac{y_{\text{Jumbo}}}{x_{\text{Jumbo}}} t_x.$$

With the ratio estimator, the total weight of the elephants from three years ago is multiplied by the ratio of (weight now)/(weight 3 years ago) for the selected elephant.

6.8
Chapter Summary

Unequal-probability samples occur naturally in many situations, particularly in cluster sampling when the psus have unequal sizes. If the psu population totals t_i are highly correlated with ψ_i, then an unequal-probability sampling design can greatly increase efficiency. All estimators studied so far in the book can be viewed as special cases of the estimators used in unequal-probability sampling.

We can draw an unequal-probability sample either with or without replacement. Selecting a with-replacement sample with unequal probabilities is simple; on each of the n draws, select one of the N psus with specified probability ψ_i, where $\sum_{i=1}^{N} \psi_i = 1$. Since any psu can be selected on each of the n draws, a psu can appear more than once in the sample.

Estimation is also simple in a with-replacement probability sample, and the estimators have the same form for either a one-stage or a multi-stage sample. The population total is estimated by $\hat{t}_\psi$ given in (6.5) and (6.13):

$$\hat{t}_\psi = \frac{1}{n} \sum_{i \in \mathcal{R}} \frac{\hat{t}_i}{\psi_i} = \sum_{i \in \mathcal{R}} \sum_{j \in \mathcal{S}_i} w_{ij} y_{ij},$$

where $\mathcal{R}$ denotes the set of psus selected for the sample (including psus as many times as they are selected). If an SRS of m_i of the M_i ssus is taken at stage 2, then $w_{ij} = [1/(n\psi_i)](M_i/m_i)$. An unbiased estimator of $V(\hat{t}_\psi)$ is given by

$$\hat{V}(\hat{t}_\psi) = \frac{1}{n} \frac{1}{n-1} \sum_{i \in \mathcal{R}} \left(\frac{\hat{t}_i}{\psi_i} - \hat{t}_\psi \right)^2,$$

which is simply the sample variance of the values of $\hat{t}_i/\psi_i$ divided by n. If a psu appears more than once in the sample, each time a different probability subsample of ssus is selected for estimating t_i. In with-replacement sampling, the population mean $\bar{y}_U$ is estimated using (6.10) by $\hat{\bar{y}}_\psi = \hat{t}_\psi/\hat{M}_{0\psi}$, where $\hat{M}_{0\psi} = \sum_{i \in \mathcal{R}} \sum_{j \in \mathcal{S}_i} w_{ij}$. Equation (6.15) gives $\hat{V}(\hat{\bar{y}}_\psi)$ using ratio estimation methods.

Although the estimators of means and totals in without-replacement unequal-probability sampling have simple form, variance estimation and sample selection methods can be complicated. We recommend using software such as SAS PROC SURVEYSELECT to select an unequal-probability sample without replacement when n/N is large.

With $\pi_i = P(\text{psu } i \text{ is included in the sample, } \mathcal{S})$, the Horvitz–Thompson estimator of the population total for a without-replacement sample is

$$\hat{t}_{\text{HT}} = \sum_{i \in \mathcal{S}} \frac{\hat{t}_i}{\pi_i},$$

where $\hat{t}_i$ is an unbiased estimator of the psu total t_i. The sampling weight for ssu j of psu i is

$$w_{ij} = \frac{1}{\pi_i} \frac{1}{P(\text{ssu } j \text{ of psu } i \text{ is in sample} \mid \text{psu } i \text{ is in sample})};$$

in terms of the weights,

$$\hat{t}_{\text{HT}} = \sum_{i \in S} \sum_{j \in S_i} w_{ij} y_{ij}$$

and

$$\bar{y}_{\text{HT}} = \frac{\displaystyle\sum_{i \in S} \sum_{j \in S_i} w_{ij} y_{ij}}{\displaystyle\sum_{i \in S} \sum_{j \in S_i} w_{ij}}.$$

The variance of the Horvitz–Thompson estimator is given in (6.20) and (6.21) for one-stage cluster sampling and in (6.26) and (6.27) for two-stage cluster sampling. The SYG unbiased estimator of $V(\hat{t}_{\text{HT}})$, given in (6.29), requires knowledge of the joint inclusion probabilities $\pi_{ik} = P(\text{psus } i \text{ and } j \text{ are included in the sample})$ and is often difficult to compute. In many situations, we recommend using the with-replacement variance estimators in (6.30) and (6.34), which do not require knowledge of the π_{ik}'s; if n/N is small, a with-replacement variance estimator performs well.

Unequal-probability sampling is used in many large-scale government surveys to improve efficiency. It is also frequently used in telephone surveys and natural resource surveys.

Key Terms

Horvitz–Thompson estimator: The Horvitz–Thompson estimator of a population total t is $\hat{t}_{\text{HT}} = \sum_{i \in S} \hat{t}_i / \pi_i$. This is the most general form of the estimator of t in without-replacement samples with inclusion probabilities π_i.

Inclusion probability: π_i is the probability that psu i is included in the sample.

Joint inclusion probability: π_{ij} is the probability that psus i and j are both included in the sample.

Poisson sampling: A sampling process in which independent Bernoulli trials determine whether each unit in the population is to be included in the sample.

Probability proportional to size (pps) sampling: Unequal-probability sampling method in which the probability of sampling a unit is proportional to the number of elements in the unit.

Random digit dialing: A method used in telephone surveys in which a probability sample of telephone numbers is selected from the set of possible telephone numbers.

For Further Reading

Overton and Stehman (1995) give a clearly written overview and examples of unequal-probability sampling. Chapter 9 of Brewer (2002) discusses the Horvitz–Thompson estimator and methods for approximating its variance. Brewer and Hanif (1983) present more than 50 methods for drawing with- and without-replacement samples with unequal probabilities. Tillé (2006) presents algorithms for selecting unequal-probability samples. Tillé also describes how to select balanced samples, in which a sample is designed so that estimated population totals of auxiliary variables equal the true population totals of those variables. Programs in the R statistical programming language that will select balanced samples are given in Matei and Tillé (2005).

Rao (2005) outlines the history of how practical problems have spurred development of survey methods, with an interesting section on the history of unequal-probability sampling. Hansen and Hurwitz (1943) develop the theory of pps sampling with replacement. Horvitz and Thompson (1952) extend the work of Hansen and Hurwitz to unequal-probability sampling without replacement.

6.9 Exercises

A. Introductory Exercises

1 For each of the following situations, say what unit might be used as psu. Do you believe there would be a strong clustering effect? Would you sample psus with equal or unequal probabilities?

a You want to estimate the percentage of patients of U.S. Air Force optometrists and ophthalmologists who wear contact lenses.

b Human taeniasis is acquired by ingesting larvae of the pork tapeworm in inadequately cooked pork. You have been asked to design a survey to estimate the percentage of inhabitants of a village who have taeniasis. A medical examination is required to diagnose the condition.

c You wish to estimate the total number of cows and heifers on all Ontario dairy farms; in addition, you would like to find estimates of the birth rate and stillbirth rate.

d You want to estimate the percentages of undergraduate students at U.S. universities who are registered to vote, and who are affiliated with each political party.

e A fisheries agency is interested in the distribution of carapace width of snow crabs. A trap hauled from a fishing boat has a limit of 30 crabs.

f You wish to conduct a customer satisfaction survey of persons who have taken guided bus tours of the Grand Canyon rim area. Tour groups range in size from 8 to 44 persons.

2 An investigator wants to take an unequal-probability sample of 10 of the 25 psus in the population listed below and in file exercise0602.dat, and wishes to sample units with replacement.

psu	ψ_i	psu	ψ_i
1	0.000110	14	0.014804
2	0.018556	15	0.005577
3	0.062998	16	0.070784
4	0.078216	17	0.069635
5	0.075245	18	0.034650
6	0.073983	19	0.069492
7	0.076580	20	0.036590
8	0.038981	21	0.033853
9	0.040772	22	0.016959
10	0.022876	23	0.009066
11	0.003721	24	0.021795
12	0.024917	25	0.059186
13	0.040654		

a Adapt the cumulative-size method to draw a sample of size 10 with replacement with probabilities ψ_i. Instead of randomly selecting integers between 1 and $M_0 = \sum_{i=1}^{N} M_i$, select 10 random numbers between 0 and 1.

b Adapt Lahiri's method to draw a sample of size 10 with replacement with probabilities ψ_i.

3 For the supermarket example in Section 6.1, suppose that the ψ_i's are the same, but that each store has $t_i = 75$. What is $E[\hat{t}_\psi]$? $V[\hat{t}_\psi]$?

4 For the supermarket example in Section 6.1, suppose that the ψ_i's are 7/16 for store A, and 3/16 for each of stores B, C, and D. Show that $\hat{t}_\psi$ is unbiased, and find its variance. Do you think that the sampling scheme with these ψ_i's is a good one?

5 Return to the supermarket example of Section 6.1. Now let's select two supermarkets with replacement. List the 16 possible samples (A,A), (A,B), etc., and find the probability with which each sample would be selected. Calculate $\hat{t}_\psi$ for each sample. What is $E[\hat{t}_\psi]$? $V[\hat{t}_\psi]$?

6 The file azcounties.dat gives data from the 2000 U.S. Census on population and housing unit counts for the counties in Arizona (excluding Maricopa County and Pima County, which are much larger than the other counties and would be placed in a separate stratum). For this exercise, suppose that year 2000 population (M_i) is known and you want to take a sample of counties to estimate the total number of housing units ($t = \sum_{i=1}^{13} t_i$). The file has the value of t_i for every county so you can calculate the population total and variance.

a Calculate the selection probabilities ψ_i for a sample of size 1 with probability proportional to 2000 population. Find $\hat{t}_\psi$ for each possible sample, and calculate the theoretical variance $V(\hat{t}_\psi)$.

b Repeat (a) for an equal probability sample of size 1. How do the variances compare? Why do you think one design is more efficient than the other?

c Now take a with-replacement sample of size 3. Find $\hat{t}_\psi$ and $\hat{V}(\hat{t}_\psi)$ for your sample.

7 For a simple random sample with replacement, with $\psi_i = 1/N$, show that (6.6) simplifies to

$$\hat{V}(\hat{t}_\psi) = \frac{N^2}{n} \frac{1}{n-1} \sum_{i \in \mathcal{R}} (t_i - \bar{t})^2,$$

where the sum is over all n units in the sample (including units as many times as they appear in the sample).

8 Let's return to the situation in Exercise 6 of Chapter 2, in which we took an SRS to estimate the average and total numbers of refereed publications of faculty and research associates. Now, consider a pps sample of faculty. The 27 academic units range in size from 2 to 92. We used Lahiri's method to choose 10 psus with probabilities proportional to size and with replacement, and took an SRS of four (or fewer, if $M_i < 4$) members from each psu. Note that academic unit 14 appears three times in the sample; each time it appears, a different subsample was collected.

Academic Unit	M_i	ψ_i	y_{ij}
14	65	0.0805452	3, 0, 0, 4
23	25	0.0309789	2, 1, 2, 0
9	48	0.0594796	0, 0, 1, 0
14	65	0.0805452	2, 0, 1, 0
16	2	0.0024783	2, 0
6	62	0.0768278	0, 2, 2, 5
14	65	0.0805452	1, 0, 0, 3
19	62	0.0768278	4, 1, 0, 0
21	61	0.0755886	2, 2, 3, 1
11	41	0.0508055	2, 5, 12, 3

Find the estimated total number of publications, along with its standard error.

B. Working with Survey Data

9 The file statepps.dat lists the number of counties, land area, and 1992 population for the 50 states plus the District of Columbia.

   a Use the cumulative-size method to draw a sample of size 10 with replacement, with probabilities proportional to land area. What is ψ_i for each state in your sample?

   b Use the cumulative-size method to draw a sample of size 10 with replacement, with probabilities proportional to population. What is ψ_i for each state in your sample?

   c How do the two samples differ? Which states tend to be in each sample?

10 Use your sample of states drawn with probability proportional to population, from Exercise 9, for this problem.

   a Using the sample, estimate the total number of counties in the United States, and find the standard error of your estimate. How does your estimate compare with the true value of total number of counties (which you can calculate, since the file statepps.dat contains the data for the whole population)?

b Now suppose that your friend Tom finds the ten values of numbers of counties in your sample, but does not know that you selected these states with probabilities proportional to population. Tom then estimates the total number of counties using formulas for an SRS. What values for the estimated total and its standard error are calculated by Tom? How do these values differ from yours? Is Tom's estimator unbiased for the population total?

11 In Example 2.5, we took an SRS to estimate the total acreage devoted to farming in the United States in 1992. Now, use the sample of states drawn with probability proportional to land area in Exercise 9, and then subsample five counties randomly from each state using file agpop.dat. Estimate the total acreage devoted to farming in 1992, along with its standard error.

12 The file statepop.dat, used in Example 6.5, also contains information on total number of farms, number of veterans, and other items.

a Plot the total number of farms versus the probabilities of selection ψ_i. Does your plot indicate that unequal-probability sampling will be helpful here?

b Estimate the total number of farms in the United States, along with its standard error.

13 Use the file statepop.dat for this problem.

a Plot the total number of veterans versus the probabilities of selection ψ_i. Does your plot indicate that unequal-probability sampling will be helpful here?

b Estimate the total number of veterans in the United States, and find the standard error for your estimate.

c Estimate the total number of Vietnam veterans in the United States, and find the standard error for your estimate.

14 In Example 6.11, we calculated the with-replacement variance for $\hat{t}_{HT}$. In this example, the sampling fraction n/N is 1/3, so the with-replacement variance is likely to overestimate the without-replacement variance. The joint inclusion probabilities for the psus are given in file classppsjp.dat, and can also be obtained by running the SAS program given on the website.

a Calculate $\hat{V}_{HT}(\hat{t}_{HT})$ and $\hat{V}_{SYG}(\hat{t}_{HT})$ for this data set.

b SAS software approximates the without-replacement variance in unequal-probability sampling using

$$\left(1 - \frac{n}{N}\right) \hat{V}_{WR}(\hat{t}_{HT}).$$

Calculate this approximation for the class data.

c How do these estimates compare, and how do they compare with the with-replacement variance calculated in Example 6.11?

C. Working with Theory

All of the problems in this section require knowledge of probability.

15 **a** Prove that Lahiri's method results in a probability proportional to size sample with replacement. HINT: Let J be an integer with $J \geq \max\{M_i\}$. Let $U_1, U_2, \ldots$ be discrete uniform $\{1, \ldots, N\}$ random variables, let $V_1, V_2, \ldots$ be discrete uniform $\{1, \ldots, J\}$ random variables, and assume all U_i and V_j are independent. To select the first psu, we generate pairs $(U_1, V_1), (U_2, V_2), \ldots$ until $V_j \leq M_{U_j}$.

b Suppose the population has N psus, with sizes $M_1, M_2, \ldots, M_N$. Let X represent the number of pairs of random numbers that must be generated to obtain a sample of size n. Find $E[X]$.

16 Note that the random variables $Q_1, \ldots, Q_N$ in Section 6.3 have a joint multinomial distribution with probabilities $\psi_1, \psi_2, \ldots, \psi_N$. Use properties of the multinomial distribution to show that $\hat{t}_\psi$ in (6.13) is an unbiased estimator of t with variance given by

$$V(\hat{t}_\psi) = \frac{1}{n}\sum_{i=1}^{N}\psi_i\left(\frac{t_i}{\psi_i} - t\right)^2 + \frac{1}{n}\sum_{i=1}^{N}\frac{V_i}{\psi_i}. \tag{6.46}$$

Also show that (6.14) is an unbiased estimator of the variance in (6.46). HINT: Use properties of conditional expectation in Appendix A, and write

$$V(\hat{t}_\psi) = V(E[\hat{t}_\psi \mid Q_1, \ldots, Q_N]) + E(V[\hat{t}_\psi \mid Q_1, \ldots, Q_N]).$$

17 Show that (6.28) and (6.29) are equivalent when an SRS of psus is selected as in Chapter 5. Are they equal if psus are selected with unequal probabilities?

18 Show that the formulas for stratified random sampling in (3.1), (3.3), and (3.5) follow from the formulas for the Horvitz–Thompson estimator in Section 6.4.3. For a stratified random sample, we sample from every stratum in the population. Thus, if we treat strata as if they were psus, $\pi_i = 1$ for every stratum in the population.

19 Use the population in Exercise 2 of Chapter 4 for this exercise. Let ψ_i be proportional to x_i.

a Using the draw-by-draw method illustrated in Example 6.8, calculate π_i for each unit and π_{ij} for each pair of units, for a without-replacement sample of size two.

b What is $V(\hat{t}_{HT})$? How does it compare with the with-replacement variance using (6.46)?

20 *Covariance of estimated population totals in a cluster sample.* Suppose a one-stage cluster sample is taken from a population of N psus, with inclusion probabilities π_i. Let t_x and t_y be the population totals for response variables x and y, and let t_{ix} and t_{iy} be the totals of variables x and y in psu i.

a Show that

$$\mathrm{Cov}\,(\hat{t}_x, \hat{t}_y) = \sum_{i=1}^{N}\frac{1 - \pi_i}{\pi_i}t_{ix}t_{iy} + \sum_{i=1}^{N}\sum_{\substack{k=1 \\ k \neq i}}^{N}(\pi_{ik} - \pi_i\pi_k)\frac{t_{ix}\,t_{ky}}{\pi_i\,\pi_k}.$$

b If an SRS of n of the N psus is selected, with $\pi_i = n/N$ and $\pi_{ik} = (n/N)[(n-1)/(N-1)]$, show using part (a) that

$$\text{Cov}\,(\hat{t}_x, \hat{t}_y) = \frac{N^2}{(N-1)n}\left(1 - \frac{n}{N}\right)\left[\sum_{i=1}^{N} t_{ix}t_{iy} - \frac{t_x t_y}{N}\right].$$

21 *Comparing two domain means in a cluster sample.* In Exercise 24 of Chapter 4, you showed that in an SRS where $\bar{y}_1$ and $\bar{y}_2$ estimate respective population domain means $\bar{y}_{U1}$ and $\bar{y}_{U2}$, $V(\bar{y}_1 - \bar{y}_2) \approx V(\bar{y}_1) + V(\bar{y}_2)$ because $\text{Cov}\,(\bar{y}_1, \bar{y}_2) \approx 0$. Now let's explore what happens when a one-stage cluster sample is selected from a population of N psus. For simplicity, assume that each psu has M ssus and that an SRS of n psus is selected. Let $\hat{\bar{y}}_1$ and $\hat{\bar{y}}_2$ be the estimators of the domain means from the cluster sample. Similarly to Exercise 24 of Chapter 4, let $x_{ij} = 1$ if ssu j of psu i is in domain 1 and $x_{ij} = 0$ if ssu j of psu i is in domain 2, and let $u_{ij} = x_{ij}y_{ij}$.

  a Find

$$\text{Cov}\left[\left(\hat{\bar{u}} - \frac{t_u}{t_x}\hat{\bar{x}}\right),\ \left\{\hat{\bar{y}} - \hat{\bar{u}} - \frac{t_y - t_u}{N - t_x}(1 - \hat{\bar{x}})\right\}\right].$$

  HINT: Use Exercise 20.

  b Show that the covariance in (a) is 0 if for each psu, all of the elements in that psu belong to the same domain—that is, either $t_{ix} = 0$ or $t_{ix} = M$ for each psu i. [If the covariance in (a) is 0, then (4.26) implies that $\text{Cov}\,(\hat{\bar{y}}_1, \hat{\bar{y}}_2) \approx 0$.]

  c Give an example in which the covariance in (a) is not 0.

22 *Indirect sampling.* Suppose you want to take a sample of students in a university but your sampling frame is a list of all classes offered by the university. A student may be in more than one class, so a probability sample of classes, which includes all students in those classes, may contain some students multiple times. Lavallée (2007) describes a generalized weight share method for such situations, and this exercise is adapted from results in his book.

  Let $\mathcal{U}^A$ be the sampling frame population with N units. Let $Z_i = 1$ if unit i is in the sample $\mathcal{S}^A$ and 0 otherwise, with $\pi_i = P(Z_i = 1)$. The target population $\mathcal{U}^B$ has M elements. Each element in $\mathcal{U}^B$ is linked with one or more of the units in $\mathcal{U}^A$; let

$$\ell_{ik} = \begin{cases} 1 & \text{if element } k \text{ from } \mathcal{U}^B \text{ is linked to unit } i \text{ from } \mathcal{U}^A \\ 0 & \text{otherwise} \end{cases}$$

and let $L_k = \sum_{i=1}^{N} \ell_{ik}$. We assume $L_k \geq 1$ for each k and that L_k is known. In our example, $\ell_{ik} = 1$ if student k is in class i and L_k is the number of classes taken by student k. Let y_k be a characteristic associated with element k of $\mathcal{U}^B$. We want to estimate $t_y = \sum_{k=1}^{M} y_k$.

  a Let $u_i = \sum_{k=1}^{M} (\ell_{ik}y_k/L_k)$ and let

$$\hat{t}_y = \sum_{i=1}^{N} \frac{Z_i}{\pi_i}u_i.$$

Show that $\hat{t}_y$ is an unbiased estimator of t_y, with

$$V(\hat{t}_y) = \sum_{i=1}^{N} \frac{1 - \pi_i}{\pi_i} u_i^2 + \sum_{i=1}^{N} \sum_{j \neq i}^{N} \frac{\pi_{ij} - \pi_i \pi_j}{\pi_i \pi_j} u_i u_j.$$

b Let $\mathcal{S}^B$ be the set of distinct units sampled from $\mathcal{U}^B$ using this procedure. Show that $\hat{t}_y$ can be rewritten as

$$\hat{t}_y = \sum_{k \in \mathcal{S}^B} \frac{1}{L_k} \sum_{i=1}^{N} \frac{Z_i}{\pi_i} \ell_{ik} y_k.$$

We can view $w_k^* = \frac{1}{L_k} \sum_{i=1}^{N} \frac{Z_i}{\pi_i} \ell_{ik}$ as a "weight" for y_k.

c If $L_k = 1$ for all k, show that $\hat{t}_{HT}$ in (6.19) is a special case of $\hat{t}_y$. What is w_k^* in this case?

d Suppose $\mathcal{U}^A = \{1, 2, 3\}, \mathcal{U}^B = \{1, 2\}$ and the values of ℓ_{ik} are given in the following table:

		Element k from $\mathcal{U}^B$	
ℓ_{ik}		1	2
Unit i	1	1	0
from $\mathcal{U}^A$	2	1	1
	3	0	1

Suppose $y_1 = 4$ and $y_2 = 6$, so that $t_y = 10$. Find the value of $\hat{t}_y$ for each of the three possible SRSs of size 2 from $\mathcal{U}^A$. Using the sampling distribution of $\hat{t}_y$, show that $\hat{t}_y$ is unbiased but that $V(\hat{t}_y) > 0$. Even though each possible SRS from $\mathcal{U}^A$ contains both units from $\mathcal{U}^B$ (so in effect, a census is taken of $\mathcal{U}^B$), the variance of $\hat{t}_y$ is not zero.

e Data file wtshare.dat contains information from a hypothetical SRS of size $n = 100$ from a population of $N = 40,000$ adults. Each adult in the sample is asked about his or her children: how many children between ages 0 and 5, whether those children attend preschool, and how many other adults in the population claim the child as part of their household. Estimate the total number of children in the population who attend preschool. Use the with-replacement variance estimator

$$\frac{1}{n} \frac{1}{n-1} \sum_{i \in \mathcal{S}^A} \left(\frac{n u_i}{\pi_i} - \hat{t}_y \right)^2$$

to construct an approximate 95% CI for the total number of children who attend preschool.

23 In simple random sampling, we know that a without-replacement sample of size n has smaller variance than a with-replacement sample of size n. The same result is not always true for unequal-probability sampling designs (Raj, 1968, p. 56). Consider a with-replacement design with selection probabilities ψ_i, and a corresponding without-replacement design with inclusion probabilities $\pi_i = n\psi_i$; assume $n\psi_i < 1$ for $i = 1, \ldots, N$.

a Consider a population with $N = 4$ and $t_1 = -5$, $t_2 = 6$, $t_3 = 0$, and $t_4 = -1$. The joint inclusion probabilities for a without-replacement sample of size 2 are $\pi_{12} = 0.004$, $\pi_{13} = \pi_{23} = \pi_{24} = 0.123$, $\pi_{14} = 0.373$, and $\pi_{34} = 0.254$. Find the value of π_i for each unit. Show that for this design and population, $V(\hat{t}_\psi) < V(\hat{t}_{HT})$.

b Show that for $\pi_i = n\psi_i$ and $V(\hat{t}_\psi)$ in (6.8),

$$V(\hat{t}_\psi) = \frac{1}{2n} \sum_{i=1}^{N} \sum_{k=1}^{N} \pi_i \pi_k \left(\frac{t_i}{\pi_i} - \frac{t_k}{\pi_k} \right)^2.$$

c Using $V(\hat{t}_{HT})$ in (6.21), show that if

$$\pi_{ik} \geq \frac{n-1}{n} \pi_i \pi_k \quad \text{for all } i \text{ and } k,$$

then $V(\hat{t}_{HT}) \leq V(\hat{t}_\psi)$.

d Gabler (1984) shows that if

$$\sum_{i=1}^{N} \min_k \left(\frac{\pi_{ik}}{\pi_k} \right) \geq n-1,$$

then $V(\hat{t}_{HT}) \leq V(\hat{t}_\psi)$. Show that if $\pi_{ik} \geq (n-1)\pi_i\pi_k/n$ for all i and k, then Gabler's condition is met.

e (Requires knowledge of linear algebra.) Show that if $V(\hat{t}_{HT}) \leq V(\hat{t}_\psi)$, then

$$\pi_{ik} \leq 2\frac{n-1}{n}\pi_i\pi_k \quad \text{for all } i \text{ and } k.$$

HINT: Use the results in Theorem 6.1 to simplify $V(\hat{t}_\psi) - V(\hat{t}_{HT})$ so that it may be written as $\sum_{i=1}^{N} \sum_{k=1}^{N} a_{ik} t_i t_k$. Then $\mathbf{A}$, the matrix with elements a_{ik}, must be nonnegative definite and therefore all principal 2×2 submatrices must have determinant ≥ 0.

24 Consider a without-replacement sample of size 2 from a population of size 4, with joint inclusion probabilities $\pi_{12} = \pi_{34} = 0.31$, $\pi_{13} = 0.20$, $\pi_{14} = 0.14$, $\pi_{23} = 0.03$, and $\pi_{24} = 0.01$.

a Calculate the inclusion probabilities π_i for this design.

b Suppose $t_1 = 2.5$, $t_2 = 2.0$, $t_3 = 1.1$, and $t_4 = 0.5$. Find $\hat{V}_{HT}(\hat{t}_{HT})$ and $\hat{V}_{SYG}(\hat{t}_{HT})$ for each possible sample.

25 *Brewer's (1963, 1975) procedure for without-replacement unequal-probability sampling.* For a sample of size $n = 2$, let π_i be the desired probability of inclusion for psu i, with the usual constraint that $\sum_{i=1}^{N} \pi_i = n$. Let $\psi_i = \pi_i/2$ and

$$a_i = \frac{\psi_i(1 - \psi_i)}{1 - \pi_i}.$$

Draw the first psu with probability $a_i / \sum_{k=1}^{N} a_k$ of selecting psu i. Supposing psu i is selected at the first draw, select the second psu from the remaining $N - 1$ psus with probabilities $\psi_j/(1 - \psi_i)$.

a   Show that

$$\pi_{ij} = \frac{\psi_i \psi_j}{\sum\limits_{k=1}^{N} a_k} \left(\frac{1}{1-\pi_i} + \frac{1}{1-\pi_j} \right).$$

b   Show that P(psu i selected in sample) $= \pi_i$. HINT: First show that

$$2 \sum_{k=1}^{N} a_k = 1 + \sum_{k=1}^{N} \frac{\psi_k}{1-\pi_k}.$$

c   The SYG estimator of $V(\hat{t}_{HT})$ for one-stage sampling is given in (6.23). Show that $\pi_i \pi_j - \pi_{ij} \geq 0$ for Brewer's method, so that the SYG estimator of the variance is always nonnegative.

26   The following table gives population values for a small population of clusters:

psu, i	M_i	Values, y_{ij}	t_i
1	5	3, 5, 4, 6, 2	20
2	4	7, 4, 7, 7	25
3	8	7, 2, 9, 4, 5, 3, 2, 6	38
4	5	2, 5, 3, 6, 8	24
5	3	9, 7, 5	21

You wish to select two psus without replacement with probabilities of inclusion proportional to M_i. Using Brewer's method from Exercise 25, construct a table of π_{ij} for the possible samples. What is the variance of the one-stage Horvitz–Thompson estimator?

27   Rao (1963) discusses the following rejective method for selecting a pps sample without replacement: Select n psus with probabilities ψ_i and with replacement. If any psu appears more than once in the sample, reject the whole sample and select another n psus with replacement. Repeat until you obtain a sample of n psus with no duplicates. Find π_{ij} and π_i for this procedure, for $n = 2$.

28   *The Rao-Hartley-Cochran (1962) method for selecting psus with unequal probabilities.* To take a sample of size n, divide the population into n random groups of psus, $U_1, U_2, \ldots, U_n$. Then select one psu from each group with probability proportional to size. Let N_k be the number of psus in group k. If psu i is in group k, it is selected with probability $x_{ki} = M_i / \sum_{j \in U_k} M_j$; the estimator is

$$\hat{t}_{RHC} = \sum_{k=1}^{n} \frac{t_i}{x_{ki}}.$$

Show that $\hat{t}_{RHC}$ is unbiased for t, and find its variance. HINT: Use two sets of indicator variables. Let $I_{ki} = 1$ if psu i is in group k and 0 otherwise, and let $Z_i = 1$ if psu i is selected to be in the sample.

29   The estimators of $V(\hat{t}_{HT})$ in (6.22) and (6.23) require knowledge of the joint inclusion probabilities π_{ik}. To use these formulas, the data file must contain an $n \times n$ matrix of the π_{ik}'s, which can dramatically increase the size of the data file; in addition, computing the variance estimator is complicated. If the joint inclusion probabilities π_{ik} could be approximated as a function of the π_i's, estimation would be simplified. Let $c_i = \pi_i(1 - \pi_i)$. Hájek (1964) (see Berger, 2004, for extensions) suggested approximating π_{ik} by

$$\tilde{\pi}_{ik} = \pi_i \pi_k \left[1 - (1 - \pi_i)(1 - \pi_k)/\sum_{j=1}^{N} c_j \right].$$

a Does the set of $\tilde{\pi}_{ik}$'s satisfy condition (6.18)? Can they be joint inclusion probabilities?

b What is $\tilde{\pi}_{ik}$ if an SRS is taken? Show that if N is large, $\tilde{\pi}_{ik}$ is close to π_{ik}.

c Show that if $\tilde{\pi}_{ik}$ is substituted for π_{ik} in (6.21), the expression for the variance can be written as

$$V_{\text{Haj}}(\hat{t}_{\text{HT}}) = \sum_{i=1}^{N} c_i e_i^2,$$

where $e_i = t_i/\pi_i - A$ and

$$A = \frac{1}{N} \sum_{j=1}^{N} c_j \frac{t_j}{\pi_j}.$$

HINT: Write (6.21) as

$$\frac{1}{2} \sum_{i=1}^{N} \sum_{k=1}^{N} (\pi_i \pi_k - \tilde{\pi}_{ik}) \left(\frac{t_i}{\pi_i} - \frac{t_k}{\pi_k} \right)^2.$$

d We can estimate $V_{\text{Haj}}(\hat{t}_{\text{HT}})$ by

$$\hat{V}_{\text{Haj}}(\hat{t}_{\text{HT}}) = \sum_{i \in \mathcal{S}} \tilde{c}_i \hat{e}_i^2,$$

where $\tilde{c}_i = (1 - \pi_i)n/(n - 1)$, $\hat{e}_i = t_i/\pi_i - \hat{A}$, and $\hat{A} = \sum_{j \in \mathcal{S}} \tilde{c}_j \frac{t_j}{\pi_j} / \sum_{j \in \mathcal{S}} \tilde{c}_j$. Show that if an SRS of size n is taken, then $\hat{V}_{\text{Haj}}(\hat{t}_{\text{HT}}) = N^2(1 - n/N)s_t^2/n$.

30 This exercise is based on results in Brewer and Donadio (2003).

a Show, using the results in Theorem 6.1, that the variance in (6.21) can be rewritten as:

$$V(\hat{t}_{\text{HT}}) = \sum_{i=1}^{N} \pi_i \left(\frac{t_i}{\pi_i} - \frac{t}{n} \right)^2 - \sum_{i=1}^{N} \pi_i^2 \left(\frac{t_i}{\pi_i} - \frac{t}{n} \right)^2$$

$$+ \sum_{i=1}^{N} \sum_{\substack{k=1 \\ k \neq i}}^{N} (\pi_{ik} - \pi_i \pi_k) \left(\frac{t_i}{\pi_i} - \frac{t}{n} \right) \left(\frac{t_k}{\pi_k} - \frac{t}{n} \right). \qquad (6.47)$$

HINT: Write $t_i/\pi_i - t_k/\pi_k = t_i/\pi_i - t/n + t/n - t_k/\pi_k$.

b The first term in (6.47) is the variance that would result if a with-replacement sample with selection probabilities $\psi_i = \pi_i/n$ were taken. Brewer and Donadio (2003) suggest that the second term may be viewed as a finite population correction for unequal-probability sampling, so that the first two terms in (6.47) approximate $V(\hat{t}_{HT})$ without depending on the joint inclusion probabilities π_{ik}. Calculate the three terms in (6.47) for an SRS of size n.

c Suppose that there exist constants c_i such that $\pi_{ik} \approx \pi_i\pi_k(c_i + c_k)/2$. Show that with this substitution, the third term in (6.47) can be approximated by

$$\sum_{i=1}^{N} \pi_i^2 (1 - c_i) \left(\frac{t_i}{\pi_i} - \frac{t}{n} \right)^2$$

so that

$$V(\hat{t}_{HT}) \approx \sum_{i=1}^{N} \pi_i(1 - c_i\pi_i) \left(\frac{t_i}{\pi_i} - \frac{t}{n} \right)^2. \tag{6.48}$$

Two choices suggested for c_i are $c_i = (n-1)/(n-\pi_i)$ or (following Hartley and Rao, 1962),

$$c_i = \frac{n-1}{\left(1 - 2\pi_i + \frac{1}{n}\sum_{k=1}^{N}\pi_k^2 \right)}.$$

Calculate the variance approximation in (6.48) for an SRS with each of these choices of c_i.

31 (Requires calculus.) Suppose in (6.46) that the variance of the estimator of the total in psu i is $V(\hat{t}_i) = M_i^2 S_i^2/m_i$. If you can only subsample $C = \sum_{i=1}^{n} m_i$ ssus, what values of m_i minimize (6.46)?

32 Consider the Mitofsky–Waksberg method, discussed in Example 6.12. Show that the probability that psu i is selected as the first psu in the sample is

$$P(\text{select psu } i) = \frac{M_i}{M_0}.$$

HINT: See Exercise 15 and argue that the Mitofsky–Waksberg method for selecting psus is a special case of Lahiri's method.

33 In Example 6.12 and Exercise 32, it was shown that the Mitofsky–Waksberg method produces a self-weighting sample if any psu in the sample has at least k residential telephone numbers. Suppose a psu in the sample has $x < k$ residential numbers. What is the relative weight for a telephone number in that psu?

34 One drawback of the Mitofsky–Waksberg method as described in Example 6.12 is that the sequential sampling procedure of selecting numbers in the psu until one has a total of k residential numbers can be cumbersome to implement. Suppose in the second stage you dial an additional $(k-1)$ numbers whether they are residential or not, and let x be the number of residential lines among the $(k-1)$. What are the relative weights for the residential telephone numbers?

35   The Mitofsky–Waksberg method, described in Example 6.12, gives a self-weighting sample of telephone numbers under ideal circumstances. Does it give a self-weighting sample of adults? Why or why not? If not, what relative weights should be used?

36   Suppose a three-stage cluster sample is taken from a population with N psus, M_i ssus in the ith psu, and L_{ij} tsus (tertiary sampling units) in the jth ssu of the ith psu. To draw the sample, n psus are randomly selected, then m_i ssus from the selected psus, then l_{ij} tsus from the selected ssus.

   a   Show that the sample weights are

$$w_{ijk} = \frac{N}{n} \frac{M_i}{m_i} \frac{L_{ij}}{l_{ij}}.$$

   b   Let $\hat{t} = \sum_{i \in S} \sum_{j \in S_i} \sum_{k \in S_{ij}} w_{ijk} y_{ijk}$. Show that $E[\hat{t}] = t = \sum_{i=1}^{N} \sum_{j=1}^{M_i} \sum_{k=1}^{L_{ij}} y_{ijk}$.

   c   Using the properties of conditional expectation in Section A.4, find an expression for $V(\hat{t})$.

37   (Model-based.) Suppose the entire population is observed in the sample, so that $n = N$ and $m_i = M_i$. Examine the three estimators $\hat{T}_{unb}$, $\hat{T}_{ratio}$ (from Section 5.6) and $\hat{T}_P$ (from Section 6.7). If the entire population is observed, which of these estimators equal $T = \sum_{i=1}^{N} \sum_{j=1}^{M_i} Y_{ij}$?

D. Projects and Activities

38   *Rectangles.* Use the population of rectangles in Exercise 30 of Chapter 2 for the exercise. The file rectlength.dat contains information on the vertical length of each of the 100 rectangles in the population.

   a   Select a sample of 10 rectangles with replacement from the 100 rectangles, with probability proportional to the length of the rectangle.

   b   For your sample, plot t_i, the area of the rectangle, vs. the selection probability ψ_i. What is the correlation between t_i and ψ_i?

   c   Estimate the total area of all 100 rectangles, and find a 95% confidence interval for the total area. Compare your answers with the estimate and confidence interval from the SRS in Exercise 30 of Chapter 2. Did unequal-probability sampling result in a smaller variance estimate?

39   Repeat Exercise 38(a), using a without-replacement sample of 10 rectangles selected with probability proportional to the length of the rectangle. You will need to use a program such as SAS PROC SURVEYSELECT to select the sample.

   a   What are the inclusion probabilities π_i for the rectangles in your sample?

   b   Estimate the total area of all 100 rectangles using the Horvitz–Thompson estimator $\hat{t}_{HT}$.

   c   Find the with-replacement variance estimate for $\hat{t}_{HT}$.

   d   (Requires knowledge of the joint inclusion probabilities.) Find the SYG variance estimate for $\hat{t}_{HT}$. How does this compare with the estimate in (b)?

40   Historians wanting to use data from United States Censuses collected in the pre-computer age faced the daunting task of poring over reels of handwritten records on microfilm, arranged in geographical order. The Public Use Microdata Samples (PUMS) were constructed by taking samples of the records and typing those records into the computer. Ruggles (1995, p. 44) described the PUMS construction for the 1940 Census:

> The population schedules of the 1940 census are preserved on 4,576 microfilm reels. Each census page contains information on forty individuals. Two lines on each page were designated as "sample lines" by the Census Bureau: the individuals falling on those lines—5 percent of the population—were asked a set of supplemental questions that appear at the bottom of the census page.
>
> Two of every five census pages were systematically selected for examination. On each selected census page, one of the two designated sample lines was then randomly selected. Data-entry personnel then counted the size of the sample unit containing the targeted sample line. Units size six or smaller were included in the sample in inverse proportion to their size. Thus, every one-person unit was included in the sample, every second two-person unit, every third three-person unit, and so on. Units with seven or more persons were included with a probability of 1-in-7: every seventh household of size seven or more was selected for the sample.

  a Explain why this is a cluster sample. What are the psus? The ssus?

  b What effect do you think the clustering will have on estimates of race? age? occupation?

  c Construct a table for the inclusion probabilities for persons in one-person units, two-person units, and so on.

  d What happens if you estimate the mean age of the population by the average age of all persons in the sample? What estimator should you use?

  e Do you think that taking a systematic sample was a good idea for this sample? Why or why not?

  f Does this method provide a representative sample of households? Why or why not?

  g What type of sample is taken of the individuals with supplementary information? Explain.

41 Ruggles (1995, p. 45) also describes the 1950 PUMS:

> The 1950 census schedules are contained on 6,278 microfilm reels. Each census page contains information on thirty individuals. Every fifth line on the census page was designated as a sample line, and additional questions for the sample-line individuals appear at the bottom of the form. For the last sample-line individual on each page, there was a block of additional supplemental questions. Thus, 20 percent of individuals were asked a basic set of supplemental questions, and 3.33 percent of individuals were asked a full set of supplemental questions.
>
> One-in-eleven pages within enumeration districts was selected randomly. On each selected census page, the sixth sample-line individual (the one with the full set of questions) was selected for inclusion in the sample. Any other members of the sample unit containing the selected individual were also included.

Answer the same questions from Exercise 40 for the 1950 PUMS.

42 In Exercise 35 of Chapter 2, you estimated the size of an audience by taking an SRS. Explain how this is a special case of cluster sampling. Obtain a seating chart for an auditorium in which the rows have different numbers of seats. Using the seating chart, select an unequal-probability sample of 10 or 20 rows, with probabilities proportional to the numbers of seats in each row. Why might you expect the unequal-probability sample to have a smaller variance for the estimated audience size than an SRS of the same number of rows?

Estimate the audience size for this auditorium using your unequal-probability sample; count the number of people in each selected row. Give a 95% CI for the total number of people in the auditorium.

43   *Create your own stock market index fund.* The data file sp500.dat contains a listing of the stocks in the S&P 500 Index, along with the market capitalization of each company, as of April 2006. The market capitalization of a company is the market value of its outstanding shares, calculated as (price per share) × (number of shares outstanding).

There are several ways you could own a self-weighting sample of dollars represented by all the companies in this index. You could take an SRS of the individual dollars in the stock market, buying shares in each company for which you have at least one dollar in your SRS. This can be cumbersome, however, and would mean buying shares of a large (and random) number of companies.

An easier way is to take a sample of companies with probability proportional to market capitalization. Suppose you have $300,000 to invest. Select a sample of 30 companies from the list of 498 companies in the file with probability proportional to market capitalization. Create a file of the companies in your sample; for each company, state how much money you will invest in that company so that you have a self-weighting sample of dollars in the index.

44   *Baseball data.*

a   Use the population in the file baseball.dat to take a two-stage cluster sample (without replacement) with the teams as the psus, with probabilities proportional to the total number of runs scored for the teams. Your sample should have approximately 150 players altogether, as in the SRS from Exercise 32 of Chapter 2. Describe how you selected your sample.

b   Construct the sampling weights for your sample.

c   Let $\hat{t}_i$ be the estimated total of the variable *logsal* for team i in your sample, and let π_i be the inclusion probability for team i. Plot $\hat{t}_i$ vs. π_i.

d   Use your sample to estimate the mean of the variable *logsal*, and give a 95% CI.

e   Estimate the proportion of players in the data set who are pitchers, and give a 95% CI.

f   Do you think that unequal-probability sampling resulted in more efficiency for your estimators? Why, or why not?

45   *IPUMS exercises.* Exercise 37 of Chapter 2 described the IPUMS data.

a   Select an unequal-probability sample of 10 psus, with probability proportional to number of persons. Take a subsample of 20 persons in each of the selected psus.

b   Using the sample you selected, estimate the population mean and total of *inctot* and give the standard errors of your estimates.

7

Complex Surveys

There is no more effective medicine to apply to feverish public sentiment than figures. To be sure, they must be properly prepared, must cover the case, not confine themselves to a quarter of it, and they must be gathered for their own sake, not for the sake of a theory. Such preparation we get in a national census.

—Ida Tarbell, *The Ways of Woman* (1915)

Most large surveys involve several of the ideas we have discussed: A survey may be stratified with several stages of clustering and rely on ratio and regression estimation to adjust for other variables. The formulas for estimating standard errors can become complicated, especially if there are several stages of clustering without replacement. Sampling weights and design effects are commonly used in complex surveys to simplify matters. These, and plots for complex survey data, are discussed in this chapter. The chapter concludes with a description of the National Crime Victimization Survey (NCVS) design, and with parallels between survey samples and designed experiments.

7.1
Assembling Design Components

We have seen most of the components of a complex survey: random sampling, ratio estimation, stratification, and clustering. Now, let's see how to assemble them into one sampling design. Although in practice weights (Section 7.2) are often used to find point estimates and computer-intensive methods (Chapter 9) are used to calculate variances of the estimates, understanding the basic principles of how the components work together is important. Here are the concepts you already know, in a modular form ready for assembly.

7.1.1 Building Blocks for Surveys

1 *Cluster sampling with replacement.* Select a sample of n clusters with replacement; primary sampling unit (psu) i is selected with probability ψ_i on a draw. Estimate the total for psu i using an unbiased estimator $\hat{t}_i$. Then treat the n values (the sample is with replacement, so some of the values in the set may be from the same psus) of $u_i = \hat{t}_i/\psi_i$ as observations: Estimate the population total by $\bar{u}$, and estimate the variance of the estimated total by s_u^2/n.

2   *Cluster sampling without replacement.* Select a sample of n psus without replacement; π_i is the probability that psu i is included in the sample. Estimate the total for psu i using an unbiased estimator $\hat{t}_i$, and calculate an unbiased estimator of the variance of $\hat{t}_i$, $\hat{V}(\hat{t}_i)$. Then estimate the population total with the Horvitz-Thompson estimator[1] from (6.19):

$$\hat{t}_{HT} = \sum_{i \in \mathcal{S}} \frac{\hat{t}_i}{\pi_i}.$$

Use an exact formula from Chapters 5 or 6 or a method from Chapter 9 to estimate the variance. We often estimate the variance assuming that psus were selected with replacement, as discussed in Section 6.4.

3   *Stratification.* Let $\hat{t}_1, \ldots, \hat{t}_H$ be unbiased estimators of the stratum totals $t_1, \ldots, t_H$, and let $\hat{V}(\hat{t}_1), \ldots, \hat{V}(\hat{t}_H)$ be unbiased estimators of the variances. Then estimate the population total by

$$\hat{t} = \sum_{h=1}^{H} \hat{t}_h$$

and its variance by

$$\hat{V}(\hat{t}) = \sum_{h=1}^{H} \hat{V}(\hat{t}_h).$$

Stratification usually forms the coarsest classification: Strata may be, for example, areas of the country, different area codes, or types of habitat. Clusters (sometimes several stages of clusters) are sampled from each stratum in the design, and additional stratification may occur within clusters. Many surveys have a *stratified multistage survey design*, in which a stratified sample is taken of psus, and subsamples of secondary sampling units (ssus), are selected within each selected psu. With several stages of clustering and stratification, it helps to draw a diagram or construct a table of the survey design, as illustrated in the following example.

EXAMPLE 7.1   Malaria has long been a serious health problem in the Gambia. Malaria morbidity can be reduced by using bed nets that are impregnated with insecticide, but this is only effective if the bed nets are in widespread use. In 1991, a nationwide survey was designed to estimate the prevalence of bed net use in rural areas. The survey is described and results reported in D'Alessandro et al. (1994).

The sampling frame consisted of all rural villages of fewer than 3000 people in the Gambia. The villages were stratified by three geographic regions (eastern, central,

---

[1]Recall that the Horvitz-Thompson estimator encompasses the other without-replacement, unbiased estimators of the total as special cases, as discussed in Section 6.4.4.

and western) and by whether the village had a public health clinic (PHC) or not. In each region five districts were chosen with probability proportional to the district population as estimated in the 1983 national census. In each district four villages were chosen, again with probability proportional to census population: two PHC villages and two non-PHC villages. Finally, six compounds were chosen more or less randomly from each village, and a researcher recorded the number of beds and nets, along with other information, for each compound.

In summary, the sample design is the following:

Stage	Sampling Unit	Stratification
1	District	Region
2	Village	PHC/non-PHC
3	Compound	

To calculate estimates or standard errors using formulas from the previous chapters, you would start at Stage 3 and work up. The following are steps you would follow to estimate the total number of bed nets (without using ratio estimation):

1 Record the total number of nets for each compound.

2 Estimate the total number of nets for each village by (number of compounds in the village) × (average number of nets per compound). Find the estimated variance of the total number of nets, for each village.

3 Estimate the total number of nets for the PHC villages in each district. Villages were sampled from the district with probabilities proportional to population, so formulas from Chapter 6 need to be used to estimate the total and the variance of the estimated total. Repeat for the non-PHC villages in each district.

4 Add the estimates from the two strata (PHC and non-PHC) to estimate the number of nets in each district; sum the estimated variances from the two strata to estimate the variance for the district.

5 At this point you have the estimated total number of nets and its estimated variance, for each district. Now use two-stage cluster sampling formulas to estimate the total number of nets for each region.

6 Finally, add the estimated totals for each region to estimate the total number of bed nets in the Gambia. Add the region variances as called for in stratified sampling.

Sounds a little complicated, doesn't it? And we have not even included ratio estimation, which would almost certainly be incorporated here because we know approximate population numbers for the numbers of beds at each stage. Fortunately, we do not always have to go to this much work in complex surveys. As we shall see later in this chapter and in Chapter 9, we can use sampling weights and computer-intensive methods to avoid much of this effort. Using a with-replacement variance estimator allows us to estimate a variance using only the weighting, stratification, and psu information. ∎

7.1.2 Ratio Estimation in Complex Surveys

Ratio estimation may be used at almost any level of the survey, although it is usually used near the top. We discussed ratio estimation with stratified random sampling in Section 4.5. The principles are the same for any probability sampling design used within the strata in a stratified multistage sample. Suppose that the population total t_x is known for an auxiliary variable x, and that $\hat{t}_y$ and $\hat{t}_x$ are unbiased estimators for t_y and t_x, respectively, from the sample. The **combined ratio estimator** of the population total for variable y is

$$\hat{t}_{yrc} = \hat{B}t_x,$$

where

$$\hat{B} = \frac{\hat{t}_y}{\hat{t}_x};$$

in Section 9.1 we show that the mean squared error (MSE) of $\hat{t}_{yrc}$ can be estimated by

$$\hat{V}(\hat{t}_{yrc}) = \left(\frac{t_x}{\hat{t}_x}\right)^2 \left[\hat{V}(\hat{t}_y) + \hat{B}^2\hat{V}(\hat{t}_x) - 2\hat{B}\widehat{\text{Cov}}(\hat{t}_y, \hat{t}_x)\right].$$

The **separate ratio estimator** applies ratio estimation within each stratum first, then combines the strata:

$$\hat{t}_{yrs} = \sum_{h=1}^{H}\hat{t}_{yhr} = \sum_{h=1}^{H}t_{xh}\frac{\hat{t}_{yh}}{\hat{t}_{xh}},$$

with

$$\hat{V}(\hat{t}_{yrs}) = \sum_{h=1}^{H}\hat{V}(\hat{t}_{yhr}).$$

As we saw in Section 5.2.3, we often use ratio estimation for estimating means, letting the auxiliary variable x_i be 1 if unit i is in the sample and 0 otherwise. Then $\hat{t}_x$ estimates the population size, and the ratio $\hat{B}=\hat{t}_y/\hat{t}_x$ divides the estimated population total by the estimated population size.

Other ratios are often of interest as well. One quantity of interest in the bed net survey was the proportion of beds that have nets. In this case, x refers to beds and y refers to nets. Then, t_x is the total number of beds in the population and t_y is the total number of bed nets in the population. We estimate the proportion of beds that have nets by $\hat{B}=\hat{t}_y/\hat{t}_x$. Alternatively, the ratio can be estimated separately for each region if it is desired to compare the bed net coverage for the regions.

7.1.3 Simplicity in Survey Design

All these design components have been shown to increase efficiency in survey after survey. Sometimes, though, an inexperienced survey designer is tempted to use a complex sampling design simply because it is there or has been used in the past, not because it has been demonstrated to be more efficient. Make sure you know from pretests or previous research that a complex design really is more efficient and practical. A simpler design giving the same amount of information per dollar spent is almost always to be preferred to a more complicated design: It is often easier to administer and analyze, and data from the survey are less likely to be analyzed incorrectly by subsequent analysts. A complex design should be efficient for estimating *all* quantities of primary interest—an optimal allocation in stratified sampling for estimating the total amount U.S. businesses spend on health care benefits may be very inefficient for estimating the percentage of businesses that declare bankruptcy in a year.

7.2
Sampling Weights

7.2.1 Constructing Sampling Weights

In most large sample surveys, weights are used to calculate point estimates. We have already seen how sampling weights are used in stratified sampling and in cluster sampling. In without-replacement sampling, the sampling weight for an observation unit is always the reciprocal of the probability that the observation unit is included in the sample.

Recall that for stratified random sampling,

$$\hat{t}_{str} = \sum_{h=1}^{H} \sum_{j \in S_h} w_{hj} y_{hj},$$

where the sampling weight $w_{hj} = (N_h/n_h)$ can be thought of as the number of observations in the population represented by the sample observation y_{hj}. The probability of selecting the jth unit in the hth stratum to be in the sample is $\pi_{hj} = n_h/N_h$, so the sampling weight is simply the inverse of the probability of selection: $w_{hj} = 1/\pi_{hj}$.

The sum of the sampling weights in stratified random sampling equals the population size N; each sampled unit "represents" certain number of units in the population, so the whole sample "represents" the whole population. The stratified sampling estimator of $\bar{y}_U$ is

$$\bar{y}_{str} = \frac{\displaystyle\sum_{h=1}^{L} \sum_{j \in S_h} w_{hj} y_{hj}}{\displaystyle\sum_{h=1}^{L} \sum_{j \in S_h} w_{hj}}.$$

The same forms of the estimators were used in cluster sampling in Section 5.3, and in the general form of weighted estimators in Section 6.4.4. In cluster sampling with equal probabilities, for example,

$$w_{ij} = \frac{NM_i}{nm_i} = \frac{1}{\text{probability that the } j\text{th ssu in the } i\text{th psu is in the sample}}.$$

Again,

$$\hat{t} = \sum_{i\in S}\sum_{j\in S_i} w_{ij}y_{ij}$$

and the estimator of the population mean is

$$\frac{\hat{t}}{\displaystyle\sum_{i\in S}\sum_{j\in S_i} w_{ij}}.$$

For cluster sampling with unequal probabilities, when π_i is the probability that the ith psu is in the sample, and $\pi_{j|i}$ is the probability that the jth ssu is in the sample given that the ith psu is in the sample, the sampling weights are $w_{ij} = 1/(\pi_i\pi_{j|i})$.

For three-stage cluster sampling, the principle extends: Let w_p be the weight for the psu, $w_{s|p}$ be the weight for the ssu, and $w_{t|s,p}$ be the weight associated with the tsu (tertiary sampling unit). Then the overall sampling weight for an observation unit is

$$w = w_p \times w_{s|p} \times w_{t|s,p}.$$

All the information needed to construct point estimates is contained in the sampling weights; when computing point estimates, the sometimes cumbersome probabilities with which psus, ssus, and tsus are selected appear only through the weights. But the sampling weights give no information on how to find standard errors of the estimates, and thus knowing the sampling weights alone will not allow you to do inferential statistics. Variances of estimates depend on the probabilities that any pair of observation units is selected to be in the sample, and requires more knowledge of the sampling design than given by weights alone.

Very large weights are often truncated or smoothed, so that no single observation has a very large contribution to the overall estimate. While this biases the estimators, it can reduce the MSE (Elliott and Little, 2000). Truncation is often used when weights are used to adjust for nonresponse, as described in Chapter 8.

Since we consider stratified multistage designs in the remainder of this book, from now on we will adopt a unified notation for estimators of population totals. We consider y_i to be a measurement on observation unit i, and w_i to be the sampling weight of observation unit i. Thus, for a stratified random sample, y_i is an observation unit within a particular stratum, and $w_i = N_h/n_h$, where unit i is in stratum h. This allows us to write the general estimator of the population total as

$$\hat{t}_y = \sum_{i\in S} w_iy_i, \tag{7.1}$$

where all measurements are at the observation unit level. The general estimator of the population mean is

$$\hat{\bar{y}} = \frac{\hat{t}_y}{\displaystyle\sum_{i\in S} w_i}; \tag{7.2}$$

$\sum_{i\in S} w_i$ estimates the number of observation units in the population.

EXAMPLE **7.2**   The Gambia bed net survey in Example 7.1 was designed so that within each region each compound would have almost the same probability of being included in the survey; probabilities varied only because different districts had different numbers of persons in PHC villages and because number of compounds might not always be exactly proportional to village population. For the central region PHC villages, for example, the probability that a given compound would be included in the survey was

$$P(\text{district selected}) \times P(\text{village selected} \mid \text{district selected})$$
$$\times P(\text{compound selected} \mid \text{district and village selected})$$
$$\propto \frac{D1}{R} \times \frac{V}{D2} \times \frac{1}{C},$$

where
C = number of compounds in the village
V = number of people in the village
$D1$ = number of people in the district
$D2$ = number of people in the district in PHC villages
R = number of people in PHC villages in all central districts

Since the number of compounds in a village will be roughly proportional to the number of people in a village, V/C should be approximately the same for all compounds. The value of R is also the same for all compounds within a region. The weights for each region, the reciprocals of the inclusion probabilities, differ largely because of the variability in $D1/D2$. As R varies from stratum to stratum, though, compounds in more populous strata have higher weights than those in less populous strata.   ■

7.2.2   Self-Weighting and Non-Self-Weighting Samples

Sampling weights for all observation units are equal in self-weighting surveys. Self-weighting samples can, in the absence of nonsampling errors, be considered representative of the population because each observed unit represents the same number of unobserved units in the population. Standard statistical methods may then be applied to the sample to obtain point estimates. A histogram of the sample values displays the approximate frequencies of occurrence in the population; the sample mean, median, and other sample statistics estimate the corresponding population quantities. In addition, self-weighting samples often yield smaller variances, and sample statistics are more robust (Kish, 1992).

Most large self-weighting samples used in practice are not simple random samples (SRSs), however. Stratification is used to reduce variances and obtain separate estimates for domains of interest; clustering, often with unequal probabilities, is used to reduce costs. You need to use statistical software that is specifically designed for survey data to obtain valid statistics from complex survey data. If you instead use statistical software that is intended for data fulfilling the usual statistical assumption that observations are independent and identically distributed, the standard errors, hypothesis test results, and confidence intervals (CIs) produced by the software will be wrong. If the sample is not self-weighting, estimates of means and percentiles produced by standard statistical software will also be biased. When you read a paper or book in which the authors analyze data from a complex survey, see whether they accounted for the data structure in the analysis, or whether they simply ran the raw data through a non-survey statistical package procedure and reported the results. If the latter, their inferential results must be viewed with suspicion; it is possible that they only find statistical significance because they fail to account for the survey design in the standard errors.

Many surveys, of course, purposely sample observation units with different probabilities. The disproportionate sampling probabilities often occur in the stratification: a higher sampling fraction is used for a stratum of large business establishments than for a stratum of small business establishments. The United States National Health and Nutrition Examination Survey (NHANES) purposely oversamples areas containing large black and Mexican-American populations (Ezzati-Rice and Murphy, 1995; National Center for Health Statisitcs, 2005); oversampling these populations allows comparison of the health of racial and ethnic minorities.

7.2.3   Weights and a Model-Based Analysis of Survey Data

You might think that a statistician taking a model-based perspective could ignore the weights altogether. After all, to a model-based survey statistician, the sample design is irrelevant and the important part of the analysis is finding a model that summarizes the population structure; as sampling weights are functions of the probabilities of selection in the design, perhaps they too are irrelevant.

But the model-based and randomization-based approaches are not as far apart as some of the literature debating the issue would have you believe. Remember, a statistician designing a survey to be analyzed using weights implicitly visualizes a model for the data; NHANES is stratified and subpopulations oversampled precisely because researchers believe there will be a difference among the subpopulations. Such differences also need to be included in the model. If you ignore the weights in analyzing data from NHANES, for example, you implicitly assume that whites, blacks, and Mexican Americans are largely interchangeable in health status. Ignoring the clustering in the inference assumes that observations in the same cluster are uncorrelated, which is not generally true. A data analyst who ignores stratification variables and dependence among observations is not fitting a good model to the data but is simply being lazy. A good analysis of survey data using models is difficult, and requires extensive validation of the model. The books edited by Skinner, Holt, and Smith (1989a) and Chambers and Skinner (2003) contain several chapters on modeling data from complex surveys.

Many researchers have found that sampling weights contain information that can be used in a model-based analysis. Little (1991) develops a class of models that result in estimators that behave like estimators obtained using survey weights. Pfeffermann (1993, 1996) describes a framework for deciding on whether to use sampling weights in regression models of survey data. Thompson (1997) and Binder and Roberts (2003) discuss differences between model-based and design based inference.

7.3
Estimating a Distribution Function

So far, we have concentrated on estimating population means, totals, and ratios. Historically, sampling theory was developed primarily to find these basic statistics and to answer questions such as "What percentage of adult males are unemployed?" or "What is the total amount of money spent on health care in the United States?" or "What is the ratio of the numbers of exotic to native birds in an area?"

But statistics other than means or totals may be of interest. You may want to estimate the median income in Canada, find the 95th percentile of test scores, or construct a histogram to show the distribution of fish lengths. An insurance company may set reimbursements for a medical procedure using the 75th percentile of charges for the procedure. We can estimate any of these quantities (but not their standard errors, however) with sampling weights. The sampling weights allow us to construct an empirical distribution for the population.

Suppose the values for the entire population of N units are known. Then any quantity of interest may be calculated from the **probability mass function**,

$$f(y) = \frac{\text{number of units whose value is } y}{N}$$

or the **cumulative distribution function** (cdf)

$$F(y) = \frac{\text{number of units with value } \leq y}{N} = \sum_{x \leq y} f(x).$$

In probability theory, these are the probability mass function and cdf for the random variable Y, where Y is the value obtained from a random sample of size one from the population. Then $f(y) = P\{Y = y\}$ and $F(y) = P\{Y \leq y\}$. Of course, $\sum f(y) = F(\infty) = 1$.

Any population quantity can be calculated from the probability mass function or cdf. The population mean is

$$\bar{y}_U = \sum_{\substack{\text{values of } y \\ \text{in population}}} y f(y).$$

The population variance, too, can be written using the probability mass function:

$$S^2 = \frac{1}{N-1} \sum_{i=1}^{N} (y_i - \bar{y}_U)^2$$

$$= \frac{N}{N-1} \sum_y f(y) \left[y - \sum_x x f(x) \right]^2$$

$$= \frac{N}{N-1} \left[\sum_y y^2 f(y) - \left(\sum_x x f(x) \right)^2 \right].$$

If the cdf F were continuous, we would define the population median to be the value m satisfying $F(m) = 1/2$. Because F has jumps at the values of y in the population, however, it is possible that the function $F(y)$ does not attain the value $1/2$. We define the finite population median to be the value m satisfying $F(m) = 1/2$ if such a value exists; otherwise, a population median is any value in the interval $[m_1, m_2]$, where m_1 is the largest value of y in the population with $F(y) < 1/2$ and m_2 is the smallest value of y with $F(y) > 1/2$. In general, θ_q is a $100q$th quantile (percentile) if $F(\theta_q) = q$ if such a value exists; otherwise, $\theta_q \in [a, b]$ where a is the largest population

FIGURE   7.1
The function $F(y)$ for the population of heights.

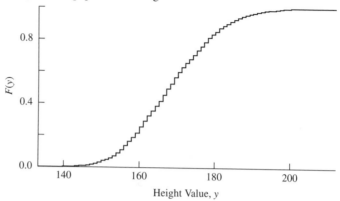

value of y with $F(y) < q$ and b is the smallest value of y with $F(y) > q$. If $q < 1/N$, θ_q is the smallest value of y and if $q > 1 - 1/N$, θ_q is the largest value of y.

EXAMPLE 7.3   Consider an artificial population of 1000 men and 1000 women in file htpop.dat. Each person's height is measured to the nearest centimeter (cm). The frequency table in file htcdf.dat gives the probability mass function and cdf for the 2000 persons in the population. Figures 7.1 and 7.2 show the graphs of $F(y)$ and $f(y)$. The population mean is $\bar{y}_U = \sum yf(y) = 168.6$.

Now let's take an SRS of size 200 from the population (file htsrs.dat). An SRS is self-weighting; each person in the sample represents $w_i = 10$ persons in the population. Hence, the histogram of the sample should resemble $f(y)$ from the population; Figure 7.3 shows that it does.

But suppose a stratified sample of 160 women and 40 men (file htstrat.dat) is taken instead of a self-weighting sample. In the stratified sample, each woman has weight $1000/160 = 6.25$ and each man has weight $1000/40 = 25$. A histogram of the raw data will distort the population distribution, as illustrated in Figure 7.4. The sample mean and median are too low because men are underrepresented in the sample.   ∎

Sampling weights allow us to construct empirical probability mass and cdfs for the data. Any statistics can then be calculated. Define the **empirical probability mass function** (epmf) to be the sum of the weights for all observations taking on the value y, divided by the sum of all the weights:

$$\hat{f}(y) = \frac{\displaystyle\sum_{i \in S: y_i = y} w_i}{\displaystyle\sum_{i \in S} w_i}.$$

The **empirical cumulative distribution function** (empirical cdf) $\hat{F}(y)$ is the sum of all weights for observations with values $\leq y$, divided by the sum of all weights:

$$\hat{F}(y) = \sum_{x \leq y} \hat{f}(x).$$

FIGURE 7.2

The function $f(y)$ for the population of heights.

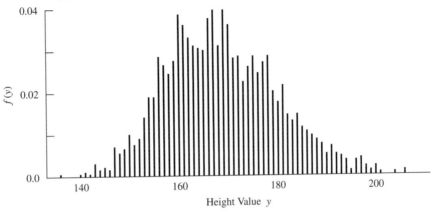

FIGURE 7.3

A histogram of raw data from an SRS of size 200. The general shape is similar to that of $f(y)$ for the population because the sample is self-weighting.

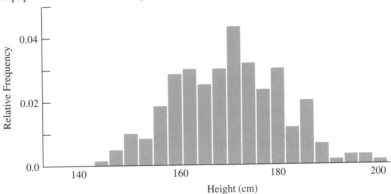

FIGURE 7.4

A histogram of raw data (not using weights) from a stratified sample of 160 women and 40 men. Tall persons are underrepresented in the sample, so this histogram distorts the population distribution.

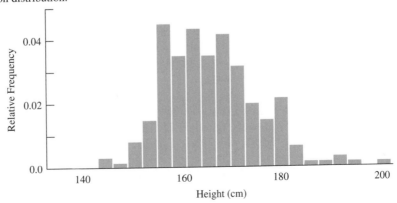

FIGURE   7.5

The estimate $\hat{f}(y)$ for the stratified sample of 160 women and 40 men.

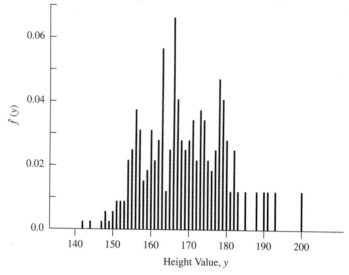

For a self-weighting sample, $\hat{f}(y)$ reduces to the relative frequency of y in the sample. For a non-self-weighting sample, $\hat{f}(y)$ and $\hat{F}(y)$ are attempts to reconstruct the population functions f and F from the sample. The weight w_i is the number of population units represented by unit i, so $\displaystyle\sum_{i \in S: y_i = y} w_i$ estimates the total number of units in the population that have value y. If all weights are integers, we can view $\hat{F}(y)$ as the cdf of a "pseudo-population" constructed by repeating observation y_i w_i times (see Exercise 6). Consider a probability sample of size 3 from a population of size 10, with sampled values given in the following table.

Sample value y_i	4	6	7
Weight w_i	2	3	5

Using the weights, we can construct a pseudo-population with values $\{4, 4, 6, 6, 6, 7, 7, 7, 7, 7\}$; each value of y_i is replicated w_i times. This is not the true population, of course, but it represents an attempt to estimate the population from the sample. For this sample of size 3, $\hat{F}(4) = 2/10$, $\hat{F}(6) = 5/10$, and $\hat{F}(7) = 1$. In most surveys, weights are not integers and the population size is too large to permit constructing a pseudo-population, but it is sometimes helpful to think of $\hat{F}(y)$ as an estimator of the population cdf $F(y)$.

EXAMPLE 7.4   Each woman in the stratified sample in Example 7.3 has sampling weight 6.25; each man has sampling weight 25. The empirical probability mass function from the stratified sample is in Figure 7.5. The weights correct the underrepresentation of taller people found in the histogram in Figure 7.4. The scarcity of men in the sample, however, demands a price: The right tail of $\hat{f}(y)$ has a few spikes of size $0.0125 = 25/2000$, each spike coming from one man in the stratified sample, rather than a number of values tapering off. ∎

The epmf $\hat{f}(y)$ can be used to find estimates of population quantities. First express the population characteristic in terms of $f(y)$: $\bar{y}_U = \sum yf(y)$ or

$$S^2 = \frac{N}{N-1}\left[\sum_y f(y)\left\{y - \sum_x xf(x)\right\}^2\right] = \frac{N}{N-1}\left[\sum_y y^2 f(y) - \left\{\sum_y yf(y)\right\}^2\right].$$

Then, substitute $\hat{f}(y)$ for every appearance of $f(y)$ to obtain an estimate of the population characteristic. Using this method, then,

$$\hat{\bar{y}} = \sum_y y\hat{f}(y) = \frac{\displaystyle\sum_{i\in S} w_i y_i}{\displaystyle\sum_{i\in S} w_i}$$

and

$$\hat{S}^2 = \frac{N}{N-1}\left[\sum_y y^2 \hat{f}(y) - \left\{\sum_y y\hat{f}(y)\right\}^2\right]. \tag{7.3}$$

Population quantiles are estimated similarly. Recall that θ_q is a $100q$th quantile if $\theta_q \in [a, b]$ where a is the largest population value of y with $F(y) < q$ and b is the smallest value of y with $F(y) > q$. Since the empirical cdf $\hat{F}$ is a step function, we usually interpolate to find a unique value for the quantile. Let y_1 be the largest value in the sample for which $\hat{F}(y_1) \le q$ and let y_2 be the smallest value in the sample for which $\hat{F}(y_2) \ge q$. Then

$$\hat{\theta}_q = y_1 + \frac{q - \hat{F}(y_1)}{\hat{F}(y_2) - \hat{F}(y_1)}(y_2 - y_1). \tag{7.4}$$

Table 7.1 shows the difference in the estimates when weights for the stratified sample are incorporated through the function $\hat{f}(y)$. The statistics calculated using weights are much closer to the population quantities. We note that estimators calculated using this method are not necessarily unbiased, however, or numerically stable. In particular, the estimator of S^2 in (7.3) is sensitive to roundoff error and in practice a different estimator such as those studied by Courbois and Urquhart (2004) may be preferable.

This simple example only involved stratification, but the method is the same for any survey design. You need to know only the sampling weights to estimate almost anything through the empirical cdf. If desired, you can smooth the empirical cdf before estimating quantiles. Nusser et al. (1996) use a semiparametric approach for estimating daily dietary intakes of various nutrients from the Continuing Survey of Food Intakes by Individuals, a stratified multistage survey.

Although the weights may be used to find point estimates through the empirical cdf, calculating standard errors is much more complicated, and requires knowledge of the sampling design. Typically, in a stratified multistage sample, we calculate variances assuming that psus were selected with replacement in each stratum. This

TABLE 7.1

Estimates from samples in Example 7.3

Quantity	Population	SRS	Stratified, No Weights	Stratified with Weights
Mean	168.6	168.9	164.6	169.0
Median	167.3	168.8	162.8	167.6
25th percentile	159.9	159.7	156.6	160.7
90th percentile	183.2	183.4	177.5	181.5
Variance, S^2	124.5	122.6	93.4	116.8

simplifies the analysis considerably, since we do not need to know the joint inclusion probabilities of the psus or any information about the subsampling design to calculate the with-replacement variance. In most surveys, the with-replacement variance estimates are larger than the without-replacement variance estimates, but the increase is small if the first-stage sampling fractions are small. Variances of statistics calculated from the empirical cdf will be discussed in Chapter 9.

7.4
Plotting Data from a Complex Survey

Simple plots reveal much information about data from a small SRS or representative systematic sample. Histograms or smoothed density estimates display the shape of the data; scatterplots and scatterplot matrices show relationships between variables; other plots discussed in Chambers et al. (1983), Cleveland (1994), and Robbins (2005) emphasize other features of the data. In a complex sampling design, however, a single plot will not display the richness of the data. As seen in Figure 7.4, plots commonly used for SRSs can mislead when applied to raw data from non-self-weighting samples. Clustering causes numerous difficulties in plotting data from a complex survey, as noted in Example 5.7, because we may want to display the clustering structure as well as possible unequal weighting in the graphs; the problems are compounded because data sets from surveys are often very large and involve several levels of clustering.

Data should be plotted both with and without weights to see the effect of the weights. In addition, data should be plotted separately for each stratum, and for each psu if possible to examine variability in the responses. You already know how to plot the raw data without weights; in this section we provide some examples of incorporating the weights into the graphics.

7.4.1 Univariate Plots

7.4.1.1 Histograms

One of the simplest plots for displaying the shape of data is the histogram. To construct a relative frequency histogram for an SRS of size n, divide the range of the data into

k bins with each bin having width b. Then the height of the histogram in the jth bin is

$$\text{height}(j) = \frac{\text{relative frequency for bin } j}{b} = \frac{\displaystyle\sum_{i \in S} u_i(j)}{bn},$$

where $u_i(j) = 1$ if observation i is in bin j and 0 otherwise. If a sample is self-weighting, as with an SRS, a regular histogram of the sample data will estimate the population probability mass function.

We saw in Figure 7.4, though, that if a sample is not self-weighting a histogram of the raw data may underrepresent some parts of the population in the display. We can use the sampling weights to construct a histogram that estimates the population histogram. As before, divide the range of the data into k bins with each bin having width b. Now use the sampling weights w_i to find the height of the histogram in bin j:

$$\text{height}(j) = \frac{\displaystyle\sum_{i \in S} w_i u_i(j)}{b \displaystyle\sum_{i \in S} w_i}. \tag{7.5}$$

Dividing by the quantity $b \sum_{i \in S} w_i$ ensures that the total area under the histogram equals 1.

EXAMPLE 7.5   To construct a histogram of the height data from the stratified sample in file htstrat.dat (Example 7.3), first decide on a bin width, b. We decide to use $b = 3$ as in Figure 7.4. This choice gives 20 histogram bins. The cutpoints for the histogram bins are at 141, 144, 147, 150, ..., 198, and 201. The first histogram bar includes persons in the sample whose heights are in the interval (141, 144]; the sample contains one woman with height 142 and one woman with height 144. Each woman in the sample has sampling weight 6.25, so the height of the first histogram bar is

$$\frac{2(6.25)}{b \displaystyle\sum_{i \in S} w_i} = \frac{12.5}{(3)(2000)} = 0.00208.$$

The biggest histogram bar includes persons in the sample with heights in the interval (165, 168]; the sample contains 19 women and 6 men with heights in this range, so the height for the bar corresponding to (165, 168] is

$$\frac{19(6.25) + 6(25)}{b \displaystyle\sum_{i \in S} w_i} = \frac{268.75}{(3)(2000)} = 0.04479.$$

The heights for the other histogram bars are computed similarly. The histogram for the stratified sample, incorporating the weights, is in Figure 7.6. The histogram with the weights shows higher relative frequencies for heights over 165 than does the histogram without weights in Figure 7.4; Figure 7.6 gives a better picture of the shape of the population distribution. SAS code for creating histograms that incorporate the sampling weights is provided on the website. ∎

FIGURE   **7.6**

Histogram of height data from stratified sample, incorporating the sampling weights.

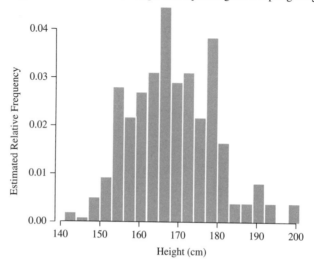

7.4.1.2   Boxplots

Side-by-side boxplots, sometimes called box-and-whisker plots, are a useful way to display the distribution of a population or to compare domain distributions visually. The box in a boxplot has lines at the 25th, 50th, and 75th quantiles, and whiskers that extend to the extremes of the data (or, alternatively, to a multiple of the interquartile range). If the sample is not self-weighting, the weights should be used to calculate the quantiles in the display.

EXAMPLE 7.6   Consider again the height data in file htstrat.dat. We use the sampling weights to estimate the quantiles of the data. Using all 200 observations, we note that $\hat{F}(167) = \sum_{y_i \leq 167} w_i / \sum_{i \in S} w_i = 0.4844$ and $\hat{F}(168) = \sum_{y_i \leq 168} w_i / \sum_{i \in S} w_i = 0.5125$. Thus, any value between 167 and 168 is a median. Several methods can be used to choose one value to estimate the median. We interpolate between the two bounds according to the empirical cdf probabilities, and use

$$m = 167 + \frac{0.5 - 0.4844}{0.5125 - 0.4844}(168 - 167) = 167.6.$$

We find the 25th and 75th percentiles similarly. For the 25th percentile, $\hat{F}(160) = 0.2344$, $\hat{F}(161) = 0.2563$, and

$$\text{estimated 25th percentile} = 160 + \frac{0.25 - 0.2344}{0.2563 - 0.2344}(161 - 160) = 160.7.$$

For the 75th percentile, $\hat{F}(176) = 0.7344$, $\hat{F}(177) = 0.7594$, and

$$\text{estimated 75th percentile} = 176 + \frac{0.75 - 0.7344}{0.7594 - 0.7344}(177 - 176) = 176.6.$$

SAS PROC SURVEYMEANS will calculate percentiles using the weights; the SAS code for calculating percentiles is given on the website with output below. We'll

FIGURE   7.7

Boxplots of height data from stratified sample, incorporating the sampling weights. The first box uses data from the entire sample, the second box uses data from the women, and the third box uses data from the men.

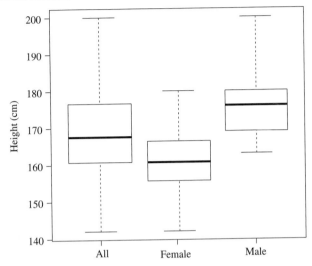

discuss how these standard errors are calculated in Section 9.5.2; for boxplots, we just use the point estimates.

Quantiles

Variable	Percentile	Estimate	Std Error	95% Confidence Limits	
height	25% Q1	160.714286	0.693338	158.759271	161.493819
	50% Median	167.555556	1.011620	165.569707	169.559572
	75% Q3	176.625000	1.330767	172.910731	178.159325

Quantiles for a domain are estimated in similar fashion, using an empirical cdf that includes only observations in that domain. Define $x_i = 1$ if observation i is in domain d, and 0 otherwise. Then the empirical cdf restricted to domain d, is

$$\hat{F}_d(y) = \frac{\sum\limits_{i \in S: y_i \leq y} x_i w_i}{\sum\limits_{i \in S} x_i w_i}.$$

For the women, $\hat{F}_F(155) = 0.2$, $\hat{F}_F(156) = 0.275$, so the 25th percentile for women is estimated by $155 + (0.25 - 0.2)/(0.275 - 0.2) = 155.7$. Similarly, the median for women is 160.7 and the 75th percentile for women is 166.4. The 25th, 50th, and 75th percentiles for men are 169, 176, and 180, respectively. Side-by-side boxplots of the data, using these estimated quantiles and extending the whiskers to the range of the data, are shown in Figure 7.7. Similar boxplots can be constructed from the estimated quantiles using SAS code on the website. ∎

7.4.1.3   Density estimates*

Smoothed density estimates are useful for displaying the shape of the estimated population data for a variable that takes on a wide range of values. The books by Scott (1992), Wand and Jones (1995), and Simonoff (1996) are useful references on smoothing with data from an SRS. The idea of smoothing methods is to create a smooth version of a histogram. Instead of having bars in a histogram, one could create a plot by connecting the heights at the midpoints of the histogram bins. Such a plot would not be particularly smooth, however, and could be improved by using each possible value of y as the midpoint of a histogram bin of width b, finding the height for that bin, and then drawing a line through those values. In essence, the histogram bars slide continuously along the horizontal axis; as points enter and leave the bar, the height corresponding to the midpoint changes. A symmetric density function K, called a kernel function, is used to allow more flexibility in the smoothing method. Bellhouse and Stafford (1999) and Buskirk and Lohr (2005) adapted kernel density estimation to survey data by incorporating the weights, with

$$\hat{f}(y; b) = \frac{1}{b \sum_{i \in S} w_i} \sum_{i \in S} w_i K \left[\frac{y - y_i}{b} \right].$$

Commonly used kernel functions include the normal kernel function $K_N(t) = \exp(-t^2/2)/\sqrt{2\pi}$ and the quadratic kernel function $K_Q(t) = \frac{3}{4}(1 - t^2)$ for $|t| < 1$. The sliding histogram described above corresponds to a box kernel with $K_B(t) = 1$ for $|t| \leq 1/2$ and $K_B(t) = 0$ for $|t| > 1/2$; in that case, $\hat{f}(y; b)$ corresponds to the histogram height given in (7.5) for a point y in the middle of a bin of width b.

Figure 7.8 shows a smoothed density estimate for the height data from Examples 7.3 and 7.5. The website gives the code for constructing this plot. As with the histogram in Figure 7.6, using the sampling weights increases the estimated density in the right tail despite the paucity of data in that region.

The choice of b, called the bandwidth, determines the amount of smoothing to be used. Small values of b use little smoothing since the sliding window is small. A large value of b provides much smoothing since each point in the plot represents the weighted average of many points from the data. One problem with density estimation in survey data is that respondents may round their answers. For example, some respondents may round their height to 165 or 170 cm, causing spikes at those values. You may want to choose b large to increase the amount of smoothing, or you may want to adopt a model for the effect of rounding by the respondent.

7.4.1.4   Displaying stratification and clustering information

The histograms, boxplots, and density estimates for survey data use the sampling weights to approximate the corresponding plots that would be constructed if we knew the data values for the entire population.

EXAMPLE 7.7   The 1987 Survey of Youth in Custody (Beck et al., 1988; U.S. Department of Justice, 1989) sampled juveniles and young adults in long-term, state-operated juvenile institutions. Residents of facilities at the end of 1987 were interviewed about family

FIGURE   7.8

Estimated density function for the stratified sample of heights. The circles represent the data points in the sample.

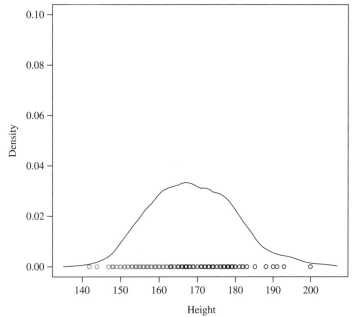

background, previous criminal history, and drug and alcohol use. Selected variables from the survey are in the file syc.dat.

The facilities form a natural cluster unit for an in-person survey; the sampling frame of 206 facilities was constructed from the 1985 Children in Custody (CIC) Census. The psus (facilities) were divided into 16 strata by number of residents in the 1985 CIC. Each of the 11 facilities with 360 or more youth formed its own stratum (strata 6–16); each of these facilities was included in the sample and residents of the 11 facilities were subsampled. In strata 1–5, facilities were sampled with probability proportional to size from the 195 remaining facilities; residents were subsampled with predetermined sampling fractions. Table 7.2 contains information about the strata.

The stratum boundaries were chosen so that the number of residents in each stratum would be comparable. It was originally intended that each resident have probability 1/8 of inclusion in the sample, which would result in a self-weighting sample with constant weight 8. The facilities in strata 14 and 16, however, had experienced a great deal of growth between 1985 and 1987, so the sampling fractions in those strata were changed to 1/11 and 1/12, respectively. In strata 1–5, weights varied from about 5 to about 15, depending on the facility's inclusion probability and the predetermined sampling fraction in that facility. The weights were further adjusted for nonresponse, and to match the sample counts with the 1987 census count of youths in long-term, state-operated facilities. After all weighting adjustments were made, weights ranged from 5 (in stratum 4) to 58 (for some youths in states that required parental permission and hence had lower response rates).

TABLE 7.2

Survey of Youth in Custody Stratum Information

Stratum	CIC Size (Number of Residents)	Number of psus in Frame	Number of Residents in CIC	Number of Eligible psus in Sample
1	1–59	99	2881	11
2	60–119	39	3525	7
3	120–179	30	4355	7
4	180–239	13	2594	7
5	240–359	14	4129	7

To estimate population quantities with standard errors from a stratified multistage sample such as the Survey of Youth in Custody, you need to know the weights, the stratification variable, and the variable describing the first-stage cluster units. In syc.dat, the weights are in variable *finalwt*, the strata are in variable *stratum* and the facilities (psus) are in variable *facility*. There is only one facility in each of strata 6–16, so that a stratified random sample of individuals is taken in each of those strata; we define the psus for those strata to be individuals rather than the facility so that they contribute to the standard errors of the estimates. SAS code for calculating the average and percentiles of *age* is on the website, with output given below. We shall discuss how to compute standard errors of quantiles in Section 9.5.2.

```
                          Std Error
Variable       Mean       of Mean       95% CL for Mean
------------------------------------------------------------
age          16.639293    0.128882    16.386326    16.892260
------------------------------------------------------------
```

```
                                 Quantiles

Variable   Percentile     Estimate    Std Error    95% Confidence Limits
------------------------------------------------------------------------
age          0% Min       11.000000       .            .           .
            25% Q1        14.805746    0.225394    14.363348    15.248145
            50% Median    15.917433    0.175991    15.572001    16.262864
            75% Q3        17.205184    0.154592    16.901754    17.508613
           100% Max       24.000000       .            .           .
------------------------------------------------------------------------
```

Let's look at some plots of the age of residents. Some youths are over age 18 because California Youth Authority facilities were included in the sample. As the survey aimed to be approximately self-weighting, the histogram of the unweighted data in Figure 7.9 and the empirical probability mass function incorporating the weights (variable *finalwt*) in Figure 7.10 are overall similar in shape. Some discrepancies appear on closer examination, though—the weights indicate that youths

FIGURE **7.9**

Histogram of all data, not incorporating weights. The histogram shows the distribution of ages in the sample, but does not necessarily reflect the distribution of ages in the population.

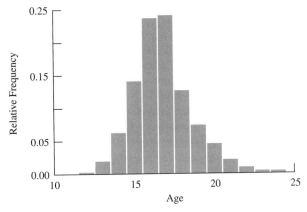

FIGURE **7.10**

Estimated probability mass function for age, $\hat{f}(y)$. The shape is similar to that of the histogram of the raw data, but there are relatively more 15-year-olds and relatively fewer 17-year-olds.

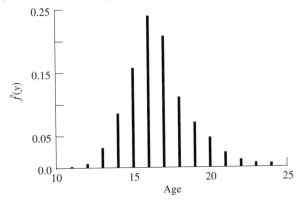

aged 15 were somewhat undersampled due to unequal selection probabilities and nonresponse, while youths aged 17 were somewhat oversampled.

If we were only interested in the distribution for the entire population, we could concentrate on plots such as those in Figures 7.9 and 7.10, and similar plots informative about univariate distributions such as probability-probability or quantile-quantile plots (see Exercises 19 and 20). But we would also like to explore stratum-to-stratum differences in age distribution. Figure 7.11 incorporates weights into boxplots of the data.

As the response variable age is discrete, we can show even more detail for each stratum. Figure 7.12 displays the sum of the weights for each age within each stratum.

FIGURE   7.11

Boxplot of age distributions for each stratum, incorporating the weights. Note the wide variability from stratum to stratum.

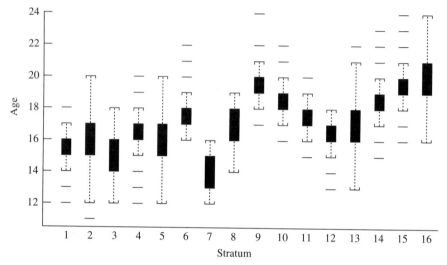

FIGURE   7.12

Age distribution for each stratum. The area of each circle is proportional to the sum of the weights for sample observations in that stratum and age class. The highest number of youths under age 18 are in strata 1 through 5.

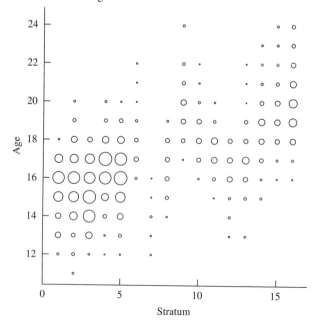

The estimated relative frequency of youths with that age in each stratum is indicated by a circle whose area is proportional to the sum of the weights.

We may also be interested in the facility-to-facility variability. Figures 7.13 and 7.14 show similar plots for the psus in stratum 5. These plots could be drawn for each stratum to show differences in psu variability among the strata. ∎

FIGURE 7.13

Boxplots of ages, incorporating weights, for the psus in stratum 5. The width of each boxplot is proportional to the number of sample observations in that facility.

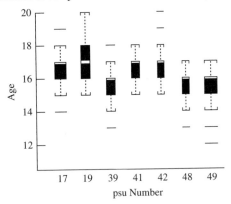

FIGURE 7.14

Age distribution for each psu in Stratum 5. The area of each circle is proportional to the sum of the weights for sample observations in that psu and age class.

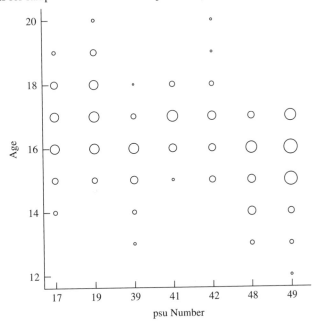

7.4.2   Bivariate Plots

You may also be interested in bivariate relationships among variables. We typically explore such relationships visually using a scatterplot. With complex survey data, unequal weights should be considered for interpreting bivariate relationships.

Since they involve two variables, scatterplots are more complicated than univariate displays. Many government surveys have large amounts of data. The U.S. Current Population Survey (CPS), for example, collects data from 60,000 households each year (U.S. Census Bureau, 2006a). A scatterplot of two continuous variables from the CPS will have so many data points that the graph may be solid black and useless for visual inspection of the data. In addition, if both variables take on integer values, for example age and years in workforce, many points will share the same x and y values.

The challenging part for scatterplots is how to incorporate the weights. In a histogram, only the horizontal axis uses the data values so the weights can be incorporated in the relative frequencies displayed on the vertical axis. But in a scatterplot, the horizontal axis displays information about the x variable and the vertical axis displays information about the y variable, so the weight information must be incorporated by some other means. Korn and Graubard (1998; 1999, Section 3.4) suggest several methods for constructing scatterplots from complex survey data. We illustrate some of these plots, and others, using the 2003–2004 NHANES data, plotting the body mass index vs. age for a stratified multistage sample of approximately 10,000 persons. It is generally a good idea to construct a variety of plots since some plots will work better with a data set than others. Body mass index is calculated as weight/height^2, in units kg/m^2. Age is topcoded at 85 to protect confidentiality of the respondents; any person with age greater than 85 is assigned age value 85. SAS code used to construct these plots from data in file nhanes.dat is given on the website. Figure 7.15 shows a plot of the raw data without weights; as you can see, the data set is so large that it is difficult to see the structure of the bivariate relationship from the graph.

7.4.2.1   Plot with circles proportional in size to observation weights

The plot in Figure 7.15 does not include information about the weights. The NHANES survey is designed so that it oversamples areas with large minority populations. The sample weights of individuals in those areas, therefore, are smaller. To get a better view of the data, we should incorporate the unequal weights. One way of doing that is to use a circle as a plotting symbol, and, for each observation, make the area of the circle proportional to the weight of the observation. This plot for the NHANES data is shown in Figure 7.16. The data are easier to see on this plot than in Figure 7.15 because the plotting symbols are smaller; however, there are still so many data points that some features may be obscured. Observations with small weights have very small circles and are nearly invisible. In larger data sets, such as the CPS, a weighted plot will still have such high data density in areas that the plot will appear to be a solid mass.

FIGURE   7.15

Plot of raw data from NHANES. There are so many data points that it is difficult to see patterns in the plot; in addition, no weighting information is used. This plot is not recommended for complex survey data with unequal weights.

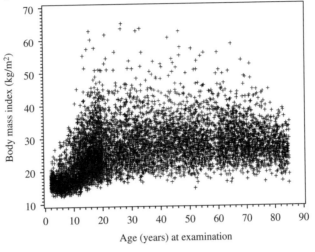

FIGURE   7.16

Weighted circle plot of NHANES data. The circle size for each point is proportional to the weight for that point.

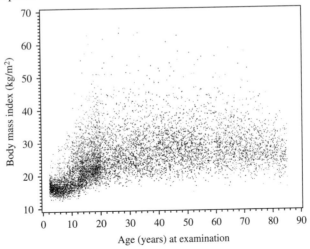

7.4.2.2   Plot a subsample of points

Instead of plotting all the data, we can plot a subset of the data. Since the sampling weight of an observation can be interpreted as the number of population units represented by that unit, a plot of a subsample selected with probabilities proportional to the weights can be interpreted much the same way as a regular scatterplot (see

FIGURE   **7.17**

Plot of subsample of NHANES data, selected with probability proportional to the weight variable.

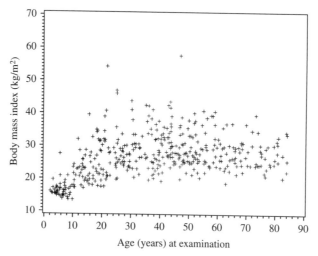

Exercise 23). Figure 7.17 shows a scatterplot of a sample of 500 points selected with replacement from the NHANES data with probability proportional to the weight variable. This plot can be repeated with different subsamples, and each plot will be different. Each plot, however, has less information than the full data set since it is based on a subsample of the data. Unusual observations such as outliers might not appear on a single plot.

7.4.2.3   Use circles to represent weights

This idea is similar to creating bins for a histogram. In a histogram, the y values are grouped into a bin, and the sum of the weights is found for the y values falling into each bin. To extend this idea to a scatterplot, divide the region into rectangles. Find the sum of the weights for the (x, y) values falling in each rectangle. Then, plot a circle with area proportional to the sum of the weights at the midpoint of the rectangle. Figure 7.18 shows the NHANES data in bins formed by rounding the x and y values to the closest multiple of 5. This type of plot is especially useful if the data set contains many points at the same values of (x, y), since the plot displays the multiplicity of points with the same values.

7.4.2.4   Use shading to represent weights

Instead of using the size of the circle to represent the sum of the weights in a bin, as in Figure 7.18, you can use shading to indicate the sum of weights. This often allows you to use more levels for the x and y values than in a plot with circles. For the plot in Figure 7.19, we form bins by rounding the x and y values to the nearest integer, creating a grid of x and y values. For each bin, calculate $z =$ sum of the weights for the points with x and y values in that bin. The shading in Figure 7.19 is proportional to the average of the z values for the four corners of each rectangle.

FIGURE   **7.18**

Circle plot of NHANES data. The area of each circle is proportional to the sum of the weights of the set of observations near the center of the circle.

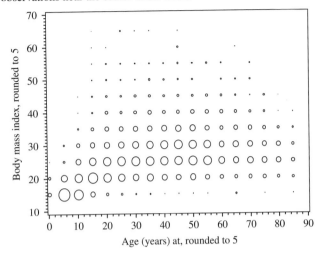

FIGURE   **7.19**

Shaded plot of NHANES data. The shading relies on the sum of the weights for each rectangle.

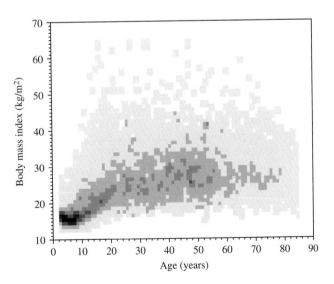

7.4.2.5   Side-by-side boxplots

Instead of creating circles at a set of gridpoints at the data or using shading, we can group the x variable into bins and draw a boxplot at the midpoint of each x bin. We then use the weights to calculate the quantiles of each bin as in Example 7.6.

FIGURE   **7.20**

Side-by-side boxplots of NHANES data. The width of each box is proportional to the sum of the weights of the set of observations used for the box. The + in each box denotes the mean.

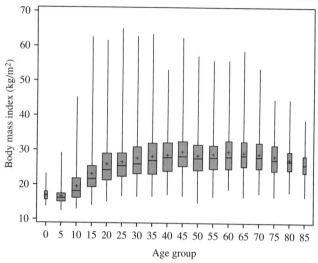

Figure 7.20 shows side-by-side boxplots of the NHANES data, where the age (x) variable is grouped in bins by rounding values to the closest multiple of 5. The width of each boxplot is proportional to the sum of the weights for observations in that bin.

7.4.2.6   Smoothed function estimates

Korn and Graubard (1998) and Bellhouse and Stafford (2001) propose using smoothed function estimates to display trends in the data. Kernel smoothing with weights, as in the smoothed density estimates in Figure 7.8, is used to obtain a trend line. Wand and Jones (1995) and Simonoff (1996) present methods for estimating trend lines for data assumed to be independent and identically distributed; Exercise 25 of Chapter 11 adapts these methods for data with unequal weights. The simplest method for estimating a trend line takes a weighted average of the y_i values that fall in the data window for x; at each point x, the function $g(x)$ is estimated by

$$\hat{g}(x;b) = \frac{\displaystyle\sum_{i\in S} \frac{w_i}{b} K\left(\frac{x - x_i}{b}\right) y_i}{\displaystyle\sum_{i\in S} \frac{w_i}{b} K\left(\frac{x - x_i}{b}\right)}.$$

Other methods fit a straight line or polynomial in each window.

The trend line can be displayed by itself, or (recommended) as an overlay on one of the other plots. Figure 7.21 displays a trend line, computed using local linear regression, with the weighted circle plot from Figure 7.16; we changed the color of the data points from black to gray so that the trend line is more visible. Note that the trend line approximately follows the line of means in the boxplots from Figure 7.20.

FIGURE   **7.21**
Weighted circle plot of NHANES data, with trend line.

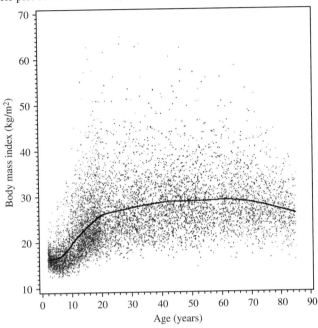

7.5
Design Effects

Cornfield (1951) suggested measuring the efficiency of a sampling plan by the ratio of the variance that would be obtained from an SRS of k observation units to the variance obtained from the complex sampling plan with k observation units. Kish (1965) named the reciprocal of Cornfield's ratio the **design effect** (abbreviated **deff**) of a sampling plan and estimator and used it to summarize the effect of the design on the variance of the estimator:

$$\text{deff(plan,statistic)} = \frac{V(\text{estimator from sampling plan})}{V(\text{estimator from an SRS with same number of observation units})}. \quad (7.6)$$

For estimating a mean from a sample with n observation units,

$$\text{deff(plan,}\hat{\bar{y}}) = \frac{V(\hat{\bar{y}})}{\left(1 - \dfrac{n}{N}\right)\dfrac{S^2}{n}}.$$

The design effect provides a measure of the precision gained or lost by use of the more complicated design instead of an SRS. Although it is a useful concept, it is not a way to avoid calculating variances: You need an estimate of the variance from the complex design to find the design effect. Of course, different quantities in the same survey may have different design effects. Kish showed how the design effect allows you to use prior knowledge for the survey design.

The SRS variance is generally easier to obtain than $V(\hat{\bar{y}})$. If estimating a proportion, the SRS variance is approximately $p(1-p)/n$; if estimating another type of mean, the SRS variance is approximately S^2/n. So if the design effect is approximately known, the variance of the estimator from the complex sample can be estimated by (deff $\times$ SRS variance). We can estimate the variance of an estimated proportion $\hat{p}$ by

$$\hat{V}(\hat{p}) = \text{deff} \times \frac{\hat{p}(1-\hat{p})}{n}.$$

We have seen design effects for several sampling plans. In Section 3.4 the design effect for stratified sampling with proportional allocation was shown to be approximately

$$\frac{V_{\text{prop}}}{V_{\text{SRS}}} \approx \frac{\sum_{h=1}^{H} \dfrac{N_h}{N} S_h^2}{S^2}$$

$$\approx \frac{\sum_{h=1}^{H} \dfrac{N_h}{N} S_h^2}{\sum_{h=1}^{H} \dfrac{N_h}{N} [S_h^2 + (\bar{y}_{Uh} - \bar{y}_U)^2]}. \tag{7.7}$$

Unless all of the stratum means are equal, the design effect for a stratified sample will usually be less than 1—stratification generally gives more precision per observation unit than an SRS.

We also looked extensively at design effects in cluster sampling, particularly in Section 5.2.2. From (5.10), the design effect for single-stage cluster sampling when all psus have M ssus is approximately

$$1 + (M - 1)\,\text{ICC}.$$

The intraclass correlation coefficient (ICC) is usually positive in cluster sampling, so the design effect is usually larger than 1; cluster samples usually give less precision per observation unit than an SRS.

In surveys with both stratification and clustering, we cannot say before calculating variances for our sample whether the design effect for a given quantity will be less than 1 or greater than 1. Stratification tends to increase precision and clustering tends to decrease it, so the overall design effect depends on whether more precision is lost by clustering than gained by stratification.

EXAMPLE **7.8**   For the bed net survey discussed in Example 7.1, the design effect for the proportion of beds with nets was calculated to be 5.89. This means that about six times as many observations are needed with the complex sampling design used in the survey to obtain the same precision that would have been achieved with an SRS. The high design effect in this survey is due to the clustering: Villages tend to be homogeneous in bed net use. If you ignored the clustering and analyzed the sample as though it were an SRS, the estimated standard errors would be much too low, and you would think you had much more precision than really existed.   ∎

7.5.1   Design Effects and Confidence Intervals

If the design effect for each statistic is known, one can use it in conjunction with standard software to obtain CIs for means and totals. If n observation units are sampled from a population of N possible observation units and if $\hat{p}$ is the survey estimate of the proportion of interest, an approximate 95% CI for p is (assuming the finite population correction is close to 1):

$$\hat{p} \pm 1.96\sqrt{\text{deff}}\sqrt{\frac{\hat{p}(1-\hat{p})}{n}}. \tag{7.8}$$

When estimating a mean rather than a proportion, if the sample is large enough to apply a central limit theorem, an approximate 95% CI is

$$\hat{\bar{y}} \pm 1.96\sqrt{\text{deff}}\sqrt{\frac{\hat{S}^2}{n}},$$

where $\hat{S}^2$ may be calculated using (7.3).

Kish (1995) and other authors sometimes use design effect to refer to the quantity

$$\text{deft}(\bar{y}) = \frac{\text{SE}(\hat{\bar{y}}_{\text{plan}})}{\dfrac{S}{\sqrt{n}}},$$

so that deft (the name *deft* is due to Tukey, 1968) will be an appropriate multiplier for a standard error or CI half-width. In practice, as Kish points out, choice of deff or deft makes little difference, but you need to pay attention to which definition a survey uses.

7.5.2   Design Effects and Sample Sizes

Design effects are extremely useful for estimating the sample size needed for a survey. That is the purpose for which it was introduced by Cornfield (1951), who used it to estimate the sample size that would be needed if the sampling unit in a survey estimating the prevalence of tuberculosis was a census tract or block rather than an individual. The maximum allowable error was specified to be 20% of the true prevalence, or $0.2 \times p$. If the prevalence of tuberculosis was 1%, the sample size for an SRS would need to be

$$n = \frac{1.96^2 p(1-p)}{(0.2p)^2} = 9508.$$

Cornfield recommended increasing the sample size for an SRS to 20,000, to give more precision in separate estimates for subpopulations. He estimated the design effect for sampling census tracts rather than individuals to be 7.4 and concluded that if census tracts, which averaged 4600 individuals, were used as a sampling unit, a sample size of 148,000 adults, rather than 20,000 adults, would be needed.

If you know the design effect for a similar survey, you only need to be able to estimate the sample size you would take using an SRS. Then multiply that sample size by deff to obtain the number of observation units you need to observe with the complex design. For sample size purposes, you may wish to use separate design effects for each stratum.

7.6
The National Crime Victimization Survey

Most crime statistics given in U.S. newspapers come from the Uniform Crime Reports, compiled by the FBI from reports submitted by law-enforcement agencies. But the Uniform Crime Reports underestimate the amount of crime in the U.S., largely because not all crimes are reported to the police.

The National Crime Victimization Survey (NCVS) is a large national survey administered by the U.S. Bureau of Justice Statistics with interviews conducted by the U.S. Census Bureau. Like the CPS, the NCVS follows a stratified, multistage cluster design. Information on the design of the CPS is found in U.S. Census Bureau (2002a); U.S. Department of Justice (2002, 2006) describe the NCVS design. The NCVS surveys households from across the United States and asks household members 12 years old and older about their experiences as victims of crime in the past six months.

We describe the design used for the 2000 NCVS.[2] The NCVS design is similar to that of many other large government surveys: Most have similar methods of stratification, clustering, and ratio estimation. We shall return to the NCVS in Chapter 8, to show how weights are adjusted for nonresponse and undercoverage in this large complex survey.

A psu in the NCVS is a county, a group of adjacent counties, or a large metropolitan area. Any psu with population about 550,000 or more (according to the 1990 census) is automatically included in the sample. Such a psu is said to be *self-representing* (SR) because it does not represent any psus other than itself. The probability this psu will be selected is one.

All other psus are grouped into strata so that each stratum group has a population of about 650,000. In the NCVS, psus are grouped into strata based on geographic location, demographic information available from the 1990 census, and on Uniform Crime Reports crime rates. One psu is selected from each of these strata with probability proportional to population size; this psu is called *non-self-representing* (NSR) because it is supposed to represent not just itself but all the psus in that stratum. Within a stratum, a psu with 100,000 population is twice as likely to be selected for the sample as a psu with population 50,000. The 2000 NCVS design had 93 SR psus and 110 NSR psus. Because victimization rates vary regionally, the large number of strata in the NCVS increases the precision of the estimates.

The second stage of sampling involves selecting enumeration districts (EDs), geographic areas used in the decennial census; an ED typically contains about 300 to

---

[2]Many structural features of the design are the same for more recent years, although there has been a drastic reduction in sample size. In 2006, the NCVS selected new psus based on the 2000 Census. Other designs are being considered for the NCVS starting after 2010 (National Research Council, 2008).

TABLE **7.3**

Sampling Stages for the 2000 NCVS

Stage	Sampling Unit	Stratification
1	psu (county, set of adjacent counties, or metropolitan area)	Location, demographic information, and crime-related characteristics
2	Enumeration District	
3	Cluster of 4 housing units	
4	Household	
5	Person within household	

400 households (750 to 1500 persons), but EDs vary considerably in population and land area. The EDs are selected with probability proportional to their 1990 census population size; the number of ED's selected within a psu is determined so that the sample of ED's will be approximately self-weighting. In the census listing, EDs are arranged by geographic location; EDs are selected using systematic sampling as described in Section 6.2, so that the sampled EDs will be distributed geographically over the selected psu. If the overall sampling rate is $1/x$, in SR psus the sampling interval is x. If using census records for the sampling frame, the addresses are numbered from 1 to the number of households in the psu. A random number k is chosen between 1 and x, and the ED's chosen to be in the sample are the ones containing addresses $k, k + x, k + 2x$, etc. In NSR psus, the sampling interval is (probability psu is selected)(x).

In the third stage of sampling, each selected ED is divided into clusters of approximately four housing units each. The census lists housing units within an ED in geographic order, and when possible, that listing is used. A sample of those clusters is taken, and each housing unit in a selected cluster of about four housing units is included in the sample. All persons aged 12 and over in the housing unit are to be interviewed for the survey.

The census listings are supplemented by *area sampling*. If the census listing of housing units were the only one used throughout that decade, there would be substantial undercoverage of the population, since no newly built housing units would be included in the sample. To allow new housing units to be included in the sample, the NCVS uses a sample of building permits for residential units and samples those. In area sampling, a field representative lists all housing units or other living quarters within a selected area of an ED, and that listing then serves as the sampling frame for that area.

In summary, the stages for the NCVS are shown in Table 7.3. Interviews for the NCVS with persons aged 12 and over are taken every month, with the housing units selected for the sample covered in a six-month period—this allows the interviewing workload to be distributed evenly throughout the year. To allow for longitudinal analyses of the data, and to be relatively certain that crimes reported for a six-month period occurred during those six months and not during an earlier time, the residents of each housing unit are interviewed every six months over a three-year period, for a total of seven interviews. For 2000, about 43,000 housing units were interviewed. Altogether, about 80,000 persons gave responses to the questionnaire.

Clearly, this is a complex survey design, and weights are used to calculate estimates of victimization rates and total numbers of crimes. The survey is designed to be approximately self-weighting, so initially each individual is assigned the same base weight of (1/probability of housing unit selection).

The NCVS is designed to be self-weighting, but sometimes a selected cluster within an Enumeration District has more housing units than originally thought; for example, an apartment building might have been erected in place of detached housing units. Then only housing units in a subsample of the cluster are interviewed. If subsampling is used, the units subsampled are assigned a weighting control factor (WCF). If only one-third are sampled, for instance, the sampled units are assigned a WCF of 3, because they will represent three times as many units. If a housing unit is in a cluster in which subsampling is not needed, it is assigned a WCF of 1. At this level, a sampled housing unit represents

$$\text{base weight} \times \text{WCF}$$

housing units in the population. This is the sampling weight for a housing unit sampled in the NCVS; as the survey attempts to interview all persons aged 12 and older in the sampled housing units, the sampling weight for a person in the sample is set equal to the weight for the housing unit.

All other weighting adjustments in the NCVS adjust for nonresponse, or are used in poststratification. Some persons selected to be in the sample are not interviewed because they are absent or refuse to participate. The interviewer gathers demographic information on the nonrespondents, and that demographic information is used to adjust the weights in an attempt to counteract the nonresponse. (This is an example of weighting class adjustments for nonresponse, as discussed in Section 8.5.) Two different weighting adjustments for nonresponse are used: the *within household noninterview adjustment factor* (WHHNAF), and the *household noninterview adjustment factor* (HHNAF). In each adjustment factor, the goal is to increase weights of interviewed units that are most similar to units that cannot be interviewed.

The WHHNAF is used to compensate for individual nonrespondents in households in which at least one member responded to the survey. It is computed separately for each of the regions (Northeast, Midwest, South, and West) of the United States. Within each region, the persons from households in which there was at least one respondent are classified into 24 cells using the race of the person designated as reference person, the age and sex of the nonresponding household member, and the nonrespondent's relationship to the reference person. Any of the 24 cells that contain fewer than 30 interviewed cases, or that produce a WHHNAF of two or more, are combined with similar cells; the collapsing of cells prevents some individuals from having weights that are too large. Then

$$\text{WHHNAF} = \frac{\text{sum of weights of all persons in cell}}{\text{sum of weights of all interviewed persons in cell}}.$$

The weights used to calculate the WHHNAF are the weights assigned to this point in the weighting procedure, that is, (base weight) × (WCF). Thus the weights of respondents in a cell are increased so that they represent the nonrespondents, and the persons in the population that the nonrespondent would represent, in addition to their original representation. After applying the WHHNAF, the weight for an individual is

$$\text{base weight} \times \text{WCF} \times \text{WHHNAF}.$$

Not all nonresponse is from nonresponding individuals in responding households. About 3 to 4 percent of households are eligible for the survey but cannot be reached or refuse to respond; the household noninterview adjustment factor is used to attempt to compensate for nonresponse at the household level. For the HHNAF, households are grouped into cells by race of the reference person and metropolitan area and urban/suburban/rural status of the residence. Then

$$\text{HHNAF} = \frac{\text{sum of weights of all persons in cell}}{\text{sum of weights of all interviewed persons in cell}}.$$

As with the WHHNAF, the weights used in calculating the HHNAF are the weights calculated so far: (base weight) $\times$ (WCF) $\times$ (WHHNAF). Cells are combined until the HHNAF is less than two.

At this point in the construction of the weights, the weight assigned to an individual is

$$\text{base weight} \times \text{WCF} \times \text{WHHNAF} \times \text{HHNAF}.$$

The sampling weights for responding individuals have been increased so that they also represent nonrespondents who are demographically similar.

Because the NCVS is a sample, the demographic information in the sample usually differs from that of the U.S. population as a whole. Two stages of ratio estimation are used to adjust the sample values so they agree better with updated census information. This adjustment is expected to reduce the variance of estimates of victimization rates.

The first stage of ratio estimation is used in NSR psus only, and is intended to reduce the variability that results from using one psu to represent the stratum. Ratio estimation is used to assign different weights to cells stratified by region, MSA status, and race. The first-stage factor,

$$\text{FSF} = \frac{\text{independent count of number of persons in cell}}{\text{sample estimate (sum of weights) of the number of persons in cell}},$$

adjusts for differences between census characteristics of sampled NSR psus and characteristics of the full set of NSR psus. The FSF equals one for SR psus, and is truncated at 1.3 for NSR psus.

The second stage of ratio estimation is applied to everyone in the sample. The persons in the sample are classified into 72 groups on the basis of their age, race, and sex. Cells need to have a count of at least 30 interviewed persons, and the SSF needs to be between 0.5 and 2.0; cells are collapsed until these conditions are met.

$$\text{SSF} = \frac{\text{independent count of number of persons in cell}}{\text{sample estimate (sum of weights) of the number of persons in cell}}.$$

The second-stage factor is a form of poststratification: it is intended to adjust the sample distribution of age, race, and sex so that the cross-classification agrees with independently taken counts that are thought to be more accurate. If the sum of weights

FIGURE   **7.22**

Boxplots of weights for the 2000 NCVS, for all persons, white males, white females, non-white males, nonwhite females, persons under age 25, persons over age 25, and victims of violent crime. The horizontal lines represent the 95th percentile, 75th percentile, median, 25th percentile, 5th percentile, and minimum. Note that the weights are much higher for nonwhite males, indicating the higher nonresponse and undercoverage in that group.

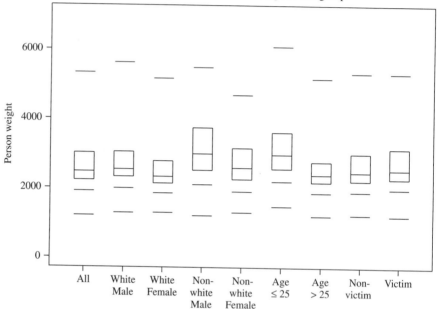

of elderly white women in the sample is larger than the current "best" estimate of the number of elderly white women in the population from updated census information, then SSF will be less than one for all elderly white women in the sample.

After all the adjustments, the final weight for person i is

$$w_i = \text{Base Weight} \times \text{WCF} \times \text{WHHNAF} \times \text{HHNAF} \times \text{FSF} \times \text{SSF}.$$

The weight w_i is used as though there were actually w_i persons in the population exactly like the one to which the weight is attached. In the 2000 NCVS, the person weights range from 1100 to 9000, with most weights between 1500 and 2500. Figure 7.22 gives boxplots for the weights for persons interviewed between July and December of 2000. The weights are included on the public use data files of the NCVS: To use them to estimate the total number of aggravated assaults reported by white females, you would define

$$y_i = \begin{cases} 1 & \text{if person } i \text{ is a white female who reported an aggravated assault} \\ 0 & \text{otherwise} \end{cases}$$

and use $\sum_{i \in S} w_i y_i$ as your estimate.

Even though the nonresponse is relatively low in the NCVS, the weights make a difference in calculating victimization rates. Estimates of victimization rates are generally higher when weights are used than when they are not used. Young black

FIGURE 7.23

Histograms of victim ages, without and with weights. The histogram constructed using weights has a higher frequency of victims in the younger age groups.

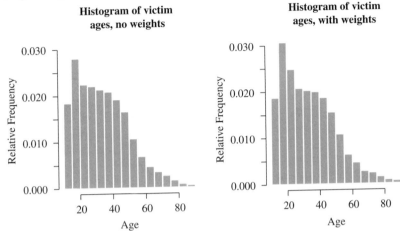

male respondents to the survey are disproportionately likely to be victims of crime, and undercoverage and nonresponse among black males is high. Figure 7.23 shows the difference that using weights makes in a histogram of ages of victims of violent crime.

The sampling design and the weighting scheme are complicated in the NCVS, so variance estimation is also complicated. Variances are now calculated by replication methods, described in Chapter 9. To protect confidentiality of respondents, the Bureau of Justice Statistics does not release the actual strata and psu variables in the public-use data sets; instead, it creates pseudo-strata and pseudo-psu variables that can be used to estimate variances. For variance estimation, we treat the psus as though they are sampled with replacement, so the subsampling information is not needed. The overall design effect for the NCVS, and for similar U.S. government surveys, is about two.

7.7
Sampling and Design of Experiments*

Numerous parallels between sample surveys and designed experiments are discussed in Fienberg and Tanur (1987) and Yates (1981). Some of these parallels are noted in this section.

Simple random sampling, in which the universe $\mathcal{U}$ has N units, is similar to the randomization approach to the comparison of two treatments using a total of N experimental units. To test the hypothesis $H_0: \mu_1 = \mu_2$, randomly assign n of the N units to treatment 1 and the remaining $N - n$ units to treatment 2. The observed value of the test statistic is compared with the reference distribution based on all $\binom{N}{n}$ possible assignments of experimental units to treatments. The p-value comes from the randomization distribution. Using randomization for inference dates back to Fisher (1925), and the theory is developed in Kempthorne (1952).

Randomization serves similar purposes in sampling and in designing experiments. In sampling, the goal is to be able to generalize our results to the whole population, and we hope that randomization gives us a representative sample. When we design an experiment, we attempt to "randomize out" all other possible influences, and we hope that we can separate the differences due to the treatments from random error. In both cases, we can quantify how often we expect to have a sample or a design that gives us a "bad" result. This quantification appears in CIs: 95% of possible samples or possible replications of an experiment are expected to yield a 95% CI that contains the true value.

The purpose of stratification is to increase the precision of our estimates by grouping similar items together. The same purpose is met in design of experiments with blocking. Cluster samples also group similar items together, but the purpose is convenience, not precision. An analogue in design of experiments is a split-plot design, which generally gives greater precision in the subplot estimate than in the whole plot estimate.

The structural similarity between surveys and designed experiments was exploited by using ANOVA tables to develop the theory of stratification and cluster sampling. We used a fixed-effects one-way ANOVA for a model-based approach to stratification and a random-effects one-way ANOVA for a model-based approach to cluster sampling. Much of the theory in cluster sampling is similar to the theory of random-effects models; in the models in Chapters 5 and 6, we relied on variance components to explain the dependence in the data.

Poststratification and ratio and regression estimation in sampling allows us to increase the precision of estimators by taking advantage of the relationship between the variable of interest and other classification variables; the same goal in designed experiments is met by using covariate adjustment, as in analysis of covariance.

Both design of experiments and sampling are involved in similar debates between using a randomization theory approach or using a model-based approach. We have touched on the different philosophical approaches for estimating functions of totals in Sections 2.9, 4.6, 3.6, 5.6, and 6.7, but much more has been said. We encourage the interested reader to start with the discussion papers by Smith (1994) and Hansen et al. (1983). Royall (1992a) succinctly summarizes a model-based approach to sampling. Brewer (2002) discusses implications of different philosophies of inference in survey sampling.

Finally, in both sample surveys and designed experiments, it is crucial that adequate effort be spent on the design of the study. No amount of statistical analysis, however sophisticated, can compensate for a poor design. Chapter 1 presented examples of disastrous results from selection bias resulting from poor survey design or execution. A call-in poll is not only useless for generalizing to a population but also harmful, as people may believe its statistics are accurate. Similarly, little can be concluded about the efficacy of treatments A and B for a medical condition if the most ill patients are assigned to treatment A; if the mean duration of symptoms is significantly less for treatment B than for treatment A, is the difference due to the treatment or to the difference in the patients?

Of course, it is sometimes possible to adjust for an imperfect design in the analysis. If a measure of the severity of the illness at the beginning of the study is available, it could be used as a covariate in comparing the two treatments, although there will still be worries about confounding with other, unmeasured quantities. Values for missing cells in a two-way ANOVA design can be estimated by a model. Similarly, available information about nonrespondents can be used to improve estimation in the presence of nonresponse, as discussed in the next chapter.

7.8
Chapter Summary

Many large surveys have a stratified multistage sampling design, in which the psus are selected by stratified sampling and then subsampled. Estimators of population quantities from a stratified multistage sample are calculated by combining the principles from Chapters 2–6. In most instances, only the stratification and information from the first stage of clustering are used to calculate standard errors of estimates.

Any population quantity can be estimated from the sample using the weights. The empirical cdf and the empirical probability mass function estimate the cdf and probability mass function of the population by incorporating the weights w_i. Since w_i can be thought of as the number of observations units in the population represented by observation unit i in the sample, the empirical cdf can be thought of as the observed cdf of a pseudo-population in which observation i in the sample is replicated w_i times.

Graphs that are commonly used for displaying data from an SRS can be adapted for complex survey data by incorporating the survey weights. Histograms, boxplot, and scatterplots that use the survey weights display features of the data that are sometimes obscured in an analysis that only reports summary statistics.

Although the survey weights can be used to find a point estimate of any population quantity, the weights are not sufficient information to calculate standard errors of statistics. Standard errors depend on the stratification and clustering in the survey design. The design effect, which is the ratio of the variance of a statistic calculated using the complex survey design to the variance that would have been obtained from an SRS with the same number of observation units, is useful for assessing the effect of design features on the variance. The design effect is often used to determine the sample size needed for a complex survey.

Key Terms

Design effect: Ratio of the variance of an estimator from the sampling plan to the variance of an estimator from an SRS with the same number of observation units.

Empirical probability mass function: An estimator of the probability mass function using sampling weights: $\hat{f}(y) = \sum_{i \in S, y_i = y} w_i / \sum_{i \in S} w_i$.

Probability mass function: $f(y) = $ (number of observation units in the population whose value is y)/N gives the distribution of the finite population.

Stratified multistage sample: A sampling design in which primary sampling units are grouped into strata; a probability sample is taken of the psus in each stratum. Secondary sampling units are then subsampled within each selected psus. In some cases, the selected ssus are also clusters and are themselves subsampled.

For Further Reading

The books edited by Skinner et al. (1989) and Chambers and Skinner (2003) are good places to start your reading about complex surveys. The volumes edited by Pfeffermann and Rao (2009a, 2009b) give a wealth of information on current topics in survey sampling. Two papers by Kish (1992, 1995) further explain the ideas behind weighting and design effects. The idea of using design effects for sample size estimation was introduced by Cornfield (1951); the paper gives an interesting example of sampling in practice.

Korn and Graubard (1999) present the theory of sampling with application to the special problems involved in health surveys. They also emphasize plotting data from surveys, and describe a number of methods for constructing scatterplots and other plots with survey data.

7.9
Exercises

A. Introductory Exercises

1   You are asked to design a survey to estimate the total number of cars without permits that park in handicapped parking places on your campus. What variables (if any) would you consider for stratification? For clustering? What information do you need to aid in the design of the survey? Describe a survey design that you think would work well for this situation.

2   Repeat Exercise 1 for a survey to estimate the total number of books in a library that need rebinding.

3   Repeat Exercise 1 for a survey to estimate the percentage of persons in your city who speak more than one language.

4   Repeat Exercise 1 for a survey to estimate the distribution of number of eggs laid by Canada geese.

5   The organization "Women tired of waiting in line" wants to estimate statistics about restroom usage. Design a survey to estimate the average amount of time spent by women in a restroom and the average time spent by men in a restroom at your university.

6   Use the data in file integerwt.dat for this exercise. The strata are constructed with $N_1 = 200$, $N_2 = 800$, $N_3 = 400$, $N_4 = 600$.

 a   Take a stratified random sample with $n_1 = 50$, $n_2 = 50$, $n_3 = 20$, and $n_4 = 25$. Calculate the sampling weight w_i for each observation in your sample (the sample sizes were selected so that each weight is an integer).

 b   Using the weights, estimate $\bar{y}_U$, S^2, and the 25th, 50th, and 75th percentiles of the population.

 c   Now create a "pseudo-population" by constructing a data set in which the data value y_i is replicated w_i times. Your pseudo-population should have $N = 2000$

observations. Estimate the same quantities in (b) using the pseudo-population and usual formulas for an SRS. How do the estimates compare with the estimates from (b)?

B. Working with Survey Data

7   Using the data in nybight.dat (see Exercise 18 of Chapter 3), find the empirical mass function of number of species caught per trawl in 1974. Be sure to use the sampling weights.

8   Using the data in teachers.dat (see Exercise 15 of Chapter 5), use the sampling weights to find the empirical mass function of the number of hours worked. What is the design effect?

9   Using the data in measles.dat (see Exercise 16 of Chapter 5), what is the design effect for percentage of parents who received a consent form? For the percentage of children who had previously had measles?

10   Using the data in file statepop.dat (see Example 6.5 of Chapter 6), draw a histogram, using the weights, of the number of veterans. How does this compare with a histogram that does not use the weights?

11   Using the data in file statepop.dat (see Example 6.5 of Chapter 6), draw one of the scatterplots, using the weights, of the number of veterans vs. number of physicians.

12   The Survey of Youth in Custody sampled youth who were residents of long-term facilities at the end of 1987. Is the sample representative of youth who have been in long-term facilities in 1987? Why, or why not?

13   The file syc.dat, used in Example 7.7 contains other information from the 1987 Survey of Youth in Custody. Draw a histogram, using the weights, for the age of the youth at first arrest. What is the average age of first arrest? The median? The 25th percentile? Use the "final weight" to estimate these quantities. How do your estimates compare to estimates obtained without using weights?

14   Using the file syc.dat and the final weights, estimate the proportion of youths who

    a   are age 14 or younger

    b   are held for a violent offense

    c   lived with both parents when growing up

    d   are male

    e   are Hispanic

    f   grew up primarily in a single-parent family

    g   have used illegal drugs.

    Give 95% CIs for your answers.

15   Use the data in file nhanes.dat for this exercise. (If you prefer, you may download the NHANES data from the website at www.cdc.gov/nchs/nhanes.htm.) Triceps skin-fold measurements are sometimes used as a gauge of body fat. We are interested in the relation between $y =$ triceps skinfold (variable *bmxtri*) and $x =$ body mass index (variable *bmxbmi*).

a   Estimate the mean value of triceps skinfold for the population, along with a 95% CI.

b   Draw a histogram of the variable triceps skinfold, using the weights. Do the data appear to be normally distributed?

c   Find the minimum, 25th, 50th, and 75th percentiles, and maximum for the variable triceps skinfold. Calculate the same quantities separately for each gender (variable *riagendr*). Use these to construct side-by-side skeletal boxplots of the data as in Figure 7.7.

d   Construct a weighted circle plot with smoothed trend line for y variable triceps skinfold and x variable body mass index. Does there appear to be a linear relationship? What other features do you see in the data?

16   Answer the questions in Exercise 15, for $y =$ waist circumference (variable *bmxwaist*) and $x =$ thigh circumference (variable *bmxthicr*).

17   The file ncvs2000.dat includes selected variables for a subset of data in the 2000 NCVS. Using the data, find estimates of the following:

a   percentage of persons who are victims of a violent crime

b   percentage of persons who have been injured in a violent crime

c   average number of crime incidents per person

d   average medical expenses for persons who are injured.

Give standard errors for your estimates.

C. Working with Theory

18   *Trimmed means.* Many statisticians recommend using trimmed means to estimate a population mean $\bar{y}_U$ if there are outliers. The procedure used to find an α-trimmed mean in an SRS of size n is to remove the largest $n\alpha$ observations and the smallest $n\alpha$ observations, and then calculate the mean of the $n(1 - 2\alpha)$ observations that remain.
Show that the α-trimmed mean for a finite population $\mathcal{U}$ of N observation units is

$$\bar{y}_{U\alpha} = \frac{\displaystyle\sum_{q_1 \leq y \leq q_2} yf(y)}{\displaystyle\sum_{q_1 \leq y \leq q_2} f(y)},$$

where q_1 and q_2 are the α and $(1-\alpha)$ quantiles, respectively. Now propose an estimator of the population α-trimmed mean for data from a complex survey using $\hat{F}(y)$ and $\hat{f}(y)$.

19   *Probability–probability plots.* A probability–probability plot (often referred to as a p–p plot) compares the empirical cdf from a sample with a specified theoretical cdf G such as the cdf of a normal distribution with specified mean and variance (Gnanadesikan, 1997). If the proposed cdf G describes the data well (including the

specification of the mean and variance), the points in a p–p plot of $\hat{F}(y)$ vs. $G(y)$ will lie approximately on a straight line with intercept 0 and slope 1.

Construct a p–p plot for the height data in htstrat.dat, used in Example 7.3. Use a normal distribution for G, with the mean and standard deviation estimated from the sample. Draw in the line with intercept 0 and slope 1. Is G a reasonable distribution to use to summarize the data?

20   *Quantile–quantile plots.* Quantile–quantile plots are often used to assess how well a theoretical probability distribution fits a data set (Chambers et al., 1983). To construct a quantile–quantile plot from an SRS of size n, order the sample values so that $y_{(1)} \leq y_{(2)} \leq \cdots \leq y_{(n)}$. Then, to compare with a continuous theoretical cdf G, calculate $x_{(i)} = G^{-1}[(i - 0.375)/(n + 0.25)]$ and plot $y_{(i)}$ vs. $x_{(i)}$ for $i = 1, \ldots, n$. If G is a good fit for the data, the quantile–quantile plot will approximate a straight line.

To use a quantile–quantile plot with survey data, let $w_{(1)}, \ldots, w_{(n)}$ be the weight values corresponding to the ordered sample $y_{(1)} \leq y_{(2)} \leq \cdots \leq y_{(n)}$. Let

$$v_{(i)} = \frac{\left(\displaystyle\sum_{j=1}^{i} w_{(j)}\right)\left(1 - \dfrac{0.375}{i}\right)}{\left(\displaystyle\sum_{j \in S} w_{(j)}\right)\left(1 + \dfrac{0.25}{n}\right)}.$$

Then plot $y_{(i)}$ vs. $G^{-1}(v_{(i)})$ and assess whether the values appear to be approximately on a straight line.

Figure 7.24 shows a histogram and quantile–quantile plot with G a standard normal cdf, for the body mass index data used in Section 7.4.2. SAS code used to produce these plots is on the website. The histogram displays a skewed distribution. The curvature in the quantile–quantile plot also indicates the skewness, since observations on the left are more compressed and observations on the right are more extreme than we would expect for normally distributed data. If a normal distribution described the data well, we would expect to see the points following a straight line.

a   Show that the plot of $y_{(i)}$ vs. $G^{-1}(v_{(i)})$ gives the SRS quantile–quantile plot when the sample is self-weighting.

b   Construct a quantile–quantile plot for the height data in htstrat.dat, used in Example 7.3. Use a standard normal cdf for G. Do you think the normal distribution describes these data well?

21   Show that in a stratified sample, $\sum y \hat{f}(y)$ produces the estimator in (3.2).

22   What is $\hat{S}^2$ in (7.3) for an SRS? How does it compare with the sample variance s^2?

23   Consider a probability sample S of n observation units from a population U of N observation units. The weights are $w_i = 1/\pi_i$, where π_i is the probability that unit i is in the sample. Now let S_2 be a subsample of S of size n_2, with units selected with probability proportional to w_i. Show that S_2 is a self-weighting sample from U.

24   In a two-stage cluster sample of rural and urban areas in Nepal, Rothenberg et al. (1985) found that the design effect for common contagious diseases was much higher

F I G U R E   7.24

Histogram and normal quantile–quantile plot of body mass index from NHANES.

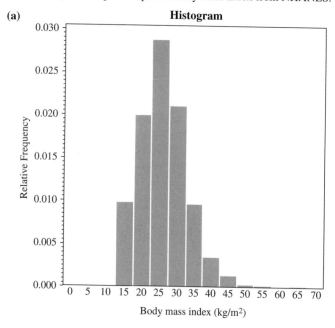

(a) Histogram

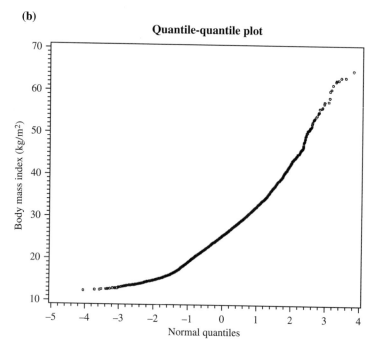

(b) Quantile-quantile plot

than for rare contagious diseases. In the urban areas measles, with an estimated incidence of 123.9 cases per 1000 children per year, had a design effect of 7.8; diphtheria, with an estimated incidence of 2.1 cases per 1000 children per year, had a design effect of 1.9.

Explain why one would expect this disparity in the design effects. (HINT: Suppose a sample of 1000 children is taken, in 50 clusters of 20 children each. Also suppose that the disease is as aggregated as possible, so if the estimated incidence were 40 per 1000, all children in two clusters would have the disease, and no children in the remaining 38 clusters would have the disease. Now calculate deff for incidences varying from 1 per 1000 to 200 per 1000.)

25 The British Crime Survey (BCS) is also a stratified, multistage survey (AyeMaung, 1995). In contrast to the NCVS, the BCS is not designed to be approximately self-weighting, as inner-city areas are sampled at about twice the rate of non-inner-city areas. In the BCS, households are selected using probability sampling, but only one adult (selected at random) is interviewed in each responding household. Set the relative sampling weight for an inner-city household to be 1.

a Consider the BCS as a sample of households. What is the relative sampling weight for a non-inner-city household?

b Consider the BCS as a sample of adults. Construct a table of relative sampling weights for the sample of adults.

Number of Adults	Inner City	Not Inner City
1		
2		
3		
4		
5		

D. Projects and Activities

26 Obtain one of the papers listed the file chapter7papers.html on the book website, or another paper employing a complex survey design, and write a short critique. Your critique should include:

a a brief summary of the design and analysis

b a discussion of the effectiveness of the design and the appropriateness of the analysis

c your recommendations for future studies of this type.

27 *Trucks.* The U.S. Vehicle Inventory and Use Survey (VIUS) was described in Exercise 34 of Chapter 3.

a Draw a histogram, using the weights, of the number of miles driven (variable *miles_annl*) for the five truck class strata.

b Draw side-by-side boxplots, using the weights, of miles per gallon (*MPG*) for each class of gross vehicle weight (*vius_gvw*).

c Draw two of the scatterplots that incorporate weights, described in Section 7.4.2, for y variable *miles_annl* and x variable model year (*adm_modelyear*). How do these differ from scatterplots that do not use the weights?

28 *IPUMS exercises.*

a Use the file ipums.dat to select a two-stage stratified cluster sample from the population. Select two psus from each stratum, with probability proportional to size. Then take a simple random sample of persons from each selected psu; use the same subsampling size within each psu. Your final sample should have about 1200 persons.

b Construct the column of sampling weights for your sample.

c Draw a histogram of the variable *inctot* for your sample, using the weights.

d Construct side-by-side boxplots of the variable *inctot* for each level of marital status (variable *marstat*).

e Draw two of the scatterplots that incorporate weights, described in Section 7.4.2, for y variable *inctot* and x variable *age*. How do these differ from scatterplots that do not use the weights?

f Using the sample you selected, estimate the population mean of *inctot* and give the standard error of your estimate. Also estimate the population total of *inctot* and give its standard error.

g Compare your results with those from an SRS with the same number of persons. Find the design effect of your response (the ratio of your variance from the unequal-probability sample to the variance from the SRS).

29 *Baseball data.* Use the two-stage sample from Exercise 37 of Chapter 5 for this exercise.

a Draw a histogram of the variable *salary* for your sample, using the weights.

b Construct side-by-side boxplots of the variable *salary* for each position.

c Draw two of the scatterplots that incorporate weights, described in Section 7.4.2, for y variable *salary* and x variable number of games played (g). How do these differ from scatterplots that do not use the weights?

d Draw two of the scatterplots that incorporate weights, described in Section 7.4.2, for y variable *salary* and x variable number of home runs (hr). What do you see in these graphs?

e Draw quantile-quantile plots (see Exercise 20) for the variable *salary* and log(*salary*). Does either variable appear to follow, approximately, a normal distribution?

30 Many governmental statistical organizations and other collectors of survey data have websites providing information on the survey design. Some of these organizations and their Internet addresses are listed in the file chapter7websites.html on the book website. The first site listed, www.fedstats.gov, provides links to U.S. government agencies that spend at least $500,000 per year on statistical activities. Many of these agencies conduct surveys. The second listing from the International Statistical Institute (ISI) provides a directory to official statistical agencies throughout the world.

Look up a website describing a complex survey. Write a summary of the purpose, design, and method used for analysis. Do you think that the design used could be improved upon? If so, how?

31 *Activity for course project.* Find a survey data set that has been collected by a federal government or large survey organization. Many of these are now available online, and contain information about stratification and clustering that you can use to calculate standard errors of survey estimates. Some examples in the United States include the NCVS, the National Health Interview Survey, the Current Population Survey, the Commercial Buildings Energy Consumption Survey, and the General Social Survey. You can find a survey by selecting a topic from www.fedstats.gov and following the links to the survey data. Many of the other organizations listed in the file chap7websites.html on the book website (see Exercise 30) also provide survey data online.

Read the documentation for the survey. What is the survey design? What stratification and clustering variables are used? (Sometimes the stratification and clustering variables are difficult to find in the documentation; look for variables containing "psu" or "str" in the name. These are often near the beginning or end of the variable listing in the codebook. Some surveys do not release stratification and clustering information to protect the confidentiality of data respondents, so make sure your survey provides that information.)

Select response variables that you are interested in. If possible, find at least one response with continuous response. Draw a histogram, using the final weight variable, for that response. Use the weights to estimate the summary statistics of the mean and 25th, 50th, and 75th percentiles. We'll return to this data set in subsequent chapters so that you will have an opportunity to study bivariate and multivariate relationships among your variables of interest.

Nonresponse

Miss Schuster-Slatt said she thought English husbands were lovely, and that she was preparing a questionnaire to be circulated to the young men of the United Kingdom, with a view to finding out their matrimonial preferences.

"But English people won't fill up questionnaires," said Harriet.

"Won't fill up questionnaires?" cried Miss Schuster-Slatt, taken aback.

"No," said Harriet, "they won't. As a nation we are not questionnaire-conscious."

—Dorothy Sayers, *Gaudy Night*

The best way to deal with nonresponse is to prevent it. After nonresponse has occurred, it is sometimes possible to construct models to predict the missing data, but predicting the missing observations is never as good as observing them in the first place. Nonrespondents often differ in critical ways from respondents; if the nonresponse rate is not negligible, inference based only upon the respondents may be seriously flawed.

We discuss two type of nonresponse in this chapter: **unit nonresponse**, in which the entire observation unit is missing, and **item nonresponse**, in which some measurements are present for the observation unit but at least one item is missing. In a survey of persons, unit nonresponse means that the person provides no information for the survey; item nonresponse means that the person does not respond to a particular item on the questionnaire. In the Current Population Survey (CPS) and the National Crime Victimization Survey (NCVS), unit nonresponse can arise for a variety of reasons: The interviewer may not be able to contact the household; the person may be ill and cannot respond to the survey; the person may refuse to participate in the survey. In these surveys, the interviewer tries to get demographic information about the non-respondent such as age, sex, and race, as well as characteristics of the dwelling unit such as urban/rural status; this information can be used later to try to adjust for the nonresponse. Item nonresponse occurs largely because of refusals: A household may decline to give information about income, for example.

In agriculture or wildlife surveys, the term *missing data* is generally used instead of *nonresponse,* but the concepts and remedies are similar. In a survey of breeding ducks, for example, some birds will not be found by the researchers; they are, in a sense, nonrespondents. The nest may be raided by predators before the investigator can determine how many eggs were laid; this is comparable to item nonresponse. Lesser and Kalsbeek (1999) discuss nonresponse and other nonsampling errors in environmental surveys.

In this chapter, we discuss four approaches to dealing with nonresponse:

1 Prevent it. Design the survey so that nonresponse is low. This is by far the best method.

2 Take a representative subsample of the nonrespondents; use that subsample to make inferences about the other nonrespondents.

3 Use a model to predict values for the nonrespondents. Weighting class adjustment methods implicitly use a model to adjust for unit nonresponse. Imputation often adjusts for item nonresponse, and parametric models may be used for either type of nonresponse.

4 Ignore the nonresponse (not recommended, but unfortunately common in practice).

8.1
Effects of Ignoring Nonresponse

EXAMPLE **8.1** Thomsen and Siring (1983) report results from a 1969 survey on voting behavior carried out by the Central Bureau of Statistics in Norway. In this survey, three calls were followed by a mail survey. The final nonresponse rate was 9.9%, which is often considered to be a small nonresponse rate. Did the nonrespondents differ from the respondents?

In the Norwegian voting register, it was possible to find out whether a person voted in the election. The percentage of persons who voted could then be compared for respondents and nonrespondents; Table 8.1 shows the results. The selected sample is all persons selected to be in the sample, including data from the Norwegian voting register for both respondents and nonrespondents.

The difference in voting rate between the nonrespondents and the selected sample was largest in the younger age groups. Among the nonrespondents, the voting rate varied with the type of nonresponse. The overall voting rate for the

TABLE **8.1**

Percentage of Persons Who Voted

	All	20–24	25–29	Age 30–49	50–69	70–79
Nonrespondents	71	59	56	72	78	74
Selected sample	88	81	84	90	91	84

SOURCE: Adapted from table 8 in Thomsen and Siring (1983).

persons who refused to participate in the survey was 81%, the voting rate for the not-at-homes was 65%, and the voting rate for the mentally and physically ill was 55%, implying that absence or illness were the primary causes of nonresponse bias. ∎

It has been demonstrated repeatedly that nonresponse can have large effects on the results of a survey—in Example 8.1, a nonresponse rate of less than 10% led to an overestimate of the voting rate in Norway. Holt and Elliot (1991, p. 334) discuss the results of a series of studies done on nonresponse in the United Kingdom indicating that "lower response rates are associated with the following characteristics: London residents; households with no car; single people; childless couples; older people; divorced/widowed people; new Commonwealth origin; lower educational attainment; self-employed."

Moreover, increasing the sample size without targeting nonresponse does nothing to reduce nonresponse bias; a larger sample size merely provides more observations from the class of persons that would respond to the survey. Increasing the sample size may actually worsen the nonresponse bias, as the larger sample size may divert resources that could have been used to reduce or remedy the nonresponse, or it may result in less care in the data collection. Recall that the infamous Literary Digest Survey of 1936, discussed in Example 1.1, had 2.4 million respondents but a response rate of less than 25%. The U.S. decennial census itself does not include the entire population, and the undercoverage rate varies for different demographic groups. Mulry (2004) discusses issues in measuring the undercoverage in the U.S. Census.

Most small surveys ignore any nonresponse that remains after callbacks and follow-ups, and report results based on complete records only. Hite (1987) did so in the survey discussed in Chapter 1, and much of the criticism of her results was based on her low response rate. Nonresponse is also ignored for many surveys reported in newspapers, both local and national.

An analysis of complete records has the underlying assumptions that the non-respondents are similar to the respondents, and that units with missing items are similar to units that have responses for every question. Much evidence indicates that this assumption does not hold true in practice. If nonresponse is ignored in the NCVS, for example, victimization rates are underestimated. Biderman and Cantor (1984) find lower victimization rates for persons who respond in three consecutive interviews than for persons who are nonrespondents in at least one of those interviews or who move before they complete the panel.

Results reported from an analysis of only complete records should be taken as representative of the population of persons who would respond to the survey, which is rarely the same as the target population. If you insist on estimating population means and totals using only the complete records and making no adjustment for nonrespondents, at the very least you should report the rate of nonresponse.

The main problem caused by nonresponse is potential bias. Think of the population as being divided into two somewhat artificial strata of respondents and nonrespondents. The population respondents are the units that would respond if they were chosen to be in the sample; the number of population respondents, N_R, is unknown.

Similarly, the N_M (M for missing) population nonrespondents are the units that would not respond. We then have the following population quantities:

Stratum	Size	Total	Mean	Variance
Respondents	N_R	t_R	$\bar{y}_{RU}$	S_R^2
Nonrespondents	N_M	t_M	$\bar{y}_{MU}$	S_M^2
Entire population	N	t	$\bar{y}_U$	S^2

The population as a whole has variance $S^2 = \sum_{i=1}^{N} (y_i - \bar{y}_U)^2/(N-1)$, mean $\bar{y}_U$, and total t. A probability sample from the population will likely contain some respondents and some nonrespondents. But, of course, on the first call we do not observe y_i for any of the units in the nonrespondent stratum. If the population mean in the nonrespondent stratum differs from that in the respondent stratum, estimating the population mean using only the respondents will produce bias.[1]

Let $\bar{y}_R$ be an approximately unbiased estimator of the mean in the respondent stratum, using only the respondents. As

$$\bar{y}_U = \frac{N_R}{N}\bar{y}_{RU} + \frac{N_M}{N}\bar{y}_{MU},$$

the bias is approximately

$$E[\bar{y}_R] - \bar{y}_U \approx \frac{N_M}{N}(\bar{y}_{RU} - \bar{y}_{MU}).$$

The bias is small if either (1) the mean for the nonrespondents is close to the mean for the respondents, or (2) N_M/N is small—there is little nonresponse. But we can never be assured of (1), as we generally have no data for the nonrespondents. Minimizing the nonresponse rate is the only sure way to control nonresponse bias.

8.2
Designing Surveys to Reduce Nonsampling Errors

A common feature of poor surveys is a lack of time spent on the design and nonresponse follow-up in the survey. Many persons new to surveys (and some, unfortunately, not new) simply jump in and start collecting data without considering potential problems in the data collection process; they mail questionnaires to everyone in the target population and analyze those that are returned. It is not surprising that such surveys have poor response rates. Some surveys reported in academic journals on purchasing, for example, have response rates between 10 and 15%. It is difficult to see how anything can be concluded about the population in such a survey.

---

[1]The variance is often too low as well. In income surveys, for example, the rich and the poor are more likely to be nonrespondents on the income questions. In that case, S_R^2, for the respondent stratum, is smaller than S^2. The point estimator of the mean may be biased, and the variance estimator may be biased, too.

A researcher who knows the target population well will be able to anticipate some of the reasons for nonresponse and prevent some of it. Most investigators, however, do not know as much about reasons for nonresponse as they think they do. They need to discover why the nonresponse occurs and resolve as many of the problems as possible before commencing the survey.

These reasons can be discovered through designed experiments and application of quality improvement methods to the data collection and processing. You do not know why previous surveys related to yours have a low response rate? Design an experiment to find out. You think errors are introduced in the data recording and processing? Use a nested design to find the sources of errors. Books on quality improvement or design of experiments such as Montgomery (2000) or Oehlert (2000) will tell you how to collect your data.

And, of course, you can rely on previous researchers' experiments to help you minimize nonsampling errors. The references on design of experiments and quality control in Chapter 15 are a good place to start; Hidiroglou et al. (1993) give a general framework for nonresponse.

EXAMPLE 8.2   The 1990 U.S. decennial census attempted to survey each of the over 100 million households in the United States. The response rate for the mail survey was 65%; households that did not mail in the questionnaire needed to be contacted in person, adding millions of dollars to the cost of the census. Increasing the mail response rate for future censuses would result in tremendous savings.

Dillman et al. (1995) report results of a factorial experiment employed in the 1992 Census Implementation Test, designed to explore the individual effects and interactions of three experimental factors on response rates. The three factors were: (1) a prenotice letter alerting the household to the impending arrival of the census form, (2) a stamped return envelope included with the census form, and (3) a reminder postcard sent a few days after the census form. The results were dramatic, as shown in Figure 8.1. The experiment established that while all three factors influenced the response rate, the letter and postcard led to greater gains in response rate than the stamped return envelope.   ■

Nonresponse can have many different causes; as a result, no single method can be recommended for every survey. Platek (1977) classifies sources of nonresponse as related to (1) survey content, (2) methods of data collection, and (3) respondent characteristics, and illustrated various sources using the diagram in Figure 8.2. Groves (1989) and Dillman et al. (2009) discuss additional sources of nonresponse. The following are some factors that may influence response rate and data accuracy.

■   *Survey content.* A survey on drug use or financial matters may have a large number of refusals. Sometimes the response rate can be increased for sensitive items by careful ordering of the questions, by using a randomized response technique (see Section 15.4), or by using a self-administered questionnaire on the computer to preserve the respondents' privacy.

■   *Time of survey.* Some calling periods or seasons of the year may yield higher response rates than others. The vacation month of August, for example, would be a bad time to take a one-time household survey in Germany.

FIGURE   8.1

Response rates achieved for each combination of the factors *letter, envelope,* and *postcard.* The observed response rate was 64.3% when all three aids were used and only 50% when none were used.

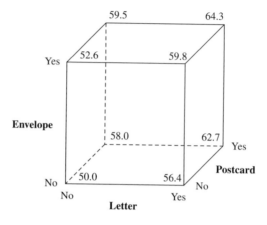

FIGURE   8.2

Factors affecting nonresponse

SOURCE: "Some Factors Affecting Non-Response," by R. Platek, 1977, *Survey Methodology, 3,* 191–214. Copyright © 1977 Survey Methodology. Reprinted with permission.

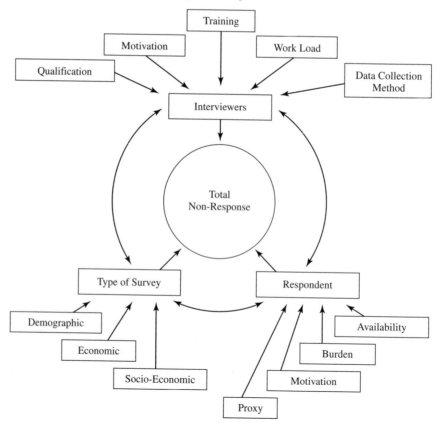

- *Interviewers.* Gower (1979) found a large variability in response rates achieved by different interviewers, with about 15% of interviewers reporting almost no nonresponse. Some field investigators in a bird survey may be better at spotting and identifying birds than others. Quality improvement methods can be applied to increase the response rate and accuracy for interviewers. The same methods can be applied to the data-coding process. These methods will be discussed in Chapter 15.

- *Data-collection method.* Generally, telephone and mail surveys have a lower response rate than in-person surveys (they also have lower costs, however). Computer-Assisted Telephone Interviewing (CATI) and Computer-Assisted Personal Interviewing (CAPI) have been demonstrated to improve accuracy of data collected in telephone and in-person surveys; with CATI and CAPI, all questions are displayed on a computer and the interviewer codes the responses in the computer as questions are asked. CATI and CAPI are especially helpful in surveys in which a respondent's answer to one question determines which question is asked next (Catlin et al., 1988). Many telephone surveys have reported a decrease in response rates in recent years (Curtin et al., 2005); some of the decline in telephone response rates is attributed to recent technology that allows people to screen calls.

  Mail, fax, and Internet surveys often have low response rates. Possible reasons for nonresponse in a mail survey should be explored before the questionnaire is mailed: Is the survey sent to the wrong address? Do recipients discard the envelope as junk mail even before opening it? Will the survey reach the intended recipient? Will the recipient believe that filling out the survey is worth the time?

  A survey conducted by an interviewer often has less item nonresponse than a self-administered questionnaire (Tourangeau et al., 2000). A person filling out a paper survey can skip questions more easily than a person who is prompted by an interviewer. Computer-assisted self-administered questionnaires can sometimes be designed so participants must provide an answer to all questions; that does not mean the answers are always truthful, however.

- *Questionnaire design.* We have already seen in Chapter 1 that question wording has a large effect on the responses received; it can also affect whether a person responds to an item on the questionnaire. Beatty and Herrmann (2002) review research on the application of cognitive research on questionnaire design. In a mail or Internet survey, a well-designed form for the respondent may increase data accuracy and reduce item nonresponse (Dillman, 2008).

- *Respondent burden.* Persons who respond to a survey are doing you an immense favor, and the survey should be as nonintrusive as possible. A shorter questionnaire, requiring less detail, may reduce the burden to the respondent. Respondent burden is a special concern in panel surveys such as the NCVS, in which sampled households are interviewed every six months for $3\frac{1}{2}$ years. DeVries et al. (1996) discuss methods used in reducing respondent burden in the Netherlands. Techniques such as stratification can reduce respondent burden because a smaller sample suffices to give the required precision. Raghunathan and Grizzle (1995) use a split-questionnaire design, in which subsets of the survey respondents are given different subsets of the questionnaire, to reduce respondent burden. With a split-questionnaire design, each individual receives a shortened questionnaire yet every question is administered to at least a subsample of respondents.

- *Survey introduction.* The survey introduction provides the first contact between the interviewer and potential respondent; a good introduction, giving the recipient motivation to respond, can increase response rates dramatically. The Nielsen Company emphasizes to households in their selected sample that their participation in the Nielsen ratings affects which television shows are aired. The respondent should be told for what purpose the data will be used (unscrupulous persons often pretend to be taking a survey when they are really trying to attract customers or converts), and assured confidentiality.

- *Incentives and disincentives.* Incentives, financial or otherwise, may increase the response rate (Singer, 2002). Disincentives may work as well: Physicians who refused to be assessed by peers after selection in a stratified sample from the College of Physicians and Surgeons of Ontario registry had their medical licenses suspended. Not surprisingly, nonresponse was low (McAuley et al., 1990).

- *Follow-up.* The initial contact of the sample is usually less costly per unit than follow-ups of the nonrespondents. If the initial survey is by mail, a reminder may increase the response rate. Not everyone responds to follow-up calls, though; some persons will refuse to respond to the survey no matter how often they are contacted. You need to decide how many follow-up calls to make before the marginal returns do not justify the money spent.

You should try to obtain at least some information about nonrespondents that can be used later to adjust for the nonresponse, and include surrogate items that can be used for item nonresponse. True, there is no complete compensation for not having the data, but partial information may be better than none. Information about the race, sex, or age of a nonrespondent may be used later to adjust for nonresponse. Questions about income may well lead to refusals, but questions about cars, employment, or education may be answered and can be used to predict income. If the pretests of the survey indicate a nonresponse problem that you do not know how to prevent, try to design the survey so that at least some information is collected for each observation unit.

The quality of survey data is largely determined at the design stage. Fisher's (1938) words about experiments apply equally well to the design of sample surveys: "To call in the statistician after the experiment is done may be no more than asking him to perform a postmortem examination: he may be able to say what the experiment died of." Any survey budget needs to allocate sufficient resources for survey design and for nonresponse follow-up. Do not scrimp on the survey design; every hour spent on design may save weeks of remorse later.

8.3
Callbacks and Two-Phase Sampling

Virtually all good surveys rely on callbacks to obtain responses from persons not at home for the first try. Analysis of callback data can provide some information about the biases that can be expected from the remaining nonrespondents.

EXAMPLE **8.3**   Traugott (1987) analyzed callback data from two 1984 Michigan polls on preference for presidential candidates. The overall response rates for the surveys were about 65%, typical for large political polls. About 21% of the interviewed sample responded on the first call; up to 30 attempts were made to reach persons who did not respond on the first call. Traugott found that later respondents were more likely to be male, older, and Republican than early respondents; while 48% of the respondents who answered the first call supported Reagan and 45% supported Mondale, 59% of the entire sample supported Reagan as opposed to 39% for Mondale. Differing procedures for nonresponse follow-up and persistence in callback may explain some of the inconsistencies among political polls.

If nonrespondents resemble late respondents, one might speculate that nonrespondents were more likely to favor Reagan. But nonrespondents do not necessarily resemble the hard-to-reach; persons who absolutely refuse to participate may differ greatly from persons who could not be contacted immediately, and nonrespondents may be more likely to have illnesses or other circumstances preventing participation. We also do not know how likely it is that nonrespondents to the surveys will vote in the election; even if we speculate that they were more likely to favor Reagan, they are not necessarily more likely to vote for Reagan. ■

Often, when the survey is designed so that callbacks will be used, the initial contact is by mail survey; the follow-up calls use a more expensive method such as a personal interview.

Hansen and Hurwitz (1946) proposed subsampling the nonrespondents and using **two-phase sampling** (also called **double sampling**) for stratification to estimate the population mean or total. The population is divided into two strata, as described in Section 8.1; the two strata are respondents and initial nonrespondents, persons who do not respond to the first call. We shall develop the theory of two-phase sampling for general survey designs in Chapter 12; here, we illustrate how it can be used for nonresponse.

In the simplest form of two-phase sampling, randomly select n units in the population. Of these, n_R respond and n_M do not respond. The values n_R and n_M, though, are random variables; they will change if a different SRS is selected. Then make a second call on a random subsample of $100\nu\%$ of the n_M nonrespondents in the sample, where the subsampling fraction ν does not depend on the data collected.

Suppose that through some superhuman effort all of the targeted nonrespondents are reached. Let $\bar{y}_R$ be the sample average of the original respondents, and $\bar{y}_M$ (M stands for "missing") be the average of the subsampled nonrespondents. The two-phase sampling estimators of the population mean and total are

$$\hat{\bar{y}} = \frac{n_R}{n}\bar{y}_R + \frac{n_M}{n}\bar{y}_M \tag{8.1}$$

and

$$\hat{t} = N\hat{\bar{y}} = \frac{N}{n}\sum_{i \in S_R} y_i + \frac{N}{n}\frac{1}{\nu}\sum_{i \in S_M} y_i, \tag{8.2}$$

where S_R represents the sampled units in the respondent stratum and S_M represents the sampled units in the nonrespondent stratum. Note that $\hat{t}$ is a weighted sum of the observed units; the weights are N/n for the respondents and $N/(nv)$ for the subsampled nonrespondents. Because only a subsample was taken in the nonrespondent stratum, each subsampled unit in that stratum represents more units in the population than does a unit in the respondent stratum.

The expected value and variance of these estimators are given in Chapter 12. From (12.8), if the finite population corrections can be ignored, we can estimate the variance by

$$\hat{V}(\hat{\bar{y}}) = \frac{n_R - 1}{n - 1}\frac{s_R^2}{n} + \frac{n_M - 1}{n - 1}\frac{s_M^2}{vn} + \frac{1}{n - 1}\left[\frac{n_R}{n}(\bar{y}_R - \hat{\bar{y}})^2 + \frac{n_M}{n}(\bar{y}_M - \hat{\bar{y}})^2\right].$$

If everyone responds in the subsample, two-phase sampling not only removes the nonresponse bias but also accounts for the original nonresponse in the estimated variance.

8.4
Mechanisms for Nonresponse

Most surveys have some residual nonresponse even after careful design and follow-up of nonrespondents. All methods for fixing up nonresponse are necessarily model-based. If we are to make any inferences about the nonrespondents, we must assume that they are related to respondents in some way.

Dividing population members into two fixed strata of would-be respondents and would-be nonrespondents, as in Section 8.1, is fine for thinking about potential nonresponse bias and for two-phase methods. To adjust for nonresponse that remains after all other measures have been taken, we need a more elaborate setup. Define the random variable

$$R_i = \begin{cases} 1 & \text{if unit } i \text{ responds} \\ 0 & \text{if unit } i \text{ does not respond.} \end{cases}$$

After sampling, the realizations of the response indicator variable are known for the units selected in the sample. A value for y_i is recorded if r_i, the realization of R_i, is 1. The probability that a unit selected for the sample will respond,

$$\phi_i = P(R_i = 1),$$

is of course unknown but assumed positive. Rosenbaum and Rubin (1983) call ϕ_i the **propensity score** for the ith unit.

Suppose that y_i is a response of interest, and that $\mathbf{x}_i$ is a vector of information known about unit i in the sample. Information used in the survey design is included in $\mathbf{x}_i$. We consider three types of missing data, using the Little and Rubin (2002) terminology of nonresponse classification.

Missing Completely at Random   If ϕ_i does not depend on $\mathbf{x}_i$, y_i, or the survey design, the missing data are **missing completely at random** (MCAR). Such a situation occurs if, for example, someone at the laboratory drops a test tube containing the blood sample of one of the survey participants—there is no reason to think that the dropping of the test tube had anything to do with the white blood cell count.[2] If data are MCAR, the respondents are representative of the selected sample.

Missing data in the NCVS would be MCAR if the probability of nonresponse is completely unrelated to region of the United States, race, sex, age, or any other variable measured for the sample, *and* if the probability of nonresponse is unrelated to any variables about victimization status. Nonrespondents would be essentially selected at random from the sample.

If the response probabilities ϕ_i are all equal and the events $\{R_i = 1\}$ are conditionally independent of each other and of the sample selection process given n_R, then the data are MCAR. If an SRS of size n is taken, then under this mechanism the respondents will be a simple random subsample of variable size n_R. The sample mean of the respondents, $\bar{y}_R$, is approximately unbiased for the population mean. The MCAR mechanism is implicitly adopted when nonresponse is ignored.

Missing at Random Given Covariates If ϕ_i depends on $\mathbf{x}_i$ but not on y_i, the data are **missing at random** (MAR); the nonresponse depends only on observed variables. We can successfully model the nonresponse, since we know the values of $\mathbf{x}_i$ for all sample units. Persons in the NCVS would be missing at random if the probability of responding to the survey depends on race, sex, and age—all known quantities—but does not vary with victimization experience within each age/race/sex class. This is sometimes termed **ignorable nonresponse**: Ignorable means that a model can explain the nonresponse mechanism and that the nonresponse can be ignored after the model accounts for it, not that the nonresponse can be completely ignored and complete-data methods used.

Not Missing at Random If the probability of nonresponse depends on the value of a missing response variable, and cannot be completely explained by values in the observed data, then the nonresponse is **not missing at random** (NMAR). This is likely the situation for the NCVS: It is suspected that a person who has been victimized by crime is less likely to respond to the survey than a nonvictim, even if they share the values of all known variables such as race, age, and sex. Crime victims may be more likely to move after a victimization, and thus not be included in subsequent NCVS interviews. Models can help in this situation, because the nonresponse probability may also depend on known variables, but cannot completely adjust for the nonresponse.

The probabilities of responding, ϕ_i, are useful for thinking about the type of nonresponse. Unfortunately, they are unknown, so we do not know for sure which type of nonresponse is present. We can sometimes distinguish between MCAR and MAR by fitting a model attempting to predict the observed probabilities of response for subgroups from known covariates; if the coefficients in a logistic regression model predicting nonresponse are significantly different from 0, the missing data are likely not MCAR. Distinguishing between MAR and NMAR is more difficult. In practice, we expect most nonresponse in surveys to be of the NMAR type. It is unreasonable to expect that we can construct a perfect model that will completely explain the

---

[2]Even here, though, the suspicious mind can create a scenario in which the nonresponse might be related to quantities of interest: Perhaps laboratory workers are less likely to drop test tubes that they believe contain HIV.

nonresponse mechanism. But we can try to reduce the bias due to nonresponse. In the next section, we discuss a method that is commonly used to estimate the ϕ_i's.

8.5
Weighting Methods for Nonresponse

In previous chapters we have seen how weights can be used in calculating estimates for various sampling schemes (see Sections 2.4, 3.3, 5.3, and 7.2). The sampling weights are the reciprocals of the inclusion probabilities, so that an estimator of the population total is $\sum_{i \in \mathcal{S}} w_i y_i$. For stratification, the weights are $w_i = N_h/n_h$ if unit i is in stratum h; for sampling units with unequal probabilities, $w_i = 1/\pi_i$.

Weights can also be used to adjust for nonresponse. Let Z_i be the indicator variable for presence in the selected sample, with $P(Z_i = 1) = \pi_i$. If R_i is independent of Z_i, then the probability that unit i will be measured is

$$P(\text{unit } i \text{ selected in sample and responds}) = \pi_i \phi_i.$$

The probability of responding, ϕ_i, is estimated for each unit in the sample, using auxiliary information that is known for all units in the selected sample. The final weight for a respondent is then $1/(\pi_i \hat{\phi_i})$. Weighting methods assume that the response probabilities can be estimated from variables known for all units; they assume MAR data.

8.5.1   Weighting Class Adjustment

Sampling weights w_i have been interpreted as the number of units in the population represented by unit i of the sample. Weighting class methods extend this approach to compensate for nonsampling errors: Variables known for all units in the selected sample are used to form weighting adjustment classes, and it is hoped that respondents and nonrespondents in the same weighting adjustment class are similar. Weights of respondents in the weighting adjustment class are increased, so that the respondents represent the nonrespondents' share of the population as well as their own.

EXAMPLE **8.4**   Suppose the age is known for every member of the selected sample and that person i in the selected sample has sampling weight $w_i = 1/\pi_i$. Then weighting classes can be formed by dividing the selected sample among different age classes, as Table 8.2 shows.

We estimate the response probability for each class by

$$\hat{\phi_c} = \frac{\text{sum of weights for respondents in class } c}{\text{sum of weights for selected sample in class } c}.$$

Then the sampling weight for each respondent in class c is multiplied by $1/\hat{\phi_c}$, the weight factor in Table 8.2. The weight of each respondent with age between 15 and 24, for example, is multiplied by 1.622. Since there was no nonresponse in the over-65 group, their weights are unchanged.   ∎

TABLE 8.2
Illustration of Weighting Class Adjustment Factors

| | \multicolumn{5}{c}{Age} | |
	15–24	25–34	35–44	45–64	65+	Total
Sample size	202	220	180	195	203	1000
Respondents	124	187	162	187	203	863
Sum of weights for sample	30,322	33,013	27,046	29,272	30,451	150,104
Sum of weights for respondents	18,693	28,143	24,371	28,138	30,451	
$\hat{\phi}_c$	0.6165	0.8525	0.9011	0.9613	1.0000	
Weight factor	1.622	1.173	1.110	1.040	1.000	

The probability of response is assumed to be the same within each weighting class, with the implication that within a weighting class, the probability of response does not depend on y. As mentioned earlier, weighting class methods assume MAR data. The weight for a respondent in weighting class c is $1/(\pi_i \hat{\phi}_c)$.

To estimate the population total using weighting class adjustments, let $x_{ci} = 1$ if unit i is in class c, and 0 otherwise. Then let the new weight for respondent i be

$$\tilde{w}_i = w_i \sum_c \frac{x_{ci}}{\hat{\phi}_c},$$

where w_i is the sampling weight for unit i; $\tilde{w}_i = w_i/\hat{\phi}_c$ if unit i is in class c. Assign $\tilde{w}_i = 0$ if unit i is a nonrespondent. Then,

$$\hat{t}_{wc} = \sum_{i \in S} \tilde{w}_i y_i$$

and

$$\hat{\bar{y}}_{wc} = \frac{\hat{t}_{wc}}{\sum\limits_{i \in S} \tilde{w}_i}.$$

In an SRS, for example, if n_c is the number of sample units in class c, n_{cR} is the number of respondents in class c, and $\bar{y}_{cR}$ is the average for the respondents in class c, then $\hat{\phi}_c = n_{cR}/n_c$ and

$$\hat{t}_{wc} = \sum_{i \in S} \sum_c \frac{N}{n} \frac{n_c}{n_{cR}} x_{ci} y_i = N \sum_c \frac{n_c}{n} \bar{y}_{cR}.$$

EXAMPLE 8.5 *The National Crime Victimization Survey.* To adjust for individual nonresponse in the NCVS, the within-household noninterview adjustment factor (WHHNAF) of Chapter 7 is used. NCVS interviewers gather demographic information on the nonrespondents, and this information is used to classify all persons into 24 weighting adjustment cells. The cells depend on the age of the person, the relation of the person to the reference person (head of household), and the race of the reference person.

For any cell, let W_R be the sum of the weights for the respondents, and W_M be the sum of the weights for the nonrespondents. Then the new weight for a respondent in a cell will be the previous weight multiplied by the weighting adjustment factor $(W_M + W_R)/W_R$. Thus the weights that would be assigned to nonrespondents are reallocated among respondents with similar (we hope) characteristics.

A problem occurs if $(W_M + W_R)/W_R$ is too large. If $(W_M + W_R)/W_R > 2$, the cell contains more nonrespondents than respondents. In this case, the variance of the estimate increases; if the number of respondents in the cell is small, the weight may not be stable. The Census Bureau collapses cells to obtain weighting adjustment factors of 2 or less. If there are fewer than 30 interviewed persons in a cell, or if the weighting adjustment factor is greater than 2, the cell is combined (collapsed) with neighboring cells until the collapsed cell has more than 30 observations and a weighting adjustment factor of 2 or less. ∎

Construction of Weighting Classes   Weighting adjustment classes should be constructed as though they were strata; as shown in the next section, weighting adjustment is similar to poststratification. If we could construct weighting classes so that in each weighting class c (a) the response variable y_i is constant in class c, (b) the response propensity ϕ_i is the same for every unit in class c, or (c) the response y_i is uncorrelated with the response propensity ϕ_i in class c, then we would largely eliminate nonresponse bias for estimating population means and totals (see Exercise 17).

Consequently, the weighting classes should be formed so that units within each class are as similar as possible with respect to the major variables of interest, and so that the response propensities vary from class to class but are relatively homogeneous within a class. At the same time, it is desirable to avoid very large weight adjustments. Eltinge and Yansaneh (1997) discuss methods for choosing the number of weighting classes to use.

8.5.2   Poststratification

Poststratification was introduced in Section 4.4; it is a form of ratio adjustment. To use poststratification to try to compensate for nonresponse, we modify the weights so that the sample is calibrated to population counts in the poststrata. Poststratification is similar to weighting class adjustment, except that population counts are used to adjust the weights. Suppose an SRS is taken. After the sample is collected, units are grouped into H different poststrata, usually based on demographic variables such as race or sex. The population has N_h units in poststratum h; of these, n_h were selected for the sample and n_{hR} responded. The poststratified estimator for $\bar{y}_U$ is

$$\bar{y}_{\text{post}} = \sum_{h=1}^{H} \frac{N_h}{N} \bar{y}_{hR};$$

the weighting class estimator for $\bar{y}_U$ is

$$\bar{y}_{wc} = \sum_{h=1}^{H} \frac{n_h}{n} \bar{y}_{hR}.$$

The two estimators are similar in form; the only difference is that in poststratification, the N_h are known while in weighting class adjustments the N_h are unknown and estimated by Nn_h/n. A variance estimator for poststratification will be given in Exercise 17 of Chapter 9.

Poststratification Using Weights

In a general survey design, the sum of the weights in a subgroup, $\sum_{i \in S_h} w_i$, is supposed to estimate the population count for that subgroup, N_h. Poststratification uses the ratio estimator within each subgroup to adjust by the true population count.

Let $x_{hi} = 1$ if unit i is a respondent in poststratum h, and 0 otherwise. Then let

$$w_i^* = w_i \sum_{h=1}^{H} x_{hi} \frac{N_h}{\sum_{j \in \mathcal{R}} w_j x_{hj}},$$

where $\mathcal{R}$ is the set of respondents in the sample. Using the modified weights,

$$\sum_{i \in \mathcal{R}} w_i^* x_{hi} = N_h,$$

and the poststratified estimator of the population total is

$$\hat{t}_{\text{post}} = \sum_{i \in \mathcal{R}} w_i^* y_i.$$

Note that the modified weights w_i^* depend on the particular sample selected.

Poststratification adjusts for undercoverage as well as nonresponse if the population count N_h includes individuals not in the sampling frame for the survey. As shown in Chapter 4, poststratification can reduce the variance of estimated population quantities by calibrating the survey to the known population counts.

EXAMPLE 8.6   The second-stage factor in the NCVS (see Section 7.6) uses poststratification to adjust the weights. After all other weighting adjustments have been done, including the weighting class adjustments for nonresponse, poststratification is used to make the sample counts agree with estimates of the population counts from the Census Bureau. Each person is assigned to one of 72 poststrata based on the person's age, race, and sex. The number of persons in the population falling in that poststratum, N_h, is known from other sources. Then, the weight for a person in poststratum h is multiplied by

$$\frac{N_h}{\text{sum of weights for all respondents in poststratum } h}.$$

With weighting classes, the weighting factor to adjust for unit nonresponse is always at least one. With poststratification, because weights are adjusted so that they sum to a known population total, the weighting factor can be any positive number, although weighting factors of two or less are desirable. ■

The poststratified estimator is approximately unbiased if within each poststratum h, (a) each unit has the same probability of responding, (b) the response propensity ϕ_i is the same for every unit, or (c) the response y_i is uncorrelated with the response propensity ϕ_i (see Exercise 18). These are big assumptions: To make them seem a little more plausible, survey researchers often use many poststrata. But a large number of poststrata may create additional problems, because poststrata with too few respondents may result in unstable estimates (Gelman and Carlin, 2002). If faced with poststrata with few observations, most practitioners collapse the poststrata with others that have similar means in key variables until they have a reasonable number of observations in each poststratum. For the CPS, a "reasonable" number means that each group has at least 20 observations and that the response rate for each group is at least 50%.

Raking Adjustments

Raking is a poststratification method that may be used when poststrata are formed using more than one variable, but only the marginal population totals are known. Raking was first used in the 1940 census to make sure that the complete census and samples taken from it gave consistent results and was introduced by Deming and Stephan (1940); Brackstone and Rao (1979) further developed the theory.

Consider the following table of sums of weights from a sample; each entry in the table is the sum of the sampling weights for persons in the sample falling in that classification (for example, the sum of the sampling weights for black females is 300).

	Black	White	Asian	Native American	Other	Sum of Weights
Female	300	1200	60	30	30	1620
Male	150	1080	90	30	30	1380
Sum of Weights	450	2280	150	60	60	3000

Now suppose we know the true population counts for the marginal totals: We know that the population has 1510 women and 1490 men, 600 blacks, 2120 whites, 150 Asians, 100 Native Americans, and 30 persons in the "Other" category. The population counts for each cell in the table, however, are unknown; we do not know the number of black females in this population and cannot assume independence. Raking allows us to adjust the weights so that the sums of weights in the margins equal the population counts.

First, adjust the rows. Multiply each entry by (true row population)/(estimated row population). Multiplying the cells in the female row by 1510/1620 and the cells in the male row by 1490/1380 results in the following table:

	Black	White	Asian	Native American	Other	Sum of Weights
Female	279.63	1118.52	55.93	27.96	27.96	1510
Male	161.96	1166.09	97.17	32.39	32.39	1490
Total	441.59	2284.61	153.10	60.35	60.35	3000

The row totals are fine now, but the column totals do not yet equal the population totals. Repeat the same procedure with the columns in the new table. The entries in the first column are each multiplied by 600/441.59. The following table results:

	Black	White	Asian	Native American	Other	Sum of Weights
Female	379.94	1037.93	54.51	46.33	13.90	1532.61
Male	220.06	1082.07	94.70	53.67	16.10	1466.60
Total	600.00	2120.00	150.00	100.00	30.00	3000

But this has thrown the row totals off again. Repeat the procedure until both row and column totals equal the population counts. The procedure converges as long as all cell counts are positive. In this example, the final table of adjusted counts is:

	Black	White	Asian	Native American	Other	Sum of Weights
Female	375.59	1021.47	53.72	45.56	13.67	1510
Male	224.41	1098.53	96.28	54.44	16.33	1490
Total	600.00	2120.00	150.00	100.00	30.00	3000

The entries in the last table may be better estimates of the cell populations (i.e., with smaller variance) than the original weighted estimates, simply because they use more information about the population. The weighting adjustment factor for each white male in the sample is 1098.53/1080; the weight of each white male is increased a little to adjust for nonresponse and undercoverage. Likewise, the weights of white females are decreased because they are overrepresented in the sample.

The assumptions for raking are the same as for poststratification, with the additional assumption that the response probabilities depend only on the row and column and not on the particular cell. Raking has some difficulties—the algorithm may not converge if some of the cell estimates are zero. There is also a danger of "overadjustment"—if there is little relation between the extra dimension in raking and the cell means, raking can increase the variance rather than decrease it.

8.5.3 Advantages and Disadvantages of Weighting Adjustments

Weighting class adjustments and poststratification can both help reduce nonresponse bias. The models for weighting adjustments for nonresponse are strong: In each weighting cell or poststratum, the respondents and nonrespondents are assumed to be similar, or each individual in a weighting class is assumed equally likely to respond to the survey or have a response propensity that is uncorrelated with y. These models never exactly describe the true state of affairs, and you should always consider their plausibility and implications. It is an unfortunate tendency of some survey practitioners to treat the weighting adjustment as a complete remedy and then act as though there was no nonresponse. Weights may improve many of the estimates, but they rarely eliminate all nonresponse bias. If weighting adjustments are made (and remember, making no adjustments is itself a model about the nature of the nonresponse), practitioners should always state the assumed response model and give evidence to justify it. Weighting adjustments are usually used for unit nonresponse, not for item nonresponse (which would require a different weight for each item).

Poststratification is a special case of **calibration** methods in survey sampling. Deville and Särndal (1992) and Särndal (2007) describe the use of calibration methods to attempt to reduce nonresponse bias. We discuss general calibration methods in Section 11.7.

8.6
Imputation

Missing items may occur in surveys for several reasons: An interviewer may fail to ask a question; a respondent may refuse to answer the question or cannot provide the information; a clerk entering the data may skip the value. Sometimes, items with responses are changed to missing when the data set is edited or cleaned—a data editor may not be able to resolve the discrepancies for an individual 3-year-old who voted in the last election, and may set both values to missing.

Imputation is commonly used to assign values to the missing items. A replacement value, often from another person in the survey who is similar to the item nonrespondent on other variables, is imputed (filled in) for the missing value. When imputation is used, an additional variable should be created for the data set that indicates whether the response was measured or imputed.

Imputation procedures are used not only to reduce the nonresponse bias but to produce a "clean," rectangular data set—one without holes for the missing values. We may want to look at tables for subgroups of the population, and imputation allows us to do that without considering the item nonresponse separately each time we construct a table.

EXAMPLE **8.7**   The CPS has an overall high household response rate, but some households refuse to answer certain questions. The nonresponse rate is about 20% on many income questions. This nonresponse would create a substantial bias in any analysis unless some corrective action were taken: Various studies suggest that the item nonresponse for the income items is highest for low-income and high-income households. Imputation for the missing data makes it possible to use standard statistical techniques such as regression without the analyst having to treat the nonresponse by using specially developed methods. For surveys such as the CPS, if imputation is to be done, the agency collecting the data has more information to guide it in filling in the missing values than does an independent analyst, because identifying information is not released on the public use tapes.

The CPS uses weighting for noninterview adjustment, and deductive and hot-deck imputation for item nonresponse. The sample is divided into classes using variables sex, age, race, and other demographic characteristics. If an item is missing, a corresponding item from another unit in that class is substituted. The classifications differ for different items; for imputing weekly earnings, several thousand classes are formed from demographic characteristics as well as occupation and education (U.S. Census Bureau, 2006a).   ■

TABLE **8.3**
Small Data Set Used to Illustrate Imputation Methods

Person	Age	Sex	Years of Education	Crime Victim?	Violent Crime Victim?
1	47	M	16	0	0
2	45	F	?	1	1
3	19	M	11	0	0
4	21	F	?	1	1
5	24	M	12	1	1
6	41	F	?	0	0
7	36	M	20	1	?
8	50	M	12	0	0
9	53	F	13	0	?
10	17	M	10	?	?
11	53	F	12	0	0
12	21	F	12	0	0
13	18	F	11	1	?
14	34	M	16	1	0
15	44	M	14	0	0
16	45	M	11	0	0
17	54	F	14	0	0
18	55	F	10	0	0
19	29	F	12	?	0
20	32	F	10	0	0

We use the small data set in Table 8.3 to illustrate some of the different methods for imputation. This artificial data set is only used for illustration; in practice, a much larger data set is needed for imputation. A "1" means the respondent answered yes to the question.

8.6.1 Deductive Imputation

Some values may be imputed in the data editing, using logical relations among the variables. Person 9 is missing the response for whether she was a victim of violent crime. But she had responded that she was not a victim of any crime, so the violent crime response should be changed to 0.

Deductive imputation may sometimes be used in longitudinal surveys. If a woman has two children in year 1 and two children in year 3, but is missing the value for year 2, the logical value to impute would be two.

8.6.2 Cell Mean Imputation

Respondents are divided into classes (cells) based on known variables, as in weighting class adjustments. Then the average of the values for the responding units in cell c,

$\bar{y}_{Rc}$, is substituted for each missing value. Cell mean imputation assumes that missing items are missing completely at random within the cells.

EXAMPLE **8.8** The four cells for our example are constructed using the variables age and sex. (In practice, of course, you would want to have many more individuals in each cell.)

		Age ≤ 34	Age ≥ 35
Sex	M	Persons 3, 5, 10, 14	Persons 1, 7, 8, 15, 16
	F	Persons 4, 12, 13, 19, 20	Persons 2, 6, 9, 11, 17, 18

Persons 2 and 6, missing the value for years of education, would be assigned the mean value for the four women aged 35 or older who responded to the question: 12.25. The mean for each cell after imputation is the same as the mean of the respondents. The imputed value, however, is not one of the possible responses to the question about education. ■

Mean imputation gives the same point estimates for means, totals, and proportions as the weighting class adjustments. Mean imputation methods fail to reflect the variability of the nonrespondents, however—all missing observations in a class are given the same imputed value. The distribution of y will be distorted because of a "spike" at the value of the sample mean of the respondents. As a consequence, the estimated variance in the subclass will be too small.

To avoid the spike, a stochastic cell mean imputation could be used. If the response variable were approximately normally distributed, the missing values could be imputed with a randomly generated value from a normal distribution with mean $\bar{y}_{cR}$ and standard deviation s_{cR}.

Mean imputation, stochastic or otherwise, distorts relationships among different variables, because imputation is done separately for each missing item. Sample correlations and other statistics are changed. Jinn and Sedransk (1989) discuss the effect of different imputation methods on secondary data analysis, for instance for estimating a regression slope.

8.6.3 Hot-Deck Imputation

In *hot-deck imputation*, as in cell mean imputation and weighting adjustment methods, the sample units are divided into classes. The value of one of the responding units in the class is substituted for each missing response. Often, the values for a set of related missing items are taken from the same donor, to preserve some of the multivariate relationships. The name *hot deck* is from the days when computer programs and data sets were punched on cards—the deck of cards containing the data set being analyzed was warmed by the card reader, so the term *hot deck* was used to refer to imputations made using the same data set. Fellegi and Holt (1976) discuss methods for data editing and hot-deck imputation with large surveys.

How is the donor unit to be chosen? Several methods are possible.

Sequential Hot-Deck Imputation Some hot-deck imputation procedures impute the value in the same subgroup that was last read by the computer. This is partly a carryover from the card days of computers (imputation could be done in one pass), and partly a belief that if the data are arranged in some geographical order, adjacent units in the same subgroup will tend to be more similar than randomly chosen units in the subgroup. One problem with using the value on the previous "card" is that often the nonrespondents also tend to occur in clusters, so one person may be a donor multiple times, in a way that the sampler cannot control. One of the other hot-deck imputation methods is usually used today for most surveys.

In our example, person 19 is missing the response for crime victimization. Person 13 had the last response recorded in her subclass, so the value 1 is imputed.

Random Hot-Deck Imputation A donor is randomly chosen from the persons in the cell with information on all the missing items. To preserve multivariate relationships, usually values from the same donor are used for all missing items of a person.

In our small data set, person 10 is missing both variables for victimization. Persons 3, 5, and 14 in his cell have responses for both crime questions, so one of the three is chosen randomly as the donor. In this case, person 14 is chosen, and his values are imputed for both missing variables.

Nearest-Neighbor Hot-Deck Imputation Define a distance measure between observations, and impute the value of a respondent who is "closest" to the person with the missing item, where closeness is defined using the distance function.

If age and sex are used for the distance function, so that the person of closest age with the same sex is selected to be the donor, the victimization responses of person 3 will be imputed for person 10.

8.6.4 Regression Imputation

Regression imputation predicts the missing value using a regression of the item of interest on variables observed for all cases. A variation is *stochastic regression imputation*, in which the missing value is replaced by the predicted value from the regression model plus a randomly generated error term.

We only have 18 complete observations for the response crime victimization (not really enough for fitting a model to our data set), but a logistic regression of the response with explanatory variable age gives the following model for predicted probability of victimization, $\hat{p}$:

$$\log \frac{\hat{p}}{1 - \hat{p}} = 2.5643 - 0.0896 \times \text{age}.$$

The predicted probability of being a crime victim for a 17-year-old is 0.74; because that is greater than a predetermined cutoff of 0.5, the value 1 is imputed for Person 10.

EXAMPLE **8.9**   Paulin and Ferraro (1994) discuss regression models for imputing income in the U.S. Consumer Expenditure Survey. Households selected for the interview component of the survey are interviewed each quarter for five consecutive quarters; in each interview, they are asked to recall expenditures for the previous three months. The data are used to relate consumer expenditures to characteristics such as family size and income; they are the source of reports that expenditures exceed income in certain income classes.

The Consumer Expenditure Survey conducts about 5,000 interviews each year, as opposed to about 60,000 for the CPS. This sample size is too small for hot-deck imputation methods, as it is less likely that suitable donors will be found for nonrespondents in a smaller sample. If imputation is to be done at all, a parametric model needs to be adopted. Paulin and Ferraro used multiple regression models to predict the log of family income (logarithms are used because the distribution of income is skewed) from explanatory variables including total expenditures and demographic variables. These models assume that income items are MAR given the covariates. ∎

8.6.5   Cold-Deck Imputation

In *cold-deck imputation*, the imputed values are from a previous survey or other information, such as from historical data. (Since the data set serving as the source for the imputation is not the one currently running through the computer, the deck is "cold.") As with hot-deck imputation, cold-deck imputation is not guaranteed to eliminate selection bias.

EXAMPLE **8.10**   Kirkman et al. (2005) describe the imputation procedures used in the 2004 Annual Survey of the Mathematical Sciences, which reports on faculty composition and degrees awarded by departments of mathematical sciences in U.S. colleges and universities. Departments are grouped into seven classes (p. 883): groups (I), (II), and (III) are doctoral-granting departments of mathematics; group (IV) consists of doctoral-granting departments of statistics, biostatistics, and biometrics; group (Va) consists of applied mathematics doctoral-granting departments; group (M) consists of departments whose highest graduate degree is a master's degree; and group (B) consists of departments granting only a baccalaureate degree. The questionnaire is sent to every institution in groups (I), (II), (III), (IV), and (Va); it is sent to a stratified sample of institutions in groups (M) and (B). The response rate in each group is generally between 90% and 100%. Before 2001, population totals were estimated by using the data from the respondents with simple projections (essentially, the weights for the respondents were increased to compensate for the nonrespondents). Beginning in 2001, the survey uses cold-deck imputation. If a doctoral department does not respond in the current year but has responded during the previous three years, the responses from the previous questionnaire are imputed for the current year's data. ∎

8.6.6   Multiple Imputation

In multiple imputation, each missing value is imputed m (≥ 2) different times. Typically, the same stochastic model is used for each imputation. These create m

different "data" sets with no missing values. Each of the m data sets is analyzed as if no imputation had been done; the different results give the analyst a measure of the additional variance due to the imputation. Multiple imputation with different models for nonresponse can give an idea of the sensitivity of the results to particular nonresponse models. See Rubin (1987, 1996, 2004) for details on implementing multiple imputation. Schenker et al. (2006) describe methods used for multiple imputation of income items in the U.S. National Health Interview Survey.

8.6.7 Advantages and Disadvantages of Imputation

Imputation creates a "clean," rectangular data set that can be analyzed by standard software. Analyses of different subsets of the data will produce consistent results. If the nonresponse is missing at random given the covariates used in the imputation procedure, imputation substantially reduces the bias due to item nonresponse. If parts of the data are confidential, the collector of the data can perform the imputation. The data collector has more information about the sample and population than is released to the public (for example, the collector may know the exact address for each sample member), and can often perform a better imputation using that information.

The foremost danger of using imputation is that future data analysts will not distinguish between the original and the imputed values. Ideally, the imputer should record which observations are imputed, how many times each nonimputed record is used as a donor, and which donor was used for a specific response imputed to a recipient. The imputed values may be good guesses, but they are not real data.

If you treat the imputed values as though they were observed in the survey, the estimated variance will be too small. This is partly because of the artificial increase in the sample size and partly because the imputed values are treated as though they were really obtained in the data collection. The true variance will be larger than that estimated from a standard software package. Rao (1996), Fay (1996), and Shao (2003) discuss methods for estimating the variances after imputation.

8.7
Parametric Models for Nonresponse*

Most of the methods for dealing with nonresponse assume that the nonresponse is *ignorable*—that is, conditionally on measured covariates, nonresponse is independent of the variables of interest. In this situation, rather than simply dividing units among different subclasses and adjusting weights, one can fit a superpopulation model. From the model, then, one predicts the values of the y's not in the sample. The model-fitting is often iterative. Often, Bayesian methods are used to fit the model.

In a completely model-based approach, we develop a model for the complete data and add components to the model to account for the proposed nonresponse mechanism. Such an approach has many advantages over other methods: the modeling approach is flexible and can be used to include any knowledge about the nonresponse mechanism, the modeler is forced to state the assumptions about nonresponse explicitly in the model, and some of these assumptions can be evaluated. In addition, variance estimates that result from fitting the model account for the nonresponse, if the model is a good one.

EXAMPLE 8.11 Many people believe that spotted owls in Washington, Oregon, and California are threatened with extinction because timber harvesting in mature coniferous forests reduces their available habitat. Good estimates of the size of the spotted owl population are needed for reasoned debate on the issue.

In the sampling plan described by Azuma et al. (1990), a region of interest is divided into N sampling regions (psus), and an SRS of n psus is selected. Let $Y_i = 1$ if psu i is occupied by a pair of owls, and 0 otherwise. Assume that the Y_i's are independent and that $P(Y_i = 1) = p$, the true proportion of occupied psus. If occupancy could be definitively determined for each psu, the proportion of psus occupied could be estimated by the sample proportion $\bar{y}$. While a fixed number of visits can establish that a psu is occupied, however, a determination that a psu is unoccupied may be wrong—some owl pairs are "nonrespondents," and ignoring the nonresponse will likely result in a too-low estimate of percentage occupancy.

Azuma et al. (1990) propose using a geometric distribution for the number of visits required to discover the owls in an occupied unit, thus modeling the nonresponse. The assumptions for the model are that: (1) the probability of determining occupancy on the first visit, η, is the same for all psus, (2) each visit to a psu is independent, and (3) visits can continue until an owl is sighted. A geometric distribution is commonly used for number of callbacks needed in surveys of people (see Potthoff et al., 1993).

Let X_i be the number of visits required to determine whether psu i is occupied or not. Under the geometric model,

$$P(X_i = x \mid Y_i = 1) = \eta(1 - \eta)^{x-1}, \text{ for } x = 1, 2, 3, \ldots.$$

The budget of the U.S. Forest Service, however, does not allow for an infinite number of visits. Suppose a maximum of s visits are to be made to each psu. The random variable Y_i cannot be observed; the observable random variables are

$$V_i = \begin{cases} k & \text{if } Y_i = 1, X_i = k \text{ and } X_i \leq s \\ 0 & \text{otherwise} \end{cases}$$

and

$$U_i = \begin{cases} 1 & \text{if } Y_i = 1 \text{ and } X_i \leq s \\ 0 & \text{otherwise.} \end{cases}$$

Here, $\sum_{i \in \mathcal{S}} U_i$ counts the number of psus observed to be occupied, and $\sum_{i \in \mathcal{S}} V_i$ counts the total number of visits made to occupied units. Using the geometric model, the probability that an owl is first observed in psu i on visit $k(\leq s)$ is

$$P(V_i = k) = \eta(1 - \eta)^{k-1}p$$

and the probability that an owl is observed on one of the s visits to psu i is

$$P(U_i = 1) = E[U_i] = [1 - (1 - \eta)^s]p.$$

Thus the expected value of the sample proportion of occupied units, $E[\overline{U}]$, is $[1 - (1 - \eta)^s]p$, and is less than the proportion of interest p if $\eta < 1$. The geometric model agrees with the intuition that owls are missed in the s visits.

We find the maximum likelihood estimates of p and η under the assumption that all psus are independent. The likelihood function

$$(\eta p)^{\sum_i u_i} (1 - \eta)^{\sum_i (v_i - u_i)} [1 - p + p(1 - \eta)^s]^{n - \sum_i u_i}$$

is maximized when

$$\hat{p} = \frac{\overline{u}}{1 - (1 - \hat{\eta})^s}$$

and when $\hat{\eta}$ solves

$$\frac{\overline{v}}{\overline{u}} = \frac{1}{\eta} - \frac{s(1 - \eta)^s}{1 - (1 - \eta)^s};$$

numerical methods are needed to calculate $\hat{\eta}$. Maximum likelihood theory also allows calculation of the asymptotic covariance matrix of the parameter estimates.

An SRS of 240 habitat psus in California had the following results:

Visit Number	1	2	3	4	5	6
Number of occupied psus	33	17	12	7	7	5

A total of 81 psus were observed to be occupied in six visits, so $\overline{u} = 81/240 = 0.3375$. The average number of visits made to occupied units was $\overline{v}/\overline{u} = 196/81 = 2.42$. Thus the maximum likelihood estimates are $\hat{\eta} = 0.334$ and $\hat{p} = 0.370$; using the asymptotic covariance matrix from maximum likelihood theory, we estimate the variance of $\hat{p}$ by 0.00137. Thus, an approximate 95% confidence interval for the proportions of units that are occupied is 0.370 ± 0.072.

Incorporating the geometric model for number of visits gave a larger estimate of the proportion of units occupied. If the model does not describe the data, however, the estimate $\hat{p}$ will still be biased; if the model is poor, $\hat{p}$ may be a worse estimate of the occupancy rate than $\overline{u}$. If, for example, field investigators were more likely to find owls on later visits because they accumulate additional information on where to look, the geometric model would be inappropriate.

We need to check whether the geometric model adequately describes the number of visits needed to determine occupancy. Unfortunately, we cannot determine whether the model would describe the situation for units in which owls are not detected in six visits, as the data are missing. We can, however, use a χ^2 goodness-of-fit test to see whether data from the six visits made are fit by the model. Under the model, we expect $n\eta(1-\eta)^{k-1}p$ of the psus to have owls observed on visit k, and we plug in our estimates of p and η to calculate expected counts:

Visit	Observed Count	Expected Count
1	33	29.66
2	17	19.74
3	12	13.14
4	7	8.75
5,6	12	9.71
Total	81	80.99

Visits 5 and 6 were combined into one category so that the expected cell count would be greater than 5. The χ^2 test statistic is 1.75, with p-value > 0.05. There is no indication that the model is inadequate for the data we have. We cannot check its adequacy for the missing data, however. The geometric model assumes observations are independent, and that an occupied psu would eventually be determined to be occupied if enough visits were made. We cannot check whether that assumption of the model is reasonable or not: If some wily owls will never be detected in any number of visits, $\hat{p}$ will still be too small. ∎

Maximum likelihood methods are often used to estimate parameters in nonresponse models. Calculation of estimates required numerical methods even for the simple model adopted for the owls, and that was a simple random sample with a simple geometric model for the response mechanism that allowed us to easily write down the likelihood function. Likelihood functions for more complex sampling designs or nonresponse mechanisms are much more difficult to construct (particularly if observations in the same cluster are considered dependent), and calculating estimates often requires intensive computations. Little and Rubin (2002) discuss likelihood-based methods for missing data in general. Stasny (1991) gives an example of using models to account for nonresponse.

8.8
What Is an Acceptable Response Rate?

Often an investigator will say, "I expect to get a 60% response rate in my survey. Is that acceptable, and will the survey give me valid results?" As we have seen in this chapter, the answer to that question depends on the nature of the nonresponse: If the nonrespondents are MCAR, then we can largely ignore the nonresponse and use the respondents as a representative sample of the population. If the nonrespondents tend to differ from the respondents, then the biases in the results from using only the respondents may make the entire survey worthless.

Many references give advice on cut-offs for acceptability of response rates. Babbie, for example, says: "A review of the published social research literature suggests that a response rate of at least 50 percent is considered adequate for analysis and reporting. A response of 60 percent is good; a response rate of 70 percent is very good"

(2007, 262). I believe that giving such absolute guidelines for acceptable response rates is dangerous and has led many survey investigators to unfounded complacency about nonresponse; many examples exist of surveys with a 70% response rate whose results are flawed. The NCVS needs corrections for nonresponse bias even with a response rate of about 95%.

Be aware that response rates can be manipulated by defining them differently. Researchers often do not say how the response rate was calculated or may use an estimate of response rate that is smaller than it should be. Many surveys inflate the response rate by eliminating units that could not be located from the denominator. Very different results for response rate accrue depending on which definition of response rate is used; all of the following have been used in surveys:

$$\frac{\text{number of completed interviews}}{\text{number of units in sample}},$$

$$\frac{\text{number of completed interviews}}{\text{number of units contacted}},$$

$$\frac{\text{completed interviews} + \text{ineligible units}}{\text{contacted units}},$$

$$\frac{\text{completed interviews}}{\text{contacted units} - (\text{ineligible units})},$$

$$\frac{\text{completed interviews}}{\text{contacted units} - (\text{ineligible units}) - \text{refusals}}.$$

Note that a "response rate" calculated using the last formula will be much higher than one calculated using the first formula because the denominator is smaller.

The American Association of Public Opinion Research (2008b) gives guidelines for classifying units in the sample as eligible, complete or partial interviews, refusals, or other categories, and gives six definitions for different response rates. They recommend that the quantities used in calculating response rate should be defined for every survey. The AAPOR guidelines are available online at www.aapor.org; these are widely accepted as the standards for reporting response rates, and using them allows response rates reported by different surveys to be compared.

The U.S. Office of Management and Budget (2006) guidelines require that a nonresponse bias assessment be performed when the expected unit response rate is below 80%, or the expected item response rate is below 70%, based on the definitions given in the document for calculating response rate. The following recommendations from the U.S. Office of Management and Budget's Federal Committee on Statistical Methodology, reported in González (1994), are helpful:

Recommendation 1. Survey staffs should compute response rates in a uniform fashion over time and document response rate components on each edition of a survey.

Recommendation 2. Survey staffs for repeated surveys should monitor response rate components (such as refusals, not-at-homes, out-of-scopes, address not locatable, postmaster returns, etc.) over time, in conjunction with routine documentation of cost and design changes.

Recommendation 3. Response rate components should be published in survey reports; readers should be given definitions of response rates used, including actual counts, and commentary on the relevance of response rates to the quality of the survey data.

Recommendation 4. Some research on nonresponse can have real payoffs. It should be encouraged by survey administrators as a way to improve the effectiveness of data collection operations.

8.9
Chapter Summary

Nonresponse and undercoverage present serious problems for survey inference. The main concern is that failure to obtain information from some units in the selected sample (nonresponse), or failure to include parts of the population in the sampling frame (undercoverage), can result in biased estimates of population quantities.

The survey design should include features to minimize nonresponse. Designed experiments can give insight into methods for increasing response rates. If possible, the survey frame should contain some information on everyone in the selected sample so that respondents and nonrespondents can be compared on those variables, and so that the auxiliary information can be used in adjusting for residual nonresponse.

Weighting adjustment methods and models can be used to try to reduce nonresponse bias. In weighting class methods, the weights of respondents in a grouping class are increased to compensate for the nonrespondents in that grouping class. In poststratification, the weights of respondents in a poststratum are increased so that they sum to an independent count of the population in that poststratum. The nonresponse mechanism can also be modeled explicitly.

Imputation methods create a "complete" data set by filling in values for data that are missing because of item nonresponse. You must be careful when analyzing imputed data sets to account for the imputation when estimating variances since the imputed values are usually derived from the data.

All surveys should report nonresponse rates. If imputation is used, the imputed values should be flagged so that data analysts know which values were observed and which values were imputed.

Key Terms

Imputation: Methods used to "fill in" values for missing items so that the data set appears complete.

Item nonresponse: Occurs when a unit has responses to some but not all of the items in the survey instrument.

Nonresponse bias: Bias that occurs because nonrespondents differ from survey respondents.

Propensity score: The probability that a unit will respond to the survey.

Raking: A weighting adjustment method in which weights are iteratively adjusted to match row and column population totals.

Respondent: A unit in the selected sample that provides data for the survey.

Selected sample: The set of population units selected to be in the sample; this includes the respondents and nonrespondents.

Two-phase sampling: A method of sampling in which, after an initial probability sample is selected, a probability subsample is selected using inclusion probabilities that may depend on results from the initial sample.

Undercoverage: Occurs when the sampling frame does not include all of the population of interest.

Unit nonresponse: A failure to obtain any information from the observation unit.

For Further Reading

Madow et al. (1983), Groves (1989), Lessler and Kalsbeek (1992), and Groves et al. (2002) cover many topics about nonresponse adjustment, from both statistical and social science viewpoints. Little and Rubin (2002) is a general reference on methods for dealing with missing data (not necessarily in surveys), and is a good reference for model-based approaches. References for more information on weighting are Oh and Scheuren (1983), Holt and Elliot (1991), and Bethlehem (2002). Särndal and Lundström (2005) give methods for adjusting for nonresponse under the unifying umbrella of calibration. Dalenius (1981) emphasizes the importance of dealing with nonsampling as well as sampling errors. References for imputation include Kalton and Kasprzyk (1986), Marker et al. (2002), and Rässler et al. (2008). The journals *Survey Methodology*, *Journal of Official Statistics*, and *Public Opinion Quarterly* publish many articles on experiments that have been done to reduce nonresponse in surveys of persons.

8.10
Exercises

A. Introductory Exercises

1   Ryan et al. (1991) report results from the Ross Laboratories Mothers' Survey, a national mail survey investigating infant feeding in the United States. Questionnaires asking mothers about type of milk fed to the infant during each of the first six months and about socioeconomic variables were mailed to a sample of mothers of six-month-old infants. The authors state that the number of questionnaires mailed increased from 1984 to 1989: "In 1984, 56,894 questionnaires were mailed and 30,694 were returned. In 1989, 196,000 questionnaires were mailed and 89,640 were returned." Low-income families were oversampled in the survey design because they had the lowest response rates. Respondents were divided into subclasses defined by region, ethnic background, age, and education; weights were computed using information from the U.S. Census Bureau.

  a   What are the advantages and drawbacks of oversampling the low-income families in this survey? What implicit model is adopted for nonresponse?

  b   Weighted counts are "comparable with those published by the U.S. Bureau of the Census and the National Center for Health Statistics" on ethnicity, maternal age, income, education, employment, birth weight, region, and participation in the Women, Infants, and Children supplemental food program. The weighted counts estimated that about 53% of mothers had one child, while the government data indicated that about 43% of mothers had one child. Does the agreement of weighted counts with official statistics indicate that the weighting corrects the nonresponse bias? Explain.

c Discuss the use of weighting in this survey. Can you think of any improvements?

2 Investigators selected an SRS of 200 high school seniors from a population of 2000 for a survey of television-viewing habits, with an overall response rate of 75%. By checking school records, they were able to find the grade point average for the non-respondents, and classify the sample accordingly:

GPA	Sample Size	Number of Respondents	Hours of TV $\bar{y}$	s_y
3.00–4.00	75	66	32	15
2.00–2.99	72	58	41	19
Below 2.00	53	26	54	25
Total	200	150		

a What is the estimate for the average number of hours of TV watched per week if only respondents are analyzed? What is the standard error of the estimate?

b Perform a χ^2 test for the null hypothesis that the three GPA groups have the same response rates. What do you conclude? What do your results say about the type of missing data: Do you think the data are MCAR? MAR? Nonignorable?

c Perform a one-way ANOVA analysis to test the null hypothesis that the three GPA groups have the same mean level of television viewing. What do you conclude? Does your ANOVA analysis indicate that GPA would be a good variable for constructing weighting cells? Why, or why not?

d Use the GPA classification to adjust the weights of the respondents in the sample. What is the weighting class estimate of the average viewing time?

e The population counts are 700 students with GPA between 3 and 4; 800 students with GPA between 2 and 3; and 500 students with GPA less than 2. Use these population counts to construct a poststratified estimate of the mean viewing time.

f What other methods might you use to adjust for the nonresponse?

g What other variables might be collected that could be used in nonresponse models?

3 The following description and assessment of nonresponse is from a study of Hamilton, Ontario, homeowners' attitudes on composting toilets:

> The survey was carried out by means of a self-administered mail questionnaire. Twelve hundred questionnaires were sent to a randomly selected sample of house-dwellers. Follow-up thank you notes were sent a week later. In total, 329 questionnaires were returned, representing a response rate of 27%. This was deemed satisfactory since many mail surveyors consider a 15 to 20% response rate to be a good return (Wynia et al., 1993, p. 362).

Do you agree that the response rate of 27% is satisfactory? Suppose the investigators came to you for statistical advice on analyzing these data and designing a follow-up survey. What would you tell them?

4   Kosmin and Lachman (1993) had a question on religious affiliation included in 56 consecutive weekly household surveys; the subject of household surveys varied from week to week from cable TV use, to preference for consumer items, to political issues. After four callbacks, the unit nonresponse rate was 50%; an additional 2.3% refused to answer the religion question. The authors say:

> Nationally, the sheer number of interviews and careful research design resulted in a high level of precision ... Standard error estimates for our overall national sample show that we can be 95% confident that the figures we have obtained have an error margin, plus or minus, of less than 0.2%. This means, for example, that we are more than 95% certain that the figure for Catholics is in the range of 25.0% to 26.4% for the U.S. population. (p. 286):

a   Critique the preceding statement.

b   If you anticipated item nonresponse, do you think it would be better to insert the question of interest in different surveys each week, as was done here, or to use the same set of additional questions in each survey? Explain your answer. How would you design an experiment to test your conjecture?

B. Working with Survey Data

5   The issue of nonresponse in the Winter Break Closure Survey (in file winter.dat) was briefly mentioned in Exercise 19 of Chapter 3. What model is adopted for nonresponse when the formulas from stratified sampling are used to estimate the proportion of university employees who would answer "yes" to the question "Would you want to have Winter Break Closure again?" Do you think this is a reasonable model? How else might you model the effects of nonresponse in this survey? What additional information could be collected to adjust for unit nonresponse?

6   The American Statistical Association (ASA) studied whether it should offer a certification designation for its members, so that statisticians meeting the qualifications could be designated as "Certified Statisticians." In 1994, the ASA surveyed its membership about this issue, with data in file certify.dat. The survey was sent to all 18,609 members; 5001 responses were obtained. Results from the survey were reported in the October 1994 issue of *Amstat News.*

Assume that in 1994, the ASA membership had the following characteristics: 55% have Ph.D.'s and 38% have Master's degrees; 29% work in industry, 34% work in academia, and 11% work in government. The cross-classification between education and workplace was unavailable.

a   What are the response rates for the various subclasses of ASA membership? Are the nonrespondents MCAR? Do you think they are MAR?

b   Use raking to adjust the weights for the six cells defined by education (Ph.D. or non-Ph.D.) and workplace (industry, academia, or other). Start with an initial weight of 18,609/5001 for each respondent. What assumptions must you make to use raking?

c Can you conclude from this survey that a majority of the ASA membership opposed certification in 1994? Why, or why not?

7 The ACLS survey in Example 3.4 had nonresponse. Calculate the response rate in each stratum for the survey. What model was adopted for the nonresponse in Example 3.4? Is there evidence that the nonresponse rate varies among the strata, or that it is related to the percentage female membership?

8 Weights are used in the Survey of Youth in Custody (discussed in Example 7.7) to adjust for unit nonresponse. Use a hot-deck procedure to impute values for the variable measuring with whom the youth lived when growing up. What variables will you use to group the data into classes?

9 Repeat Exercise 8, using a regression imputation model.

10 Repeat Exercise 8, for the variable "have used illegal drugs."

11 Repeat Exercise 9, for the variable "have used illegal drugs."

12 Gnap (1995) conducted a survey on teacher workload which was used in Exercise 15 of Chapter 5.

a The original survey was intended as a one-stage cluster sample. What was the overall response rate?

b Would you expect nonresponse bias in this study? If so, in which direction would you expect the bias to be? Which teachers do you think would be less likely to respond to the survey?

c Gnap also collected data on a random subsample of the nonrespondents in the "large" stratum, in file teachnr.dat. How do the respondents and nonrespondents differ?

d Is there evidence of nonresponse bias, when you compare the subsample of non-respondents to the respondents in the original survey?

13 Not all of the parents surveyed in the study discussed in Exercise 16 of Chapter 5 returned the questionnaire. In the original sampling design, 50 questionnaires were mailed to parents of children in each school, for a total planned sample size of 500. We know that of the 9962 children who were not immunized during the campaign, the consent form had not been returned for 6698 of the children, the consent form had been returned but immunization refused for 2061 of the children, and 1203 children whose parents had consented were absent on immunization day.

a Calculate the response rate for each cluster. What is the correlation of the response rate and the percentage of respondents in the school who returned the consent form? Of the response rate and the percentage of respondents in each school who refused consent?

b Overall, about 67% (6698/9962) of the parents in the target population did not return the consent form. Using the data from the respondents, calculate a 95% CI for the proportion of parents in the sample who did not return the consent form. Calculate two additional interval estimates for this quantity: one assuming that the missing values are all 0's, and one assuming that the missing values are all 1's. What is the relation between your estimates and the population quantity?

c Repeat part (b), examining the percentage of parents that returned the form but refused to have their children immunized.

d Do you think nonresponse bias is a problem for this survey?

14 Use the data in file agpop.dat for this exercise. Let y_i be the value of *acres92* for unit i and x_i be the value of *acres87* for unit i. Draw an SRS of size 400. Now generate missing data from your sample by generating a standard uniform random variable U_i for each observation and deleting the observation if $16U_i \geq \ln(x_i)$. (Sample SAS code for this is on the website.)

a If you ignore the missing data, do you expect the mean of y to be too large or too small?

b Compute the mean from the data set with missing values. Does a 95% CI, computed ignoring the nonresponse, contain the true mean from the population?

c Now poststratify the sample by region, using the stratum population sizes in Example 3.2 to adjust the sampling weight in each region. Does this appear to reduce the bias?

d Try a different poststratification. This time, form 4 groups based on the value of x_i in the population, and find the number of population counties in each group. How do the results of this poststratification compare with the poststratification by region?

15 Repeat Exercise 14, using weighting class methods instead of poststratification. With weighting class methods, you adjust the weights using counts from the selected sample rather than the population.

C. Working with Theory

16 Let $Z_i = 1$ if unit i is included in the sample and 0 otherwise, with $P(Z_i = 1) = \pi_i$. Let $R_i = 1$ if unit i responds to the survey and 0 otherwise, with $P(R_i = 1) = \phi_i$ and $\bar{\phi}_U = \sum_{i=1}^{N} \phi_i/N$. Assume R_i is independent of Z_i for each $i = 1, \ldots, N$. Let $\hat{\bar{y}}_R$ estimate the population mean $\bar{y}_U = \sum_{i=1}^{N} y_i/N$ using only the respondents:

$$\hat{\bar{y}}_R = \frac{\displaystyle\sum_{i=1}^{N} Z_i R_i w_i y_i}{\displaystyle\sum_{i=1}^{N} Z_i R_i w_i},$$

where $w_i = 1/\pi_i$. Show that the bias of $\hat{\bar{y}}_R$ is approximately

$$E[\hat{\bar{y}}_R] - \bar{y}_U \approx \frac{1}{N} \sum_{i=1}^{N} \frac{\phi_i y_i}{\bar{\phi}_U} \approx \frac{\text{Cov}(\phi, y)}{\bar{\phi}_U},$$

where $\text{Cov}(\phi, y) = \sum_{i=1}^{N} (\phi_i - \bar{\phi}_U)(y_i - \bar{y}_U)/(N-1)$. As a consequence, the nonresponse bias is approximately zero if either (1) $\phi_i = \bar{\phi}_U$ for all i, that is, the response propensity is the same for all units, or (2) the propensity to respond is uncorrelated with the response y_i (Tremblay, 1986).

17 Let Z_i and R_i be as defined in Exercise 16. Divide the sample into C weighting classes and define $x_{ci} = 1$ if unit i is in class c and 0 otherwise. Let $\bar{\phi}_c = \sum_{i=1}^{N} \phi_i x_{ci} / \sum_{i=1}^{N} x_{ci}$,

$$\hat{\phi}_c = \frac{\sum\limits_{i=1}^{N} Z_i R_i w_i x_{ci}}{\sum\limits_{i=1}^{N} Z_i w_i x_{ci}},$$

and

$$\hat{t}_{wc} = \sum_{i=1}^{N} Z_i R_i w_i y_i \sum_{c=1}^{C} \frac{x_{ci}}{\hat{\phi}_c}.$$

Show that if the weighting classes are sufficiently large,

$$\text{Bias}(\hat{t}_{wc}) \approx \sum_{c=1}^{C} \sum_{i=1}^{N} x_{ci} \phi_i (y_i - \bar{y}_{cu}) / \bar{\phi}_c,$$

where $\bar{y}_{cu} = \sum_{i=1}^{N} y_i x_{ci} / \sum_{i=1}^{N} x_{ci}$. Thus, the weighting class adjustments for nonresponse in Section 8.5 produce an approximately unbiased estimator if (a) $\phi_i = \bar{\phi}_c$ for all units in class c, (b) $y_i = \bar{y}_{cu}$ for all units in class c, or (c) within each class c, the propensity to respond is uncorrelated with y_i.

18 Let Z_i and R_i be as defined in Exercise 16. Divide the sample into H poststrata. Let N_h be the number of population units in poststratum h, obtained from an independent source such as a population register or census. Show that if the poststrata are sufficiently large,

$$\text{Bias}(\hat{t}_{\text{post}}) \approx \sum_{h=1}^{H} \frac{N_h}{N} \text{Cov}_h(\phi, y) / \bar{\phi}_h,$$

where $\bar{\phi}_h = \sum_{i=1}^{N} x_{hi} \phi_i / N_h$ and $\text{Cov}_h(\phi, y)$ is the population covariance of the y_i's and ϕ_i's for population units in poststratum h.

19 *Effect of weighting class adjustment on variances.* Suppose that an SRS of size n is taken. Let $Z_i = 1$ if unit i is included in the sample and 0 otherwise, with $P(Z_i = 1) = n/N$. Two weighting classes are used to adjust for nonresponse; define $x_i = 1$ if unit i is in class 1 and 0 if unit i is in class 2. Let $R_i = 1$ if unit i responds to the survey and 0 otherwise. Assume that the R_i's are independent Bernoulli random variables with $P(R_i = 1) = x_i \phi_1 + (1 - x_i) \phi_2$, and that R_i is independent of $Z_1, \ldots, Z_N$. The sample sizes in the two classes are $n_1 = \sum_{i=1}^{N} Z_i x_i$ and $n_2 = \sum_{i=1}^{N} Z_i (1 - x_i)$; note that n_1 and n_2 are random variables. Similarly, the number of respondents in the two classes are $n_{1R} = \sum_{i=1}^{N} Z_i R_i x_i$ and $n_{2R} = \sum_{i=1}^{N} Z_i R_i (1 - x_i)$. Assume the number of respondents in each group is sufficiently large so that $E[n_c/n_{cR}] \approx 1/\phi_c$ for $c = 1, 2$. With these assumptions, the weighting class adjusted estimator of the mean,

$$\hat{\bar{y}}_{wc} = \frac{n_1}{n} \frac{1}{n_{1R}} \sum_{i=1}^{N} Z_i R_i x_i y_i + \frac{n_2}{n} \frac{1}{n_{2R}} \sum_{i=1}^{N} Z_i R_i (1 - x_i) y_i$$

is approximately unbiased for the population mean $\bar{y}_\mathcal{U}$ (see Exercise 17). Find the approximate variance of $\hat{\bar{y}}_{wc}$. Hint: Use Property A.4 of Conditional Expectation in Section A.4.

20 *The Hartley (1946) and Politz–Simmons (1949) method.* Suppose that all calls are made during Monday through Friday evenings. Each respondent is asked whether he or she was at home at the time of the interview, on each of the four preceding weeknights. Respondent i replies that she was home k_i of the 4 nights. It is then assumed that the probability of response is proportional to the number of nights at home during interviewing hours, so the probability of response is estimated by $\hat{\phi}_i = (k_i + 1)/5$. Let

$$\hat{\bar{y}}_{\text{HPS}} = \frac{\displaystyle\sum_{i \in \mathcal{S}} w_i y_i / \hat{\phi}_i}{\displaystyle\sum_{i \in \mathcal{S}} w_i / \hat{\phi}_i}.$$

a Under what circumstances would you expect the method to reduce bias due to non-response? What assumptions must be made for the estimator to be approximately unbiased?

b What are some potential drawbacks of the method for use in practice? How does it adjust for persons who were not at home during any of the five nights, or who refused to participate in the survey?

D. Projects and Activities

21 Find a recent poll on a website. How do they describe the sources of error in the survey? Do they give the nonresponse rate, or reference a document that details the treatment of nonresponse? How do they adjust for nonresponse in their estimates?

22 Find an example of a survey in a popular newspaper or magazine. Is the nonresponse rate given? If so, how was it calculated? How do you think the nonresponse might have affected the conclusions of the survey? Give suggestions for how the journalist could discuss nonresponse problems in the article.

23 Find an example of a survey in a scholarly journal. How did the authors calculate the nonresponse rate? How did the survey deal with nonresponse? How do you think the nonresponse might have affected the conclusions of the study? Do you think the authors adequately account for potential nonresponse biases? What suggestions do you have for future studies?

24 The U.S. National Science Foundation Division of Science Resources Studies published results from the 2003 Survey of Doctorate Recipients in "Characteristics of Doctoral Scientists and Engineers in the United States: 2003."[3] How does this survey deal with nonresponse, discussed on page 153 of the report? Do you think that nonresponse bias is a problem for this survey?

---

[3]NSF Publication 06-320. Available at www.nsf.gov/statistics/nsf06320/pdf/nsf06320.pdf.

25   How did the survey you critiqued in Exercise 26 of Chapter 7 deal with nonresponse? In your opinion, did the investigators adequately address the problems of nonresponse? What suggestions do you have for improvement?

26   Answer the questions in Exercise 25 for the survey you examined in Exercise 30 of Chapter 7.

27   *Activity for course project.*   Return to the data you chose in Exercise 31 of Chapter 7. What kinds of nonresponse occur in your data set? How does the survey define nonresponse rate, and what are the nonresponse rates for the survey? What methods are used to try to adjust for the nonresponse?

9

Variance Estimation in Complex Surveys

Rejoice that under cloud and star
  The planet's more than Maine or Texas.
Bless the delightful fact there are
  Twelve months, nine muses, and two sexes;
And infinite in earth's dominions
Arts, climates, wonders, and opinions

—Phyllis McGinley, "In Praise of Diversity"[1]

Population means and totals are easily estimated using weights. Estimating variances is more intricate: In Chapter 7 we noted that in a complex survey with several levels of stratification and clustering, variances for estimated means and totals are calculated at each level and then combined as the survey design is ascended. Poststratification and nonresponse adjustments also affect the variance.

In previous chapters, we have presented and derived variance formulas for a variety of sampling plans. Some of the variance formulas, such as those for simple random samples (SRSs), are relatively simple. Other formulas, such as $\hat{V}(\hat{t})$ from a two-stage cluster sample without replacement, are more complicated. All work for estimating variances of estimated totals. But we often want to estimate other quantities from survey data for which we have presented no variance formula. For example, in Chapter 4 we derived an approximate variance for a ratio of two means when an SRS is taken. What if you want to estimate a ratio, but the survey is not an SRS? How would you estimate the variance?

This chapter describes several methods for estimating variances of estimated totals and other statistics from complex surveys. Section 9.1 describes the commonly used linearization method for calculating variances of estimators. Sections 9.2 and 9.3 present random group and resampling methods for calculating variances of linear and nonlinear statistics. Section 9.4 describes the calculation of generalized variance functions, and Section 9.5 describes constructing confidence intervals (CIs).

---

[1]From *The Love Letters of Phyllis McGinley*, by Phyllis McGinley. Copyright 1951, 1952, 1953, 1954 by Phyllis McGinley. Copyright renewed © 1979, 1980, 1981, 1982 by Phyllis Hayden Blake. Used by permission of Viking Penguin, a division of Penguin Books USA Inc.

9.1
Linearization (Taylor Series) Methods

Most of the variance formulas in Chapters 2 through 6 were for estimators of means and totals. Those formulas can be used to find variances for any linear combination of estimated means and totals. Let y_{ij} be the response of unit i to item j. Suppose $\hat{t}_1, \ldots, \hat{t}_k$ are unbiased estimators of the k population totals $t_1, \ldots, t_k$, with $\hat{t}_j = \sum_{i \in S} w_i y_{ij}$. Then, for any constants $a_1, \ldots, a_k$, we can define a new variable

$$q_i = \sum_{j=1}^{k} a_j y_{ij}$$

so that

$$\hat{t}_q = \sum_{i \in S} w_i q_i = \sum_{j=1}^{k} a_j \hat{t}_j$$

and

$$V\left(\sum_{j=1}^{k} a_j \hat{t}_j\right) = V(\hat{t}_q) = \sum_{j=1}^{k} a_j^2 V(\hat{t}_j) + 2 \sum_{j=1}^{k-1} \sum_{l=j+1}^{k} a_j a_l \text{Cov}(\hat{t}_j, \hat{t}_l). \quad (9.1)$$

Thus, if t_1 is the total number of dollars robbery victims reported stolen, t_2 is the number of days of work they missed because of the crime, and t_3 is their total medical expenses, one measure of financial consequences of robbery (assuming \$150 per day of work lost) might be $\hat{t}_1 + 150\hat{t}_2 + \hat{t}_3$. By (9.1), the variance is

$$V(\hat{t}_1 + 150\hat{t}_2 + \hat{t}_3) = V(\hat{t}_q)$$
$$= V(\hat{t}_1) + 150^2 V(\hat{t}_2) + V(\hat{t}_3)$$
$$+ 300 \, \text{Cov}(\hat{t}_1, \hat{t}_2) + 2 \, \text{Cov}(\hat{t}_1, \hat{t}_3) + 300 \, \text{Cov}(\hat{t}_2, \hat{t}_3),$$

where $q_i = y_{i1} + 150 y_{i2} + y_{i3}$ is the financial loss from robbery for person i.

Suppose, though, that we are interested in the proportion of total loss accounted for by the stolen property, t_1/t_q. This is not a linear statistic, as t_1/t_q cannot be expressed in the form $a_1 t_1 + a_2 t_q$ for constants a_1 and a_2. But Taylor's theorem from calculus allows us to **linearize** a smooth nonlinear function $h(t_1, t_2, \ldots, t_k)$ of the population totals; Taylor's theorem gives the constants $a_0, a_1, \ldots, a_k$ so that

$$h(t_1, \ldots, t_k) \approx a_0 + \sum_{j=1}^{k} a_j t_j.$$

Then $V[h(\hat{t}_1, \ldots, \hat{t}_k)]$ may be approximated by $V(\sum_{j=1}^{k} a_j \hat{t}_j)$, which we know how to calculate using (9.1).

Taylor series approximations have long been used in statistics to calculate approximate variances. Woodruff (1971) illustrates their use in complex surveys. Binder (1983) gives a more rigorous treatment of Taylor series methods for complex surveys and tells how to use linearization when the parameter of interest θ solves $h(\theta, t_1, \ldots, t_k) = 0$, but θ is an implicit function of $t_1, \ldots, t_k$.

FIGURE 9.1

The function $h(x) = x(1 - x)$, along with the tangent to the function at point p. If $\hat{p}$ is close to p, then $h(\hat{p})$ will be close to the tangent line. The slope of the tangent line is $h'(p) = 1 - 2p$.

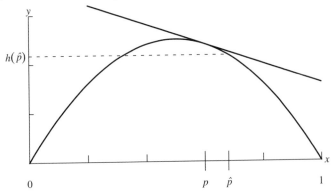

EXAMPLE 9.1   The quantity $\theta = p(1 - p)$, where p is a population proportion, may be estimated by $\hat{\theta} = \hat{p}(1 - \hat{p})$. Assume that $\hat{p}$ is an unbiased estimator of p and that $V(\hat{p})$ is known. Let $h(x) = x(1 - x)$, so $\theta = h(p)$ and $\hat{\theta} = h(\hat{p})$. Now h is a nonlinear function of x, but the function can be approximated at any nearby point a by the tangent line to the function; the slope of the tangent line is given by the derivative, as illustrated in Figure 9.1.

The first-order version of Taylor's theorem states that if the second derivative of h is continuous, then

$$h(x) = h(a) + h'(a)(x - a) + \int_{a}^{x} (x - t)h''(t)dt;$$

under conditions commonly satisfied in statistics, the last term is small relative to the first two and we use the approximation

$$h(\hat{p}) \approx h(p) + h'(p)(\hat{p} - p)$$
$$= p(1 - p) + (1 - 2p)(\hat{p} - p).$$

Then,

$$V[h(\hat{p})] \approx (1 - 2p)^2 V(\hat{p} - p),$$

and $V(\hat{p})$ is known, so the approximate variance of $h(\hat{p})$ can be estimated by

$$\hat{V}[h(\hat{p})] = (1 - 2\hat{p})^2 \hat{V}(\hat{p}). \quad \blacksquare$$

The following are the basic steps for constructing a linearization estimator of the variance of a nonlinear function of means or totals:

1   Express the quantity of interest as a function of means or totals of variables measured or computed in the sample. In general, $\theta = h(t_1, t_2, \dots, t_k)$ or $\theta = h(\bar{y}_{1U}, \dots, \bar{y}_{kU})$. In Example 9.1, $\theta = h(\bar{y}_U) = h(p) = p(1 - p)$ and $\hat{\theta} = h(\hat{p})$.

2   Find the partial derivative of h with respect to each argument. The partial derivatives, evaluated at the population quantities, form the linearizing constants a_j.

3 Apply Taylor's theorem to linearize the estimate:

$$h(\hat{t}_1, \hat{t}_2, \ldots, \hat{t}_k) \approx h(t_1, t_2, \ldots, t_k) + \sum_{j=1}^{k} a_j(\hat{t}_j - t_j),$$

where

$$a_j = \left. \frac{\partial h(c_1, c_2, \ldots, c_k)}{\partial c_j} \right|_{t_1, t_2, \ldots, t_k}.$$

4 Define the new variable q by

$$q_i = \sum_{j=1}^{k} a_j y_{ij}.$$

Now find the estimated variance of $\hat{t}_q = \sum_{i \in S} w_i q_i$, substituting estimators for unknown population quantities. This will generally approximate the variance of $\hat{\theta} = h(\hat{t}_1, \ldots, \hat{t}_k)$.

EXAMPLE 9.2 We used linearization methods to approximate the variance of the ratio and regression estimators in Chapter 4. In Chapter 4, we used an SRS, estimator $\hat{B} = \bar{y}/\bar{x} = \hat{t}_y/\hat{t}_x$, and the approximation

$$\hat{B} - B = \frac{\bar{y} - B\bar{x}}{\bar{x}} \approx \frac{\bar{y} - B\bar{x}}{\bar{x}_U} = \sum_{i \in S} \frac{y_i - Bx_i}{n\bar{x}_U}.$$

In (4.10), we estimated the variance by

$$\hat{V}(\hat{B}) = \left(1 - \frac{n}{N}\right) \frac{s_e^2}{n\bar{x}^2},$$

where s_e^2 is the sample variance of the residuals $e_i = y_i - \hat{B}x_i$. Essentially, we used Taylor's theorem to obtain this estimator. The steps below give the same result.

1 Express B as a function of the population totals. Let $h(c, d) = d/c$, so

$$B = h(t_x, t_y) = \frac{t_y}{t_x} \quad \text{and} \quad \hat{B} = h(\hat{t}_x, \hat{t}_y) = \frac{\hat{t}_y}{\hat{t}_x}.$$

Assume that the estimators $\hat{t}_x$ and $\hat{t}_y$ are unbiased.

2 The partial derivatives are

$$\frac{\partial h(c, d)}{\partial c} = -\frac{d}{c^2} \quad \text{and} \quad \frac{\partial h(c, d)}{\partial d} = \frac{1}{c};$$

evaluated at $c = t_x$ and $d = t_y$, these are $-t_y/t_x^2$ and $1/t_x$.

3 By Taylor's theorem,

$$\hat{B} = h(\hat{t}_x, \hat{t}_y)$$

$$\approx h(t_x, t_y) + \left. \frac{\partial h(c, d)}{\partial c} \right|_{t_x, t_y} (\hat{t}_x - t_x) + \left. \frac{\partial h(c, d)}{\partial d} \right|_{t_x, t_y} (\hat{t}_y - t_y).$$

Using the partial derivatives from Step 2,

$$\hat{B} - B \approx -\frac{t_y}{t_x^2}(\hat{t}_x - t_x) + \frac{1}{t_x}(\hat{t}_y - t_y).$$

4   The approximate mean squared error (MSE) of $\hat{B}$ is

$$E[(\hat{B} - B)^2] \approx E\left[\left\{ -\frac{t_y}{t_x^2}(\hat{t}_x - t_x) + \frac{1}{t_x}(\hat{t}_y - t_y) \right\}^2 \right] \tag{9.2}$$

$$= \frac{1}{t_x^2}\{B^2 V(\hat{t}_x) + V(\hat{t}_y) - 2B\,\mathrm{Cov}(\hat{t}_x, \hat{t}_y)\}. \tag{9.3}$$

Substitute estimators of the unknown quantities into (9.2) to define

$$q_i = \frac{1}{\hat{t}_x}[y_i - \hat{B}x_i] = \frac{1}{\hat{t}_x}e_i,$$

and find $\hat{V}(\hat{B}) = \hat{V}(\hat{t}_q) = \hat{V}(\hat{t}_e)/\hat{t}_x^2$ using the survey design. For an SRS, this results in the variance estimator in (4.10). ∎

The method in Example 9.2 requires substituting estimators for quantities such as B. Note that alternative variance estimators can be derived from (9.2). In particular, if t_x is known, it can be used in place of an estimator $\hat{t}_x$ in the denominator of q_i, giving $\hat{V}_2(\hat{B}) = \hat{V}(\hat{t}_e)/t_x^2$. The estimators $\hat{V}(\hat{B})$ and $\hat{V}_2(\hat{B})$ are asymptotically equivalent, since we expect $t_x/\hat{t}_x \approx 1$ for large sample sizes. For small samples, $\hat{V}(\hat{B})$ works slightly better than $\hat{V}_2(\hat{B})$ in many situations (see Exercise 19 of Chapter 4). An alternative procedure for deriving linearization variance estimators that results in a unique estimator is discussed in Exercise 23.

Advantages: If the partial derivatives are known, linearization almost always gives a variance estimate for a statistic and can be applied in general sampling designs. Linearization methods have been used for a long time in statistics, and the theory is well developed. Software exists for calculating linearization variance estimates for many nonlinear functions of interest such as ratios and regression coefficients; some software will be discussed in Section 9.6.

Disadvantages: Calculations can be messy, and the method is difficult to apply for complex functions involving weights. You must either find analytical expressions for the partial derivatives of h or calculate the partial derivatives numerically. A separate variance formula is needed for each nonlinear statistic that is estimated, and that can require much special programming; a different method is needed for each statistic. In addition, not all statistics can be expressed as a smooth function of the population totals—the median and other quantiles, for example, do not fit into this framework. The accuracy of the linearization approximation depends on the sample size—the variance estimator is often biased downwards if the sample is not large enough.

9.2
Random Group Methods

9.2.1 Replicating the Survey Design

Suppose the basic survey design is replicated independently R times. *Independently* here means that each of the R sets of random variables used to select the sample is independent of the other sets—after each sample is drawn, the sampled units are replaced in the population so they are available for later samples. Then the R replicate samples produce R independent estimates of the quantity of interest; the variability among those estimates can be used to estimate the variance of $\hat{\theta}$. Mahalanobis (1939, 1946) describes early uses of the method, which he calls "replicated networks of sample units" and "interpenetrating sampling."

Let

$$\theta = \text{parameter of interest}$$
$$\hat{\theta}_r = \text{estimate of } \theta \text{ calculated from } r\text{th replicate}$$
$$\tilde{\theta} = \frac{1}{R} \sum_{r=1}^{R} \hat{\theta}_i.$$

If $\hat{\theta}_r$ is an unbiased estimator of θ, so is $\tilde{\theta}$, and

$$\hat{V}_1(\tilde{\theta}) = \frac{1}{R} \frac{1}{R-1} \sum_{r=1}^{R} (\hat{\theta}_r - \tilde{\theta})^2 \qquad (9.4)$$

is an unbiased estimator of $V(\tilde{\theta})$. Note that $\hat{V}_1(\tilde{\theta})$ is the sample variance of the R independent estimators of θ divided by R—the usual estimator of the variance of a sample mean.

EXAMPLE **9.3**   The 1991 Information Please Almanac listed enrollment, tuition, and room-and-board costs for every four-year college in the United States. Suppose we want to estimate the ratio of nonresident tuition to resident tuition for public colleges and universities in the United States. In a typical implementation of the random group method, independent samples would be chosen using the same design, and $\hat{\theta}$ found for each sample. Let's take four SRSs of size 10 each. The four SRSs are without replacement, but the same college can appear in more than one of the four SRSs. The data are in file college91.dat, with summary statistics for the four SRSs in Table 9.1.

For this example,

$$\hat{\theta}_i = \frac{\text{average of nonresident tuitions for sample } i}{\text{average of resident tuitions for sample } i},$$

so $\hat{\theta}_1 = 2.3288$, $\hat{\theta}_2 = 2.5802$, $\hat{\theta}_3 = 2.4591$, and $\hat{\theta}_4 = 3.1110$. The sample average of the four independent estimates of θ is $\tilde{\theta} = 2.6198$. The sample standard deviation of the four estimates is 0.343, so the standard error of $\tilde{\theta}$ is $0.343/\sqrt{4} = 0.172$. The variance is estimated from four independent observations, so a 95% CI for the ratio is

$$2.6198 \pm 3.18(0.172) = [2.07, 3.17],$$

TABLE 9.1

Summary Statistics for Four SRSs of Colleges, Used in Example 9.3

Sample Number	Average Enrollment	Average Resident Tuition	Average Nonresident Tuition
Sample 1	6934.2	1559.0	3630.6
Sample 2	6968.6	1505.2	3883.7
Sample 3	4790.2	1527.5	3756.3
Sample 4	8613.0	1527.1	4750.8

where 3.18 is the t critical value with 3 degrees of freedom (df). Note that the small number of replicates causes the CI to be wider than it would be if more replicate samples were taken, because the estimate of the variance with 3 df is not very stable. ∎

9.2.2 Dividing the Sample into Random Groups

In practice, subsamples are not usually drawn independently, but the complete sample is selected according to the survey design. The complete sample is then divided into R groups, so that each group forms a miniature version of the survey, mirroring the sample design. The groups are then treated as though they are independent replicates of the basic survey design. This method was first described by Hansen et al. (1953, p. 440).

If the sample is an SRS of size n, the groups are formed by randomly apportioning the n observations into R groups, each of size n/R. These pseudo-random groups are not quite independent replicates because an observation unit can only appear in one of the groups; if the population size is large relative to the sample size, however, the groups can be treated as though they are independent replicates. In a cluster sample, the psus are randomly divided among the R groups. The psu takes all its observation units with it to the random group, so each random group is still a cluster sample. In a stratified multistage sample, a random group contains a sample of psus from each stratum. Note that if k psus are sampled in the smallest stratum, at most k random groups can be formed.

If θ is a nonlinear quantity, $\tilde{\theta}$ will not, in general, be the same as $\hat{\theta}$, the estimator calculated directly from the complete sample. For example, in ratio estimation, $\tilde{\theta} = (1/R) \sum_{r=1}^{R} \hat{\bar{y}}_r / \hat{\bar{x}}_r$, while $\hat{\theta} = \hat{\bar{y}}/\hat{\bar{x}}$. Usually, $\hat{\theta}$ is a more stable estimator than $\tilde{\theta}$. Sometimes $\hat{V}_1(\hat{\theta})$ is used to estimate $V(\hat{\theta})$, although it is an overestimate. Another estimator of the variance is slightly larger, but is often used:

$$\hat{V}_2(\hat{\theta}) = \frac{1}{R}\frac{1}{R-1}\sum_{r=1}^{R}(\hat{\theta}_r - \hat{\theta})^2 \tag{9.5}$$

EXAMPLE 9.4   The 1987 Survey of Youths in Custody, discussed in Example 7.7, was divided into seven random groups. The survey design had 16 strata. Strata 6–16 each consisted of one facility ($=$ psu), and these facilities were sampled with probability one. In strata 1–5, facilities were selected with probability proportional to number of residents in the 1985 Children in Custody census.

It was desired that each random group be a miniature of the sampling design. For each self-representing facility in strata 6–16, random group numbers were assigned as follows: The first resident selected from the facility was assigned a number between 1 and 7. Let's say the first resident was assigned number 6. Then the second resident in that facility would be assigned number 7, the third resident 1, the fourth resident 2, and so on. In strata 1–5, all residents in a facility (psu) were assigned to the same random group. Thus for the seven facilities sampled in stratum 2, all residents in facility 33 were assigned random group number 1, all residents in facility 9 were assigned random group number 2, and so on. Seven random groups were formed because strata 2 through 5 each have seven psus.

After all random group assignments were made, each random group had the same basic design as the original sample. Random group 1, for example, forms a stratified sample in which a (roughly) random sample of residents is taken from the self-representing facilities in strata 6–16, and an unequal-probability sample of facilities is taken from each of strata 1–5.

To use the random group method to estimate a variance, $\hat{\theta}$ is calculated for each random group. The following table shows estimates of mean age of residents for each random group (SAS code for these calculations is given on the website); each estimate was calculated using

$$\hat{\theta}_r = \frac{\sum w_i y_i}{\sum w_i},$$

where w_i is the final weight for resident i, and the summations are over observations in random group r.

Random Group Number	Estimate of Mean Age, $\hat{\theta}_r$
1	16.55
2	16.66
3	16.83
4	16.06
5	16.32
6	17.03
7	17.27

The seven estimates of θ are treated as independent observations, so

$$\tilde{\theta} = \frac{1}{7} \sum_{r=1}^{7} \hat{\theta}_r = 16.67$$

and

$$\hat{V}_1(\tilde{\theta}) = \frac{1}{7} \left\{ \frac{1}{6} \sum_{r=1}^{7} (\hat{\theta}_r - \tilde{\theta})^2 \right\} = \frac{0.1704}{7} = 0.024.$$

Using the entire data set, we calculate $\hat{\theta} = 16.64$ with

$$\hat{V}_2(\tilde{\theta}) = \frac{1}{7}\left\{ \frac{1}{6}\sum_{r=1}^{7}(\hat{\theta}_r - \hat{\theta})^2 \right\} = \frac{0.1716}{7} = 0.025.$$

We can use either $\tilde{\theta}$ or $\hat{\theta}$ to calculate CIs; using $\hat{\theta}$, a 95% CI for mean age is

$$16.64 \pm 2.45\sqrt{0.025} = [16.3, 17.0]$$

(2.45 is the t critical value with 6 df). ∎

Advantages: No special software is necessary to estimate the variance, and it is very easy to calculate the variance estimate. The method is well-suited to multiparameter or nonparametric problems. It can be used to estimate variances for percentiles and nonsmooth functions as well as variances of smooth functions of the population totals. Random group methods are easily used after weighting adjustments for nonresponse and undercoverage.

Disadvantages: The number of random groups is often small—this gives imprecise estimates of the variances (see Exercise 18). If $\hat{\theta}$ is a nonlinear statistic, $\tilde{\theta}$ can have large bias if the number of observations in each group is small. Generally one would like at least ten random groups to obtain a more stable estimate of the variance and to avoid inflating the CI by using a critical value from a t distribution with few df. Setting up the random groups can be difficult in complicated designs, as each random group must have the same design structure as the complete survey. The survey design may limit the number of random groups that can be constructed; if two psus are selected in each stratum, then only two random groups can be formed.

9.3
Resampling and Replication Methods

Random group methods are easy to compute and explain but are unstable if a complex sample can only be split into a small number of groups. Resampling methods treat the sample as if it were itself a population; we take different samples from this new "population" and use the subsamples to estimate the variance. All of the methods in this section calculate variance estimates for a sample in which psus are sampled with replacement. If psus are sampled without replacement, these methods may still be used, but are expected to overestimate the variance and result in conservative CIs, as discussed in Section 6.4.3.

9.3.1 Balanced Repeated Replication (BRR)

Some surveys are stratified to the point that only two psus are selected from each stratum. This gives the highest degree of stratification possible while still allowing calculation of variance estimates in each stratum.

TABLE   9.2

A Small Stratified Random Sample, Used to Illustrate BRR

Stratum	$\dfrac{N_h}{N}$	y_{h1}	y_{h2}	$\bar{y}_h$	$y_{h1} - y_{h2}$
1	0.30	2,000	1,792	1,896	208
2	0.10	4,525	4,735	4,630	−210
3	0.05	9,550	14,060	11,805	−4,510
4	0.10	800	1,250	1,025	−450
5	0.20	9,300	7,264	8,282	2,036
6	0.05	13,286	12,840	13,063	446
7	0.20	2,106	2,070	2,088	36

9.3.1.1   BRR in a Stratified Random Sample

We illustrate BRR for a problem we already know how to solve—calculating the variance for $\bar{y}_{\text{str}}$ from a stratified random sample. More complex statistics from stratified multistage samples are discussed in Section 9.3.1.2.

Suppose an SRS of two observation units is chosen from each of seven strata. We arbitrarily label one of the sampled units in stratum h as y_{h1}, and the other as y_{h2}. The sampled values are given in Table 9.2.

The estimated population mean is

$$\bar{y}_{\text{str}} = \sum_{h=1}^{H} \frac{N_h}{N} \bar{y}_h = 4451.7.$$

Ignoring the finite population corrections (fpcs) in (3.5) gives the variance estimator

$$\hat{V}_{\text{str}}(\bar{y}_{\text{str}}) = \sum_{h=1}^{H} \left(\frac{N_h}{N} \right)^2 \frac{s_h^2}{n_h};$$

when $n_h = 2$, as here, $s_h^2 = (y_{h1} - y_{h2})^2/2$, so

$$\hat{V}_{\text{str}}(\bar{y}_{\text{str}}) = \sum_{h=1}^{H} \left(\frac{N_h}{N} \right)^2 \frac{(y_{h1} - y_{h2})^2}{4}.$$

Here, $\hat{V}_{\text{str}}(\bar{y}_{\text{str}}) = 55{,}892.75$. This may overestimate the variance if sampling is without replacement.

To use the random group method, we would randomly select one of the observations in each stratum for group 1 and assign the other to group 2. The groups in this situation are half-samples. For example, group 1 might consist of $\{y_{11}, y_{22}, y_{32}, y_{42}, y_{51}, y_{62}, y_{71}\}$ and group 2 of the other seven observations. Then,

$$\hat{\theta}_1 = (0.3)(2000) + (0.1)(4735) + \cdots + (0.2)(2106) = 4824.7$$

and

$$\hat{\theta}_2 = (0.3)(1792) + (0.1)(4525) + \cdots + (0.2)(2070) = 4078.7.$$

The random group estimate of the variance—in this case, 139,129—has only 1 df for a two-psu-per-stratum design and is unstable in practice. If a different assignment of observations to groups had been made—had, for example, group 1 consisted of y_{h1} for strata 2, 3, and 5 and y_{h2} for strata 1, 4, 6, and 7—then $\hat{\theta}_1 = 4508.6$, $\hat{\theta}_2 = 4394.8$, and the random group estimate of the variance would have been 3238.

McCarthy (1966, 1969) notes that altogether 2^H possible half-samples could be formed, and suggests using a balanced sample of the 2^H possible half-samples to estimate the variance. **Balanced repeated replication** uses the variability among R replicate half-samples that are selected in a balanced way to estimate the variance of $\hat{\theta}$.

To define balance, let's introduce the following notation. Half-sample r can be defined by a vector $\boldsymbol{\alpha}_r = (\alpha_{r1}, \ldots, \alpha_{rH})$: Let

$$y_h(\boldsymbol{\alpha}_r) = \begin{cases} y_{h1} & \text{if } \alpha_{rh} = 1 \\ y_{h2} & \text{if } \alpha_{rh} = -1. \end{cases}$$

Equivalently,

$$y_h(\boldsymbol{\alpha}_r) = \frac{\alpha_{rh} + 1}{2} y_{h1} - \frac{\alpha_{rh} - 1}{2} y_{h2}.$$

If group 1 contains observations $\{y_{11}, y_{22}, y_{32}, y_{42}, y_{51}, y_{62}, y_{71}\}$ as above, then $\boldsymbol{\alpha}_1 = (1, -1, -1, -1, 1, -1, 1)$. Similarly, $\boldsymbol{\alpha}_2 = (-1, 1, 1, 1, -1, 1, -1)$. The set of R replicate half-samples is **balanced** if

$$\sum_{r=1}^{R} \alpha_{rh} \alpha_{rl} = 0 \quad \text{for all } l \neq h.$$

For replicate r, calculate $\hat{\theta}(\boldsymbol{\alpha}_r)$ the same way as $\hat{\theta}$ but using only the observations in the half-sample selected by $\boldsymbol{\alpha}_r$. For estimating the mean of a stratified random sample, $\hat{\theta}(\boldsymbol{\alpha}_r) = \sum_{h=1}^{H} (N_h/N) y_h(\boldsymbol{\alpha}_r)$. Define the BRR variance estimator to be

$$\hat{V}_{\text{BRR}}(\hat{\theta}) = \frac{1}{R} \sum_{r=1}^{R} [\hat{\theta}(\boldsymbol{\alpha}_r) - \hat{\theta}]^2.$$

If the set of half-samples is balanced, then for stratified random sampling $\hat{V}_{\text{BRR}}(\bar{y}_{\text{str}}) = \hat{V}_{\text{str}}(\bar{y}_{\text{str}})$. (The proof of this is left as Exercise 19.) If, in addition, $\sum_{r=1}^{R} \alpha_{rh} = 0$ for $h = 1, \ldots, H$, then $\frac{1}{R} \sum_{r=1}^{R} \bar{y}_{\text{str}}(\boldsymbol{\alpha}_r) = \bar{y}_{\text{str}}$.

For our example, the set of α's in the following table meets the balancing condition $\sum_{r=1}^{8} \alpha_{rh} \alpha_{rl} = 0$ for all $l \neq h$. The 8×7 matrix of -1's and 1's has orthogonal columns; in fact, it is the design matrix (excluding the column of ones) for a fractional factorial design (Box et al., 1978), called a Hadamard matrix. Wolter (1985) gives more detail on constructing these matrices.

		Stratum (h)						
		1	2	3	4	5	6	7
	α_1	-1	-1	-1	1	1	1	-1
	α_2	1	-1	-1	-1	-1	1	1
	α_3	-1	1	-1	-1	1	-1	1
Half-sample	α_4	1	1	-1	1	-1	-1	-1
(r)	α_5	-1	-1	1	1	-1	-1	1
	α_6	1	-1	1	-1	1	-1	-1
	α_7	-1	1	1	-1	-1	1	-1
	α_8	1	1	1	1	1	1	1

The estimate from each half-sample, $\hat{\theta}_r = \bar{y}_{str}(\alpha_r)$, is calculated from the data in Table 9.2.

Half-sample	$\hat{\theta}(\alpha_r)$	$[\hat{\theta}(\alpha_r) - \hat{\theta}]^2$
1	4732.4	78,792.5
2	4439.8	141.6
3	4741.3	83,868.2
4	4344.3	11,534.8
5	4084.6	134,762.4
6	4592.0	19,684.1
7	4123.7	107,584.0
8	4555.5	10,774.4
average	4451.7	55,892.8

The average of $[\hat{\theta}(\alpha_r) - \hat{\theta}]^2$ for the eight replicate half-samples is 55,892.75, which is the same as $\hat{V}_{str}(\bar{y}_{str})$ for sampling with replacement. Note that we can calculate the BRR variance estimate by creating a new variable of weights for each replicate half-sample. The sampling weight for observation i in stratum h is $w_{hi} = N_h/n_h$, and

$$\bar{y}_{str} = \frac{\displaystyle\sum_{h=1}^{H}\sum_{i=1}^{2} w_{hi} y_{hi}}{\displaystyle\sum_{h=1}^{H}\sum_{i=1}^{2} w_{hi}}.$$

Define

$$w_{hi}(\alpha_r) = \begin{cases} 2w_{hi} & \text{if observation } i \text{ of stratum } h \text{ is in} \\ & \text{the half-sample selected by } \alpha_r \\ 0 & \text{otherwise.} \end{cases}$$

Then

$$\bar{y}_{str}(\alpha_r) = \frac{\displaystyle\sum_{h=1}^{H}\sum_{i=1}^{2} w_{hi}(\alpha_r) y_{hi}}{\displaystyle\sum_{h=1}^{H}\sum_{i=1}^{2} w_{hi}(\alpha_r)}.$$

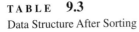

TABLE 9.3

Data Structure After Sorting

Observation Number	Stratum Number	psu Number	ssu Number	Weight, w_i	Response Variable 1	Response Variable 2	Response Variable 3
1	1	1	1	w_1	y_1	x_1	u_1
2	1	1	2	w_2	y_2	x_2	u_2
3	1	1	3	w_3	y_3	x_3	u_3
4	1	1	4	w_4	y_4	x_4	u_4
5	1	2	1	w_5	y_5	x_5	u_5
6	1	2	2	w_6	y_6	x_6	u_6
7	1	2	3	w_7	y_7	x_7	u_7
8	1	2	4	w_8	y_8	x_8	u_8
9	1	2	5	w_9	y_9	x_9	u_9
10	2	1	1	w_{10}	y_{10}	x_{10}	u_{10}
11	2	1	2	w_{11}	y_{11}	x_{11}	u_{11}
Etc.							

Similarly, for any statistic $\hat{\theta}$ calculated using the weights w_{hi}, $\hat{\theta}(\alpha_r)$ is calculated exactly the same way, but using the new weights $w_{hi}(\alpha_r)$. Using the new weight variables instead of selecting the subset of observations simplifies calculations for surveys with many response variables—the same column $w(\alpha_r)$ can be used to find the rth half-sample estimate for all quantities of interest. The modified weights also make it easy to extend the method to stratified multistage samples. SAS software will print the Hadamard matrix defining the half-samples and construct replicate weights; code for analyzing the data in Table 9.2 using BRR is on the website.

9.3.1.2   BRR in a Stratified Multistage Survey

When $\bar{y}_U$ is the only quantity of interest in a stratified random sample, BRR is simply a fancy method of calculating the variance in (3.5) and adds little extra to the procedure in Chapter 3. BRR's value in a complex survey comes from its ability to estimate the variance of a general population quantity θ, where θ may be a ratio of two variables, a correlation coefficient, a quantile, or another quantity of interest.

Suppose the population has H strata, and two psus are selected from stratum h with unequal probabilities and with replacement. (In replication methods, we like sampling with replacement because the subsampling design does not affect the variance estimator, as we saw in Section 6.3.) The same method may be used when sampling is done without replacement in each stratum, but the estimated variance is expected to be larger than the without-replacement variance. The data file for a complex survey with two psus per stratum often resembles that shown in Table 9.3, after sorting by stratum and psu.

The vector α_r defines the half-sample r: If $\alpha_{rh} = 1$, then all observation units in psu 1 of stratum h are in half-sample r. If $\alpha_{rh} = -1$, then all observation units in psu 2 of stratum h are in half-sample r. The vectors $\boldsymbol{\alpha}_r$ are selected in a balanced way, exactly as in stratified random sampling. Now, for half-sample r, create a new column of weights $w(\boldsymbol{\alpha}_r)$:

$$w_i(\boldsymbol{\alpha}_r) = \begin{cases} 2w_i & \text{if observation unit } i \text{ is in half-sample } r \\ 0 & \text{otherwise.} \end{cases} \tag{9.6}$$

For the data structure in Table 9.3 with $\alpha_{r1} = -1$ and $\alpha_{r2} = 1$, the column $w(\boldsymbol{\alpha}_r)$ will be

$$(0, 0, 0, 0, 2w_5, 2w_6, 2w_7, 2w_8, 2w_9, 2w_{10}, 2w_{11}, \dots).$$

Now use the column $w(\boldsymbol{\alpha}_r)$ instead of w to estimate quantities for half-sample r. The estimate of the population total of y for the full sample is $\sum_{i \in \mathcal{S}} w_i y_i$; the estimate of the population total of y for half-sample r is $\sum_{i \in \mathcal{S}} w_i(\boldsymbol{\alpha}_r) y_i$. If $\theta = t_y / t_x$, then $\hat{\theta} = \sum_{i \in \mathcal{S}} w_i y_i / \sum_{i \in \mathcal{S}} w_i x_i$, and $\hat{\theta}(\boldsymbol{\alpha}_r) = \sum_{i \in \mathcal{S}} w_i(\boldsymbol{\alpha}_r) y_i / \sum_{i \in \mathcal{S}} w_i(\boldsymbol{\alpha}_r) x_i$. We saw in Section 7.3 that the empirical distribution function is calculated using the weights:

$$\hat{F}(y) = \frac{\text{sum of } w_i \text{ for all observations with } y_i \leq y}{\text{sum of } w_i \text{ for all observations}}.$$

Then the empirical distribution using half-sample r is

$$\hat{F}_r(y) = \frac{\text{sum of } w_i(\boldsymbol{\alpha}_r) \text{ for all observations with } y_i \leq y}{\text{sum of } w_i(\boldsymbol{\alpha}_r) \text{ for all observations}}.$$

If θ is the population median, then $\hat{\theta}$ may be defined as the smallest value of y for which $\hat{F}(y) \geq 1/2$, and $\hat{\theta}(\boldsymbol{\alpha}_r)$ is the smallest value of y for which $\hat{F}_r(y) \geq 1/2$.
    For any quantity θ, we define

$$\hat{V}_{\text{BRR}}(\hat{\theta}) = \frac{1}{R} \sum_{r=1}^{R} [\hat{\theta}(\boldsymbol{\alpha}_r) - \hat{\theta}]^2. \tag{9.7}$$

BRR can also be used to estimate covariances of statistics: If θ and η are two quantities of interest, then

$$\widehat{\text{Cov}}_{\text{BRR}}(\hat{\theta}, \hat{\eta}) = \frac{1}{R} \sum_{r=1}^{R} [\hat{\theta}(\boldsymbol{\alpha}_r) - \hat{\theta}][\hat{\eta}(\boldsymbol{\alpha}_r) - \hat{\eta}].$$

Other BRR variance estimators, variations of (9.7), are described in Exercise 20.
    While the exact equivalence of $\hat{V}_{\text{BRR}}(\bar{y}_{\text{str}}(\boldsymbol{\alpha}))$ and $\hat{V}_{\text{str}}(\bar{y}_{\text{str}})$ does not extend to nonlinear statistics, Rao and Wu (1985) show that if θ is a smooth function of the population totals, the variance estimator from BRR is asymptotically equivalent to that from linearization. BRR also provides a consistent estimator of the variance for quantiles when a stratified random sample is taken (Shao and Wu, 1992).
    When a replication method such as BRR is used, data analysts can calculate variances from data files without needing to know the stratification and clustering information. The public-use data set can consist of the response variables, original weights, and the columns of replicate weights. The statistic $\hat{\theta}$ is calculated by using the

TABLE 9.4
NHANES Data with Replicate Weights

Stratum	psu	BMI	Weight	Replicate Weight 1	Replicate Weight 2	Replicate Weight 3	··· ···	Replicate Weight 15	Replicate Weight 16
39	2	50.85	5,824.78	11,649.56	0	0	···	11,649.56	0
41	1	20.78	5,564.04	11,128.08	0	11,128.08	···	11,128.08	0
33	2	19.60	12,947.34	25,894.68	0	25,894.68	···	0	25,894.68
37	1	21.64	7,304.95	14,609.89	0	14,609.89	···	0	14,609.89
33	1	15.76	8,385.25	0	16,770.49	0	···	16,770.49	0
33	2	28.32	19,994.16	39,988.32	0	39,988.32	···	0	39,988.32
41	1	38.03	15,876.72	31,753.44	0	31,753.44	···	31,753.44	0
33	2	26.76	40,061.77	80,123.54	0	80,123.54	···	0	80,123.54

original weights w_i with the data vector of y_i's. Then the columns of replicate weights are used to perform the variance estimation: We calculate $\hat{\theta}(\alpha_r)$, for $r = 1, \ldots, R$, by performing the same calculations used to find $\hat{\theta}$, with weights $w_i(\alpha_r)$ substituted for the original weights w_i. Equation (9.7) is then applied to estimate the variance of $\hat{\theta}$. Weighting adjustments for nonresponse, such as those discussed in Section 8.5, can be incorporated into the replicate weights so that the BRR estimate of variance includes the effects of the nonresponse and calibration adjustments (Canty and Davison, 1999).

EXAMPLE 9.5   Let's use BRR to estimate variances from the NHANES data studied in Section 7.4.2. The public-use data set includes variables for pseudo-stratum and pseudo-psu that can be used for variance estimation. (The original strata and psu variables are not released to the public to preserve confidentiality of the respondents' data.) Each pseudo-stratum has two pseudo-psus, so BRR can be used. We generate replicate weights for these data using SAS software. The replicate weights can then be used to calculate standard errors for any statistic, not just those in the software package. For example, some software packages do not yet calculate medians from survey data; the original weight variable can be used to estimate a median, and the replicate weights can then be used to estimate the variance of the estimated median.

Since our replicate weights are based on the final weight variable included in the NHANES data, however, they do not incorporate effects of nonresponse adjustment on the variance. Many data sets that are made available to the public have replicate weights that account for the nonresponse adjustments, and those are preferred if available.

The data set has 15 pseudo-strata, so we use a 16×16 Hadamard matrix (16 is the first multiple of 4 after 15 for which a Hadamard matrix exists). The replicate weights for a few of the observations are given in Table 9.4. Each entry in the replicate weight columns is either 0 or $2w_i$. Note that the pattern of replicate weights is the same for each of the three observations from pseudo-psu 2 of pseudo-stratum 33.

Using the replicate weights, and the SAS code given on the website, we estimate the mean and the median first using the original weight vector and then using each of the 16 vectors of replicate weights. Using the original weight vector, we estimate the mean body mass index (variable *bmxbmi*) as $\hat{\bar{y}} = 26.19$ and the median as $\hat{m} = 25.60$. The values calculated using the replicate weights are:

	Replicate Weight							
	1	2	3	4	5	6	7	8
Mean	26.3364	25.9977	26.0700	26.0741	26.4049	26.2483	25.9767	26.3310
Median	25.64	25.31	25.49	25.39	25.67	25.56	25.4	25.76
	9	10	11	12	13	14	15	16
Mean	26.3255	25.9909	26.1574	26.1584	26.1406	26.0324	26.2851	26.4078
Median	25.83	25.38	25.53	25.50	25.62	25.56	25.68	25.88

Using (9.7) we estimate $\hat{V}(\hat{\bar{y}}) = 0.0215$ and $\hat{V}(\hat{m}) = 0.026$. The estimates from the replicate weights tend to be close to each other; to avoid roundoff error, it is best to do these calculations on the computer. ∎

Advantages: BRR gives a variance estimator that is asymptotically equivalent to that from linearization methods for smooth functions of population totals. It can also be used for estimating variances of quantiles. The data analyst only needs the columns of replicate weights, and does not need the original sampling design information, to calculate variances. It requires relatively few computations (and relatively few columns of replicate weights) when compared with the jackknife and the bootstrap.

Disadvantages: As defined above, BRR can only be used in situations in which there are two psus per stratum. In practice, though, it is often extended to other sampling designs by using more complicated balancing schemes (see Fay, 1989 and Judkins, 1990). BRR, like the jackknife and bootstrap, estimates the with-replacement variance, and may overestimate the without-replacement variance.

9.3.2 Jackknife

The **jackknife** method, like BRR, extends the random group method by allowing the replicate groups to overlap. The jackknife was introduced by Quenouille (1956) as a method of reducing bias; Tukey (1958) proposed using it to estimate variances and calculate CIs. In this section, we describe the *delete-1 jackknife*; Shao and Tu (1995) discuss other forms of the jackknife and give theoretical results.

For an SRS, let $\hat{\theta}_{(j)}$ be the estimator of the same form as $\hat{\theta}$, but not using observation j. Thus, if $\hat{\theta} = \bar{y}$, then $\hat{\theta}_{(j)} = \bar{y}_{(j)} = \sum_{i \neq j} y_i/(n-1)$. For an SRS, define the delete-1 jackknife estimator (so called because we delete one observation in each replicate) as

$$\hat{V}_{JK}(\hat{\theta}) = \frac{n-1}{n} \sum_{j=1}^{n} (\hat{\theta}_{(j)} - \hat{\theta})^2. \tag{9.8}$$

Why the multiplier $(n-1)/n$? Let's look at $\hat{V}_{JK}(\hat{\theta})$ when $\hat{\theta} = \bar{y}$. When $\hat{\theta} = \bar{y}$,

$$\bar{y}_{(j)} = \frac{1}{n-1}\sum_{i\neq j} y_i = \frac{1}{n-1}\left(\sum_{i=1}^{n} y_i - y_j\right) = \bar{y} - \frac{1}{n-1}(y_j - \bar{y}).$$

TABLE 9.5

Jackknife Calculations for Example 9.6

j	x	y	$\bar{x}_{(j)}$	$\bar{y}_{(j)}$	$\hat{B}_{(j)}$
1	1365	3747	1580.6	3617.7	2.2889
2	1677	4983	1545.9	3480.3	2.2513
3	1500	1500	1565.6	3867.3	2.4703
4	1080	2160	1612.2	3794.0	2.3533
5	1875	2475	1523.9	3759.0	2.4667
6	3071	5135	1391.0	3463.4	2.4899
7	1542	3950	1560.9	3595.1	2.3032
8	930	4050	1628.9	3584.0	2.2003
9	1340	4140	1583.3	3574.0	2.2573
10	1210	4166	1597.8	3571.1	2.2350

Then,

$$\sum_{j=1}^{n}(\bar{y}_{(j)} - \bar{y})^2 = \frac{1}{(n-1)^2}\sum_{j=1}^{n}(y_j - \bar{y})^2 = \frac{1}{n-1}s_y^2,$$

so $\hat{V}_{JK}(\bar{y}) = s_y^2/n$, the with-replacement estimator of the variance of $\bar{y}$.

EXAMPLE 9.6 Let's use the jackknife to estimate the ratio of nonresident tuition to resident tuition for the first group of colleges in Example 9.3. Here, $\hat{\theta} = \bar{y}/\bar{x}$, $\hat{\theta}_{(j)} = \hat{B}_{(j)} = \bar{y}_{(j)}/\bar{x}_{(j)}$, and

$$\hat{V}_{JK}(\hat{B}) = \frac{n-1}{n}\sum_{j\in\mathcal{S}}(\hat{B}_{(j)} - \hat{B})^2.$$

For each jackknife group in Table 9.5, omit one observation. Thus, $\bar{x}_{(1)}$ is the average of all x's except for x_1: $\bar{x}_{(1)} = (1/9)\sum_{i=2}^{9}x_i$. Here, $\hat{B} = 2.3288$, $\sum(\hat{B}_{(j)} - \hat{B})^2 = 0.1043$, and $\hat{V}_{JK}(\hat{B}) = .09377$. ■

How can we extend this to a cluster sample? One might think that you could just delete one observation unit at a time, but that will not work—deleting one observation unit at a time destroys the cluster structure and gives an estimate of the variance that is only correct if the intraclass correlation coefficient is zero. In any resampling method and in the random group method, keep observation units within a psu together while constructing the replicates—this preserves the dependence among observation units within the same psu. For a cluster sample, then, we would apply the jackknife variance estimator in (9.8) by letting n be the number of psus, and letting $\hat{\theta}_{(j)}$ be the estimate of θ that we would obtain by deleting all the observations in psu j.

In a stratified multistage cluster sample, the jackknife is applied separately in each stratum at the first stage of sampling, with one psu deleted at a time. Suppose there are H strata, and n_h psus are chosen for the sample from stratum h. Assume these psus are chosen with replacement.

To apply the jackknife, delete one psu at a time. Let $\hat{\theta}_{(hj)}$ be the estimator of the same form as $\hat{\theta}$ when psu j of stratum h is omitted. To calculate $\hat{\theta}_{(hj)}$, define a new weight variable: Let

$$
w_{i(hj)} = \begin{cases} w_i & \text{if observation unit } i \text{ is not in stratum } h \\ 0 & \text{if observation unit } i \text{ is in psu } j \text{ of stratum } h \\ \dfrac{n_h}{n_h - 1} w_i & \text{if observation unit } i \text{ is in stratum } h \text{ but not in psu } j. \end{cases}
$$

Then use the weights $w_{i(hj)}$ to calculate $\hat{\theta}_{(hj)}$, and

$$
\hat{V}_{\text{JK}}(\hat{\theta}) = \sum_{h=1}^{H} \frac{n_h - 1}{n_h} \sum_{j=1}^{n_h} (\hat{\theta}_{(hj)} - \hat{\theta})^2. \tag{9.9}
$$

EXAMPLE 9.7   Here we use the jackknife to calculate the variance of the mean egg volume from Example 5.7. We calculated $\hat{\theta} = \bar{y}_r = 4375.947/1757 = 2.49$. In that example, since we did not know the number of clutches in the population, we calculated the with-replacement variance.

First, we find the weight vector for each of the 184 jackknife iterations. We have only one stratum, so $h = 1$ for all observations. For $\hat{\theta}_{(1,1)}$, delete the first psu. Thus the new weights for the observations in the first psu are 0; the weights in all remaining psus are the previous weights times $n_h/(n_h - 1) = 184/183$. Using the weights from Example 5.7, the new jackknife weight columns are shown in Table 9.6.

Note that the sums of the jackknife weights vary from column to column because the original sample is not self-weighting. We calculated $\hat{\theta}$ as $(\sum w_i y_i)/\sum w_i$;

TABLE   9.6

Jackknife Weights for Example 5.7. The values w_i are the relative weights; $w_{i(k)}$ is the set of jackknife weights for the replication omitting psu k.

clutch	csize	w_i	$w_{i(1)}$	$w_{i(2)}$	...	$w_{i(184)}$
1	13	6.5	0	6.535519	...	6.535519
1	13	6.5	0	6.535519	...	6.535519
2	13	6.5	6.535519	0	...	6.535519
2	13	6.5	6.535519	0	...	6.535519
3	6	3	3.016393	3.016393	...	3.016393
3	6	3	3.016393	3.016393	...	3.016393
4	11	5.5	5.530055	5.530055	...	5.530055
4	11	5.5	5.530055	5.530055	...	5.530055
$\vdots$	$\vdots$	$\vdots$	$\vdots$	$\vdots$		$\vdots$
183	13	6.5	6.535519	6.535519	...	6.535519
183	13	6.5	6.535519	6.535519	...	6.535519
184	12	6	6.032787	6.032787	...	0
184	12	6	6.032787	6.032787	...	0
Sum	3514	1757	1753.53	1753.53	...	1754.54

to find $\hat{\theta}_{(hj)}$, we follow the same procedure but use $w_{i(hj)}$ in place of w_i. Thus, $\hat{\theta}_{(1,1)} = 4349.348/1753.53 = 2.48034; \hat{\theta}_{(1,2)} = 4345.036/1753.53 = 2.47788; \hat{\theta}_{(1,184)} = 4357.819/1754.54 = 2.48374$. Using (9.9), then, we calculate $\hat{V}_{JK}(\hat{\theta}) = 0.00373$. This results in a standard error of 0.061, the same as calculated in Example 5.7. SAS code for constructing jackknife weights and finding the jackknife estimate of the variance is given on the website. ∎

EXAMPLE 9.8 We used the random group method to estimate the variance of mean age of residents for the Survey of Youth in Custody in Example 9.4. The jackknife can also be used to estimate the variance. SAS code on the website results in the following output:

```
                  Data Summary

Number of Strata                     16
Number of Clusters                  861
Number of Observations             2621
Sum of Weights                    25012


            Variance Estimation

Method                         Jackknife
Number of Replicates                 861


                  Statistics

                    Std Error
Variable    Mean    of Mean         95% CL for Mean
-----------------------------------------------------------
age       16.639293  0.130106   16.3839236   16.8946626
-----------------------------------------------------------
```

The standard error from the jackknife method differs from that given by the random group method; the random group standard error is based on only 7 groups and is less stable than the jackknife standard error. ∎

Advantages: The jackknife is an all-purpose method. The same procedure is used to estimate the variance for every statistic for which jackknife can be used. The jackknife works in stratified multistage samples in which BRR does not apply because more than two psus are sampled in each stratum. The jackknife provides a consistent estimator of the variance when θ is a smooth function of population totals (Krewski and Rao, 1981). Replication methods such as the jackknife can be used to account for some of the effects of imputation on the variance estimates (Rao and Shao, 1992).

Disadvantages: For some sampling designs, the jackknife may require a large amount of computation. The jackknife performs poorly for estimating the variances of some statistics that are not smooth functions of population totals. For example, the jackknife does not give a consistent estimator of the variance of quantiles in an SRS.

9.3.3 Bootstrap

As with the jackknife, theoretical results for the **bootstrap** were first developed for areas of statistics other than survey sampling; Shao and Tu (1995) summarize theoretical results for the bootstrap in complex survey samples. We first describe the bootstrap for an SRS with replacement, as developed by Efron (1979, 1982) and described in Davison and Hinkley (1997). Suppose S is an SRS with replacement of size n. We hope, in drawing the sample, that it reproduces properties of the whole population. We then treat the sample S as if it were a population, and take resamples from S. If the sample really is similar to the population—if the empirical probability mass function of the sample is similar to the probability mass function of the population—then samples generated from the empirical probability mass function should behave like samples taken from the population.

EXAMPLE 9.9   Let's use the bootstrap to estimate the variance of the median height, θ, in the height population from Example 7.3, using the sample in the file ht.srs. The population median height is $\theta = 168$; the sample median from ht.srs is $\hat{\theta} = 169$. Figure 7.2, the probability mass function for the population, and Figure 7.3, the histogram of the sample, are similar in shape (largely because the sample size for the SRS is large), so we would expect that taking an SRS of size n with replacement from S would be like taking an SRS with replacement from the population. A resample from S, though, will not be exactly the same as S because the sample is with replacement—some observations in S may occur twice or more in the resample, while other observations in S may not occur at all.

We take an SRS of size 200 with replacement from S to form the first resample. The first resample from S has an empirical probability mass function similar to but not identical to that of S; the resample median is $\hat{\theta}_1^* = 170$. Repeating the process, the second resample from S has median $\hat{\theta}_2^* = 169$. We take a total of $R = 2000$ resamples from S and calculate the sample median from each sample, obtaining $\hat{\theta}_1^*, \hat{\theta}_2^*, \dots, \hat{\theta}_R^*$. We obtain the following frequency table for the 2000 resample medians:

Median of Resample	165.0	166.0	166.5	167.0	167.5	168.0	168.5	169.0	169.5	170.0	170.5	171.0	171.5	172.0
Frequency	1	5	2	40	15	268	87	739	111	491	44	188	5	4

The sample mean of these 2000 values is 169.3 and the sample variance of these 2000 values is 0.9148; this is the bootstrap estimate of the variance of the sample median. An approximate 95% CI may be constructed using the bootstrap variance as $169.3 \pm 1.96\sqrt{.9148} = [167.4, 171.2]$. Alternatively, the bootstrap distribution may be used to calculate a CI directly. The bootstrap distribution estimates the sampling distribution of $\hat{\theta}$, so a 95% percentile CI may be calculated by finding the 2.5 percentile and the 97.5 percentile of the bootstrap distribution. For this example, a 95% percentile CI for the median is [167.5, 171]. Manly (1997) describes other methods for finding CIs using the bootstrap.   ∎

If the original SRS is without replacement, Gross (1980) proposes creating N/n copies of the sample to form a "pseudopopulation," then drawing R SRSs without replacement from the pseudopopulation. If n/N is small, the with-replacement and without-replacement bootstrap distributions should be similar.

Shao (2003) describes methods for using the bootstrap with data from a complex survey. In all of the methods, we take bootstrap resamples of the psus within each stratum. As with BRR and the jackknife, observations within a psu are always kept together in the bootstrap iterations.

Here are steps for using the rescaling bootstrap of Rao and Wu (1988) for a stratified multistage sample. Let n_h be the number of psus sampled from stratum h. Let R be the number of bootstrap replicates to be created. Typically, $R = 500$ or 1,000, although some statisticians use smaller values of R.

1. For bootstrap replicate r $(r = 1, \ldots, R)$, select an SRS of $n_h - 1$ psus with replacement from the n_h sample psus in stratum h. Do this independently for each stratum. Let $m_{hj}(r)$ be the number of times psu j of stratum h is selected in replicate r.

2. Create the replicate weight vector for replicate r as

$$w_i(r) = w_i \times \frac{n_h}{n_h - 1} m_{hj}(r), \text{ for observation } i \text{ in psu } j \text{ of stratum } h.$$

   The result is R vectors of replicate weights.

3. Use the vectors of replicate weights to estimate $V(\hat{\theta})$. Let $\hat{\theta}_r^*$ be the estimator of θ, calculated the same way as $\hat{\theta}$ but using weights $w_i(r)$ instead of the original weights w_i. Then,

$$\hat{V}_B(\hat{\theta}) = \frac{1}{R - 1} \sum_{i=1}^{R} (\hat{\theta}_r^* - \hat{\theta})^2.$$

EXAMPLE 9.10 We use the bootstrap to estimate variances from the data in file htstrat.dat, discussed in Example 7.3. The bootstrap weights are constructed by taking 1000 stratified random samples with replacement from the data set; we select 159 women and 39 men with replacement in each resample. The average height is estimated by $\bar{y}_{str} = 169.02$ with bootstrap standard error 0.737; the standard error calculated using the stratified sampling formula in (3.5), ignoring the fpc, is 0.739. The SAS macro on the website uses SAS PROC SURVEYSELECT to construct the replicate bootstrap weights. ∎

EXAMPLE 9.11 We noted in Section 8.6 that if a data set has imputed values and then is analyzed as if those imputed values were real, the resulting variance estimate is too low. Replication methods such as bootstrap can be used to account for some of the effects of imputation on the variance estimates. Zhang et al. (1998) use the bootstrap with the 1993–94 Schools and Staffing Survey to estimate the amount that imputation inflated the variance estimates. The survey uses several types of imputation, including hot-deck imputation. They found that the standard errors calculated accounting for the imputation could be up to twice as large as if the imputation were ignored. ∎

Advantages: The bootstrap will work for smooth functions of population means and for some nonsmooth functions such as quantiles in general sampling designs. The bootstrap is well suited for finding CIs directly: To calculate a 90% CI, one can merely take the 5th and 95th percentiles from $\hat{\theta}_1^*, \hat{\theta}_2^*, \ldots, \hat{\theta}_R^*$, or can use a bootstrap-t method such as that described in Efron (1982).

Disadvantages: In some settings, the bootstrap may require more computations than BRR or jackknife, since R is typically a very large number. In other large surveys, however, for example if a stratified random sample is taken, the bootstrap may require fewer computations than the jackknife. The bootstrap variance estimate differs when a different set of bootstrap samples is taken.

9.4
Generalized Variance Functions

In many large government surveys such as the U.S. Current Population Survey (CPS) or the Canadian Labour Force Survey, hundreds or thousands of estimates are calculated and published. The agencies analyzing the survey results could calculate standard errors for each published estimate and publish additional tables of the standard errors, but that would add greatly to the labor involved in publishing timely estimates from the surveys. In addition, other analysts of the public-use data files may wish to calculate additional estimates, and the public-use files may not provide enough information to allow calculation of standard errors.

Generalized variance functions (GVFs) are provided in a number of surveys to calculate standard errors. They have been used for the CPS since 1947. Wolter (2007, Chapter 7) describes the theory underlying GVFs.

Criminal Victimization in the United States, 1990 (U.S. Department of Justice, p. 146), gives GVF formulas for calculating standard errors in the 1990 National Crime Victimization Survey (NCVS). If $\hat{t}$ is an estimated number of persons or households victimized by a particular type of crime, or if $\hat{t}$ estimates a total number of victimization incidents,

$$\hat{V}(\hat{t}) = a\hat{t}^2 + b\hat{t}.$$

(9.10)

If $\hat{p}$ is an estimated proportion,

$$\hat{V}(\hat{p}) = \frac{b}{\hat{t}}\hat{p}(1 - \hat{p}),$$

(9.11)

where $\hat{t}$ is the estimated base population for the proportion. For the 1990 NCVS, the values of a and b were $a = -0.00001833$ and $b = 3725$. For example, for 1990 it was estimated that 1.23% of persons aged 20 to 24 were robbed, and it was also estimated that there were 18,017,100 persons in that age group. Thus the GVF estimate of SE($\hat{p}$) is

$$\sqrt{\frac{3725}{18,017,100}.0123(1 - .0123)} = 0.0016.$$

Assuming that asymptotic results apply, this gives an approximate 95% CI of $0.0123\pm$ $(1.96)(0.0016)$, or $[0.0091, 0.0153]$. There were an estimated 800,510 completed robberies in 1990. Using (9.10), the standard error of this estimate is

$$\sqrt{(-0.00001833)(800{,}510)^2 + 3725(800{,}510)} = 54{,}499.$$

Where do these formulas come from? Suppose t_i is the total number of observation units belonging to a class, say the total number of persons in the United States who were victims of violent crime in 1990. Let $p_i = t_i/N$, the proportion of persons in the population belonging to that class. If d_i is the design effect (deff) in the survey for estimating p_i (see Section 7.5), then

$$V(\hat{p}_i) \approx d_i\frac{p_i(1-p_i)}{n} = \frac{b_i}{N}p_i(1-p_i), \tag{9.12}$$

where $b_i = d_i \times (N/n)$. Similarly,

$$V(\hat{t}_i) \approx d_iN^2\frac{p_i(1-p_i)}{n} = a_it_i^2 + b_it_i,$$

where $a_i = -d_i/n$. If estimating a proportion in a domain, say the proportion of persons in the 20–24 age group who were robbery victims, the denominator in (9.12) is changed to the estimated population size of the domain (see Section 4.2).

If the deffs are similar for different estimates so that $a_i \approx a$ and $b_i \approx b$, then constants a and b can be estimated that give (9.10) and (9.11) as approximations to the variance for a number of quantities. The general procedure for constructing a generalized variance function is as follows:

1 Using replication or some other method, estimate variances for k population totals of special interest, $\hat{t}_1, \hat{t}_2, \ldots, \hat{t}_k$. Let v_i be the relative variance for $\hat{t}_i$, $v_i = \hat{V}(\hat{t}_i)/\hat{t}_i^2$, for $i = 1, 2, \ldots, k$.

2 Postulate a model relating v_i to $\hat{t}_i$. The 1990 NCVS and many other surveys use the model

$$v_i = \alpha + \frac{\beta}{\hat{t}_i}. \tag{9.13}$$

This is a linear regression model with response variable v_i and explanatory variable $1/\hat{t}_i$. Valliant (1987) found that this model produces consistent estimators of the variances for the class of superpopulation models he studied.

3 Use regression techniques to estimate α and β by a and b. Valliant (1987) suggests using weighted least squares to estimate the parameters, giving higher weight to items with small v_i.

4 Use the estimated regression equation to predict the relative variance of an estimated total $\hat{t}_{\text{new}}$: $\hat{v}_{\text{new}} = a + b/\hat{t}_{\text{new}}$. Since $\hat{v}_{\text{new}}$ is the predicted value of the relative variance $\hat{V}(\hat{t}_{\text{new}})/\hat{t}_{\text{new}}^2$, the GVF estimate of $V(\hat{t}_{\text{new}})$ is $\hat{V}(\hat{t}_{\text{new}}) = a\hat{t}_{\text{new}}^2 + b\hat{t}_{\text{new}}$. The GVF model can also be used to estimate the variance of a percentage, with $\hat{V}(\hat{p}) = \hat{p}(1-\hat{p})b/\hat{t}$, where $\hat{t}$ is the estimated number of units in the base of the percentage (see Exercise 25).

The a_i and b_i for individual items are replaced by quantities a and b which are calculated from all k items. For the 1990 NCVS, $b = 3725$. Most weights in the 1990

NCVS are between 1500 and 2500; $\hat{\beta}$ approximately equals the (average weight) $\times$ (deff), if the overall deff is about two.

The model used in (9.13) is relatively simple; if the deffs vary greatly among different responses, the simple model may give inaccurate estimates of variances for some responses. Krenzke (1995) describes alternative models considered for the NCVS, and provides a good example of the process used to develop a GVF model that takes account of nonconstant deffs.

Valliant (1987) found that if design effects for the k estimated totals are similar, the GVF variances are often more stable than the direct estimates of variance, as they smooth out some of the fluctuations from item to item. If a quantity of interest does not follow the model in Step 2, however, the GVF estimate of the variance is likely to be poor, and you can only know that it is poor by calculating the variance directly.

Advantages: The GVF may be used when insufficient information is provided in the public-use data files to allow direct calculation of standard errors. The data collector can calculate the GVF, and often has more information for estimating variances than is released to the public. A GVF saves a great deal of time and speeds production of annual reports. It is also useful for designing similar surveys in the future.

Disadvantages: The model relating v_i to $\hat{t}_i$ may not be appropriate for the quantity you are interested in, resulting in an unreliable estimate of the variance. You must be careful about using GVFs for estimates not included when calculating the regression parameters. If a subpopulation has an unusually high degree of clustering (and hence a high deff), the GVF estimate of the variance may be much too small.

9.5
Confidence Intervals

9.5.1   Confidence Intervals for Smooth Functions of Population Totals

Theoretical results exist for most of the variance estimation methods discussed in this chapter, stating that under certain assumptions $(\hat{\theta} - \theta)/\sqrt{\hat{V}(\hat{\theta})}$ asymptotically follows a standard normal distribution. These results and conditions are given in Binder (1983), for linearization estimates; in Krewski and Rao (1981) and Rao and Wu (1985), for jackknife and BRR; in Rao and Wu (1988) and Sitter (1992), for bootstrap. Consequently, when the assumptions are met, an approximate 95% CI for θ may be constructed as

$$\hat{\theta} \pm 1.96\sqrt{\hat{V}(\hat{\theta})}.$$

Alternatively, a t_{df} percentile may be substituted for 1.96, with df = (number of groups $-$ 1) for the random group method, and df = (number of psus $-$ number of strata) for the other methods. Rust and Rao (1996) give guidelines for appropriate dfs. The bootstrap method may also be used to calculate CIs directly.

Roughly speaking, the assumptions for linearization, jackknife, BRR, and bootstrap are as follows:

1   The quantity of interest θ can be expressed as a smooth function of the population totals; more precisely, $\theta = h(t_1, t_2, \ldots, t_k)$, where the second-order partial derivatives of h are continuous.

2   The sample sizes are large: Either the number of psus sampled in each stratum is large, or the survey contains a large number of strata. (See Rao and Wu, 1985, for the precise technical conditions needed.) Also, to construct a CI using the normal distribution, the sample sizes must be large enough so that the sampling distribution of $\hat{\theta}$ is approximately normal.

Furthermore, a number of simulation studies indicate that these CIs behave well in practice. Wolter (1985) summarizes some of the simulation studies; others are found in Kovar et al. (1988) and Rao et al. (1992). These studies indicate that the jackknife and linearization tend to give similar estimates of the variance, while the bootstrap and BRR give slightly larger estimates. Sometimes a transformation may be used so that the sampling distribution of a statistic is closer to a normal distribution: if estimating total income, for example, a log transformation may be used because the distribution of income is extremely skewed.

9.5.2   Confidence Intervals for Population Quantiles

The theoretical results described above for BRR, jackknife, bootstrap, and linearization methods do not apply to population quantiles, however, because they are not smooth functions of population totals. Special methods have been developed to construct CIs for quantiles; McCarthy (1993) compares several CIs for the median, and his discussion applies to other quantiles as well.

Let q be between 0 and 1. Then define the quantile θ_q as $\theta_q = F^{-1}(q)$, where $F^{-1}(q)$ is defined to be the smallest value y satisfying $F(y) \geq q$. Similarly, define $\hat{\theta}_q = \hat{F}^{-1}(q)$. (Alternatively, as in Example 7.6, interpolation can be used to define quantiles.) Now F^{-1} and $\hat{F}^{-1}$ are *not* smooth functions, but we assume the population and sample are large enough that they can be well approximated by continuous functions.

Some of the methods already discussed work quite well for constructing CIs for quantiles. The random group method works well if the number of random groups, R, is moderate. Let $\hat{\theta}_q(r)$ be the estimated quantile from random group r. Then, a CI for θ_q is

$$\hat{\theta}_q \pm t \sqrt{\frac{1}{R(R-1)} \sum_{r=1}^{R} [\hat{\theta}_q(r) - \hat{\theta}_q]^2},$$

where t is the appropriate percentile from a t distribution with $R - 1$ degrees of freedom. Similarly, studies by McCarthy (1993), Kovar et al. (1988), Sitter (1992), Rao et al. (1992), and Shao and Chen (1998) indicate that in certain designs CIs can be formed using

$$\hat{\theta}_q \pm 1.96 \sqrt{\hat{V}(\hat{\theta}_q)},$$

where the variance estimate is calculated using BRR or bootstrap.

FIGURE   9.2

Woodruff's confidence interval for the quantile θ_q if the empirical distribution function is continuous. Since $F(y)$ is a proportion, we can easily calculate a confidence interval for any value of y, shown on the vertical axis. We then look at the corresponding points on the horizontal axis to form a confidence interval for θ_q.

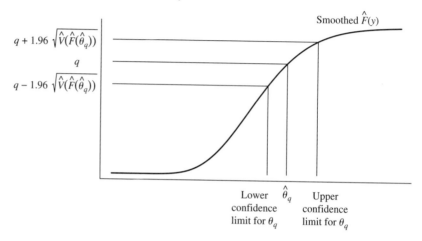

An alternative interval can be constructed based on a method introduced by Woodruff (1952). For any y, $\hat{F}(y)$ is a function of population totals: $\hat{F}(y) = \sum_{i \in S} w_i u_i / \sum_{i \in S} w_i$, where $u_i = 1$ if $y_i \le y$ and $u_i = 0$ if $y_i > y$. Thus, a method in this chapter can be used to estimate $V[\hat{F}(y)]$ for any value y, and an approximate 95% CI for $F(y)$ is given by

$$\hat{F}(y) \pm 1.96 \sqrt{\hat{V}[\hat{F}(y)]}.$$

Now let's use the CI for $q = F(\theta_q)$ to obtain an approximate CI for θ_q. Since we have a 95% CI,

$$0.95 \approx P\left\{ \hat{F}(\theta_q) - 1.96\sqrt{\hat{V}[\hat{F}(\hat{\theta}_q)]} \le q \le \hat{F}(\theta_q) + 1.96\sqrt{\hat{V}[\hat{F}(\hat{\theta}_q)]} \right\}$$

$$= P\left\{ q - 1.96\sqrt{\hat{V}[\hat{F}(\hat{\theta}_q)]} \le \hat{F}(\theta_q) \le q + 1.96\sqrt{\hat{V}[\hat{F}(\hat{\theta}_q)]} \right\}$$

$$= P\left(\hat{F}^{-1}\left\{ q - 1.96\sqrt{\hat{V}[\hat{F}(\hat{\theta}_q)]} \right\} \le \theta_q \le \hat{F}^{-1}\left\{ q + 1.96\sqrt{\hat{V}[\hat{F}(\hat{\theta}_q)]} \right\} \right).$$

So an approximate 95% CI for the quantile θ_q is

$$\left[\hat{F}^{-1}\left\{ q - 1.96\sqrt{\hat{V}[\hat{F}(\hat{\theta}_q)]} \right\}, \hat{F}^{-1}\left\{ q + 1.96\sqrt{\hat{V}[\hat{F}(\hat{\theta}_q)]} \right\} \right].$$

The derivation of this CI is illustrated in Figure 9.2. An appropriate t critical value may be substituted for 1.96 if desired.

We need several technical assumptions to use the Woodruff-method interval. These assumptions are stated by Rao and Wu (1987) and Francisco and Fuller (1991), who studied a similar CI. Essentially, the problem is that both F and $\hat{F}$ are step functions;

they have jumps at the values of y in the population and sample. The technical conditions basically say that the jumps in F and in $\hat{F}$ should be small, and that the sampling distribution of $\hat{F}(y)$ is approximately normal. Sitter and Wu (2001) show that the Woodruff method gives CIs with approximately correct coverage probabilities even when q is large or small.

EXAMPLE 9.12 Let's use Woodruff's method to construct a 95% CI for the median height in the file htstrat.dat, discussed in Example 7.3. The following values were obtained for the empirical distribution function:

y	165	166	167	168	169	170	171
$\hat{F}(y)$	0.3781	0.4438	0.4844	0.5125	0.5375	0.5656	0.6000

In Example 7.6, we estimated the population median by

$$\hat{\theta}_{0.5} = 167 + \frac{0.5 - 0.4844}{0.5125 - 0.4844}(168 - 167) = 167.6.$$

Note that

$$\hat{F}(\hat{\theta}_q) = \frac{\sum\limits_{h=1}^{2}\sum\limits_{i\in S_h} w_{hi} u_{hi}}{\sum\limits_{h=1}^{2}\sum\limits_{i\in S_h} w_{hi}} = \frac{\sum\limits_{h=1}^{2}\sum\limits_{i\in S_h} w_{hi} u_{hi}}{2000}$$

where $u_{hi} = 1$ if $y_{hi} \leq \hat{\theta}_{0.5}$ and 0 otherwise, so, using the variance for the combined ratio estimator in Section 4.5,

$$\hat{V}[\hat{F}(\hat{\theta}_q)] = \frac{1}{(2000)^2}\left[(1000)^2 \sum\left(1 - \frac{160}{1000}\right)\frac{s_{e1}^2}{160} + (1000)^2 \sum\left(1 - \frac{40}{1000}\right)\frac{s_{e2}^2}{40}\right],$$

where s_{eh}^2 is the sample variance of the values $e_{hi} = u_{hi} - 0.5$ for stratum h. Using the values $s_{e1}^2 = 0.1789$ and $s_{e2}^2 = 0.1641$ results in $\hat{V}[\hat{F}(\hat{\theta}_{0.5})] = 0.00121941$. Thus, for this sample, $1.96\sqrt{\hat{V}[\hat{F}(\hat{\theta}_{0.5})]} = 0.0684$.

The lower confidence bound for the median is then $\hat{F}^{-1}(0.5 - 0.0684)$, and the upper confidence bound for the median is $\hat{F}^{-1}(0.5 + 0.0684)$. We again use linear interpolation to obtain

$$\hat{F}^{-1}(0.4316) = 165 + \frac{0.4316 - 0.3781}{0.4438 - 0.3781}(166 - 165) = 165.8$$

and

$$\hat{F}^{-1}(0.5684) = 170 + \frac{0.5684 - 0.5656}{0.6 - 0.5656}(171 - 170) = 170.1.$$

Thus, an approximate 95% CI for the median is [165.8, 170.1].

Some books and software obtain a slightly different CI for the median; these are asymptotically equivalent to the CI derived in this section if the underlying population distribution function is sufficiently smooth. SAS software calculates the CI

$$\left(\hat{F}^{-1}\left\{ \hat{F}(\hat{\theta}_q) - t_{df}\sqrt{\hat{V}[\hat{F}(\hat{\theta}_q)]} \right\}, \hat{F}^{-1}\left\{ \hat{F}(\hat{\theta}_q) + t_{df}\sqrt{\hat{V}[\hat{F}(\hat{\theta}_q)]} \right\} \right) = [165.6, 169.6].$$

SAS code for producing this CI, with output in Example 7.6, is given on the website. ∎

9.6
Chapter Summary

This chapter has briefly introduced you to some basic types of variance estimation methods that are used in practice: linearization, random groups, replication, and generalized variance functions. But this is just an introduction; you are encouraged to read some of the references mentioned in this chapter before applying these methods to your own complex survey. Much of the research done exploring properties and behavior of these methods has been done since 1980, and variance estimation methods are still a subject of research by statisticians.

Linearization methods are perhaps the most thoroughly researched in terms of theoretical properties, and have been widely used to find variance estimates in complex surveys. The main drawback of linearization, though, is that the derivatives need to be calculated for each statistic of interest, and this complicates the programs for estimating variances. If the statistic you are interested in is not handled in the software, you must write your own code.

The random group method is an intuitively appealing method for estimating variances. It is easy to explain and to compute, and can be used for almost any statistic of interest. Its main drawback is that we generally need enough random groups to have a stable estimate of the variance, and the number of random groups we can form is limited by the number of psus sampled in a stratum.

Resampling methods for stratified multistage surveys avoid partial derivatives by computing estimates for subsamples of the complete sample. They must be constructed carefully, however, so that the correlation of observations in the same cluster is preserved in the resampling. Resampling methods require more computing time than linearization but less programming time: the same method is used on all statistics. They have been shown to be equivalent to linearization for large samples when the characteristic of interest is a smooth function of population totals. Resampling methods can sometimes capture the variability in weight adjustments used for nonresponse.

The BRR method can also be used with almost any statistic, but is usually used only for two psu per stratum designs, or for designs that can be reformulated into two psu per strata. The jackknife and bootstrap can also be used for most estimators likely to be used in surveys (exception: the delete-one jackknife does not work well for estimating the variance of quantiles), and may be used in stratified multistage samples in which more than two psus are selected in each sample, but require more computing than BRR.

Generalized variance functions fit a model predicting the variance of a quantity from other characteristics. They are easy to use, but may give incorrect inferences for a statistic that does not follow the model used to develop the GVF.

Brogan (2005) reviews software packages that analyze complex survey data. SUDAAN (www.rti.org/sudaan), Stata (www.stata.com), SPSS Complex Samples (www.spss.com), and SAS (SAS Institute Inc., 2008) software use linearization methods to estimate variances of nonlinear statistics. The survey software packages WesVar (www.westat.com) and VPLX (Fay, 1990) both use resampling methods to calculate variance estimates. Recent versions of SAS and SUDAAN software also implement BRR and jackknife. Several software packages in the R language (R Development Core Team, 2008) are freely available at www.r-project.org. Lumley (2000) provides a package of R survey functions, using linearization and replication methods; Matei and Tillé (2005) give R functions for selecting samples and computing the Horvitz–Thompson estimator. The free software IVEware (www.isr.umich.edu/src/smp/ive/) uses linearization and replication methods along with multiple imputation for missing data. Software for analyzing survey data changes rapidly; the Survey Research Methods Section of the American Statistical Association (www.amstat.org/sections/SRMS) is a good resource for updated information; click on Links and Resources.

Key Terms

Balanced repeated replication: Resampling method for variance estimation used when there are two psus sampled per stratum.

Bootstrap: Resampling method for variance estimation in which samples of psus with replacement are taken within each stratum.

Generalized variance function: A formula for variance estimation constructed based on a regression model for the variances.

Jackknife: Resampling method for variance estimation in which each psu is deleted in turn.

Linearization: A method for estimating the variance of a nonlinear function of estimated population totals by using a Taylor series expansion.

For Further Reading

The methods discussed in this chapter are described in more detail by Rao (1988), Rao (1997), Rust and Rao (1996), Shao (2003), Wolter (2007), and Brogan (2005).

Binder (1983, 1996) presents a general theory for using the linearization method of estimating the variance, even when the quantities of interest are defined implicitly. Demnati and Rao (2004) derive linearization variance estimators using the weights.

Rao and Wu (1985) give theory (and references to earlier work) showing the asymptotic equivalence of different variance estimators. Canty and Davison (1999) review resampling methods for variance estimation and illustrate how they can account for nonresponse adjustments. Chapter 6 of Shao and Tu (1995) presents theory for the jackknife and bootstrap used in complex surveys.

The CIs presented in this chapter are developed under the design-based approach. A 95% CI may be interpreted in the repeated sampling sense that if samples were repeatedly taken from the finite population, we would expect 95% of the resulting CIs to include the true value of the quantity in the population. In some situations, you may want to consider constructing a conditional CI instead. In poststratification, for example, the sample sizes in each poststratum are random variables; a conditional CI estimates the variance conditionally on the poststrata sample sizes [see (4.22)]. Särndal et al. (1992, section 7.10) and Casady and Valliant (1993) discuss conditional CIs and give a bibliography of other work.

9.7
Exercises

A. Introductory Exercises

1 Which of the variance estimation methods in this chapter would be suitable for estimating the proportion of beds that have bednets for the Gambia bednet survey in Example 7.1? Explain why each method is or is not appropriate.

2 Use the jackknife to estimate $V(\bar{y})$ for the data in srs30.dat, and verify that $\hat{V}_{JK}(\bar{y}) = s^2/30$ for these data. What are the jackknife weights for jackknife replicate j?

3 Use Woodruff's method to construct a 95% CI for the median of the data in file srs30.dat.

4 Estimate the 25th percentile, median, and 75th percentile for the variable *acres92* in file agstrat.dat, used in Example 3.2. Give a 95% CI for each parameter.

B. Working with Survey Data

5 Use the random groups in the data file syc.dat to estimate the variances for the estimates of the proportion of youth who:

a Are age 14 or younger

b Are held for a violent offense

c Lived with both parents when growing up

d Are male

e Are Hispanic

f Grew up primarily in a single-parent family

g Have used illegal drugs.

6 Calculate the jackknife estimate of the variance for the regression estimate of the population mean age of trees in a stand for the data in Exercise 3 of Chapter 4. How does the jackknife variance compare with the variance calculated using linearization methods?

7 Use the jackknife to estimate the variances of your estimates in parts (b) and (c) of Exercise 16 of Chapter 5.

8   Use the jackknife to estimate the variance of the ratio estimator used in Example 4.2. How does it compare with the linearization estimator?

9   Use Woodruff's method to construct a 95% CI for the median weekday greens fee for nine holes, using the SRS in file golfsrs.dat.

10 Use the data in nhanes.dat along with the BRR method to estimate the variance of the ratio of triceps skinfold to body mass index (see Exercise 15 of Chapter 7).

11 Use the data in nhanes.dat along with the BRR method to estimate the variance of the estimated median value of waist circumference (see Exercise 16 of Chapter 7).

12 Use the data in file ncvs2000.dat for this exercise. The public use NCVS files list a total of 184 pseudo-strata, each with two pseudo-psus (as with the NHANES data in Example 9.5, the original stratification and clustering information is altered to preserve confidentiality). Construct replicate weights variables from the pseudo-stratum and pseudo-psu information, using the BRR method. Then use the replicate weights to estimate the percentage of persons who are victims of a violent crime, along with its standard error. Compare your results to those of Exercise 17 in Chapter 7.

C. Working with Theory

All of the problems in this section require probability and calculus.

13 As in Example 9.1, let $h(p) = p(1 - p)$.

a   Find the remainder term in the Taylor expansion, $\int_a^x (x - t)h''(t)dt$, and use it to find an exact expression for $h(\hat{p})$.

b   Is the remainder term likely to be smaller than the other terms? Explain.

c   Find an exact expression for $V[h(\hat{p})]$ for a simple random sample with replacement. How does it compare with the approximation in Example 9.1? HINT: Use moments of the Binomial distribution to find $E(\hat{p}^4)$.

14 The straight-line regression slope for the population is

$$B_1 = \frac{\sum\limits_{i=1}^{N} (x_i - \bar{x}_U)(y_i - \bar{y}_U)}{\sum\limits_{i=1}^{N} (x_i - \bar{x}_U)^2}.$$

a   Express B_1 as a function of population totals $t_1 = \sum_{i=1}^{N} x_i y_i$, $t_2 = \sum_{i=1}^{N} x_i$, $t_3 = \sum_{i=1}^{N} y_i$, $t_4 = \sum_{i=1}^{N} x_i^2$, and $t_5 = \sum_{i=1}^{N} 1 = N$, so that $B_1 = h(t_1, t_2, t_3, t_4, t_5)$.

b   Let $\hat{B}_1 = h(\hat{t}_1, \hat{t}_2, \hat{t}_3, \hat{t}_4, \hat{t}_5)$, and suppose that $E[\hat{t}_i] = t_i$ for $i = 1, 2, 3, 4, 5$. Use the linearization method to find an approximation to the variance of $\hat{B}_1$. Express your answer in terms of $V(\hat{t}_i)$ and Cov $(\hat{t}_i, \hat{t}_j)$.

c   What is the linearization approximation to the variance for an SRS of size n?

d   Find a linearized variate q_i so that $\hat{V}(\hat{B}_1) = \hat{V}(\hat{t}_q)$.

15   The variance of a population is

$$S^2 = \frac{1}{N-1} \sum_{i=1}^{N} (y_i - \bar{y}_U)^2.$$

a   Express S^2 as a function h of population totals $t_1 = \sum_{i=1}^{N} y_i^2$, $t_2 = \sum_{i=1}^{N} y_i$, and $t_3 = \sum_{i=1}^{N} (1)$.

b   Find an estimator $\hat{S}^2$ by substituting estimators for t_1, t_2, and t_3.

c   Find the linearization variance estimator of $\hat{S}^2$.

16   The correlation coefficient for the population is

$$R = \frac{\displaystyle\sum_{i=1}^{N} (x_i - \bar{x}_U)(y_i - \bar{y}_U)}{\sqrt{\displaystyle\sum_{i=1}^{N} (x_i - \bar{x}_U)^2 \sum_{i=1}^{N} (y_i - \bar{y}_U)^2}}.$$

a   Express R as a function of population totals $t_1 = \sum_{i=1}^{N} x_i$, $t_2 = \sum_{i=1}^{N} y_i$, $t_3 = \sum_{i=1}^{N} x_i^2$, $t_4 = \sum_{i=1}^{N} x_i y_i$, and $t_5 = \sum_{i=1}^{N} y_i^2$, so that $R = h(t_1, t_2, t_3, t_4, t_5)$.

b   Let $r = h(\hat{t}_1, \ldots, \hat{t}_5)$, and suppose that $E[\hat{t}_i] = t_i$ for $i = 1, \ldots, 5$. Use the linearization method to find an approximation to the variance of r. Express your answer in terms of $V(\hat{t}_i)$ and $\text{Cov}\,(\hat{t}_i, \hat{t}_j)$.

c   What is the linearization approximation to the variance for an SRS of size n?

17   *Variance estimation with poststratification.* Suppose we poststratify the sample into L poststrata, with population counts $N_1, N_2, \ldots, N_L$. Then the poststratified estimator for the population total is

$$\hat{t}_{\text{post}} = \sum_{l=1}^{L} \frac{N_l}{\hat{N}_l} \hat{t}_l = h(\hat{t}_1, \ldots, \hat{t}_L, \hat{N}_1, \ldots, \hat{N}_L),$$

where

$$\hat{t}_l = \sum_{i \in S} w_i x_{li} y_i, \quad \hat{N}_l = \sum_{i \in S} w_i x_{li},$$

and $x_{li} = 1$ if unit i is in poststratum l and 0 otherwise. Show, using linearization, that

$$V(\hat{t}_{\text{post}}) \approx V \left\{ \sum_{l=1}^{L} \left(\hat{t}_l - \frac{t_l}{N_l} \hat{N}_l \right) \right\}.$$

We can thus let $q_i = \sum_{l=1}^{L} x_{li} \left(y_i - \frac{\hat{t}_l}{\hat{N}_l} \right)$ and estimate $V(\hat{t}_{\text{post}})$ by $\hat{V}(\hat{t}_{\text{post}}) = \hat{V} \left(\sum_{i \in S} w_i q_i \right)$.

18   Consider the random group estimator of the variance from Section 9.2.2. The parameter of interest is $\theta = \bar{y}_U$. A simple random sample with replacement of size n is taken from the population. The sample is divided into R random groups, each of size m. Let $\hat{\theta}_r$ be the sample mean of the m observations in random group r, let $\hat{\theta} = \bar{y} = \sum_{r=1}^{R} \hat{\theta}_r / R$, and let $\hat{V}_2(\hat{\theta})$ be the variance estimator defined in (9.5). Show that

$$\mathrm{CV}\,[\hat{V}_2(\hat{\theta})] = \left[\frac{\kappa}{m} + 3\frac{m-1}{m} - \frac{R-3}{R-1}\right]^{1/2}\frac{1}{\sqrt{R}}$$

where $\kappa = \sum_{i=1}^{N}(y_i - \bar{y}_U)^4/[(N-1)S^4]$.

19   Suppose a stratified random sample is taken with two observations per stratum. Show that if $\sum_{r=1}^{R}\alpha_{rh}\alpha_{rl} = 0$ for $l \neq h$, then

$$\hat{V}_{\mathrm{BRR}}(\bar{y}_{\mathrm{str}}) = \hat{V}_{\mathrm{str}}(\bar{y}_{\mathrm{str}}).$$

HINT: First note that

$$\bar{y}_{\mathrm{str}}(\alpha_i) - \bar{y}_{\mathrm{str}} = \sum_{h=1}^{H}\frac{N_h}{N}\alpha_{ih}\frac{y_{h1}-y_{h2}}{2}.$$

Then express $\hat{V}_{\mathrm{BRR}}(\bar{y}_{\mathrm{str}})$ directly using y_{h1} and y_{h2}.

20   Other BRR estimators of the variance are

$$\frac{1}{4R}\sum_{r=1}^{R}[\hat{\theta}(\alpha_r) - \hat{\theta}(-\alpha_r)]^2$$

and

$$\frac{1}{2R}\sum_{r=1}^{R}[\{\hat{\theta}(\alpha_r) - \hat{\theta}\}^2 + \{\hat{\theta}(-\alpha_r) - \hat{\theta}\}^2].$$

For a stratified random sample with two observations per stratum, show that if $\sum_{r=1}^{R}\alpha_{rh}\alpha_{rl} = 0$ for $l \neq h$, then each of these variance estimators is equivalent to $\hat{V}_{\mathrm{str}}(\bar{y}_{\mathrm{str}})$.

21   Suppose the parameter of interest is $\theta = h(t)$, where $h(t) = at^2 + bt + c$ and t is the population total. Let $\hat{\theta} = h(\hat{t})$. Show, in a stratified random sample with two observations per stratum, that if $\sum_{r=1}^{R}\alpha_{rh}\alpha_{rl} = 0$ for $l \neq h$, then

$$\frac{1}{4R}\sum_{r=1}^{R}\left[\hat{\theta}(\alpha_r) - \hat{\theta}(-\alpha_r)\right]^2 = \hat{V}_L(\hat{\theta}),$$

the linearization estimator of the variance (see Rao and Wu, 1985).

22   The linearization method in Section 9.1 is the one historically used to find variances. Binder (1996) proposes proceeding directly to the estimate of the variance by evaluating the partial derivatives at the sample estimates rather than at the population quantities. What is Binder's estimate for the variance of the ratio estimator? Does it differ from that in Section 9.1?

23   *An alternative approach to linearization variance estimators.* Demnati and Rao (2004) derive a unified theory for linearization variance estimation using weights. Let θ be the population quantity of interest, and define the estimator $\hat{\theta}$ to be a function of the vector of sampling weights and the population values:

$$\hat{\theta} = g(\mathbf{w}, \mathbf{y}_1, \mathbf{y}_2, \ldots, \mathbf{y}_k),$$

where $\mathbf{w} = (w_1, \ldots, w_N)^T$ with w_i the sampling weight of unit i ($w_i = 0$ if i is not in the sample), and $\mathbf{y}_j$ is the vector of population values for the jth response variable. Then a linearization variance estimator can be found by taking the partial derivatives of the function with respect to the *weights*. Let

$$z_i = \frac{\partial g(\mathbf{w}, \mathbf{y}_1, \mathbf{y}_2, \ldots, \mathbf{y}_k)}{\partial w_i}$$

evaluated at the sampling weights w_i. Then we can estimate $V(\hat{\theta})$ by

$$\hat{V}(\hat{\theta}) = \hat{V}(\hat{t}_z) = \hat{V}\left(\sum_{i \in S} w_i z_i\right).$$

For example, considering the ratio estimator of the population total,

$$\hat{\theta} = g(\mathbf{w}, \mathbf{x}, \mathbf{y}) = \frac{\hat{t}_y}{\hat{t}_x} t_x = \frac{\displaystyle\sum_{k \in S} w_k y_k}{\displaystyle\sum_{k \in S} w_k x_k} t_x.$$

The partial derivative of $\hat{\theta} = g(\mathbf{w}, \mathbf{x}, \mathbf{y})$ with respect to w_i is

$$z_i = \frac{\partial g(\mathbf{w}, \mathbf{x}, \mathbf{y})}{\partial w_i} = \frac{y_i}{\displaystyle\sum_{k \in S} w_k x_k} t_x - \frac{x_i \displaystyle\sum_{k \in S} w_k y_k}{\left(\displaystyle\sum_{k \in S} w_k x_k\right)^2} t_x = (y_i - \hat{B} x_i)\frac{t_x}{\hat{t}_x}.$$

For an SRS, finding the estimated variance of $\hat{t}_z$ gives (4.11).

Consider the poststratified estimator in Exercise 17.

a   Write the estimator as $\hat{t}_{\text{post}} = g(\mathbf{w}, \mathbf{y}, \mathbf{x}_1, \ldots, \mathbf{x}_L)$, where $x_{li} = 1$ if observation i is in poststratum l and 0 otherwise.

b   Find an estimator of $V(\hat{t}_{\text{post}})$ using the Demnati–Rao (2004) approach.

24   Consider the model sometimes adopted for GVFs in (9.13). Consider the one-stage cluster design studied in Section 5.2.2, in which each psu has size M and an SRS of n psus is selected from the N psus in the population. Assume that N is large and n/N is small.

a   For a binary response with $p = \bar{y}_U$, show that

$$V(\hat{t}) \approx (NM)^2 \frac{p(1-p)}{nM}[1 + (M-1)\text{ICC}].$$

b   Show that the relative variance $v = V(\hat{t})/t^2$ can be written as $v \approx \alpha + \beta/t$, and give α and β. Consequently, if the intraclass correlation coefficient is similar for responses in the survey, the GVF method should work well.

25   Let b be an estimator for β in the model for GVFs in (9.13). Let $B = t_y/t_x$ and $\hat{B} = \hat{t}_y/\hat{t}_x$. Suppose that $\hat{B}$ and $\hat{t}_x$ are independent.

a Using the model in (9.13) and the result in (9.2), show that we can estimate $V(\hat{B})$ by

$$\hat{V}(\hat{B}) = \hat{B}^2 \left[\frac{b}{\hat{t}_y} - \frac{b}{\hat{t}_x} \right].$$

b Now let B be a proportion for a subpopulation, where t_x is the size of the subpopulation and t_y is the number of units in that subpopulation having a certain characteristic. Show that $\hat{V}(\hat{B}) = b\hat{B}(1 - \hat{B})/\hat{t}_x$ and that $\hat{V}(\hat{B}) = \hat{V}(1 - \hat{B})$.

D. Projects and Activities

26 *Forest data.* Use the forest data from Exercise 36 of Chapter 2 for this exercise. Construct the jackknife weights for your SRS of size 2000, and use the jackknife weights to estimate the variance of the ratio of hillshade index at 9 am to hillshade index at noon. Compare your answer with the linearization variance estimate you calculated in Exercise 41 of Chapter 4.

27 *Index fund.* In Exercise 43 of Chapter 6, you selected a sample of size 30 from the S&P 500 companies with probability proportional to market capitalization. Construct jackknife weights for this sample.

28 *Trucks.* Use the data from the Vehicle Inventory and Use Survey (VIUS), described in Exercise 34 of Chapter 3, for this problem. The survey design is stratified random sampling, with a sample size of 136,113 trucks.

   a Which of the variance estimation methods in this chapter can be used to estimate the variance of the estimated ratio of miles driven in 2002 (*miles_annl*) to lifetime miles driven (*miles_life*)? What are the advantages and drawbacks of each method?

   b Which of the variance estimation methods can be used to estimate the variance of the estimated median number of miles driven in 2002? What are the advantages and drawbacks of each method?

   c Use the bootstrap with 500 replications to estimate the variances of the estimates in (a) and (b).

29 *Baseball data.* Construct jackknife weights for your dataset from Exercise 35 of Chapter 3. Use these weights to estimate the variance of the estimated mean of the variable *logsal*, and of the ratio (total number of home runs)/(number of runs scored).

30 *IPUMS exercises.* Construct the jackknife weights for your dataset from Exercise 38 of Chapter 5. Use these weights to estimate the variances of the estimated population mean and total of *inctot*.

31 Find a survey on the Internet that releases replicate weights for variance estimation. What method was used to construct the replicate weights? How are nonresponse adjustments incorporated into the replicate weights?

32 *Activity for course project.* Describe the method used for variance estimation for the survey you looked at in Exercise 31 of Chapter 7. If the survey releases replicate weight variables, what method was used to construct them? Do the replicate weights incorporate nonresponse adjustments? If so, how? Now do either (a) or (b) for your survey:

a   If the survey releases replicate weight variables, use them to estimate the variance of the estimated means you found in Exercise 31 of Chapter 7. If the replicate weights are formed by BRR or bootstrap, also estimate the variances of the estimated quantiles.

b   If the survey releases stratification and clustering information, use these to construct replicated weights using one of the resampling methods described in this chapter. Use the replicate weights to estimate the variance of the estimated means you found in Exercise 31 of Chapter 7.

10

Categorical Data Analysis in Complex Surveys

But Statistics must be made otherwise than to prove a preconceived idea.

—Florence Nightingale, Annotation in *Physique Sociale* by A. Quetelet

Up to now we have mostly been looking at how to estimate summary quantities such as means, totals, and percentages in different sampling designs. Totals and percentages are important for many surveys to provide a description of the population: for instance, the percentage of the population victimized by crime or the total number of unemployed persons in the United States. Often, though, researchers are interested in multivariate questions: Is race associated with criminal victimization, or can we predict unemployment status from demographic variables? Such questions are typically answered in statistics using techniques in categorical data analysis or regression. The techniques you learned in an introductory statistics course, though, assumed that observations were all independent and identically distributed from some population distribution. These assumptions are no longer met in data from complex surveys; in this and the following chapter we examine the effects of the complex sampling design on commonly used statistical analyses.

Since much information from sample surveys is collected in the form of percentages, categorical data methods are extensively used in the analysis. In fact, many of the data sets used to illustrate the chi-square test in introductory statistics textbooks originate in complex surveys. Our greatest concern is with the effects of clustering on hypothesis tests and models for categorical data, since clustering usually decreases precision. We begin by reviewing various chi-square tests when a simple random sample (SRS) is taken from a large population.

10.1
Chi-Square Tests with Multinomial Sampling

EXAMPLE 10.1 Each couple in an SRS of 500 married couples from a large population is asked whether (1) the household owns at least one personal computer and (2) the household subscribes to cable television. The following contingency table presents the outcomes:

Observed Count		Computer? Yes	No	
Cable?	Yes	119	188	307
	No	88	105	193
		207	293	500

Are households with a computer more likely to subscribe to cable? A chi-square test for independence is often used for such questions. Under the null hypothesis that owning a computer and subscribing to cable are independent, the expected counts for each cell in the contingency table are the following:

Expected Count		Computer? Yes	No	
Cable?	Yes	127.1	179.9	307
	No	79.9	113.1	193
		207	293	500

Pearson's chi-square test statistic is

$$X^2 = \sum_{\text{all cells}} \frac{(\text{observed count} - \text{expected count})^2}{\text{expected count}} = 2.281.$$

The **likelihood ratio chi-square test statistic is**

$$G^2 = 2 \sum_{\text{all cells}} (\text{observed count}) \ln \left(\frac{\text{observed count}}{\text{expected count}} \right) = 2.275.$$

The two test statistics are asymptotically equivalent; for large samples, each approximately follows a chi-square (χ^2) distribution with 1 degree of freedom (df) under the null hypothesis. The p-value for each statistic is 0.13, giving no reason to doubt the null hypothesis that owning a computer and subscribing to cable television are independent.

If owning a computer and subscribing to cable are independent events, the odds that a cable subscriber will own a computer should equal the odds that a non-cable-subscriber will own a computer. We estimate the odds of owning a computer if the household subscribes to cable as 119/188 and estimate the odds of owning a computer if the household does not subscribe to cable as 88/105. The **odds ratio** is therefore estimated as

$$\frac{\dfrac{119}{188}}{\dfrac{88}{105}} = 0.755.$$

If the null hypothesis of independence is true, we expect the odds ratio to be close to one. Equivalently, we expect the logarithm of the odds ratio to be close to zero. The log odds is -0.28 with asymptotic standard error

$$\sqrt{\frac{1}{119} + \frac{1}{88} + \frac{1}{188} + \frac{1}{105}} = 0.186;$$

an approximate 95% confidence interval (CI) for the log odds is $-0.28 \pm 1.96(0.186) = [-0.646, 0.084]$. This CI includes 0, and confirms the result of the hypothesis test that there is no evidence against independence. ∎

Chi-square tests are commonly used in three situations: testing independence of factors, testing homogeneity of proportions, and testing goodness of fit. Each assumes a form of random sampling. These tests are discussed in more detail in Agresti (2002) and Simonoff (2006).

10.1.1 Testing Independence of Factors

Each of n independent observations is cross-classified by two factors: row factor R with r levels and column factor C with c levels. Each observation has probability p_{ij} of falling into row category i and column category j, giving the following table of true probabilities. Here, $p_{i+} = \sum_{j=1}^{c} p_{ij}$ is the probability that a randomly selected unit will fall in row category i, and $p_{+j} = \sum_{i=1}^{r} p_{ij}$ is the probability that a randomly selected unit will fall in column category j.

		C				
		1	2	$\cdots$	c	
	1	p_{11}	p_{12}	$\cdots$	p_{1c}	p_{1+}
	2	p_{21}	p_{22}	$\cdots$	p_{2c}	p_{2+}
R	$\vdots$	$\vdots$	$\vdots$		$\vdots$	$\vdots$
	r	p_{r1}	p_{r2}	$\cdots$	p_{rc}	p_{r+}
		p_{+1}	p_{+2}	$\cdots$	p_{+c}	1

The observed count in cell (i, j) from the sample is x_{ij}. If all units in the sample are independent, the x_{ij}'s are from a multinomial distribution with rc categories; this sampling scheme is known as **multinomial sampling**. In surveys, the assumptions for multinomial sampling are met in an SRS with replacement; they are approximately met in an SRS without replacement when the sample size is small compared with the population size. The latter situation occurred in Example 10.1: Independent multinomial sampling means we have a sample of 500 (approximately) independent households, and we observe to which of the four categories each household belongs.

The null hypothesis of independence is

$$H_0 : p_{ij} = p_{i+}p_{+j} \quad \text{for } i = 1, \ldots, r \quad \text{and} \quad j = 1, \ldots, c. \tag{10.1}$$

Let $m_{ij} = np_{ij}$ represent the expected counts. If H_0 is true, $m_{ij} = np_{i+}p_{+j}$, and m_{ij} can be estimated by

$$\hat{m}_{ij} = n\hat{p}_{i+}\hat{p}_{+j} = n\frac{x_{i+}}{n}\frac{x_{+j}}{n},$$

where $\hat{p}_{ij} = x_{ij}/n$, $\hat{p}_{+j} = \sum_{i=1}^{r} \hat{p}_{ij}$, and $\hat{p}_{i+} = \sum_{j=1}^{c} \hat{p}_{ij}$. Pearson's chi-square test statistic is

$$X^2 = \sum_{i=1}^{r}\sum_{j=1}^{c} \frac{(x_{ij} - \hat{m}_{ij})^2}{\hat{m}_{ij}} = n\sum_{i=1}^{r}\sum_{j=1}^{c} \frac{(\hat{p}_{ij} - \hat{p}_{i+}\hat{p}_{+j})^2}{\hat{p}_{i+}\hat{p}_{+j}}. \tag{10.2}$$

The likelihood ratio test statistic is

$$G^2 = 2\sum_{i=1}^{r}\sum_{j=1}^{c} x_{ij} \ln\left(\frac{x_{ij}}{\hat{m}_{ij}}\right) = 2n\sum_{i=1}^{r}\sum_{j=1}^{c} \hat{p}_{ij} \ln\left(\frac{\hat{p}_{ij}}{\hat{p}_{i+}\hat{p}_{+j}}\right). \tag{10.3}$$

If multinomial sampling is used with a sufficiently large sample size, X^2 and G^2 are approximately distributed as a χ^2 random variable with $(r-1)(c-1)$ degrees of freedom (df) under the null hypothesis. How large is "sufficiently large" depends on the number of cells and expected probabilities; Fienberg (1979) argues that p-values will be approximately correct if (a) the expected count in each cell is greater than 1 and (b) $n \geq 5 \times$ (number of cells).

An equivalent statement to (10.1) is that all odds ratios equal 1:

$$H_0 : \frac{p_{11}p_{ij}}{p_{1j}p_{i1}} = 1 \qquad \text{for all } i \geq 2 \text{ and } j \geq 2.$$

We may estimate any odds ratio $(p_{ij}p_{kl})/(p_{il}p_{kj})$ by substituting in estimated proportions: $(\hat{p}_{ij}\hat{p}_{kl})/(\hat{p}_{il}\hat{p}_{kj})$. If the sample is sufficiently large, the *logarithm* of the estimated odds ratio is approximately normally distributed with estimated variance (see Exercise 15)

$$\hat{V}\left[\ln\left(\frac{\hat{p}_{ij}\hat{p}_{kl}}{\hat{p}_{il}\hat{p}_{kj}}\right)\right] = \frac{1}{x_{ij}} + \frac{1}{x_{kl}} + \frac{1}{x_{il}} + \frac{1}{x_{kj}}.$$

10.1.2 Testing Homogeneity of Proportions

The Pearson and likelihood ratio test statistics in (10.2) and (10.3) may also be used when independent random samples from r populations are each classified into c categories. Multinomial sampling is done within each population, so the sampling scheme is called **product-multinomial sampling**. Product-multinomial sampling is equivalent to stratified random sampling when the sampling fraction for each stratum is small or when sampling is with replacement.

The difference between product-multinomial sampling and multinomial sampling is that the row totals p_{i+} and x_{i+} are fixed quantities in product-multinomial sampling—x_{i+} is the predetermined sample size for stratum i. The null hypothesis that the proportion of observations falling in class j is the same for all strata is

$$H_0 : \frac{p_{1j}}{p_{1+}} = \frac{p_{2j}}{p_{2+}} = \cdots = \frac{p_{rj}}{p_{r+}} = p_{+j} \qquad \text{for all } j = 1, \ldots, c. \tag{10.4}$$

If the null hypothesis in (10.4) is true, again $m_{ij} = np_{i+}p_{+j}$ and the expected counts under H_0 are $\hat{m}_{ij} = np_{i+}\hat{p}_{+j}$, exactly as in the test for independence.

EXAMPLE **10.2** The sample sizes used in Exercise 14 of Chapter 3, the stratified sample of nursing students and tutors, were the sample sizes for the respondents. Let's use a chi-square test for homogeneity of proportions to test the null hypothesis that the nonresponse rate is the same for each stratum. The four strata form the rows in the following contingency table.

	Nonrespondent	Respondent	
General student	46	222	268
General tutor	41	109	150
Psychiatric student	17	40	57
Psychiatric tutor	8	26	34
	112	397	509

The two chi-square test statistics are $X^2 = 8.218$, with p-value 0.042 and $G^2 = 8.165$, with p-value 0.043. There is thus evidence of different nonresponse rates among the four groups. However, the following table shows that the difference is not attributable to either the main effect of general/psychiatric or student/tutor:

	Nonresponse rate	
	Student	Tutor
General	17%	27%
Psychiatric	30%	24%

Further investigation would be needed to explore the nonresponse pattern. ∎

10.1.3 Testing Goodness of Fit

In the classical goodness of fit test, multinomial sampling is again assumed, with independent observations classified into k categories. The null hypothesis is

$$H_0 : p_i = p_i^{(0)} \qquad \text{for } i = 1, \dots, k,$$

where $p_i^{(0)}$ is prespecified or is a function of parameters θ to be estimated from the data.

EXAMPLE 10.3 Webb (1955) examined the safety records for 17,952 Air Force pilots for an 8-year period around World War II and constructed the following frequency table.

Number of Accidents	Number of Pilots
0	12,475
1	4,117
2	1,016
3	269
4	53
5	14
6	2
7	2

If accidents occur randomly—if no pilots are more or less "accident-prone" than others—a Poisson distribution should fit the data well. We estimate the mean of the Poisson distribution by the mean number of accidents per pilot in the sample, 0.40597. The observed and expected probabilities under the null hypothesis that the data follow a Poisson distribution are given in the following table. The expected probabilities are computed using the Poisson probabilities $e^{-\lambda}\lambda^x/x!$ with $\lambda = 0.40597$.

Number of Accidents	Observed Proportion, $\hat{p}_i$	Expected Probability Under H_0, $\hat{p}_i^{(0)}$
0	0.6949	0.6663
1	0.2293	0.2705
2	0.0566	0.0549
3	0.0150	0.0074
4	0.0030	0.0008
5+	0.0012	0.0001

The two chi-square test statistics are

$$X^2 = \sum_{\text{all cells}} \frac{(\text{observed count} - \text{expected count})^2}{\text{expected count}}$$

$$= \sum_{i=1}^{k} \frac{(n\hat{p}_i - n\hat{p}_i^{(0)})^2}{n\hat{p}_i^{(0)}} \tag{10.5}$$

$$= n\sum_{i=1}^{k} \frac{(\hat{p}_i - \hat{p}_i^{(0)})^2}{\hat{p}_i^{(0)}}$$

and

$$G^2 = 2n\sum_{i=1}^{k} \hat{p}_i \ln\left(\frac{\hat{p}_i}{\hat{p}_i^{(0)}}\right). \tag{10.6}$$

For the pilots, $X^2 = 756$ and $G^2 = 400$. If the null hypothesis is true, both statistics follow a χ^2 distribution with 4 df (2 df are spent on n and $\hat{\lambda}$). Both p-values are less than 0.0001, providing evidence that a Poisson model does not fit the data. More pilots have no accidents, or more than two accidents, than would be expected under the Poisson model. There is thus evidence that some pilots are more accident-prone than would occur under the Poisson model. ∎

All of the chi-square test statistics in (10.2), (10.3), (10.5), and (10.6) grow with n. If the null hypothesis is not exactly true in the population—if households with cable are even infinitesimally more likely to own a personal computer than households without cable—we can almost guarantee rejection of the null hypothesis by taking a large enough random sample. This property of the hypothesis test means that it will be sensitive to artificially inflating the sample size by ignoring clustering.

10.2
Effects of Survey Design on Chi-Square Tests

The survey design can affect both the estimated cell probabilities and the tests of association or goodness of fit. In complex survey designs, we no longer have the random sampling that gives both X^2 and G^2 an approximate χ^2 distribution. Thus, if we ignore the survey design and use the chi-square tests described in Section 10.1, the significance levels and p-values will be wrong. Clustering, especially, can have a strong effect on the p-values of chi-square tests. In a cluster sample with a positive intraclass correlation coefficient (ICC), the true p-value will often be much larger than the p-value reported by a statistical package using the assumption of independent multinomial sampling. Let's see what can happen to hypothesis tests if the survey design is ignored in a cluster sample.

EXAMPLE 10.4 Suppose that both husband and wife are asked about the household's cable and computer status for the survey discussed in Example 10.1, and both give the same answer. While the assumptions of multinomial sampling were met for the SRS of couples, they are not met for the cluster sample of persons—far from being independent units, the husband and wife from the same household agree completely in their answers. The ICC for the cluster sample is 1.

What happens if we ignore the clustering? The contingency table for the observed frequencies is as follows:

		Computer?		
Observed Count		Yes	No	
Cable?	Yes	238	376	614
	No	176	210	386
		414	586	1000

The estimated proportions and odds ratio are identical to those in Example 10.1: $\hat{p}_{11} = 238/1000 = 119/500$ and the odds ratio is

$$\frac{\dfrac{238}{376}}{\dfrac{176}{210}} = 0.755.$$

But $X^2 = 4.562$ and $G^2 = 4.550$ are twice the values of the test statistics in Example 10.1. If you ignored the clustering and compared these statistics to a χ^2 distribution with 1 df, you would report a "p-value" of 0.033 and conclude that the data provided evidence that having a computer and subscribing to cable are not independent. If playing this game, you could lower the "p-value" even more by interviewing both children in each household as well, thus multiplying the original test statistics by 4.

Can you attain an arbitrarily low p-value by observing more ssus per psu? Absolutely not. The statistics X^2 and G^2 have a null χ_1^2 distribution *when multinomial sampling is used*. When a cluster sample is taken instead, and when the intraclass correlation coefficient is positive, X^2 and G^2 do *not* follow a χ_1^2 distribution under the null hypothesis. For the 1000 husbands and wives, $X^2/2$ and $G^2/2$ follow a χ_1^2 distribution under H_0—this gives the same *p*-value found in Example 10.1. ∎

10.2.1 Contingency Tables for Data from Complex Surveys

The observed counts x_{ij} do not necessarily reflect the relative frequencies of the categories in the population unless the sample is self-weighting. Suppose an SRS of elementary school classrooms in Denver is taken, and each of ten randomly selected students in each classroom is evaluated for self-concept (high or low) and clinical depression (present or not). Students are selected for the sample with unequal probabilities—students in small classes are more likely to be in the sample than students from large classes. A table of observed counts from the sample, ignoring the inclusion probabilities, would not give an accurate picture of the association between self-concept and depression in the population if the degree of association differs with class size. Even if the association between self-concept and depression is the same for different class sizes, the estimates of numbers of depressed students using the margins of the contingency table may be wrong.

Remember, though, that sampling weights can be used to estimate any population quantity. Here, they can be used to estimate the cell proportions. Estimate p_{ij} by

$$\hat{p}_{ij} = \frac{\displaystyle\sum_{k \in S} w_k y_{kij}}{\displaystyle\sum_{k \in S} w_k}, \tag{10.7}$$

where

$$y_{kij} = \begin{cases} 1 & \text{if observation unit } k \text{ is in cell } (i,j) \\ 0 & \text{otherwise} \end{cases}$$

and w_k is the weight for observation unit k. Thus,

$$\hat{p}_{ij} = \frac{\text{sum of weights for observation units in cell } (i,j)}{\text{sum of weights for all observation units in sample}}.$$

If the sample is self-weighting, $\hat{p}_{ij}$ will be the proportion of observation units falling in cell (i,j). Using the estimates $\hat{p}_{ij}$, construct the table

		C				
		1	2	$\cdots$	c	
	1	$\hat{p}_{11}$	$\hat{p}_{12}$	$\cdots$	$\hat{p}_{1c}$	$\hat{p}_{1+}$
	2	$\hat{p}_{21}$	$\hat{p}_{22}$	$\cdots$	$\hat{p}_{2c}$	$\hat{p}_{2+}$
R	$\vdots$	$\vdots$	$\vdots$		$\vdots$	$\vdots$
	r	$\hat{p}_{r1}$	$\hat{p}_{r2}$	$\cdots$	$\hat{p}_{rc}$	$\hat{p}_{r+}$
		$\hat{p}_{+1}$	$\hat{p}_{+2}$	$\cdots$	$\hat{p}_{+c}$	1

to examine associations, and estimate odds ratios by $(\hat{p}_{ij}\hat{p}_{kl})/(\hat{p}_{il}\hat{p}_{kj})$. A CI for p_{ij} may be constructed by using any method of variance estimation discussed so far, or a design effect (deff) may be used to modify the SRS CI, as in (7.8).

Do not throw the observed counts away, however. If the odds ratios calculated using the $\hat{p}_{ij}$ differ appreciably from the odds ratios calculated using the observed counts x_{ij}, you should explore why they differ. Perhaps the odds ratio for depression and self-concept differs for larger classes or depends on socioeconomic factors related to class size. If that is the case, you should include these other factors in a model for the data or perhaps test the association separately for large and small classes.

10.2.2 Effects on Hypothesis Tests and Confidence Intervals

We can estimate contingency table proportions and odds ratios using weights. The weights, however, are not sufficient for constructing hypothesis tests and CIs—these depend on the clustering and (sometimes) stratification of the survey design.

Let's look at the effect of stratification first. If the strata in a stratified random sample are the row categories, the stratification poses no problem—we essentially have product-multinomial sampling as described in Section 10.1 and can test for homogeneity of proportions the usual way.

Often, though, we want to study association between factors that are not stratification variables. In general, stratification increases precision of the estimates. For an SRS, (10.2) gives

$$X^2 = n \sum_{i=1}^{r} \sum_{j=1}^{c} \frac{(\hat{p}_{ij} - \hat{p}_{i+}\hat{p}_{+j})^2}{\hat{p}_{i+}\hat{p}_{+j}}.$$

A stratified sample with n observation units provides the same precision for estimating p_{ij} as an SRS with n/d_{ij} observation units, where d_{ij} is the deff for estimating p_{ij}. If the stratification is worthwhile, the deffs will generally be less than 1. Consequently, if we use the SRS test statistics in (10.2) or (10.3) with the $\hat{p}_{ij}$ from the stratified sample, X^2 and G^2 will be smaller than they should be to follow a null $\chi^2_{(r-1)(c-1)}$ distribution; "p-values" calculated ignoring the stratification will be too large and H_0 will not be rejected as often as it should be. Thus, while SAS PROC FREQ or another statistics program for non-survey data may give you a p-value of 0.04, the actual p-value may be 0.02. Ignoring the stratification results in a conservative test. Similarly, a CI constructed for a log odds ratio is generally too large if the stratification is ignored. Your estimates are really more precise than the SRS CI indicates.

Clustering usually has the opposite effect. Design effects for $\hat{p}_{ij}$ with a cluster sample are usually greater than 1—a cluster sample with n observation units gives the same precision as an SRS with fewer than n observations. If the clustering is ignored, X^2 and G^2 are expected to be larger than if the equivalently sized SRS were taken, and "p-values" calculated ignoring the clustering are likely to be too small. An analysis ignoring the survey design may give you a p-value of 0.04, while the actual p-value may be 0.25. If you ignore the clustering, you may well declare an association to be statistically significant when it is really just due to random variation in the data. CIs for log odds ratios will be narrower than they should be—the estimates are not as precise as the CIs from an SRS-based analysis would lead you to believe.

Ignoring clustering in chi-square tests is often more dangerous than ignoring stratification. An SRS-based chi-square test using stratified data will still indicate strong associations; it just will not uncover all weaker associations. Ignoring clustering, however, will lead to declaring associations statistically significant that really are not. Ignoring the clustering in goodness-of-fit tests may lead to adopting an unnecessarily complicated model to describe the data.

An investigator ignorant of sampling theory will often analyze a stratified sample correctly, using the strata as one of the classification variables. But the investigator may not even record the clustering, and too often simply runs the observed counts through SAS PROC FREQ or SPSS CROSSTABS and accepts the printed-out p-value as truth. To see how this could happen, consider an investigator wanting to replicate Basow and Silberg's (1987) study on whether male and female professors are evaluated differently by college students. (The original study was discussed in Example 5.1.) The investigator selects a stratified sample of male and female professors at the college and asks each student in those professors' classes to evaluate the professor's teaching. Over 2000 student responses are obtained, and the investigator cross-classifies those responses by professor gender and by whether the student gives the professor a high or low rating. The investigator, comparing Pearson's X^2 statistic on the observed counts to a χ_1^2 distribution, declares a statistically significant association between professor gender and student rating. The stratification variable *professor gender* is one of the classification variables, so no adjustments need be made for the stratification. But the reported p-value is almost certainly incorrect, for a number of reasons: (1) The clustering of students within a class is ignored—indeed, the investigator does not even record which professor is evaluated by a student, but only records the professor's gender, so the investigation cannot account for the clustering. If student evaluations reflect teaching quality, students of a "good" professor would be expected to give higher ratings than students of a "bad" professor. The ICC for students is positive, and the equivalent sample size in an SRS is less than 2000. The p-value reported by the investigator is then much too small, and the investigator may be wrong in concluding faculty women receive a different mean level on student evaluations. (2) A number of students may give responses for more than one professor in the sample. It is unclear what effect these multiple responses would have on the test of independence. (3) Not all students attend class or turn in the evaluation. Some of the nonresponse may be missing completely at random (a student was ill the day of the study), but some may be related to perceived teaching quality (the student skips class because the professor is confusing).

The societal implications of reporting false positive results because clustering is ignored can be expensive. A university administrator may decide to give female faculty an unnecessary handicap when determining raises that are based in part on student evaluations. A medical researcher may conclude that a new medication with more side effects than the standard treatment is more effective for combating a disease, even though the statistical significance is due to the cluster inflation of the sample size. A government official may decide that a new social program is needed to remedy an "inequity" demonstrated in the hypothesis test. The same problem occurs outside of sample surveys as well, particularly in biostatistics. Clusters may correspond to pairs of eyes, to patients in the same hospital, or to repeated measures on the same person.

Is the clustering problem serious in surveys taken in practice? A number of studies have found that it can be. Holt et al. (1980) found that the actual significance levels for tests nominally conducted at the $\alpha = 0.05$ level ranged from 0.05 to 0.50. Fay (1985) references a number of studies demonstrating that the SRS-based test statistics "may give extremely erroneous results when applied to data arising from a complex sample design." The simulation study in Thomas et al. (1996) calculated actual significance levels attained for X^2 and G^2 when the nominal significance level was set at $\alpha = 0.05$—they found actual significance levels of about 0.30 to 0.40.

10.3
Corrections to χ^2 Tests

In this section, we outline some of the basic approaches for testing independence with data from a complex survey. The theory for goodness of fit tests and tests for homogeneity of proportions is similar. In complex surveys, though, unlike in multinomial and product multinomial sampling, the tests for independence and homogeneity of proportions are not necessarily the same. Holt et al. (1980) note that often (but not always) clustering has less effect on tests for independence than on tests for goodness of fit or homogeneity of proportions.

Recall from (10.1) that the null hypothesis of independence is

$$H_0 : p_{ij} = p_{i+}p_{+j} \qquad \text{for } i = 1,\ldots,r \quad \text{and} \quad j = 1,\ldots,c.$$

For a 2×2 table, $p_{ij} = p_{i+}p_{+j}$ for all i and j is equivalent to $p_{11}p_{22} - p_{12}p_{21} = 0$, so the null hypothesis reduces to a single equation. In general, the null hypothesis can be expressed as $(r-1)(c-1)$ distinct equations, which leads to $(r-1)(c-1)$ df for the χ^2 tests used for multinomial sampling. Let

$$\theta_{ij} = p_{ij} - p_{i+}p_{+j}.$$

Then, the null hypothesis of independence is

$$H_0 : \theta_{11} = 0, \ \theta_{12} = 0, \ \ldots, \ \theta_{r-1,c-1} = 0.$$

10.3.1 Wald Tests

The Wald (1943) test was the first to be used for testing independence in complex surveys (Koch et al., 1975). For the 2×2 table, the null hypothesis involves one quantity,

$$\theta = \theta_{11} = p_{11} - p_{1+}p_{+1} = p_{11}p_{22} - p_{12}p_{21},$$

and θ is estimated by

$$\hat{\theta} = \hat{p}_{11}\hat{p}_{22} - \hat{p}_{12}\hat{p}_{21}.$$

The quantity θ is a smooth function of population totals, so we estimate $V(\hat{\theta})$ using one of the methods in Chapter 9. If the sample sizes are sufficiently large and $H_0: \theta = 0$ is true, then $\hat{\theta}/\sqrt{\hat{V}(\hat{\theta})}$ approximately follows a standard normal distribution. Equivalently, under H_0, the **Wald statistic**

$$X_W^2 = \frac{\hat{\theta}^2}{\hat{V}(\hat{\theta})} \tag{10.8}$$

approximately follows a χ^2 distribution with 1 df. In practice, we often compare X_W^2 to an F distribution with 1 and κ df, where κ is the df associated with the variance estimator. If the random group method is used to estimate the variance, then κ equals (number of groups) $-$ 1; if another method is used, κ equals (number of psus) $-$ (number of strata).

EXAMPLE 10.5 Let's look at the association between variables "Was anyone in your family ever incarcerated?" (variable *famtime*) and "Have you ever been put on probation or sent to a correctional institution for a violent offense?" (variable *everviol*) using data from the Survey of Youths in Custody. A total of $n = 2588$ youths in the survey had responses for both items. The following table gives the sum of the weights for each category.

		Ever Violent?		
		No	Yes	Total
Family Member	No	4,761	7,154	11,915
Incarcerated?	Yes	4,838	7,946	12,784
	Total	9,599	15,100	24,699

This results in the following table of estimated proportions:

		Ever Violent?		
		No	Yes	Total
Family Member	No	0.1928	0.2896	0.4824
Incarcerated?	Yes	0.1959	0.3217	0.5176
	Total	0.3886	0.6114	1.0000

Thus,

$$\hat{\theta} = \hat{p}_{11}\hat{p}_{22} - \hat{p}_{12}\hat{p}_{21} = \hat{p}_{11} - \hat{p}_{1+}\hat{p}_{+1} = 0.0053.$$

We can write $\theta = h(p_{11}, p_{12}, p_{21}, p_{22})$ and $\hat{\theta} = h(\hat{p}_{11}, \hat{p}_{12}, \hat{p}_{21}, \hat{p}_{12})$ for $h(a, b, c, d) = ad - bc$, so we can use linearization (see Exercise 14) or a resampling method to estimate $V(\hat{\theta})$. The random group method can also be used, although it does not give variance estimates that are as accurate as the other methods (see Exercise 5). Using linearization in SAS PROC SURVEYFREQ (SAS code is on the website), we obtain $X_W^2 = (0.0053)^2 / \hat{V}(\hat{\theta}) = 0.995$ with p-value $= 0.32$.

This test gives no evidence of an association between the two factors, when we look at the population as a whole. But of course the hypothesis test does not say anything about possible associations among the two variables in subpopulations—it could occur, for example, that violence and incarceration of a family member are positively associated among older youth, and negatively associated among younger youth—we would need to look at the subpopulations separately or fit a loglinear model to see if this was the case. ∎

For larger tables, let $\theta_{ij} = p_{ij} - p_{i+}p_{+j}$ and let $\boldsymbol{\theta} = [\theta_{11} \, \theta_{12} \, \ldots \, \theta_{r-1,c-1}]^T$ be the $(r-1)(c-1)$-vector of θ_{ij}'s, so that the null hypothesis is

$$H_0 : \boldsymbol{\theta} = \mathbf{0}.$$

The Wald statistic is then

$$X_W^2 = \hat{\boldsymbol{\theta}}^T \, \hat{V}(\hat{\boldsymbol{\theta}})^{-1} \hat{\boldsymbol{\theta}},$$

where $\hat{V}(\hat{\boldsymbol{\theta}})$ is the estimated covariance matrix of $\hat{\boldsymbol{\theta}}$. In very large samples, X_W^2 approximately follows a $\chi^2_{(r-1)(c-1)}$ distribution under H_0. But "large" in a complex survey refers to a large number of psus, not necessarily to a large number of observation units. In a 4×4 contingency table, $\hat{V}(\hat{\boldsymbol{\theta}})$ is a 9×9 matrix, and requires calculation of 45 different variances and covariances. If a cluster sample has only 50 psus, the estimated covariance matrix will be very unstable. In practice, the Wald test for large contingency tables often performs poorly, and we do not recommend its use. Some modifications of the Wald test perform better; see Thomas et al. (1996).

Thomas (1989) suggested using the Bonferroni correction for larger tables. In an $R \times C$ table the null hypothesis of independence,

$$H_0 : \theta_{11} = 0, \; \theta_{12} = 0, \; \ldots, \; \theta_{r-1,c-1} = 0,$$

has $m = (r-1)(c-1)$ components:

$$H_0(1) : \theta_{11} = 0$$
$$H_0(2) : \theta_{12} = 0$$
$$\vdots$$
$$H_0(m) : \theta_{(r-1)(c-1)} = 0.$$

Instead of using the estimated covariance of all $\hat{\theta}_{ij}$'s as in the full multivariate Wald test, we can use the Bonferroni inequality to test each component $H_0(k)$ separately with significance level α/m. H_0 will be rejected at level α if any of the $H_0(k)$ is rejected at level α/m—that is, if

$$\frac{\hat{\theta}_{ij}^2}{\hat{V}(\hat{\theta}_{ij})} > F_{1,\kappa,\alpha/m}$$

for $i = 1, \ldots, (r-1)$ and $j = 1, \ldots, (c-1)$. Each test statistic is compared to an $F_{1,\kappa}$ distribution, where the estimator of the variance has κ df. Since the Bonferroni adjustment for multiple testing is used, this is a conservative test, although it appears to work well in practice.

10.3.2 Rao–Scott Tests

The test statistics X^2 and G^2 do not follow a $\chi^2_{(r-1)(c-1)}$ distribution in a complex survey under the null hypothesis of independence. But both statistics have a skewed distribution, and a multiple of X^2 or G^2 may approximately follow a χ^2 distribution.

We can obtain a **first-order correction** by matching the mean of the test statistic to the mean of the $\chi^2_{(r-1)(c-1)}$ distribution (Rao and Scott, 1981, 1984). The mean of a $\chi^2_{(r-1)(c-1)}$ distribution is $(r-1)(c-1)$; we can calculate $E[X^2]$ or $E[G^2]$ under the complex sampling design when H_0 is true and compare the test statistic

$$X_F^2 = \frac{(r-1)(c-1)X^2}{E[X^2]}$$

or

$$G_F^2 = \frac{(r-1)(c-1)G^2}{E[G^2]}$$

to a $\chi^2_{(r-1)(c-1)}$ distribution. Bedrick (1983) and Rao and Scott (1984) show that under H_0,

$$E[X^2] \approx E[G^2] \approx \sum_{i=1}^{r}\sum_{j=1}^{c}(1-p_{ij})d_{ij} - \sum_{i=1}^{r}(1-p_{i+})d_i^R - \sum_{j=1}^{c}(1-p_{+j})d_j^C, \quad (10.9)$$

where d_{ij} is the deff for estimating p_{ij}, d_i^R is the deff for estimating p_{i+}, and d_j^C is the deff for estimating p_{+j}. In practice, if the estimator of the cell variances has κ df, it works slightly better to compare $X_F^2/(r-1)(c-1)$ or $G_F^2/(r-1)(c-1)$ to an F distribution with $(r-1)(c-1)$ and $(r-1)(c-1)\kappa$ df.

The first-order correction can often be used with published tables because you need to estimate only variances of the proportions in the contingency table—you need not estimate the full covariance matrix of the $\hat{p}_{ij}$ as is required for the Wald test. But we are only adjusting the test statistic so that its mean under H_0 is $(r-1)(c-1)$; p-values of interest come from the tail of the reference distribution, and it does not necessarily follow that the tail of the distribution of X_F^2 matches the tail of the $\chi^2_{(r-1)(c-1)}$ distribution. Rao and Scott (1981) show that X_F^2 and G_F^2 have a null χ^2 distribution if and only if all the deffs for the variances and covariances of the $\hat{p}_{ij}$ are equal. Otherwise, the variance of X_F^2 is larger than the variance of a $\chi^2_{(r-1)(c-1)}$ distribution, and p-values from X_F^2 are often a bit smaller than they should be (but closer to the actual p-values than if no correction was done at all).

EXAMPLE 10.6 In the Survey of Youth in Custody, let's look at the relationship between age and whether the youth was sent to the institution for a violent offense (using variable *crimtype, currviol* was defined to be 1 if *crimtype* = 1 and 0 otherwise). Using the weights, we estimate the proportion of the population falling in each cell:

		Age Class			
		≤ 15	16 or 17	≥ 18	Total
	No	0.1698	0.2616	0.1275	0.5589
Violent Offense?					
	Yes	0.1107	0.1851	0.1453	0.4411
	Total	0.2805	0.4467	0.2728	1.0000

First, let's look at what happens if we ignore the clustering and pretend that the test statistic in (10.2) follows a χ^2 distribution with 2 df. With $n = 2621$ youths in the table, Pearson's X^2 statistic is

$$X^2 = n\sum_{i=1}^{2}\sum_{j=1}^{3}\frac{(\hat{p}_{ij}-\hat{p}_{i+}\hat{p}_{+j})^2}{\hat{p}_{i+}\hat{p}_{+j}} = 34.12.$$

Comparing this to a χ_2^2 distribution yields an incorrect "*p*-value" of 3.9×10^{-8}.

The following design effects were estimated, using the stratification and clustering information in the survey:

Design Effects		Age Class			
		≤ 15	16 or 17	≥ 18	Total
Violent Offense?	No	14.9	4.0	3.5	6.8
	Yes	4.7	6.5	3.8	6.8
	Total	14.5	7.5	6.6	

Several of the deffs are very large, as might be expected because some facilities have mostly violent or mostly nonviolent offenders. All of the residents of facility 31, for example, are there for a violent offense. In addition, the facilities with primarily nonviolent offenders tend to be larger. We would expect the clustering, then, to have a substantial effect on the hypothesis test.

Using (10.9), we estimate $E[X^2]$ by 4.9 and use $X_F^2 = 2X^2/4.9 = 14.0$. Comparing 14.0/2 to an $F_{2,1690}$ distribution gives an approximate *p*-value of 0.001. SAS code for calculating the Rao–Scott test directly is given on the website. This *p*-value is probably still a bit too small, though, because of the wide disparity in the deffs. ∎

Rao and Scott (1981, 1984) also proposed a **second-order correction**—matching the mean and variance of the test statistic to the mean and variance of a χ^2 distribution, as done for ANOVA model tests by Satterthwaite (1946). Satterthwaite compared a test statistic T with skewed distribution to a χ^2 reference distribution by choosing a constant k and df ν so that $E[kT] = \nu$ and $V[kT] = 2\nu$ (ν and 2ν are the mean and variance of a χ^2 distribution with ν df). Here, letting $m = (r-1)(c-1)$, we know that $E[kX_F^2] = km$ and

$$V[kX_F^2] = V\left[\frac{kmX^2}{E(X^2)}\right] = \frac{V[X^2]k^2m^2}{[E(X^2)]^2},$$

so matching the moments gives

$$\nu = 2\frac{[E(X^2)]^2}{V[X^2]} \quad \text{and} \quad k = \frac{\nu}{m}.$$

Then,

$$X_S^2 = \frac{\nu X_F^2}{(r-1)(c-1)} \tag{10.10}$$

is compared to a χ^2 distribution with ν df. The statistic G_S^2 is formed similarly. Again, if the estimator of the variances of the $\hat{p}_{ij}$ has κ df, it works slightly better to compare X_S^2/ν or G_S^2/ν to an F distribution with ν and $\nu\kappa$ df.

In general, estimating $V[X^2]$ is somewhat involved, and requires the complete covariance matrix of the $\hat{p}_{ij}$'s, so the second-order correction often cannot be used when the data are only available in published tables. If the deffs are all similar, the first- and second-order corrections will behave similarly. When the deffs vary appreciably, however, *p*-values using X_F^2 may be too small, and X_S^2 may perform better. Exercise 18 tells how the second-order correction can be calculated.

10.3.3   Model-Based Methods for χ^2 Tests

All the methods above use the covariance estimates of the proportions to adjust the χ^2 tests. A model-based approach may also be used. We describe a model due to Cohen (1976) for a cluster sample with two observation units per cluster. Extensions and other models that have been used for cluster sampling are described in Altham (1976), Brier (1980), Rao and Scott (1981), Wilson and Koehler (1991), and Chowdhury and McGilchrist (2001). Many of these models assume that the deff is the same for each cell and margin.

EXAMPLE 10.7 Cohen (1976) presents an example exploring the relationship between gender and diagnosis with schizophrenia. The data consisted of 71 hospitalized pairs of siblings. Many mental illnesses tend to run in families, so we might expect that if one sibling is diagnosed with schizophrenia, the other sibling is more likely to be diagnosed with schizophrenia. Thus, any analysis that ignores the dependence among siblings is likely to give *p*-values that are much too small. If we just categorize the 142 patients by gender and diagnosis and ignore the correlation between siblings, we get the following table. Here, *S* means the patient was diagnosed with schizophrenia, and *N* means the patient was not diagnosed with schizophrenia.

	S	N	
Male	43	15	58
Female	32	52	85
	75	67	142

If analyzed using the assumption of multinomial sampling, $X^2 = 17.89$ and $G^2 = 18.46$. Such an analysis, however, assumes that all the observations are independent, so the "*p*-value" of 0.00002 is incorrect.

We know the clustering structure for the 71 clusters, though. You can see in Table 10.1 that most of the pairs fall in the diagonal blocks: If one sibling has schizophrenia, the other is more likely to have it. In 52 of the sibling pairs, either both siblings are diagnosed as having schizophrenia, or both siblings are diagnosed as not having schizophrenia.

Let q_{ij} be the probability that a pair falls in the (i, j) cell in the classification of the pairs. Thus, q_{11} is the probability that both siblings are schizophrenic and male, q_{12} is the probability that the younger sibling is a schizophrenic female and the older sibling is a schizophrenic male, etc. Then model the q_{ij}'s by

$$q_{ij} = \begin{cases} aq_i + (1-a)q_i^2 & \text{if } i = j \\ (1-a)q_i q_j & \text{if } i \neq j \end{cases} \tag{10.11}$$

where a is a clustering effect and q_i is the probability that an individual is in class i ($i =$ SM, SF, NM, NF). If $a = 0$, members of a pair are independent, and we can just do the regular chi-square test using the individuals—the usual Pearson's X^2, calculated ignoring the clustering, would be compared to a $\chi^2_{(r-1)(c-1)}$ distribution. If $a = 1$, the two siblings are perfectly correlated so we essentially have only one piece of information from each pair—$X^2/2$ would be compared to a $\chi^2_{(r-1)(c-1)}$ distribution.

TABLE **10.1**

Cluster Information for the 71 Pairs of Siblings

| | | \multicolumn{4}{c}{Younger Sibling} | | | |
		SM	SF	NM	NF	
	SM	13	5	1	3	22
Older	SF	4	6	1	1	12
Sibling	NM	1	1	2	4	8
	NF	3	8	3	15	29
		21	20	7	23	71

For a between 0 and 1, if the model holds, $X^2/(1 + a)$ approximately follows a $\chi^2_{(r-1)(c-1)}$ if the null hypothesis is true.

The model may be fit by maximum likelihood (see Cohen, 1976 for details). Then, $\hat{a} = .3006$, and the estimated probabilities for the four cells are the following:

	S	N	Total
Male	0.2923	0.1112	0.4035
Female	0.2330	0.3636	0.5966
Total	0.5253	0.4748	1.0000

We can check the model by using a goodness-of-fit test for the clustered data in Table 10.1. This model does not exhibit significant lack of fit, while the model assuming independence does. For testing whether gender and schizophrenia are independent in the 2 × 2 table, $X^2/1.3006 = 13.76$, which we compare to a χ^2_1 distribution. The resulting p-value is 0.0002, about 10 times as large as the p-value from the analysis that pretended siblings were independent. ∎

10.4
Loglinear Models

If there are more than two classification variables, we are often interested in seeing if there are more complex relationships in the data. Loglinear models are commonly used to study these relationships.

10.4.1 Loglinear Models with Multinomial Sampling

In a two-way table, if the row variable and the column variable are independent, then $p_{ij} = p_{i+}p_{+j}$. Equivalently,

$$\ln p_{ij} = \ln (p_{i+}) + \ln (p_{+j})$$
$$= \mu + \alpha_i + \beta_j,$$

where

$$\sum_{i=1}^{r} \alpha_i = 0 \quad \text{and} \quad \sum_{j=1}^{c} \beta_j = 0.$$

This is called a **loglinear model** because the logarithms of the cell probabilities follow a linear model. The model for independence in a 2×2 table may be written as

$$\mathbf{y} = \mathbf{X}\boldsymbol{\beta},$$

where

$$\mathbf{y} = \begin{bmatrix} \ln(p_{11}) \\ \ln(p_{12}) \\ \ln(p_{21}) \\ \ln(p_{22}) \end{bmatrix}, \quad \mathbf{X} = \begin{bmatrix} 1 & 1 & 1 \\ 1 & 1 & -1 \\ 1 & -1 & 1 \\ 1 & -1 & -1 \end{bmatrix}, \quad \text{and } \boldsymbol{\beta} = \begin{bmatrix} \mu \\ \alpha_1 \\ \beta_1 \end{bmatrix}.$$

The parameters $\boldsymbol{\beta}$ are estimated using the estimated probabilities $\hat{p}_{ij}$. For the data in Example 10.1, the estimated probabilities are as follows:

		Computer?		
		Yes	No	
Cable?	Yes	0.238	0.376	0.614
	No	0.176	0.210	0.386
		0.414	0.586	1.000

The parameter estimates are $\hat{\mu} = -1.428$, $\hat{\alpha}_1 = 0.232$, and $\hat{\beta}_1 = -0.174$. The fitted values of $\hat{p}_{ij}$ for the model of independence are then

$$\hat{p}_{ij} = \exp(\hat{\mu} + \hat{\alpha}_i + \hat{\beta}_j),$$

and are given in the following table:

		Computer?		
		Yes	No	
Cable?	Yes	0.254	0.360	0.614
	No	0.160	0.226	0.386
		0.414	0.586	1.000

We would also like to see how well this model fits the data. We can do that in two ways:

1 Test the goodness of fit of the model using either X^2 in (10.5) or G^2 in (10.6): For a two-way contingency table, these statistics are equivalent to the statistics for testing independence. For the computer/cable example, the likelihood ratio statistic for goodness of fit is 2.27. In multinomial sampling, X^2 and G^2 approximately follow a $\chi^2_{(r-1)(c-1)}$ distribution if the model is correct.

2 A full, or saturated, model for the data can be written as

$$\log p_{ij} = \mu + \alpha_i + \beta_j + (\alpha\beta)_{ij}$$

with $\sum_{i=1}^{r} (\alpha\beta)_{ij} = \sum_{j=1}^{c} (\alpha\beta)_{ij} = 0$. The last term is analogous to the interaction term in a two-way ANOVA model. This model will give a perfect fit to the observed cell probabilities because it has rc parameters. The null hypothesis of independence is equivalent to

$$H_0 : (\alpha\beta)_{ij} = 0 \text{ for } i = 1, \ldots, r-1; j = 1, \ldots, c-1.$$

Standard statistical software packages give estimates of the $(\alpha\beta)_{ij}$'s and their asymptotic standard errors under multinomial sampling. For the saturated model in the computer/cable example, SAS PROC CATMOD gives the following:

Effect	Parameter	Estimate	Standard Error	Chi-Square	Prob
CABLE	1	0.2211	0.0465	22.59	0.0000
COMP	2	-0.1585	0.0465	11.61	0.0007
CABLE*COMP	3	-0.0702	0.0465	2.28	0.1313

The values in the column "Chi-Square" are the Wald test statistics for testing whether that parameter is zero. Thus the p-value, under multinomial sampling, for testing whether the interaction term is zero is 0.1313—again, for this example, this is exactly the same as the p-value from the test for independence.

10.4.2 Loglinear Models in a Complex Survey

What happens in a complex survey? We obtain point estimates of the model parameters like we always do, by using weights. Thus, we estimate p_{ij} by (10.7), and calculate pseudo-maximum likelihood estimates of the loglinear model parameters incorporating the weights. If we use software that does not account for the survey design, however, the test statistics for goodness of fit and the asymptotic standard errors for the parameter estimates will be wrong.

Many of the same corrections used for χ^2 tests of independence can also be used for hypothesis tests in loglinear models. Rao and Thomas (2003) describe various tests of goodness of fit for contingency tables from complex surveys; these include Wald tests, jackknife, and Rao–Scott corrections to X^2 and G^2.

The Bonferroni inequality may also be used to compare nested loglinear models. For testing independence in a two-way table, for example, we compare the saturated model with the reduced model of independence and test each of the $m = (r-1)(c-1)$ null hypotheses

$$H_0(1) : (\alpha\beta)_{11} = 0$$

$$\vdots$$

$$H_0(m) : (\alpha\beta)_{(r-1)(c-1)} = 0$$

separately at level α/m.

More generally, we can compare any two nested loglinear models using this method. For a three-dimensional $r \times c \times d$ table, let

$$\mathbf{y} = [\ln(p_{111}) \ln(p_{112}) \ldots \ln(p_{rcd})]^T$$

Suppose the smaller model is

$$\mathbf{y} = X\boldsymbol{\beta}$$

and the larger model is

$$\mathbf{y} = X\boldsymbol{\beta} + Z\boldsymbol{\theta},$$

where $\boldsymbol{\theta}$ is a vector of length m. Then we can fit the larger model, and perform m separate hypothesis tests of the null hypotheses

$$H_0 : \theta_i = 0,$$

each at level α/m, by comparing $\hat{\theta}_i/\text{SE}(\hat{\theta}_i)$ to a t distribution.

EXAMPLE 10.8 Let's look at a three-dimensional table from the Survey of Youths in Custody, to examine relationships among the variables "Was anyone in your family ever incarcerated?" "Have you ever been put on probation or sent to a correctional institution for a violent offense?" and age. The cell probabilities are p_{ijk}. The estimated probabilities $\hat{p}_{ijk}$, estimated using weights, are in the following table:

		Family Member Incarcerated?				
		No		Yes		
		Ever Violent?		Ever Violent?		
		No	Yes	No	Yes	Total
Age Class	≤ 15	0.0588	0.0698	0.0659	0.0856	0.2801
	16–17	0.0904	0.1237	0.0944	0.1375	0.4461
	≥ 18	0.0435	0.0962	0.0355	0.0986	0.2738
	Total	0.1928	0.2896	0.1959	0.3217	1.0000

The saturated model for the three-way table is

$$\log p_{ijk} = \mu + \alpha_i + \beta_j + \gamma_k + (\alpha\beta)_{ij} + (\alpha\gamma)_{ik} + (\beta\gamma)_{jk} + (\alpha\beta\gamma)_{ijk}.$$

SAS PROC CATMOD, using the weights, gives the following parameter estimates for the saturated model:

```
                                          Standard     Chi-
Effect              Parameter  Estimate     Error     Square    Prob
---------------------------------------------------------------------
AGECLASS                    1   -0.1149   0.00980     137.45  0.0000
                            2    0.3441   0.00884    1515.52  0.0000
EVERVIOL                    3   -0.2446   0.00685    1275.26  0.0000
AGECLASS*EVERVIOL           4    0.1366   0.00980     194.27  0.0000
                            5    0.0724   0.00884      67.04  0.0000
FAMTIME                     6    0.0242   0.00685      12.51  0.0004
AGECLASS*FAMTIME            7    0.0555   0.00980      32.03  0.0000
                            8    0.0128   0.00884       2.10  0.1473
EVERVIOL*FAMTIME            9   -0.0317   0.00685      21.42  0.0000
AGECLAS*EVERVIOL*FAMTIME   10    0.00888  0.00980       0.82  0.3646
                           11    0.0161   0.00884       3.33  0.0680
```

Because this is a complex survey, and because SAS PROC CATMOD acts as though the sample size is $\sum w_i$ when the weights are used, the standard errors and p-values given for the parameters are completely wrong. But we can estimate the variance of each parameter by refitting the loglinear model on each of the random groups, and use the random group estimate of the variance to perform hypothesis tests on individual parameters. The random group standard errors for the 11 parameters are:

Parameter	Estimate	Standard error	Test statistic
1	−0.1149	0.1709	−0.67
2	0.3441	0.0953	3.61
3	−0.2446	0.0589	−4.15
4	0.1366	0.0769	1.78
5	0.0724	0.0379	1.91
6	0.0242	0.0273	0.89
7	0.0555	0.0191	2.91
8	0.0128	0.0218	0.59
9	−0.0317	0.0233	−1.36
10	0.0089	0.0191	0.47
11	0.0161	0.0167	0.96

The null hypothesis of no interactions among variables is

$$H_0 : (\alpha\beta)_{ij} = (\alpha\gamma)_{ik} = (\beta\gamma)_{jk} = (\alpha\beta\gamma)_{ijk} = 0;$$

or, using the parameter numbering in the output,

$$H_0 : \beta_4 = \beta_5 = \beta_7 = \beta_8 = \beta_9 = \beta_{10} = \beta_{11} = 0.$$

This null hypothesis has seven components; to use the Bonferroni test, we test each individual parameter at the $0.05/7$ level. The $(1 - .05/14)$ percentile of a t_6 distribution is 4.0; none of the test statistics $\hat{\beta}_i/\text{SE}(\hat{\beta}_i)$, for $i = 4, 5, 7, 8, 9, 10, 11$, exceed that critical value, so we would not reject the null hypothesis that all three variables are independent. We might want to explore the *ageclass* × *famtime* interaction further, however. ∎

10.5
Chapter Summary

Since many surveys collect categorical data, we often want to perform chi-square tests to explore association among variables. We can estimate probabilities in contingency tables using the sampling weights. Pearson and likelihood-ratio chi-square tests for association must be modified to account for the stratification and clustering in the survey design. Wald tests use the design-based variance so that the Wald test statistic approximately follows a chi-square distribution. The Rao–Scott test modifies the usual Pearson or likelihood-ratio test statistics by the average design effect to obtain corrected p-values.

Key Terms

Loglinear model: A model used for associations in categorical data.

Rao-Scott correction: A modification to a chi-square test statistic to account for the complex survey design.

Wald test statistic: A statistic for testing $H_0 : \theta = \theta_0$ of the form $(\hat{\theta} - \theta_0)^T [\hat{V}(\hat{\theta})]^{-1} (\hat{\theta} - \theta_0)$, in which the variance estimator accounts for the complex survey design.

For Further Reading

Agresti (2002) and Simonoff (2006) are good references on the analysis of categorical data in non-survey situations. The books edited by Skinner et al. (1989a) and Chambers and Skinner (2003) contain chapters on categorical data analysis on complex survey data.

A number of methods have been proposed to account for the survey design when testing for goodness of fit, homogeneity of populations, and independence of variables. Thomas et al. (1996) describe more than 25 methods that have been developed for testing independence in two-way tables and provide a useful bibliography. Some of these methods, and variations, are described in more detail in Rao and Thomas (1988, 1989, 2003). Fay (1985) describes an alternative method that involves jackknifing the test statistic itself. Scott (2007) reviews Rao–Scott corrections and outlines other areas of application.

10.6
Exercises

A. Introductory Exercises

1   Find an example or exercise in an introductory statistics textbook that performs a chi-square test on data from a survey. What design do you think was used for the survey? Is a chi-square test for multinomial sampling appropriate for the data? Why, or why not?

2   Read one of the articles listed in the file chapter10papers.html on the book website, or another research article in which a categorical data analysis is performed on survey data. Describe the sampling design and the method of analysis. Did the authors account for the design in their data analysis? Should they have analyzed the data differently?

3   Schei and Bakketeig (1989) took an SRS of 150 women between 20 and 49 years of age from Trondheim, Norway. Their goal was to investigate the relationship between sexual and physical abuse by a spouse and certain gynecological symptoms in the women. Of the 150 women selected to be in the sample, 15 had moved, 1 had died, 3 were excluded because they were not eligible for the study, and 13 refused to participate.

Of the 118 women who participated in the study, 20 reported some type of sexual or physical abuse from their spouse: eight reported being hit, two being kicked or bitten, seven being beaten up, and three being threatened or cut with a knife. Seventeen of the women in the study reported a gynecological symptom of irregular bleeding or pelvic pain. The numbers of women falling into the four categories of gynecological symptom and abuse by spouse are given in the following contingency table:

		Abuse		
		No	Yes	
	No	89	12	101
Gynecological Symptom Present?				
	Yes	9	8	17
		98	20	118

a If abuse and presence of gynecological symptoms are not associated, what are the expected probabilities in each of the four cells?

b Perform a χ^2 test of association for the variables abuse and presence of gynecological symptoms.

c What is the response rate for this study? Which definition of response rate did you use? Do you think that the nonresponse might affect the conclusions of the study? Explain.

4 Samuels (1996) collected data to examine how well students do in follow-up courses if the prerequisite course is taught by a part-time or full-time instructor. The following table gives results for students in Math I and Math II.

Instructor for Math I	Instructor for Math II	Grade in Math II		Total
		A, B, or C	D, F, or Withdraw	
Full Time	Full Time	797	461	1258
Full Time	Part Time	311	181	492
Part Time	Full Time	570	480	1050
Part Time	Part Time	909	449	1358
Total		2587	1571	4158

a The null hypothesis here is that the proportion of students receiving an A, B, or C is the same for each of the four combinations of instructor type. Is this a test of independence, homogeneity, or goodness of fit?

b Perform a hypothesis test for the null hypothesis in (a), assuming students are independent.

c Do you think the assumption that students are independent is valid? Explain.

B. Working with Survey Data

5 In Example 10.5 we used linearization to estimate $V(\hat{\theta})$. We can alternatively use the random group method to estimate the variance of $\hat{\theta}$. Calculate $\hat{p}_{11}\hat{p}_{22} - \hat{p}_{12}\hat{p}_{21}$ for each of the seven random groups, as discussed in Example 9.4, and find the variance of the seven nearly independent estimates of θ. Form the Wald statistic based on your estimated variance. Since the estimate of the variance from the random groups method has only six df, the test statistic should be compared to an $F_{1,6}$ distribution rather than to a χ_1^2 distribution. Do you reach the same conclusion as in Example 10.5?

6 Use the file winter.dat for this exercise. The data were first discussed in Exercise 19 of Chapter 3.

    a Test the null hypothesis that *class* is not associated with *breakaga*. In the context of Section 10.1, what type of sampling was done?

    b Now construct a 2×2 contingency table for the variables *breakaga* and *work*. Use the sampling weights to estimate the probabilities p_{ij} for each cell.

    c Calculate the odds ratio using the $\hat{p}_{ij}$ from (b). How does this compare with an odds ratio calculated using the observed counts (and ignoring the sampling weights)?

    d Estimate $\theta = p_{11}p_{22} - p_{21}p_{12}$ using the $\hat{p}_{ij}$ you calculated in (b).

    e Test the null hypothesis $H_0 : \theta = 0$.

    f How did the stratification affect the hypothesis test?

7 Use the file teachers.dat for this exercise. The data were first discussed in Exercise 15 of Chapter 5.

    a Construct a new variable *zassist*, which takes on the value 1 if a teacher's aide spends any time assisting the teacher, and 0 otherwise. Construct another new variable *zprep*, which takes on values Low, Medium, and High based on the amount of time the teacher spends in school on preparation.

    b Construct a 2×3 contingency table for the variables *zassist* and *zprep*. Use the sampling weights to estimate the probabilities p_{ij} for each cell.

    c Using the Rao–Scott method, test the null hypothesis that *zassist* is not associated with *zprep*.

8 The following data are from the Canada Health Survey, and given in Rao and Thomas (1989, p. 107). They relate smoking status (current smoker, occasional smoker, never smoked) to fitness level for 2505 persons. Smokers who had quit were not included in the analysis. The estimated proportions in the table below were found by applying the sample weights to the sample. The design effects are in brackets. We would like to test whether smoking status and fitness level are independent.

Smoking Status	Recommended	Fitness level Minimum acceptable	Unacceptable	
Current	0.220 [3.50]	0.150 [4.59]	0.170 [1.50]	0.540 [1.44]
Occasional	0.023 [3.45]	0.010 [1.07]	0.011 [1.09]	0.044 [2.32]
Never	0.203 [3.49]	0.099 [2.07]	0.114 [1.51]	0.416 [2.44]
Total	0.446 [4.69]	0.259 [5.96]	0.295 [1.71]	1.000

    a What is the value of X^2 if you assume the 2505 observations were collected in a multinomial sample? Of G^2? What is the p-value for each statistic under multinomial sampling, and why are these p-values incorrect?

    b Using (10.9) find the approximate expected value of X^2 and G^2.

    c Calculate the corrected statistics X_F^2 and G_F^2 for these data, and find p-values for the hypothesis tests. Does the clustering in the Canada Health Survey make a difference in the p-value you obtain?

9 The following data are from Rao and Thomas (1988), and were collected in the Canadian Class Structure Survey, a stratified multistage sample collected in 1982–83 to study employment and social structure. Canada was divided into 35 strata by region and population size; two psus were sampled in 34 of the strata, and one psu sampled in the 35th stratum. Variances were estimated using balanced repeated replication using the 34 strata with two psus. Estimated design effects are in brackets behind the estimated proportion for each cell.

	Males	Females	Total
Decision-making managers	0.103 [1.20]	0.038 [1.31]	0.141 [1.09]
Advisor-managers	0.018 [0.74]	0.016 [1.95]	0.034 [1.95]
Supervisors	0.075 [1.81]	0.043 [0.92]	0.118 [1.30]
Semi-autonomous workers	0.105 [0.71]	0.085 [1.85]	0.190 [1.44]
Workers	0.239 [1.42]	0.278 [1.15]	0.516 [1.86]
Total	0.540 [1.29]	0.460 [1.29]	

a What is the value of X^2 if you assume the 1463 persons were surveyed in a simple random sample? Of G^2? What is the p-value for each statistic under multinomial sampling, and why are these p-values incorrect?

b Using (10.9), find the approximate expected value of X^2 and G^2.

c How many df are associated with the BRR variance estimates?

d Calculate the first-order corrected statistics X_F^2 and G_F^2 for these data, and find approximate p-values for the hypothesis tests. Does the clustering in the survey make a difference in the p-value you obtain?

e The second-order Rao–Scott correction gave test statistic $X_S^2 = 38.4$, with 3.07 df. How does the p-value obtained using the X_S^2 compare with the p-value from X_F^2?

10 Using the data in syc.dat, define the variable *currprop* as 1 if *crimtype* = 2 and 0 otherwise. Perform a Rao–Scott test of whether *currprop* is associated with age group, for the groups given in Example 10.6. Also give the table of estimated probabilities for the cross-classification.

11 Using the NHANES data in nhanes.dat, categorize the respondents in three groups: normal, with body mass index (*bmxbmi*) less than 25; overweight, with $25 \le bmxbmi < 30$, and obese, with $bmxbmi \ge 30$. Also create two age groups, of persons under age 30 and persons at least 30 years old. Create a table of the estimated probabilities for the cross-classification of these two categorical variables, and perform a Rao–Scott test of association.

12 Using the data in ncvs2000.dat, test whether being a victim of violent crime is associated with gender.

C. Working with Theory

13 Some researchers have used the following method to perform tests of association in two-way tables. Instead of using the original observation weights w_k, define

$$w_k^* = \frac{n w_k}{\sum_{i \in S} w_i},$$

where n is the number of observation units in the sample. The sum of the new weights w_k^*, then, is n. The "observed" count for cell (i,j) is

$$x_{ij} = \text{sum of the } w_k^* \text{'s for observations in cell } (i,j)$$

and the "expected" count for cell (i,j) is $\hat{m}_{ij} = (x_{i+}x_{+j})/n$. Then compare the test statistic

$$\sum_{i=1}^{r}\sum_{j=1}^{c}\frac{(x_{ij}-\hat{m}_{ij})^2}{\hat{m}_{ij}}$$

to a $\chi^2_{(r-1)(c-1)}$ distribution.

Does this test give correct p-values for data from a complex survey? Why, or why not? HINT: Try it out on the data in Examples 10.1 and 10.4.

14 (Requires calculus.) Consider X_W^2 in (10.8).

   a Use the linearization method of Section 9.1 to approximate $V(\hat{\theta})$ in terms of $V(\hat{p}_{ij})$ and $\text{Cov}(\hat{p}_{ij},\hat{p}_{kl})$. Show that if we let $y_{ijk} = 1$ if observation k is in cell (i,j) and 0 otherwise, then $\hat{V}(\hat{\theta}) = \hat{V}(\hat{q})$, where $q_k = \hat{p}_{22}y_{11k} + \hat{p}_{11}y_{22k} - \hat{p}_{12}y_{21k} - \hat{p}_{21}y_{12k}$.

   b What is the Wald statistic, using the linearization estimate of $V(\hat{\theta})$ in (a), when multinomial sampling is used? (Under multinomial sampling, $V(\hat{p}_{ij}) = p_{ij}(1-p_{ij})/n$ and $\text{Cov}(\hat{p}_{ij},\hat{p}_{kl}) = -p_{ij}p_{kl}/n$.) Is this the same as Pearson's X^2 statistic?

15 (Requires calculus.) *Estimating the log odds ratio in a complex survey.* Let

$$\theta = \log\left(\frac{p_{11}p_{22}}{p_{12}p_{21}}\right) \quad \text{and} \quad \hat{\theta} = \log\left(\frac{\hat{p}_{11}\hat{p}_{22}}{\hat{p}_{12}\hat{p}_{21}}\right).$$

   a Use the linearization method of Section 9.1 to approximate $V(\hat{\theta})$ in terms of $V(\hat{p}_{ij})$ and $\text{Cov}(\hat{p}_{ij},\hat{p}_{kl})$.

   b Using (a), show that $\hat{V}(\hat{\theta}) = \dfrac{1}{x_{11}} + \dfrac{1}{x_{12}} + \dfrac{1}{x_{21}} + \dfrac{1}{x_{22}}$ under multinomial sampling.

16 In Section 10.3.1, we used a Wald test for $H_0 : \theta = 0$, where $\theta = \theta_{11} = p_{11}p_{22} - p_{12}p_{21}$. An equivalent null hypothesis is $H_0 : \eta = 0$, where $\eta = \log\left[(p_{11}p_{22})/(p_{12}p_{21})\right]$. Using the result of Exercise 15, derive the Wald test statistic for $H_0 : \eta = 0$.

17 Show that for multinomial sampling, $X_F^2 = X^2$. HINT: What is $E[X^2]$ in (10.9) for a multinomial sample?

18 (Requires mathematical statistics and theory of linear models.) *Deriving the first- and second-order corrections to Pearson's X^2 (see Rao and Scott, 1981).*

   a Suppose the random vector $\mathbf{Y}$ is normally distributed with mean $\mathbf{0}$ and covariance matrix Σ. Then, if $\mathbf{C}$ is symmetric and positive definite, show that $\mathbf{Y}^T\mathbf{C}\mathbf{Y}$ has the same distribution as $\sum \lambda_i W_i$, where the W_i's are independent χ_1^2 random variables and the λ_i's are the eigenvalues of $\mathbf{C}\Sigma$.

   b Let $\hat{\boldsymbol{\theta}} = (\hat{\theta}_{11},\ldots,\hat{\theta}_{1,(c-1)},\ldots,\hat{\theta}_{(r-1),1},\ldots,\hat{\theta}_{(r-1),(c-1)})^T$, where $\hat{\theta}_{ij} = \hat{p}_{ij} - \hat{p}_{i+}\hat{p}_{+j}$. Let $\mathbf{A}$ be the covariance matrix of $\hat{\boldsymbol{\theta}}$ if a multinomial sample of size n is taken and the null hypothesis is true. Using (a), argue that $\hat{\boldsymbol{\theta}}^T\mathbf{A}^{-1}\hat{\boldsymbol{\theta}}$ asymptotically has the same distribution as $\sum \lambda_i W_i$, where the W_i are independent χ_1^2 random variables, and the λ_i's are the eigenvalues of $\mathbf{A}^{-1} V(\hat{\boldsymbol{\theta}})$.

c What are $E[\hat{\boldsymbol{\theta}}^T \mathbf{A}^{-1} \hat{\boldsymbol{\theta}}]$ and $V[\hat{\boldsymbol{\theta}}^T \mathbf{A}^{-1} \hat{\boldsymbol{\theta}}]$ in terms of the λ_i's?

d Find $E[\hat{\boldsymbol{\theta}}^T \mathbf{A}^{-1} \hat{\boldsymbol{\theta}}]$ and $V[\hat{\boldsymbol{\theta}}^T \mathbf{A}^{-1} \hat{\boldsymbol{\theta}}]$ for a 2 × 2 table. You may want to use your answer in Exercise 14.

19 We know the clustering structure for the data in Example 10.7. Use results from Chapter 5 (assume one-stage cluster sampling) to estimate the proportion for each cell and margin in the 2 × 2 table, and find the variance for each estimated proportion. Now use estimated design effects to perform a hypothesis test of independence using X_F^2. How do the results compare to the model-based test?

D. Projects and Activities

20 *Trucks.* Use the VIUS data described in Exercise 34 of Chapter 3. Define the variable *heavy* to be 1 if the gross vehicle weight is higher than 10,000 pounds and 0 otherwise, and define the variable *autotran* to be 1 if the vehicle has automatic transmission and 0 otherwise. Using the sample weights, construct a 2 × 2 table of estimated probabilities for the cross-classification of *heavy* and *autotran*. What is the design effect for each estimated proportion? Carry out a Rao–Scott test for independence. How do the results compare with a Wald test for independence?

21 *Baseball data.* For the sample you selected in Exercise 37 of Chapter 5, define the variable *pitcher* to be 1 if the player is a pitcher and 0 otherwise, and the variable *million* to be 1 if the salary is greater than $1 million and 0 otherwise.

a Test whether the variables *pitcher* and *million* are associated, using the first-order Rao–Scott test.

b Using the sampling weights, estimate the log odds ratio.

22 *IPUMS exercises.* Use your sample from Exercise 28 of Chapter 7 for this problem.

a Create a categorical variable from *inctot* with two categories: low income and high income. Use the median income as the dividing point for the categories in the new variable *catinc*.

b Conduct hypothesis tests to explore whether *catinc* is associated with (i) *race* or (ii) *sex*. What method did you use to account for the complex sampling design?

23 *Activity for course project.* Return to the survey you explored in Exercise 31 of Chapter 7. Now consider two categorical responses in the survey. Construct a two-way table of estimated proportions, using the weight variable. Conduct a hypothesis test to explore whether these variables are associated. What method did you use to account for the complex sampling design?

11

Regression with Complex Survey Data

Now he knew that he knew nothing fundamental and, like a lone monk stricken with a conviction of sin, he mourned, "If I only knew more! ...Yes, and if I could only remember statistics!"

—Sinclair Lewis, *It Can't Happen Here*

EXAMPLE **11.1** How are maternal drug use and smoking related to birth weight and infant mortality? What variables are the best predictors of neonatal mortality? How is the birth weight of an infant related to that of older siblings?

In most of this book, we have emphasized estimating population means and totals—for example, how many low-birth-weight babies are born in the United States each year? Questions on the relation between variables, however, are often answered in statistics by using some form of a regression analysis. A response variable (for example, birth weight) is related to a number of explanatory variables (for example, maternal smoking, family income, and maternal age). We would like to be able to use the resulting regression equation not only to identify the relationship among variables for our data, but also to predict the value of the response for future infants, or for infants not included in the sample.

You know how to fit regression models if the "usual assumptions," reviewed in Section 11.1, are met. These assumptions are often not met for data from complex surveys, however. To answer the questions above, for example, you might want to use data from the 1988 Maternal and Infant Health Survey (MIHS) in the United States. The survey, collected by the U.S. Census Bureau for the National Center for Health Statistics, provides data on a number of factors related to pregnancy and infant health, including weight gain, smoking, and drug use during pregnancy; maternal exposure to toxic wastes and hazards; and complications during pregnancy and delivery (Sanderson et al., 1991). But, like most large-scale surveys, the MIHS is not a simple random sample (SRS). Stratified random samples were drawn from the 1988 vital records from the contiguous 48 states and the District of Columbia. The samples included 10,000 certificates of live birth from the 3,909,510 live births in 1988, 4000 reports of fetal death from the estimated 15,000 fetal deaths of 28 weeks' or more gestation, and 6000 certificates of death for infants under 1 year of age from the population of 38,910 such deaths. Because black infants have higher incidence of low birth weight and infant mortality than white infants, black infants had a higher sampling fraction than nonblack infants. Low-birth-weight infants were also over-sampled. Mothers in the sampled records were mailed a questionnaire asking about prenatal care; smoking, drinking, and drug use; family income; hospitalization; health of the baby; and a number of other related variables. After receiving permission from the mother, investigators also sent questionnaires to the prenatal care providers and hospitals, asking about the mother's and baby's health before and after birth. ■

As we found for analysis of contingency tables in the previous chapter, unequal probabilities of selection and the clustering and stratification of the sample complicate a statistical analysis. In the MIHS, the unequal inclusion probabilities for infants in different strata may need to be considered when fitting regression models. If a survey involves clustering, as does the National Crime Victimization Survey (NCVS), then standard errors for the regression coefficients calculated under the assumption that observations are independent will be incorrect.

In this chapter, we explore how to do regression in complex sample surveys. We review the traditional model-based approach to regression analysis, as taught in intro-ductory statistics courses, in Section 11.1. In Section 11.2, we discuss a design-based approach to regression, and present methods for calculating standard errors of regres-sion coefficients. Section 11.4 contrasts design-based and model-based approaches, Section 11.5 discusses a model-based approach, and Section 11.6 applies these ideas to logistic regression.

We already used regression estimation in Chapter 4. In Chapter 4, though, the emphasis was on using information in an auxiliary variable to increase the precision of the estimate of the population total, $t_y = \sum_{i=1}^{N} y_i$. In Sections 11.1 to 11.6 of this chapter, our primary interest is in exploring the relation among different variables, and thus in estimating the regression coefficients. In Section 11.7, we return to the use of regression for improving the precision of estimated totals.

11.1
Model-Based Regression in Simple Random Samples

As usually exposited in areas of statistics other than sampling, regression inference is based on a model that is assumed to describe the relationship between the explanatory variable, x, and the response variable, y. The straight-line model commonly used for a single explanatory variable is

$$Y_i \mid x_i = \beta_0 + \beta_1 x_i + \varepsilon_i, \tag{11.1}$$

where Y_i is a random variable for the response, x_i is an explanatory variable, and β_0 and β_1 are unknown parameters. The Y_i's are random variables; the data collected in the sample of size n are one realization of those n random variables, $\{y_i, i \in S\}$. The ε_i's, the deviations of the response variable about the line described by the model, are assumed to satisfy conditions (A1) through (A3):

(A1) $E[\varepsilon_i] = 0$ for all i. In other words, $E[Y_i|x_i] = \beta_0 + \beta_1 x_i$.

(A2) $V[\varepsilon_i] = \sigma^2$ for all i. The variance about the regression line is the same for all values of x.

(A3) $\text{Cov}[\varepsilon_i, \varepsilon_j] = 0$ for $i \neq j$. Observations are uncorrelated.

Often, (A4) is also assumed: It implies (A1) through (A3), and adds the additional assumption of normally distributed ε_i's.

(A4) Conditionally on the x_i's, the ε_i's are independent and identically distributed from a normal distribution with mean 0 and variance σ^2.

The **ordinary least squares** (OLS) **estimators** of the parameters are the values $\hat{\beta}_0$ and $\hat{\beta}_1$ that minimize the residual sum of squares $\sum [y_i - (\beta_0 + \beta_1 x_i)]^2$. Estimators of the slope β_1 and intercept β_0 are obtained by solving the **normal equations**: For the model in (11.1), these are

$$\beta_0 n \quad + \beta_1 \sum x_i = \sum y_i$$
$$\beta_0 \sum x_i + \beta_1 \sum x_i^2 = \sum x_i y_i.$$

Solving the normal equations gives the parameter estimators

$$\hat{\beta}_1 = \frac{\sum x_i y_i - \frac{1}{n}\left(\sum x_i\right)\left(\sum y_i\right)}{\sum x_i^2 - \frac{1}{n}\left(\sum x_i\right)^2} \tag{11.2}$$

$$\hat{\beta}_0 = \frac{1}{n}\sum y_i - \hat{\beta}_1 \frac{1}{n}\sum x_i.$$

Both $\hat{\beta}_1$ and $\hat{\beta}_0$ are linear in y, as we can write each in the form $\sum a_i y_i$ for known constants a_i. Although not usually taught in this form, it is equivalent to (11.2) to write

$$\hat{\beta}_1 = \sum_{i \in S} \left[\frac{x_i - \frac{1}{n}\sum x_j}{\sum x_j^2 - \frac{1}{n}\left(\sum x_j\right)^2} \right] y_i$$

and

$$\hat{\beta}_0 = \sum_{i \in S} \frac{1}{n} \left[1 - \frac{x_i \sum x_j - \frac{1}{n}\left(\sum x_j\right)^2}{\sum x_j^2 - \frac{1}{n}\left(\sum x_j\right)^2} \right] y_i.$$

If assumptions (A1) to (A3) are satisfied, then $\hat{\beta}_0$ and $\hat{\beta}_1$ are the **best linear unbiased estimators**—among all linear estimators that are unbiased under model (11.1), $\hat{\beta}_0$ and $\hat{\beta}_1$ have the smallest variance. If assumption (A4) is met, we can use the t distribution to construct confidence intervals (CIs) and hypothesis tests for the slope and intercept of the "true" regression line. Under assumption (A4),

$$\frac{\hat{\beta}_1 - \beta_1}{\sqrt{\hat{V}_M(\hat{\beta}_1)}}$$

follows a t distribution with $n - 2$ degrees of freedom (df). The subscript M refers to the use of the model to estimate the variance; for model (11.1), a model-unbiased estimator of the variance is

$$\hat{V}_M(\hat{\beta}_1) = \frac{\sum_{i \in S} (y_i - \hat{\beta}_0 - \hat{\beta}_1 x_i)^2 / (n - 2)}{\sum_{i \in S} (x_i - \bar{x})^2}. \tag{11.3}$$

The coefficient of determination R^2 in straight-line regression is

$$R^2 = 1 - \frac{\sum_{i \in S} (y_i - \hat{\beta}_0 - \hat{\beta}_1 x_i)^2}{\sum_{i \in S} (y_i - \bar{y})^2}.$$

EXAMPLE 11.2 To illustrate regression in the setting just discussed, we use data from Macdonell (1901), giving the length of the left middle finger (cm) and height (inches) for 3000 criminals. At the end of the nineteenth century, it was widely thought that criminal tendencies might also be expressed in physical characteristics that were distinguishable from the physical characteristics of noncriminal classes. Macdonell compared means and correlations of anthropometric measurements of the criminals to those of Cambridge men (presumed to come from a different class in society). This is an important data set in the history of statistics—it is the one Student (1908) used to demonstrate the t distribution. The entire data set for the 3000 criminals is in the file anthrop.dat.

An SRS of 200 individuals (file anthsrs.dat) was taken from the 3000 observations. Fitting a straight line model (SAS code is given on the website) with $y =$ height and $x =$ (length of left middle finger) results in the following output:

Variable	DF	Parameter Estimate	Standard Error	t Value	Pr > |t|
Intercept	1	30.31625	2.56681	11.81	<.0001
finger	1	3.04525	0.22172	13.73	<.0001

The sample data are plotted along with the OLS regression line in Figure 11.1. The model appears to be a good fit to the data ($R^2 = 0.49$), and, using the model-based analysis, a 95% CI for the slope of the line is

$$3.0453 \pm 1.972(0.2217) = [2.61, 3.48].$$

If we generated samples of size 200 from the model in (11.1) over and over again and constructed a CI for the slope for each sample, we would expect 95% of the resulting CIs to include the true value of β_1. ∎

Here are some remarks relevant to the application of regression to survey data:

1 No assumptions whatsoever are needed to calculate the estimates $\hat{\beta}_0$ and $\hat{\beta}_1$ from the data; these are simply formulas. The assumptions in (A1) to (A4) are needed to make *inferences* about the "true" but unknown parameters β_0 and β_1 and about

FIGURE 11.1

A plot of height vs. finger length for an SRS of 200 observations. The area of each circle is proportional to the number of observations at that value of (x, y). The OLS regression line, drawn in, has equation $y = 30.32 + 3.05x$.

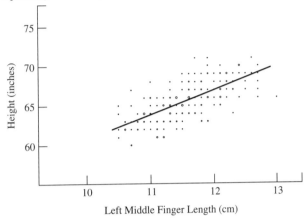

predicted values of the response variable. So the assumptions are used only when we construct a CI for β_1 or for a predicted value, or when we want to say, for example, that $\hat{\beta}_1$ is the best linear unbiased estimator of β_1.

The same holds true for other statistics we calculate. If we take a convenience sample of 100 persons, we may always calculate the average of those persons' incomes. But we cannot assess the accuracy of that statistic unless we make model assumptions about the population and sample. With a probability sample, however, we can use the sample design itself to make inferences and do not need to make assumptions about the model.

2   If the assumptions are not at least approximately satisfied, model-based inferences about parameters and predicted values will likely be wrong. For example, if observations are positively correlated rather than independent, the variance estimate from (11.3) is likely to be smaller than it should be. Consequently, regression coefficients are likely to be deemed statistically significant more often than they should be, as demonstrated in Kish and Frankel (1974).

3   We can partially check the assumptions of the model by plotting the residuals and using various diagnostic statistics as described in the regression books listed in the reference section. One commonly used plot is that of residuals versus predicted values, used to check (A1) and (A2). For the data in Example 11.2, this plot is shown in Figure 11.2, and gives no indication that the data in the sample violate assumptions (A1) or (A2). (This does not mean that the assumptions are true, just that we see nothing in the plot to indicate that they do not hold. Some of the assumptions, particularly independence, are quite difficult to check in practice.) However, we have no way of knowing whether observations not in the sample are fit by this model unless we actually see them.

4   Regression is not limited to variables related by a straight line. Let y be birth weight, and x take on the value 1 if the mother is black and 0 if the mother is

FIGURE   11.2

A plot of residuals for the model-based analysis of criminal height data, using the SRS plotted in Figure 11.1. No patterns are apparent.

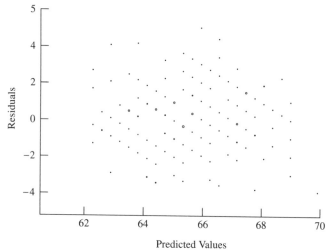

not black. Then the regression slope estimates the difference in mean birth weight for black and nonblack mothers, and the test statistic for $H_0 : \beta_1 = 0$ is the pooled t-test statistic for the null hypothesis that the mean birth weight for blacks is the same as the mean birth weight for nonblacks. Thus comparison of means for subpopulations, or domains, can be treated as a special case of regression analysis, as will be seen in Section 11.3.

11.2
Regression in Complex Surveys

Many investigators performing regression analyses on complex survey data simply run the data through standard software for the model in (11.1) and report the parameter estimates and standard errors given by the software. One may debate whether to take a model-based or design-based approach (and we shall, in Section 11.4), but the data structure needs to be taken into account in either approach.

What can happen in complex surveys?

1   Observations may have different inclusion probabilities, π_i. If π_i is related to the response variable y_i, then an analysis that does not account for the different probabilities of selection may lead to biases in the estimated regression parameters. This problem is discussed in detail by Nathan and Smith (1989), who give a bibliography of related literature.

For example, suppose that an unequal-probability sample of 200 men is taken from the population described in Example 11.2 and that the inclusion probabilities are higher for the shorter men. (For illustration purposes, I used the y_i's to set the inclusion probabilities, with π_i proportional to 24 for $y < 65$, 12 for $y = 65$, 2 for $y = 66$ or 67, and 1 for $y > 67$, with data in file anthuneq.dat.) Figure 11.3

FIGURE   **11.3**

A plot of y vs. x for an unequal-probability sample of 200 criminals. In this plot, the area of the circle is proportional to the number of observations at that data point—not to the sum of weights at the point. The OLS line, ignoring the sampling weights, is $y = 43.41 + 1.79x$. The smaller slope of this line, when compared to the slope 3.05 for the SRS in Figure 11.1, reflects the undersampling of tall men. The OLS regression estimators are biased for the population quantities because they do not incorporate the unequal sampling weights.

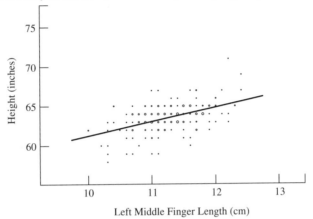

shows a scatterplot of the data from this sample, along with the ordinary least squares regression line described in Section 11.1. The OLS regression equation is $y = 43.41 + 1.79x$, compared with the equation $y = 30.32 + 3.05x$ for the SRS in Example 11.2. Ignoring the inclusion probabilities in this example leads to a very different estimate of the regression line and distorts the relationship in the population.

Nonrespondents can distort the relationship for much the same reason. If the non-respondents in the MIHS are more likely to have low-birth-weight infants, then a regression model predicting birth weight from explanatory variables may not fit the nonrespondents. Item nonresponse may have similar effects.

The stratification of the MIHS would also need to be taken into account. The survey was stratified because the investigators wanted to be sure to have an adequate sample size for blacks and low-birth-weight infants. It is certainly plausible that each stratum may have its own regression line, and postulating a single straight line to fit all the data may hide some of the information in the data.

2   Even if the estimators of the regression parameters are approximately design unbiased, the standard errors given by non-survey regression programs will likely be wrong if the survey design involves clustering. Usually, with clustering, the design effect (deff) for regression coefficients will be greater than 1.

11.2.1 Point Estimation

Traditionally, design-based sampling theory has been concerned with estimating quantities from a finite population, quantities such as $t_y = \sum_{i=1}^{N} y_i$ or $\bar{y}_U = t_y/N$.

FIGURE   11.4

A plot of the population of 3000 criminals. The area of each circle is proportional to the number of population observations at those coordinates. The population OLS regression line is $y = 30.18 + 3.06x$.

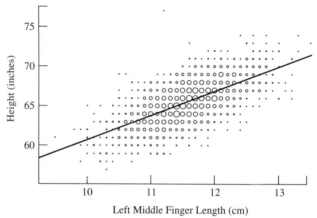

Left Middle Finger Length (cm)

In that descriptive spirit, then, the finite population quantities of interest for regression are the least squares coefficients for the population, B_0 and B_1, that minimize

$$\sum_{i=1}^{N} (y_i - B_0 - B_1 x_i)^2$$

over the entire finite population. It would be nice if the equation $y = B_0 + B_1 x$ summarizes useful information about the population (otherwise, why are you really interested in B_0 and B_1?), but no assumptions are necessary to say that these are the quantities of interest. As in Section 11.1, the normal equations are

$$B_0 N \quad + B_1 \sum_{i=1}^{N} x_i = \sum_{i=1}^{N} y_i$$

$$B_0 \sum_{i=1}^{N} x_i + B_1 \sum_{i=1}^{N} x_i^2 = \sum_{i=1}^{N} x_i y_i,$$

and B_0 and B_1 can be expressed as functions of the population totals:

$$B_1 = \frac{\sum_{i=1}^{N} x_i y_i - \frac{1}{N} \left(\sum_{i=1}^{N} x_i \right) \left(\sum_{i=1}^{N} y_i \right)}{\sum_{i=1}^{N} x_i^2 - \frac{1}{N} \left(\sum_{i=1}^{N} x_i \right)^2} = \frac{t_{xy} - \frac{t_x t_y}{N}}{t_{x^2} - \frac{(t_x)^2}{N}} \qquad (11.4)$$

$$B_0 = \frac{1}{N} \sum_{i=1}^{N} y_i - B_1 \frac{1}{N} \sum_{i=1}^{N} x_i = \frac{t_y - B_1 t_x}{N}. \qquad (11.5)$$

We know the values for the entire population for the sample drawn in Example 11.2. These population values are plotted in Figure 11.4, along with the population least squares line $y = 30.179 + 3.056x$.

As both B_0 and B_1 are functions of population totals, we can use methods derived in earlier chapters to estimate each total separately and then substitute the estimators into (11.4) and (11.5). We estimate each population total in (11.4) and (11.5) using weights, with $\hat{N} = \sum_{i \in S} w_i$, $\hat{t}_y = \sum_{i \in S} w_i y_i$, $\hat{t}_x = \sum_{i \in S} w_i x_i$, $\hat{t}_{xy} = \sum_{i \in S} w_i x_i y_i$, and $\hat{t}_{x^2} = \sum_{i \in S} w_i x_i^2$. Then,

$$
\hat{B}_1 = \frac{\sum_{i \in S} w_i x_i y_i - \dfrac{1}{\sum_{i \in S} w_i} \left(\sum_{i \in S} w_i x_i \right) \left(\sum_{i \in S} w_i y_i \right)}{\sum_{i \in S} w_i x_i^2 - \dfrac{1}{\sum_{i \in S} w_i} \left(\sum_{i \in S} w_i x_i \right)^2}
\tag{11.6}
$$

and

$$
\hat{B}_0 = \frac{\sum_{i \in S} w_i y_i - \hat{B}_1 \sum_{i \in S} w_i x_i}{\sum_{i \in S} w_i}.
\tag{11.7}
$$

Computational Note   Although (11.6) and (11.7) are correct expressions for the estimators, they are subject to roundoff error and are not as good for computation as other algorithms that have been developed. In practice, you should use professional software designed for estimating regression parameters in complex surveys. If you do not have access to such software, you can use any statistical regression package that calculates weighted least squares estimates. If you use weights w_i in the weighted least squares estimation, you will obtain the same point estimates as in (11.6) and (11.7); however, in complex surveys, the standard errors and hypothesis tests the software provides will be incorrect and should be ignored.

Plotting the Data   In any regression analysis, you *must* plot the data. Plotting multivariate data is challenging even for data from an SRS (Cook and Weisberg, 1994, discuss regression graphics in depth). Data from a complex survey design—with stratification, unequal weights, and clustering—have even more features to incorporate into plots. Some bivariate plots for survey data are discussed in Section 7.4.2. In Figure 11.5, we indicate the weighting by circle area.

EXAMPLE **11.3**   Let's estimate the finite population quantities B_0 and B_1 for the unequal-probability sample plotted in Figure 11.3. The point estimates, using the weights, are $\hat{B}_0 = 30.19$ and $\hat{B}_1 = 3.05$. If we ignored the weights and simply ran the observed data through a standard regression program such as SAS PROC REG, we get very different estimates: $\hat{\beta}_0 = 43.41$ and $\hat{\beta}_1 = 1.79$—the values in Figure 11.3.

Figure 11.5 shows why the weights, which were related to y, make a difference here. Taller men had lower inclusion probabilities and thus not as many of them appeared in the unequal-probability sample. However, the taller men that were selected had higher sampling weights; a 69-inch man in the sample represented 24 times

FIGURE 11.5

A plot of data from an unequal-probability sample. The area of each circle is proportional to the sum of the weights for observations with that value of x and y. Note that the taller men in the sample also have larger weights, so the slope of the regression line using weights is drawn upward. The regression line, calculated with the weights, is $y = 30.19 + 3.05x$.

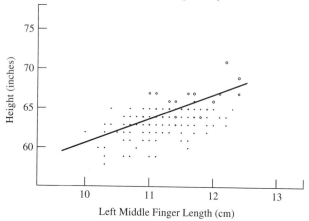

as many population units as a 60-inch man in the sample. When the weights are incorporated, estimates of the parameters are computed as though there were actually w_i data points with values (x_i, y_i). ∎

11.2.2 Standard Errors

Let's now examine the effect of the complex sampling design on the standard errors. As $\hat{B}_0$ and $\hat{B}_1$ are functions of estimated population totals, methods from Chapter 9 may be used to calculate variance estimates.

For any method of estimating the variance, under certain regularity conditions an approximate $100(1 - \alpha)\%$ CI for B_1 is given by

$$\hat{B}_1 \pm t_{\alpha/2}\sqrt{\hat{V}(\hat{B}_1)},$$

where $t_{\alpha/2}$ is the upper $\alpha/2$ point of a t distribution with df associated with the variance estimate. For linearization, jackknife, or balanced repeated replication (BRR) in a stratified multistage sample, we generally use (number of sampled psus) − (number of strata) as the df.

11.2.2.1 Standard Errors Using Linearization

The linearization variance estimator for the slope may be used because B_1 is a function of five population totals: from (11.4), $B_1 = h(t_{xy}, t_x, t_y, t_{x^2}, N)$, where

$$h(a, b, c, d, e) = \frac{a - bc/e}{d - b^2/e} = \frac{ea - bc}{ed - b^2}.$$

Using linearization, then, as you showed in Exercise 14 from Chapter 9,

$$V(\hat{B}_1) \approx V\left[\frac{\partial h}{\partial a}(\hat{t}_{xy} - t_{xy}) + \frac{\partial h}{\partial b}(\hat{t}_x - t_x) + \frac{\partial h}{\partial c}(\hat{t}_y - t_y) + \frac{\partial h}{\partial d}(\hat{t}_{x^2} - t_{x^2}) + \frac{\partial h}{\partial e}(\hat{N} - N)\right]$$

$$= V\left[\left\{t_{x^2} - \frac{(t_x)^2}{N}\right\}^{-1} \sum_{i \in S} w_i(y_i - B_0 - B_1 x_i)(x_i - \bar{x}_U)\right].$$

Define

$$q_i = (y_i - \hat{B}_0 - \hat{B}_1 x_i)(x_i - \hat{\bar{x}}),$$

where $\hat{\bar{x}} = \hat{t}_x/\hat{N}$. Then, we may use

$$\hat{V}_L(\hat{B}_1) = \frac{\hat{V}\left(\sum_{i \in S} w_i q_i\right)}{\left[\sum_{i \in S} w_i x_i^2 - \frac{\left(\sum_{i \in S} w_i x_i\right)^2}{\sum_{i \in S} w_i}\right]^2} \tag{11.8}$$

to estimate the variance of $\hat{B}_1$.

Note that the design-based variance estimator in (11.8) differs from the model-based variance estimator in (11.3), even if an SRS is taken. In an SRS of size n,

$$\hat{V}\left(\sum_{i \in S} w_i q_i\right) = \hat{V}(\hat{t}_q) = \left(1 - \frac{n}{N}\right) N^2 \frac{s_q^2}{n}$$

with

$$s_q^2 = \frac{1}{n - 1} \sum_{i \in S}(x_i - \bar{x}_S)^2 (y_i - \hat{B}_0 - \hat{B}_1 x_i)^2.$$

Thus, for an SRS, (11.8) gives

$$\hat{V}_L(\hat{B}_1) = \left(1 - \frac{n}{N}\right) \frac{n}{n - 1} \frac{\sum_{i \in S}(x_i - \bar{x}_S)^2 (y_i - \hat{B}_0 - \hat{B}_1 x_i)^2}{\left[\sum_{i \in S}(x_i - \bar{x}_S)^2\right]^2},$$

but from (11.3),

$$\hat{V}_M(\hat{\beta}_1) = \frac{\sum_{i \in S}(y_i - \hat{B}_0 - \hat{B}_1 x_i)^2}{(n - 2) \sum_{i \in S}(x_i - \bar{x})^2}.$$

Why the difference? The design-based estimator of the variance $\hat{V}_L$ comes from the inclusion probabilities of the design, while $\hat{V}_M$ comes from the average squared deviation over all possible realizations of the model. CIs constructed from the two variance estimates have different interpretations. With the design-based CI, $\hat{B}_1 \pm t_{\alpha/2}\sqrt{\hat{V}_L(\hat{B}_1)}$, the confidence level is $\sum u(S)P(S)$, where the sum is over all possible samples S that can be selected using the sampling design, $P(S)$ is the probability that sample S is selected, and $u(S) = 1$ if the CI constructed from sample S contains the population characteristic B_1 and $u(S) = 0$ otherwise. In an SRS, the design-based confidence level is the proportion of possible samples that result in a CI that includes B_1, from the set of all SRSs of size n from the finite population of fixed values $\{(x_1, y_1), (x_2, y_2), \ldots, (x_N, y_N)\}$.

For the model-based CI $\hat{\beta}_1 \pm t_{\alpha/2}\sqrt{\hat{V}_M(\hat{\beta}_1)}$, the confidence level is the expected proportion of CIs that will include β_1, from the set of all samples that could be generated from the model in (A1) to (A4). Thus the model-based estimator assumes that (A1) to (A4) hold for the infinite population mechanism that generates the data. The SRS design of the sample makes assumption (A3) (uncorrelated observations) reasonable. If a straight line model describes the relation between x and y, then (A1) is also plausible. A violation of assumption (A2) (equal variances), however, can have a large effect on inferences. The linearization design-based estimator of the variance is more robust to assumption (A2), as explored in Exercise 18.

EXAMPLE 11.4   For the SRS in Example 11.2, the model-based and design-based estimates of the variance are quite similar, as the model assumptions appear to be met for the sample and population. For these data, $\hat{B}_1 = \hat{\beta}_1$ because $w_i = 3000/200$ for all i; $\hat{V}_L(\hat{B}_1) = 0.048$ and $\hat{V}_M(\hat{\beta}_1) = (0.2217)^2 = 0.049$. In other situations, however, the estimates of the variance can be quite different; usually, if there is a difference, the linearization estimate of the variance is larger than the model-based estimate of the variance because the linearization estimate in (11.8) is valid whether the model is "correct" or not.

For the unequal-probability sample of 200 criminals in Example 11.3, define the new variable

$$q_i = (y_i - \hat{B}_0 - \hat{B}_1 x_i)(x_i - \hat{\bar{x}}) = (y_i - 30.1859 - 3.0541 x_i)(x_i - 11.51359).$$

(Note that $\hat{\bar{x}} = 11.51359$ is the estimate of $\bar{x}_U$ calculated using the unequal probabilities; the sample average of the 200 x_i's in the sample is 11.2475, which is quite a bit smaller.) Then $\hat{V}(\sum_{i \in S} w_i q_i) = 238{,}161$, and

$$\left[\sum_{i \in S} w_i x_i^2 - \frac{\left(\sum_{i \in S} w_i x_i\right)^2}{\sum_{i \in S} w_i} \right]^2 = 688{,}508,$$

so $\hat{V}_L(\hat{B}_1) = 0.35$. If the weights are ignored, then the ordinary least squares analysis gives $\hat{\beta}_1 = 1.79$ and $\hat{V}_M(\hat{\beta}_1) = 0.05$. The estimated variance is much smaller using the model, but $\hat{\beta}_1$ is biased as an estimator of B_1. Since an unequal-probability sample was taken, $\hat{V}_L(\hat{B}_1)$ should be used, giving a 95% CI of $[1.89, 4.22]$ for B_1.

These calculations can be done in SAS PROC SURVEYREG, using the code on the website. The following output is obtained:

Parameter	Estimate	Standard Error	t Value	Pr > |t|	95% Confidence Interval	
Intercept	30.1858583	6.64323949	4.54	<.0001	17.0856787	43.2860379
finger	3.0540995	0.58962334	5.18	<.0001	1.8913879	4.2168111 ∎

11.2.2.2 Standard Errors Using Jackknife

Suppose we have a stratified multistage sample, with weights w_i and H strata. A total of n_h psus are sampled in stratum h. Recall (see Section 9.3.2) that for jackknife iteration j in stratum h, we omit all observation units in psu j and recalculate the estimate using the remaining units. Define

$$w_{i(hj)} = \begin{cases} w_i & \text{if observation unit } i \text{ is not in stratum } h \\ 0 & \text{if observation unit } i \text{ is in psu } j \text{ of stratum } h \\ \dfrac{n_h}{n_h - 1} w_i & \text{if observation unit } i \text{ is in stratum } h \text{ but not in psu } j. \end{cases}$$

Then, the jackknife estimator of the with-replacement variance of $\hat{B}_1$ is

$$\hat{V}_{JK}(\hat{B}_1) = \sum_{h=1}^{H} \frac{n_h - 1}{n_h} \sum_{j=1}^{n_h} (\hat{B}_{1(hj)} - \hat{B}_1)^2, \tag{11.9}$$

where $\hat{B}_1$ is defined in (11.6) and $\hat{B}_{1(hj)}$ is of the same form but with $w_{i(hj)}$ substituted for every occurrence of w_i in (11.6).

EXAMPLE 11.5 For our two samples of size 200 from the 3000 criminals,

$$\hat{V}_{JK}(\hat{B}_1) = \frac{199}{200} \sum_{j=1}^{200} (\hat{B}_{1(j)} - \hat{B}_1)^2,$$

where $\hat{B}_{1(j)}$ is the estimated slope when observation j is deleted and the other observations reweighted accordingly. The difference between the SRS and the unequal-probability sample is in the weights. For the SRS, the original weights are $w_i = 3000/200$; consequently, $w_{i(j)} = 200 w_i / 199 = 3000/199$ for $i \neq j$. Thus for the SRS, $\hat{B}_{1(j)}$ is the OLS estimate of the slope when observation j is omitted. For the SRS, we calculate $\hat{V}_{JK}(\hat{B}_1) = 0.050$.

For the unequal-probability sample, the original weights are $w_i = 1/\pi_i$ and $w_{i(j)} = 200 w_i / 199$ for $i \neq j$. The new weights $w_{i(j)}$ are used to calculate $\hat{B}_{1(j)}$ for each jackknife iteration, giving $\hat{V}_{JK}(\hat{B}_1) = 0.461$. The jackknife estimated variance is larger than the linearization variance, as often occurs in practice. SAS code for using the jackknife to estimate the variance is on the website. ∎

11.2.3 Multiple Regression

Now let's give results for multiple regression in general. We rely heavily on matrix results found in linear models and regression books listed in the references at the end of the chapter.

Suppose we wish to find a relation between y_i and a p-dimensional vector of explanatory variables $\mathbf{x}_i$, where $\mathbf{x}_i = [x_{i1}, x_{i2}, \ldots, x_{ip}]^T$. We wish to estimate the p-dimensional vector of population parameters, $\mathbf{B}$, in the model $y = \mathbf{x}^T \mathbf{B}$. Define

$$
\mathbf{y}_U = \begin{bmatrix} y_1 \\ y_2 \\ \vdots \\ y_N \end{bmatrix} \quad \text{and} \quad \mathbf{X}_U = \begin{bmatrix} \mathbf{x}_1^T \\ \mathbf{x}_2^T \\ \vdots \\ \mathbf{x}_N^T \end{bmatrix}.
$$

The normal equations for the entire population are

$$
\mathbf{X}_U^T \mathbf{X}_U \mathbf{B} = \mathbf{X}_U^T \mathbf{y}_U,
$$

and the finite population quantities of interest are, assuming that $(\mathbf{X}_U^T \mathbf{X}_U)^{-1}$ exists,

$$
\mathbf{B} = (\mathbf{X}_U^T \mathbf{X}_U)^{-1} \mathbf{X}_U^T \mathbf{y}_U,
$$

the least squares estimates for the entire population.

Both $\mathbf{X}_U^T \mathbf{X}_U$ and $\mathbf{X}_U^T \mathbf{y}_U$ are matrices of population totals: $\mathbf{X}_U^T \mathbf{X}_U = \sum_{i=1}^{N} \mathbf{x}_i \mathbf{x}_i^T$ and $\mathbf{X}_U^T \mathbf{y}_U = \sum_{i=1}^{N} \mathbf{x}_i y_i$. The (j, k)th element of the $p \times p$ matrix $\mathbf{X}_U^T \mathbf{X}_U$ is $\sum_{i=1}^{N} x_{ij} x_{ik}$, and the kth element of the p-vector $\mathbf{X}_U^T \mathbf{y}_U$ is $\sum_{i=1}^{N} x_{ik} y_i$.

Thus, we can estimate the matrices $\mathbf{X}_U^T \mathbf{X}_U$ and $\mathbf{X}_U^T \mathbf{y}_U$ using weights. We estimate $\mathbf{X}_U^T \mathbf{X}_U = \sum_{i=1}^{N} \mathbf{x}_i \mathbf{x}_i^T$ by $\sum_{i \in \mathcal{S}} w_i \mathbf{x}_i \mathbf{x}_i^T$, and we estimate $\mathbf{X}_U^T \mathbf{y}_U = \sum_{i=1}^{N} \mathbf{x}_i y_i$ by $\sum_{i \in \mathcal{S}} w_i \mathbf{x}_i y_i$. Then, analogously to (11.6) and (11.7), define the estimator of $\mathbf{B}$ to be

$$
\hat{\mathbf{B}} = \left(\sum_{i \in \mathcal{S}} w_i \mathbf{x}_i \mathbf{x}_i^T \right)^{-1} \sum_{i \in \mathcal{S}} w_i \mathbf{x}_i y_i. \tag{11.10}
$$

Let

$$
\mathbf{q}_i = \mathbf{x}_i (y_i - \mathbf{x}_i^T \hat{\mathbf{B}}).
$$

Then, using linearization (see Exercise 20),

$$
\hat{V}\left(\hat{\mathbf{B}} \right) = \left(\sum_{i \in \mathcal{S}} w_i \mathbf{x}_i \mathbf{x}_i^T \right)^{-1} \hat{V}\left(\sum_{i \in \mathcal{S}} w_i \mathbf{q}_i \right) \left(\sum_{i \in \mathcal{S}} w_i \mathbf{x}_i \mathbf{x}_i^T \right)^{-1}. \tag{11.11}
$$

CIs for individual parameters may be constructed as

$$
\hat{B}_k \pm t \sqrt{\hat{V}(\hat{B}_k)},
$$

where t is the appropriate percentile from the t distribution.

EXAMPLE 11.6   Return to the NHANES data that we plotted in Section 7.4.2. Figure 7.21 displayed a trend line for *body mass index* plotted against *age* (see Exercise 25). We can, alternatively, fit a polynomial regression model to the sample. It appears from the data plots in Section 7.4.2 that a quadratic model might be reasonable to try. For this model, y_i = body mass index (variable *bmxbmi*) for person i and $\mathbf{x}_i = [1, x_i, x_i^2]^T$, where x_i = age for person i. SAS code to fit this model, on the website, produces the output below. Other options in SAS software for regression, and additional output, are explained in the code.

```
        Fit Statistics

R-square              0.2906
Root MSE              5.9326
Denominator DF            15
```

Estimated Regression Coefficients

Parameter	Estimate	Standard Error	t Value	Pr > |t|
Intercept	15.2978480	0.21381337	71.55	<.0001
age	0.5465938	0.01381176	39.57	<.0001
agesq	-0.0051407	0.00015865	-32.40	<.0001

The quadratic term is statistically significant in this model. (In fact, with large data sets, it is common to have many of the predictors be statistically significant because the sample size is so large.) From the output, the predicted regression model is

$$\hat{y}_i = 15.30 + 0.55 \, \text{age} - 0.005 \, \text{age}^2.$$

This model is not a perfect fit to the data. An examination of the residual plots (see Exercise 22) shows a pattern in the residuals that indicates another model might provide a better summary of the data, and other models are explored in the SAS code on the website. The values $\hat{B}_0$, $\hat{B}_1$, and $\hat{B}_2$ estimate the population quantities B_0, B_1, and B_2, which are the values that would minimize the sum of squares $\sum_{i=1}^{N} (y_i - B_0 - B_1 \text{age}_i - B_2 \text{age}_i^2)^2$ if the entire population were measured. Thus, the design-based estimates and standard errors are correct for inference about B_0, B_1, and B_2 even if the model for the population is not perfect.

SAS software estimates the value of R^2 for the data to be 0.2906. In regression with data from a random sample, R^2 is the percentage of variability in the data that is explained by the regression model. If we fit a regression model using ordinary least squares to every person in the population, we would have $R_U^2 = 1 - \text{SSW}/\text{SSTO}$, where $\text{SSW} = \sum_{i=1}^{N} (y_i - \hat{y}_i)^2$ and $\text{SSTO} = \sum_{i=1}^{N} (y_i - \bar{y}_U)^2$. We can estimate SSW and SSTO using weights as $\widehat{\text{SSW}} = \sum_{i \in S} w_i (y_i - \hat{y}_i)^2$ and $\widehat{\text{SSTO}} = \sum_{i \in S} w_i (y_i - \hat{\bar{y}})^2$, and estimate R_U^2 by $\hat{R}^2 = 1 - \widehat{\text{SSW}}/\widehat{\text{SSTO}}$. ■

11.2.4 Regression Using Weights versus Weighted Least Squares

Many regression textbooks discuss regression estimation using weighted least squares as a remedy for unequal variances. If the model generating the data is

$$Y_i = \mathbf{x}_i^T \beta + \varepsilon_i$$

with ε_i independent and normally distributed with mean 0 and variance σ_i^2, then ε_i/σ_i follows a normal distribution with mean 0 and variance 1. The weighted least squares estimator is

$$\hat{\beta}_{\text{WLS}} = (\mathbf{X}^T \mathbf{\Sigma}^{-1} \mathbf{X})^{-1} \mathbf{X}^T \mathbf{\Sigma}^{-1} \mathbf{y}$$

with $\boldsymbol{\Sigma} = \text{diag } (\sigma_1^2, \sigma_2^2, \ldots, \sigma_n^2)$. The weighted least squares estimator minimizes $\sum (y_i - x_i^T \beta)^2 / \sigma_i^2$, and gives observations with smaller variance more weight in determining the regression equation. If the model holds, then, under weighted least squares theory,

$$V_M(\hat{\beta}_{\text{WLS}}) = (\mathbf{X}^T \boldsymbol{\Sigma}^{-1} \mathbf{X})^{-1}.$$

We are **not** using weighted least squares in this sense, even though our point estimator is the same: $\hat{\mathbf{B}}$ is the value that minimizes $\sum w_i (y_i - x_i^T \mathbf{B})^2$. Our weights come from the sampling design, not from an assumed covariance structure. Our estimated variance of the coefficients is not $(\mathbf{X}^T \hat{\boldsymbol{\Sigma}}^{-1} \mathbf{X})^{-1}$, the estimated variance under weighted least squares theory, but is

$$\left(\sum_{i \in \mathcal{S}} w_i \mathbf{x}_i \mathbf{x}_i^T \right)^{-1} \hat{V} \left[\sum_{i \in \mathcal{S}} w_i \mathbf{x}_i (y_i - \mathbf{x}_i^T \hat{\mathbf{B}}) \right] \left(\sum_{i \in \mathcal{S}} w_i \mathbf{x}_i \mathbf{x}_i^T \right)^{-1}.$$

One may, of course, combine the weighted least squares approach as taught in regression courses with the finite population approach by defining the population quantities of interest to be

$$\mathbf{B} = (\mathbf{X}_U^T \boldsymbol{\Sigma}_U^{-1} \mathbf{X}_U)^{-1} \mathbf{X}_U^T \boldsymbol{\Sigma}_U^{-1} \mathbf{y}_U,$$

thus generalizing the regression model. This is essentially what is done in ratio estimation, using $\boldsymbol{\Sigma}_U = \text{diag } (x_1, x_2, \ldots, x_N)$, as will be shown in Example 11.13.

11.2.5 Software for Regression in Complex Surveys

Several software packages among those discussed in Section 9.6 will calculate regression coefficients and their standard errors for complex survey data. Before you use software written by someone else to perform a regression analysis on sample survey data, you should investigate how it deals with missing data. For example, if an observation is missing one of the x values, SAS PROC SURVEYREG excludes the observation from the analysis. If your survey has a large amount of item nonresponse on different variables, it is possible that you may end up performing your regression analysis using only twenty of the observations in your sample. You may want to consider the amount of item nonresponse as well as scientific issues when choosing covariates for your model.

Some surveys do not release enough information in the public use files to allow you to calculate estimated variances for regression coefficients. The public-use data set from the Current Population Survey, for example, contains weights for each household and person in the sample, but does not provide clustering information. Such surveys, however, often provide information on deffs for estimating population totals. In this situation, you can estimate the regression parameters using the provided weights. Then estimate the variance for the regression coefficients as though an SRS were taken, and multiply each estimated variance by an overall deff for population totals. In general, deffs for regression coefficients tend to be (but do not have to be) smaller than deffs for estimating population means and totals, so multiplying estimated variances of regression coefficients by the deff often results in a conservative estimate of the variance (see Skinner, 1989). Intuitively, this can be explained because a good regression model may control for some of the cluster-to-cluster variability in the response variable. For example, if part of the reason households in the same cluster tend to have more similar crime victimization experiences is the average income level of the neighborhood, then we would expect that adjusting for income in the regression might account for some of the cluster-to-cluster variability. The residuals from the model would then show less effect from the clustering.

11.3
Using Regression to Compare Domain Means

We often want to compare subgroups in a population. In Exercise 21 of Chapter 6, you showed that the method used to compare domain means in an SRS (namely, to form a statistic $(\hat{\bar{y}}_1 - \hat{\bar{y}}_2)/\sqrt{\hat{V}(\hat{\bar{y}}_1) + \hat{V}(\hat{\bar{y}}_2)}$ and compare that to a t distribution) can give incorrect inferences with data from a complex survey. If clusters contain units from both domains, then $\hat{\bar{y}}_1$ and $\hat{\bar{y}}_2$ are correlated so that $V(\hat{\bar{y}}_1 - \hat{\bar{y}}_2) \neq V(\hat{\bar{y}}_1) + V(\hat{\bar{y}}_2)$.

But have no fear—we can use regression to compare domain means and to fit one-way and factorial analysis of variance (ANOVA) models. To compare the means for two domains which together comprise the entire population, define a new variable x with $x_i = 1$ if observation i is in domain 1 and $x_i = 0$ if observation i is in domain 2. Then the population slope B_1 in a straight-line regression model is $B_1 = \bar{y}_{1U} - \bar{y}_{2U}$ and $\hat{B}_1 = \hat{\bar{y}}_1 - \hat{\bar{y}}_2$ (see Exercise 17). Consequently, $\hat{V}(\hat{B}_1) = \hat{V}(\hat{\bar{y}}_1 - \hat{\bar{y}}_2)$, and the 95% CI for B_1 is the 95% CI for the difference in domain means, $\bar{y}_{1U} - \bar{y}_{2U}$.

EXAMPLE 11.7 Let's compare the mean value of body mass index for men and women using the data in nhanes.dat (see Section 7.4.2 and Example 11.6). Create a variable x with $x_i = 1$ if person i is female and $x_i = 0$ if person i is male. Then fit the model $y = B_0 + B_1 x$. Partial output, from SAS code given on the website, follows:

Estimated Regression Coefficients

Parameter	Estimate	Standard Error	t Value	Pr > |t|	95% Confidence Interval	
Intercept	26.0044123	0.11636359	223.48	<.0001	25.7563891	26.2524354
x	0.3577122	0.20979752	1.71	0.1088	-0.0894606	0.8048851

The slope $\hat{B}_1$ is the difference in domain means $\hat{\bar{y}}_{\text{female}} - \hat{\bar{y}}_{\text{male}} = 26.362 - 26.004$. A 95% CI for $\bar{y}_{U,\text{female}} - \bar{y}_{U,\text{male}}$ is $[-0.089, 0.805]$. The difference in means is not significant at the 0.05 level. ■

Comparing k domain means is similar. You need to define $k-1$ indicator variables, with $x_{ij} = 1$ if observation i is in domain j and 0 otherwise, for $j = 1, \ldots, (k-1)$. The kth domain mean is estimated by $\hat{B}_0$.

FIGURE   **11.6**

Boxplot of body mass index for race-ethnicity groups defined by NHANES, incorporating the sampling weights.

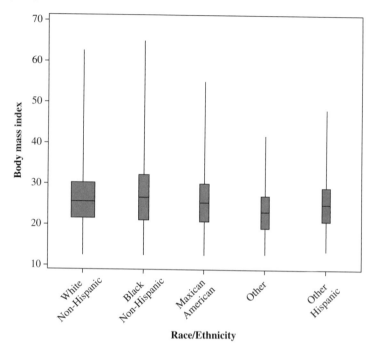

EXAMPLE 11.8   Let's compare the mean value of body mass index for the five ethnic groups (variable *ridreth2*) defined in the NHANES data. A side-by-side boxplot for these domains is shown in Figure 11.6. Output, constructed from SAS code given on the website, follows:

```
                 Tests of Model Effects

     Effect         Num DF      F Value      Pr > F

     Model             4          18.02      <.0001
     Intercept         1       25471.5       <.0001
     RIDRETH2          4          18.02      <.0001

     NOTE: The denominator degrees of freedom for the F tests
     is 15.
```

The F statistic for the null hypothesis that the mean body mass index of all five groups is the same is 18.02, indicating significant differences among the groups. You can also do pairwise comparisons of group means, and adjust the p-values for multiple testing by using a multiple comparisons method such as Bonferroni.   ∎

11.4
Should Weights Be Used in Regression?

In most areas of statistics, a regression analysis generally has one of three purposes:

1 It describes the relationship between two or more variables. Of interest may be the relationship between family income and the infant's birth weight or the relationship between education level, income, and likelihood of being a victim of violent crime. The interest is simply in a summary statistic that describes the association between the explanatory and response variables.

2 It predicts the value of y for a future observation. If we know the values for a number of demographic and health variables for an expectant mother, can we predict the birth weight of the infant, or the probability of the infant's survival?

3 It allows us to control future values of y by changing the values of the explanatory variables. For this purpose, we would like the regression equation to give us a cause-and-effect relationship between x and y.

Survey data can be used for the first and second purposes, but they generally cannot be used to establish definitive causal relationships among variables.[1] Sample surveys generally provide observational, not experimental, data. We observe a subset of possible explanatory variables, and these do not necessarily include the variables that are the root causes of changes in y. In a health survey intended to study the relationship between nutrition, exercise, and cancer incidence, survey participants may be asked about their diet and exercise habits (or the researcher may observe them) and be followed up later to see whether they have contracted cancer. Suppose that a regression analysis later indicates a significant negative association between vitamin E intake and cancer incidence, after adjusting for other variables such as age. The analysis only establishes association, not causation; you cannot conclude that cancer incidence will decrease if you start feeding people vitamin E. Although vitamin E could be the cause of the decreased cancer incidence, the cause could also be one of the unmeasured variables that is associated with both vitamin E intake and cancer incidence. To conclude that vitamin E affects cancer incidence, you need to perform an experiment: randomly assign study participants to vitamin E and no-vitamin-E groups, and observe the cancer incidence at a later time.

The purpose of a regression analysis often differs from that of an analysis to estimate population means and totals. When estimating the total number of unemployed persons from a survey, we are interested in the finite population quantity t_y; we want to estimate how many persons in the population in August 2004 were unemployed. But in a regression analysis, are you interested in B_0 and B_1, the summary statistics for the finite population? Or are you interested in uncovering a "universal truth"—to be able to say, for example, that not only do you find a positive association between

---

[1] Many statisticians would say that survey data cannot be used to make causal statements in any shape or form. Experimental units must be randomly assigned to treatments in order to infer causation. Some surveys, however, such as the study in Example 8.2, include experimentation, and for these we can often conclude that a change in the treatment caused a change in the response.

amount of fat in diet and systolic blood pressure for the population studied, but that you would expect a similar association in other populations. Cochran notes this point for comparison of domain means: "It is seldom of scientific interest to ask whether [the finite population domain means are equal], because these means would not be exactly equal in a finite population, except by rare chance. Instead, we test the null hypothesis that the two domains were drawn from *infinite* populations having the same mean" (1977, p. 39). Comparing domain means is a special case of linear regression, and Cochran's comments apply equally well to linear regression in general.

Many survey statisticians have debated whether the sampling weights are relevant for inference in regression; some of the papers discussing the issue are listed in the references at the end of this chapter. These references provide a much deeper discussion of the issues involved than we present in this section; we try to summarize the various approaches and present the contributions of each to a good analysis of survey data.

Two basic approaches have been advocated:

1   *Design-based.* The design-based approach was presented in Section 11.2. The quantities of interest are the finite population characteristics **B**, regardless of how well the model fits the population. Inferences are based on repeated sampling from the finite population, and the probability structure used for inference is that defined by the random variables indicating inclusion in the sample. A model that generates the data may exist, but we do not necessarily know what it is, so the analysis does not rely on any theoretical model. Weights are needed for estimating population means and totals, and by analogy should be used in linear regression as well.

2   *Model-based.* A stochastic model describes the relation between y_i and x_i that holds for every observation in the population. One possible model is $Y_i|\mathbf{x}_i = \mathbf{x}_i^T \boldsymbol{\beta} + \varepsilon_i$, with the ε_i's independent and normally distributed with constant variance. If the observations in the population really follow the model, then the sample design should have no effect as long as the inclusion probabilities depend on y only through the x's. The value **B** is merely the least squares estimate of β if values for the whole population were known; since only a sample is known, one should use the ordinary least squares estimators

$$\hat{\beta}_{\text{OLS}} = \left(\sum_{i \in \mathcal{S}} \mathbf{x}_i \mathbf{x}_i^T \right)^{-1} \sum_{i \in \mathcal{S}} \mathbf{x}_i y_i.$$

One searches for a model that can be thought to generate the population and then estimates the parameters for that model.

Särndal et al. (1992) adopt a *model-assisted* approach; for that approach, a model is used to specify the parameters of interest, but all inference is based on the survey design. Thus you fit a particular model because you believe it a plausible candidate for generating the population, but use the sampling weights to estimate the parameters and the sample design to estimate variances of the estimate. As inference is made using the sample design, we consider the model-assisted approach to be part of the design-based approach in this section.

The distinction among the approaches is important for the survey analyst because most regression programs use either a design-based or a model-based approach.

SAS PROC REG or the R function *lm* assume a model-based approach to regression, as exposited in Section 11.1. Survey software such as SUDAAN, WesVar, and SAS PROC SURVEYREG estimate the finite population parameters using the approach in Section 11.2. Thus it is important for you to know which approach you wish to take. Blindly running your data through software, without understanding what the software is estimating, can lead you to misinterpret the results.

Most statisticians agree that it is a good thing if a regression model describes the true state of nature. If it were known that a model would describe every possible observation involving x and y, then that model should be adopted. In the physical sciences, many models such as force = mass $\times$ acceleration can be theoretically derived. As long as you stay away from near-light velocity, any observation for which force, mass, and acceleration are accurately measured should be fit by the model. The design for how observations are sampled, then, should make little difference for finding the point estimates of regression coefficients, as every possible observation is described by the model.[2]

Unfortunately, theoretically derived models known to hold for all observations do not often exist for survey situations. An economist may conjecture a relationship between number of children, income, and amount spent on food, but there is no guarantee that this model will be appropriate for every subgroup in the population. Other variables may be related to the amount spent on food (such as educational level or amount of time away from home) but not measured in the survey. In addition, the true relation among the variables might not be exactly linear. Thus the main challenge to model-based inference is specifying the model.

If taking a model-based approach, then, you need to examine the model assumptions carefully and do everything you can to check the adequacy of the model for your data. This includes plotting the data and residuals, performing diagnostic tests, and using sampling designs that allow estimation of alternative models that may provide a better description of the relationship between variables. (Of course, you should also plot the data if adopting a design-based approach.) Inferences about observations not in the sample are based solely on the assumption that the model you have adopted applies to them, and you need to be very careful about generalizing outside of the sampled data. You must assume that the nonsampled population units can also be described by the model, and this is a very strong assumption.

Much is attractive about the model-based approach for regression: It links with sociological theories of the investigator, is consistent with other areas of statistics, and provides a mechanism for accounting for nonresponse. The model-based approach provides a framework for comparing theories about structural relationships. In addition, model-based estimates can be used with relatively small samples and with nonprobability samples. Although design-based inference does not depend on model assumptions, it does require large sample sizes in practice to be able to construct CIs. The standard errors of the model-based parameter estimates are generally lower than those of the corresponding design-based estimates.

But model misspecification and omitted covariates are of concern for a model-based analysis, and missing covariates may not show up in standard residual analyses. Moreover, in a complex survey design, the needed missing predictors may be related

---

[2]The sampling design, however, can affect the variances of the point estimates.

to the design and the survey weights. For example, for our unequal-probability sample in Figure 11.3, the inclusion probabilities we used depended on the value of y. Now, you can think of height as being determined by many, many variables $x_1, x_2, \ldots$; but the data set has only one of those possible explanatory variables. If all the other variables were included in the model, then the unequal selection probabilities would be irrelevant; because they are not, however, the inclusion probabilities π_i have useful information for estimating the regression slope.

Pfeffermann and Holmes (1985) and DuMouchel and Duncan (1983) argue that using sampling weights in regression can provide robustness to model misspecification: The weighted estimates are relatively unaffected if some independent variables are left out of the model.[3] Kott (1991) argues that sampling weights are needed in linear regression because the choice of covariates in survey data is limited to variables collected in the survey: If necessary covariates are omitted, $\hat{B}$ and $\hat{\beta}_{OLS}$ are both biased estimators of β, but the bias of $\hat{B}$ is a decreasing function of the sample size, while $\hat{\beta}_{OLS}$ is only asymptotically unbiased if the probabilities of selection are not related to the missing covariates. Rubin (1985), Smith (1988), and Little (1991) adopt a model-based perspective but argue that sampling weights are useful in model-based inference as summaries of covariates describing the mechanism by which units are included in the sample. Rubin-Bleuer and Kratina (2005) provide a rigorous mathematical framework for inference under both model-based and design-based approaches.

One point is clear: If the model you are using really does describe the mechanism generating the data, then the finite population quantity **B** should be close to the theoretical parameter β. Thus, if the model is a good one, we would expect that the point estimate of β using the model should be similar to the point estimate $\hat{\mathbf{B}}$ calculated using sampling weights. We suggest fitting a model both with and without weights. If the parameter estimates differ, then you should explore alternatives to the model you have adopted. A difference in the weighted and unweighted estimates can tell you that the proposed model does not fit well for part of the population. Lohr and Liu (1994) explore this issue for the NCVS.

EXAMPLE 11.9   Korn and Graubard (1995) illustrate the difference that including weights can make in a regression analysis, using data from the live-birth component of the 1988 MIHS. As mentioned in Example 11.1, black infants and low-birth-weight infants are over-sampled, so their sampling weights are lower than the weights for white, normal-birth-weight infants. Figure 11.7 shows a plot of the data and estimated regression line when weights are used in calculating the regression parameters; Figure 11.8 ignores the weights. The weighted regression pulls the regression line to where the population is estimated to be; in the unweighted regression, the line provides the best least-squares fit to the sample data, but does not describe the population as well. It is clear from examining the plots that the regression lines differ to such an extent because a straight-line model is not appropriate for the data; if a quadratic regression were fit instead, then the models from the weighted and unweighted regressions would show greater agreement. In this example, then, the differences between the

---

[3] But this robustness comes at a price; as mentioned earlier, the design-based variance is generally larger than the model-based variance. Kish (1992) gives a good overview of the variance inflation due to using weighted estimates rather than estimates without weights.

FIGURE **11.7**

Plot of weighted mean gestational age versus weighted mean birthweight for successive groups of approximately 500 observations. Areas of bubbles are proportional to the estimated population sizes of the groups. The straight line is the weighted linear regression fit to the original (ungrouped) data.

SOURCE: From "Examples of Differing Weighted and Unweighted Estimates from a Sample Survey," by E.L. Korn and B.I. Graubard, 1995, *The American Statistician*, 49, pp. 291–295. Copyright © 1995 American Statistical Association. Reprinted by permission.

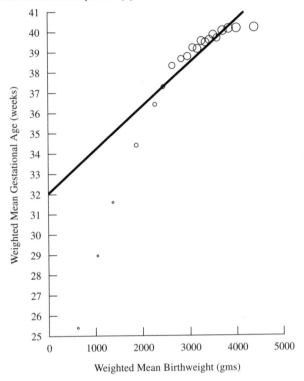

parameter estimates with weights and without weights arise because the straight-line model adopted is inappropriate. ∎

Each of the approaches to inference about regression parameters in complex surveys can be appropriate, depending upon the desired use of the regression model. You may want to consider the following questions when deciding upon your approach:

1   Are you performing a regression to generate official statistics that will be used to determine public policy? If so, you may want to use the weights to estimate parameters and the design to make inferences about the parameters. If you are using weights to estimate population and domain means, you may also want to use them to estimate regression parameters, so that the results from different analyses are consistent (see Alexander, 1991). As noted above, **B** should be close to β for a good model and large finite population, so a design-based estimate of **B** should also estimate β.

FIGURE **11.8**

Plot of weighted mean gestational age versus weighted mean birthweight for successive groups of approximately 500 observations. Areas of bubbles are proportional to the sample sizes of the groups. The straight line is the unweighted linear regression fit to the original (ungrouped) data.

SOURCE: From "Examples of Differing Weighted and Unweighted Estimates from a Sample Survey," by E.L. Korn and B.I. Graubard, 1995, *The American Statistician*, 49, pp. 291–295. Copyright © 1995 American Statistical Association. Reprinted by permission.

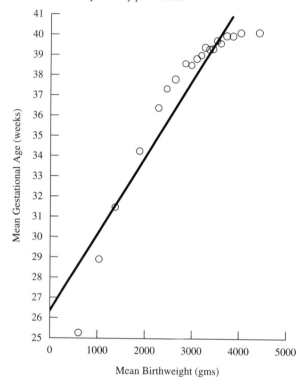

2 Was a probability sample taken? If not, then you must use a model-based approach.

3 How large is the sample size? The design-based theory relies on large sample sizes to make inferences about the parameters. If you have a small sample, you should probably use a model-based approach.

However, a mistake is often made by investigators who have heard the message that sampling weights are irrelevant in regression analysis but have ignored the rest of the discussion: They ignore the weights *and* the clustering in the data by simply running the survey data through standard regression software. This is incorrect under any approach: Whether or not weights are used to construct an estimator, the dependence in the data reflected in the clustering *must* be considered when calculating standard errors. A model-based approach that incorporates the positive correlation among observations in the same cluster is discussed in the next section.

11.5
Mixed Models for Cluster Samples

In Chapter 5 we discussed using a random effects model as a superpopulation model for cluster sampling. We can use this approach for regression analyses as well, by allowing different clusters to have their own regression equations, but relating the different regression equations for the clusters through a model.

EXAMPLE 11.10 The National Assessment of Educational Progress (NAEP) collects data on student background and achievement in the United States. It is sometimes referred to as "The Nation's Report Card," as it provides a scale for measuring student progress and comparing student achievement among different states and over time. A wealth of information is collected for each student, teacher, and participating school. In addition to proficiency scores for various subjects, the student-level data include information on the student's gender, race, ethnicity, courses taken, and variables related to socioeconomic status. School-level information includes fiscal resources, instructional methods, student-body characteristics and expectations of academic achievement.

The NAEP data can be used to identify school- and student-level variables that are associated with mathematics achievement among eighth-grade students. For simplicity, let's consider one student-level characteristic, gender; and one school-level characteristic, average amount of time spent in class on math tests. In practice, of course, you would probably include more variables in the model, as you would expect a number of characteristics to be associated with the tested mathematics achievement. Let Y_{ij} be the mathematics proficiency score of student j at school i in the sample, and let $x_{ij} = 1$ if student j at school i is female, and 0 if student j and school i is male.

We expect a clustering effect in these data—measuring all variables that might be associated with student achievement scores in mathematics is impossible, and the unmeasured characteristics of the schools, teachers, and neighborhoods induce a positive correlation in the test scores within a school. For example, the seventh- and eighth-grade mathematics teacher in one school might be superb at inspiring students to learn mathematics, but that excellence would not be recorded in the survey. The students from that class might then all perform better than average on the proficiency test, so their scores are more similar, even after adjusting for known covariates, than scores of a random sample of students from the population. When unmeasured characteristics such as these are considered over all schools, the result is a positive intraclass correlation coefficient.

Thus a model $Y_{ij} = \beta_0 + x_{ij}\beta_1 + \varepsilon_{ij}$, with ε_{ij} independent random variables with mean 0 and variance σ^2, is likely to be inappropriate for these data. If this inappropriate model is adopted, the calculated p-values for parameter estimates will be far too small. In addition, the model does not allow for different relations between gender and test score in different schools—which may occur, as some schools may encourage students of one gender more than students of the other gender.

A model that incorporates cluster effects and allows schools to have different slopes for gender is:

$$Y_{ij} = \beta_{0i} + (x_{ij} - \bar{x}_i)\beta_{1i} + \varepsilon_{ij}.$$

Here, the ε_{ij}'s are assumed to be independent $N(0, \sigma^2)$ random variables; the mean of x_{ij} for school i, $\bar{x}_i$, is subtracted from each x_{ij} so that β_{0i} can be interpreted as the average test score in school i. School i has its own straight line regression model with intercept β_{0i} and slope β_{1i}. But the slopes and intercepts from different schools are also related through a model. A simple model for the slopes and intercepts allows them to essentially be randomly distributed about a mean:

$$\beta_{0i} = \beta_0 + \delta_{0i}; \quad \beta_{1i} = \beta_1 + \delta_{1i},$$

with δ_{0i} and δ_{1i} following a bivariate normal distribution with $E_M[\delta_{0i}] = E_M[\delta_{1i}] = 0$, $V_M[\delta_{0i}] = \tau_{00}$, $V_M[\delta_{1i}] = \tau_{11}$, and Cov $(\delta_{0i}, \delta_{1i}) = \tau_{01}$. Under this situation, then the model may be written as

$$Y_{ij} = \beta_0 + (x_{ij} - \bar{x}_i)\beta_1 + \delta_{0i} + (x_{ij} - \bar{x}_i)\delta_{1i} + \varepsilon_{ij}. \tag{11.12}$$

The parameter β_0 represents the mean test score for schools; β_1 represents the mean slope for gender for schools. The random effects δ_{0i} and δ_{1i} represent the difference in the intercept and slope between school i and the average values for intercept and slope for all schools; they measure the school effect. Finally, ε_{ij} refers to additional deviation from the mean due to the individual student, after the effect of gender and school have been accounted for.

Note that if $\tau_{00} = \tau_{11} = 0$, there is no school effect on test score, and the model then reduces to a regular straight line regression model. In most applications, however, the slopes and intercepts will vary from school to school. ∎

The model in (11.12) is an example of a **mixed linear model**; it has both fixed (β_0 and β_1) and random (δ_{0i}, δ_{1i}, and ε_{ij}) coefficients. In econometrics, (11.12) is often referred to as a **random-coefficient** regression model; in the social sciences, it is called a **multilevel** or **hierarchical** linear model. Demidenko (2004) and Jiang (2007) describe the theory of mixed models. Pfeffermann et al. (1998) and Rabe-Hesketh and Skrondal (2006) discuss using mixed models with survey data. These models may be fit in SAS PROC MIXED or in specialized software packages.

The mixed model in (11.12) is a superpopulation model and is assumed to hold for all schools and students in the population. One advantage of using such a model is that it does not require that the schools be randomly selected, as long as the model describes the population. A mixed model approach is also congenial to testing different theories about mathematics education.

The model in (11.12) may also be used as a starting point for further investigation. The random effects δ_{0i} and δ_{1i} may be estimated for each school; the investigator may want to examine schools with unusually high or low values to try to conjecture why those schools might be different. The investigator may also want to include other predictor variables when estimating the intercepts and slopes for the different schools. For example, it might be conjectured that having more math tests at a school might lead to better mathematics proficiency scores, and might also lead to a smaller gender difference in the school. This extra predictor can easily be included in the mixed model. Let z_i be the average amount of time spent on math tests at school i. Then the intercept and slope at school i can be modeled as

$$\beta_{0i} = \beta_0 + \gamma_0 z_i + \delta_{0i}; \quad \beta_{1i} = \beta_1 + \gamma_1 z_i + \delta_{1i};$$

γ_0 then represents the effect of time spent on math tests on the intercept, and δ_{0i} represents the remaining school effect after adjusting for z_i.

11.6
Logistic Regression

In linear regression, the response variable is usually considered to be approximately continuous—for example, birth weight, income, or leaf area. In surveys, however, many variables of interest are dichotomous, with y_i taking only values of 1 (yes) or 0 (no). **Logistic regression** (see Hosmer and Lemeshow, 2000, for a general reference) is often used to predict probabilities of having response 1 for dichotomous variables.

First let's review logistic regression from a model-based viewpoint. Let $\mathbf{x}$ be a vector of explanatory variables and $\boldsymbol{\beta}$ be the vector of unknown parameters. Then the standard logistic regression model takes the form

$$p(\mathbf{x}) = \frac{\exp(\mathbf{x}^T \boldsymbol{\beta})}{1 + \exp(\mathbf{x}^T \boldsymbol{\beta})}, \tag{11.13}$$

where $p(\mathbf{x})$ represents the probability that a unit with covariates $\mathbf{x}$ will have a response of 1. Alternatively, the model may be expressed in logit scale, where $\text{logit}(p) = \ln[p/(1-p)]$:

$$\text{logit}[p(\mathbf{x})] = \mathbf{x}^T \boldsymbol{\beta}. \tag{11.14}$$

EXAMPLE 11.11 For the data in Example 10.1, let $y_i = 1$ if household i has a computer and $y_i = 0$ if household i does not have a computer. Let $x_i = 1$ if household i subscribes to cable and $x_i = 0$ if household i does not subscribe to cable. The fitted logistic regression model is

$$\widehat{\text{logit}}[p_i] = -0.177 - 0.281x_i.$$

Note that the slope, -0.281, is the log odds ratio from Example 10.1. It is easy to transform back to predicted conditional probabilities: When $x = 1$, then $\ln[\hat{p}/(1-\hat{p})] = -0.4573184$ so that

$$\hat{p}(1) = \frac{\exp(-0.4573184)}{1 + \exp(-0.4573184)} = 0.388 = \frac{119}{307}.$$

SAS code for calculating the parameter estimates is on the website. ∎

Much of the previous discussion in this chapter on linear regression also applies to logistic regression—a complex sample design will affect standard errors of the logistic regression coefficients, just as it affects standard errors of the linear regression coefficients. Logistic regression with one dichotomous independent variable is essentially equivalent to finding the odds ratio in a 2×2 contingency table, so the discussion in Chapter 10 about how the sampling design affects standard goodness-of-fit tests also applies to testing the significance of logistic regression coefficients.

Binder (1983), Chambless and Boyle (1985), and Roberts et al. (1987) give design-based theory for estimating logistic regression parameters. Just as the design-based theory for linear regression started with defining the population quantities of interest using the normal equations, here the quantities of interest are defined in terms of the likelihood function that would be adopted if the entire population were available for study. If there are N units in the population, this likelihood (assuming independence) is

$$L(\boldsymbol{\beta}) = \prod_{i=1}^{N} p_i^{y_i} (1 - p_i)^{1 - y_i}, \tag{11.15}$$

where $p_i = \exp(\mathbf{x}_i^T \boldsymbol{\beta})/[1 + \exp(\mathbf{x}_i^T \boldsymbol{\beta})]$ represents the probability that a unit with covariates $\mathbf{x}_i$ has a response of 1. The finite population parameter $\mathbf{B}$ is then defined to be the maximum likelihood estimate of $\boldsymbol{\beta}$ using (11.15). The parameter $\mathbf{B}$ is the solution to the system of equations

$$\sum_{i=1}^{N} x_{ij} \left[y_i - \frac{\exp(\mathbf{x}_i^T \mathbf{B})}{1 + \exp(\mathbf{x}_i^T \mathbf{B})} \right] = 0 \quad \text{for } j = 1, \ldots, p \tag{11.16}$$

if all elements in the population could be observed.

Now that $\mathbf{B}$ is defined, calculate $\hat{\mathbf{B}}$ by substituting estimators for the population totals in (11.16). A design-based estimator of $\mathbf{B}$ is given by the solution $\hat{\mathbf{B}}$ to

$$\sum_{i \in S} w_i x_{ij} \left[y_i - \frac{\exp(\mathbf{x}_i^T \hat{\mathbf{B}})}{1 + \exp(\mathbf{x}_i^T \hat{\mathbf{B}})} \right] = 0 \quad \text{for } j = 1, \ldots, p, \tag{11.17}$$

where S denotes the units included in the sample. The ith observation in the sample represents w_i observations in the population.

Variance estimation for logistic regression is discussed in the references cited above. The coefficients $\hat{\mathbf{B}}$ are defined implicitly in (11.17), so a linearization variance estimator may be obtained using methods in Binder (1983). Rao et al. (1998) present a modified version of score tests for testing the significance of logistic regression coefficients. Any of the resampling methods in Chapter 9 may be used to estimate the variance of logistic regression coefficients.

EXAMPLE 11.12 Consider using logistic regression to predict the event that body mass index > 25 from the triceps skinfold measurement (variable *bmxtri*), using the data in nhanes.dat. Partial output from SAS PROC SURVEYLOGISTIC code on the website is given below; other output given by SAS is explained in comments in the code.

Analysis of Maximum Likelihood Estimates

Parameter	DF	Estimate	Standard Error	Wald Chi-Square	Pr > ChiSq
Intercept	1	-2.6802	0.1237	469.1545	<.0001
BMXTRI	1	0.1496	0.00751	397.3564	<.0001

Odds Ratio Estimates

Effect	Point Estimate	95% Wald Confidence Limits	
BMXTRI	1.161	1.144	1.179

The Wald test used for coefficients compares $\hat{B}_j^2 / \hat{V}(\hat{B}_j)$ to a chi-square distribution with 1 df (see Section 10.3.1). ∎

Logistic regression has one important difference from linear regression. In Section 11.2, we noted the bias that can occur in estimating linear regression parameters if the inclusion probabilities are related to the response variable, but the unequal probabilities are not accounted for in the analysis. In a health survey, for example, blood pressure might be used as a stratification variable, and a higher sampling fraction used in the high-blood-pressure stratum than in the low-blood-pressure stratum. If we ignore the unequal probabilities and fit a linear regression model predicting the continuous variable blood pressure from covariates such as age, diet, and smoking history, the regression coefficients may be biased for estimating **B**.

Prentice and Pyke (1979), however, show that if a logistic regression model is valid and contains an intercept term, then the intercept is the only parameter estimate affected by a sample design that depends on the y's. Such sample designs are particularly common in epidemiology and economics, where they are referred to as case-control studies and choice-based sampling. In an epidemiology application, the population may be divided into two strata: persons with lung cancer, and persons without lung cancer. A sample is selected from each stratum; as lung cancer is rare, the stratified sample has a far greater sampling fraction (and lower sampling weights) in the cancer stratum than in the non-cancer stratum. But if the primary interest is in estimating the coefficients of age, diet and smoking history in a logistic regression, the disproportionate sampling makes no difference in a model-based analysis. We would expect that if the model is good, the only difference between a weighted and unweighted analysis would appear in the intercept terms. Of course, if a cluster sample is used, the dependence of the data induced by clustering will need to be considered in the logistic regression model for variance estimation, as discussed by Scott and Wild (2003) and Scott (2006).

11.7
Generalized Regression Estimation for Population Totals

In Chapter 4 we introduced ratio and regression estimation in the setting of SRSs, with estimators

$$\hat{t}_{yr} = \frac{\hat{t}_y}{\hat{t}_x} t_x$$

and

$$\hat{t}_{yreg} = \hat{t}_y + \hat{B}_1 (t_x - \hat{t}_x).$$

Now let's extend these estimators to complex survey samples. We want to reduce the mean squared error of the estimator $\hat{t}_y = \sum_{i \in \mathcal{S}} w_i y_i$ by including auxiliary information through the working model

$$Y_i \mid \mathbf{x}_i = \mathbf{x}_i^T \boldsymbol{\beta} + \varepsilon_i, \tag{11.18}$$

with $\mathbf{x}_i^T = (x_{i1}, x_{i2}, \ldots, x_{ip})$ and $V_M(\varepsilon_i) = \sigma_i^2$ for σ_i^2 known. We assume that the vector of true population totals $\mathbf{t_x}$ is known and thus can be used to adjust the estimator $\hat{t}_y$. We allow the variances to differ so that ratio estimation fits into this general framework. Using a working model in (11.18), but relying on the sampling design for inference, is an example of the model-assisted approach further described in Särndal et al. (1992, Chapters 6 and 7).

Define

$$\mathbf{B} = (\mathbf{X}_U^T \mathbf{\Sigma}_U^{-1} \mathbf{X}_U)^{-1} \mathbf{X}_U^T \mathbf{\Sigma}_U^{-1} \mathbf{y}_U,$$

where $\mathbf{\Sigma}_U$ is a diagonal matrix with ith diagonal element σ_i^2. The finite population parameter $\mathbf{B}$ is the weighted least squares estimate of $\boldsymbol{\beta}$ for observations in the population, using the model in (11.18). Thus the form of $\mathbf{B}$ is inspired by (11.18), but we then treat $\mathbf{B}$ as a finite population quantity to be estimated using information in the sample. Note that $\mathbf{X}_U^T \mathbf{\Sigma}_U^{-1} \mathbf{X}_U = \sum_{i=1}^N \mathbf{x}_i \mathbf{x}_i^T / \sigma_i^2$ and $\mathbf{X}_U^T \mathbf{\Sigma}_U^{-1} \mathbf{y}_U = \sum_{i=1}^N \mathbf{x}_i y_i / \sigma_i^2$. Thus, $\mathbf{B}$ may be estimated by

$$\hat{\mathbf{B}} = \left(\sum_{i \in \mathcal{S}} w_i \frac{1}{\sigma_i^2} \mathbf{x}_i \mathbf{x}_i^T \right)^{-1} \sum_{i \in \mathcal{S}} w_i \frac{1}{\sigma_i^2} \mathbf{x}_i y_i. \tag{11.19}$$

The **generalized regression** (GREG) estimator of the population total is

$$\hat{t}_{y\text{GREG}} = \hat{t}_y + (\mathbf{t_x} - \hat{\mathbf{t}}_\mathbf{x})^T \hat{\mathbf{B}}, \tag{11.20}$$

where $\hat{\mathbf{B}}$ is given in (11.19). The term $(\mathbf{t_x} - \hat{\mathbf{t}}_\mathbf{x})^T \hat{\mathbf{B}}$ in (11.20) is a regression adjustment to the Horvitz–Thompson estimator, $\hat{t}_y = \sum_{i \in \mathcal{S}} w_i y_i$. Note that $\hat{t}_{y\text{GREG}}$ is a weighted sum of the y_i values in the sample: we can write

$$\hat{t}_{y\text{GREG}} = \sum_{i \in \mathcal{S}} w_i g_i y_i, \tag{11.21}$$

where

$$g_i = 1 + (\mathbf{t_x} - \hat{\mathbf{t}}_\mathbf{x})^T \left(\sum_{j \in \mathcal{S}} w_j \frac{1}{\sigma_j^2} \mathbf{x}_j \mathbf{x}_j^T \right)^{-1} \frac{1}{\sigma_i^2} \mathbf{x}_i. \tag{11.22}$$

The values g_i are the adjustments to the weights made by using the regression estimator. For large samples, we expect $\hat{\mathbf{t}}_\mathbf{x}$ to be close to $\mathbf{t_x}$ so that g_i will be close to 1 for many observations.

For any choice of the constants σ_i^2, the GREG estimator calibrates the sample to the population total of each x variable used in the regression. To see this, look at the

GREG estimator of $\mathbf{t_x}$: from (11.21),

$$\hat{\mathbf{t}}_{\mathbf{x}\text{GREG}} = \sum_{i \in \mathcal{S}} w_i g_i \mathbf{x}_i$$

$$= \hat{\mathbf{t}}_{\mathbf{x}} + \sum_{i \in \mathcal{S}} w_i \left[(\mathbf{t_x} - \hat{\mathbf{t}}_{\mathbf{x}})^T \left(\sum_{j \in \mathcal{S}} w_j \frac{1}{\sigma_j^2} \mathbf{x}_j \mathbf{x}_j^T \right)^{-1} \frac{1}{\sigma_i^2} \mathbf{x}_i \right] \mathbf{x}_i$$

$$= \hat{\mathbf{t}}_{\mathbf{x}} + \sum_{i \in \mathcal{S}} w_i \frac{1}{\sigma_i^2} \mathbf{x}_i \left[\mathbf{x}_i^T \left(\sum_{j \in \mathcal{S}} w_j \frac{1}{\sigma_j^2} \mathbf{x}_j \mathbf{x}_j^T \right)^{-1} (\mathbf{t_x} - \hat{\mathbf{t}}_{\mathbf{x}}) \right]$$

$$= \hat{\mathbf{t}}_{\mathbf{x}} + (\mathbf{t_x} - \hat{\mathbf{t}}_{\mathbf{x}})$$

$$= \mathbf{t_x}.$$

Using linearization,

$$V(\hat{t}_{y\text{GREG}}) = V[\hat{t}_y + (\mathbf{t}_x - \hat{\mathbf{t}}_{\mathbf{x}})^T \hat{\mathbf{B}}] \approx V(\hat{t}_y - \hat{\mathbf{t}}_x^T \mathbf{B}).$$

Let $e_i = y_i - \mathbf{x}_i^T \hat{\mathbf{B}}$ be the ith residual. Then the variance may be estimated by

$$\hat{V}_1(\hat{t}_{y\text{GREG}}) = \hat{V}\left(\sum_{i \in \mathcal{S}} w_i e_i \right).$$

An alternative estimator of the variance (see Exercise 21) is

$$\hat{V}_2(\hat{t}_{y\text{GREG}}) = \hat{V}\left(\sum_{i \in \mathcal{S}} w_i g_i e_i \right).$$

If the model is a good one, we expect the variability in the residuals to be smaller than the variability in the original observations, so that the GREG estimator will be more efficient than $\hat{t}_y$. In an SRS, for example,

$$\hat{V}(\hat{t}_y) = \frac{N^2}{n}\left(1 - \frac{n}{N}\right) \frac{\sum\limits_{i \in \mathcal{S}}(y_i - \bar{y})^2}{n - 1}$$

but

$$\hat{V}(\hat{t}_{y\text{GREG}}) = \frac{N^2}{n}\left(1 - \frac{n}{N}\right) \frac{\sum\limits_{i \in \mathcal{S}} e_i^2}{n - 1};$$

if the residuals tend to be smaller than the deviations of y_i about the mean, then the estimated variance is smaller for the GREG estimator.

EXAMPLE 11.13 Ratio estimation. For ratio estimation, we adopt the working model

$$y_i = \beta x_i + \varepsilon_i, \quad V_M(\varepsilon_i) = \sigma^2 x_i.$$

The population quantity B is the weighted least squares estimate of β using the whole population. Then, using (11.19),

$$\hat{B} = \left(\sum_{i \in \mathcal{S}} w_i \frac{x_i^2}{x_i} \right)^{-1} \sum_{i \in \mathcal{S}} \frac{w_i x_i y_i}{x_i} = \frac{\displaystyle\sum_{i \in \mathcal{S}} w_i y_i}{\displaystyle\sum_{i \in \mathcal{S}} w_i x_i} = \frac{\hat{t}_y}{\hat{t}_x}.$$

The generalized regression estimator of the population total is

$$\hat{t}_{y\text{GREG}} = \hat{t}_y + (t_x - \hat{t}_x)\frac{\hat{t}_y}{\hat{t}_x} = t_x \frac{\hat{t}_y}{\hat{t}_x},$$

which is the usual ratio estimator. ∎

EXAMPLE 11.14 **Poststratification.** We discussed poststratification in Sections 4.4 and 8.5.2 as a method of calibrating estimates to population totals of subgroups and as a method of adjusting for nonresponse. Suppose we know the population counts N_c for C poststrata, $c = 1, \ldots, C$. Define the variables $x_{ic} = 1$ if observation unit i is in poststratum c and 0 otherwise, and let $\mathbf{x}_i = [x_{i1}, \ldots, x_{iC}]^T$. Consider the working model

$$Y_i = \beta_1 x_{i1} + \beta_2 x_{i2} + \cdots + \beta_C x_{iC} + \varepsilon_i$$

with $V_M(\varepsilon_i) = \sigma^2$. Then,

$$\sigma^2 \mathbf{X}_U^T \mathbf{\Sigma}_U^{-1} \mathbf{X}_U = \mathbf{X}_U^T \mathbf{X}_U = \text{diag}\,(N_1, \ldots, N_C)$$

and

$$\sigma^2 \sum_{i \in \mathcal{S}} w_i \frac{1}{\sigma^2} \mathbf{x}_i \mathbf{x}_i^T = \text{diag}\,(\hat{N}_1, \ldots, \hat{N}_C).$$

As a result, $\hat{B}_c = \hat{t}_{yc}/\hat{N}_c$, where $\hat{t}_{yc} = \sum_{i \in \mathcal{S}} w_i x_{ic} y_i$ is the estimated population total in poststratum c and $\hat{N}_c = \sum_{i \in \mathcal{S}} w_i x_{ic}$ is the estimated population count in poststratum c. The generalized regression estimator is

$$\hat{t}_{y\text{GREG}} = \hat{t}_y + \sum_{c=1}^{C} (N_c - \hat{N}_c)\frac{\hat{t}_{yc}}{\hat{N}_c} = \sum_{c=1}^{C} \frac{N_c}{\hat{N}_c} \hat{t}_{yc}. \quad \blacksquare$$

Often, the auxiliary variables are useful for many of the response variables of interest. You may want to poststratify by age, race, and gender groups when estimating every population total for your survey. This is easily implemented because the generalized regression estimator is a linear estimator in y, as seen in (11.21): $\hat{t}_{y\text{GREG}} = \sum_{i \in \mathcal{S}} w_i g_i y_i$. The weight adjustments g_i in (11.22) depend on the x's but they do not depend on values of the response variable. To estimate totals with the generalized regression estimator, form a new column in the data with values $a_i = w_i g_i$. Then use the vector of a_i as the weight vector for estimating the population total of any variable.

11.8
Chapter Summary

In regression methods with complex survey data, the population characteristics of interest are **B**, the least squares or logistic regression coefficients that would be estimated if we knew the entire population. Since **B** is a function of population totals, it is estimated by $\hat{\mathbf{B}}$ using the sampling weights. Ideally, the finite population values **B** reflect an underlying relationship between y and **x**, but inferences about **B**, using the survey design, are valid whether the regression model is a good one or not.

The generalized regression estimator provides a method for using auxiliary information to reduce the mean squared error of estimators. It can also be used to reduce bias due to nonresponse.

Key Terms

Generalized regression estimator (GREG): An estimator of a population total that uses auxiliary information through a regression model.

Model-assisted estimation: An approach to inference in which a population model motivates the form of estimators, but all inference is based on the survey design.

For Further Reading

Kutner et al. (2005) is a general reference on linear regression analysis for data assumed to be generated from a model (not survey data). Graybill (1976) and Ravishanker and Dey (2002) present theoretical results about regression models, again in the non-survey setting.

If you want to learn more about inference in sample surveys, start with the paper by Brewer and Mellor (1973), who present an insightful and entertaining debate between "Harry," a design-based survey statistician, and "Fred," who is fresh from graduate school and promotes a model-based approach. The book by Brewer (2002) also contrasts the approaches. Smith (1994) provides an interesting review of philosophies of inference, by a statistician whose previous work adhered to a model-based approach. Binder and Roberts (2003) discuss inference for regression models. Robinson (1987) studies another approach to inference in survey sampling, conditional design-based inference; references to earlier work are given in the paper. The model-assisted approach to inference that we use in this chapter is discussed in more detail by Särndal et al. (1992).

The theory for regression estimation in complex surveys has been developed by many people. Kish and Frankel (1974) is one of the first papers to show that the sample design affects estimates of regression parameters. Other references for further reading include Konijn (1962), Kalton (1983), Valliant et al. (2000), Fuller (1975, 1984, 2002), and Korn and Graubard (1999). Lehtonen and Pahkinen (2004) discuss linear and logistic regression in surveys, and present a case study of multi-level modeling in an educational survey.

Särndal (2007) gives a clear overview of generalized regression estimators and calibration in survey sampling. Estevao and Särndal (2006) discuss a functional form of calibration. Beaumont and Alavi (2004) derive a robust generalized regression estimator, and Breidt and Opsomer (2000) and Montanari and Ranalli (2005) use nonparametric regression methods in survey sampling. Gelman (2007) adopts a hierarchical Bayesian approach to weight adjustment and Valliant (2009) discusses a model-based approach to using auxiliary information when estimating population totals. Montanari (1987) and Rao (1994) present an alternative method of using regression for estimating population totals. Silva and Skinner (1997) discuss methods for selecting the *x* variables in regression estimation.

11.9
Exercises

A. Introductory Exercises

1 Read one of the articles listed in the file chapter11papers.html on the book website, or another article in which regression or logistic regression is used on data from a complex survey. Write a critique of the article. What is the purpose and design of the survey? What is the goal of the analysis? How do the authors use information from the survey design in the analysis? Do you think that the data analysis is done well? If so, why? If not, how could it have been improved? Are the conclusions drawn in the article justified?

2 An investigator wants to study the relationship between a child's age and number of siblings, and the dollar amount of the child's Christmas list presented to Santa Claus. She also wants to estimate the total number of children that visit Santa Claus, and the total dollar amount of all childrens' requests. It would be very difficult to construct a sampling frame of children who will visit Santa Claus between December 1 and December 24, but the investigator has a list of shopping malls and stores in which Santa will appear in the city, as well as the times that Santa will be at each location. The Santa sites are divided into four categories: 23 department stores, 19 discount stores, 15 toy stores, and 5 shopping malls. The investigator wants you to help design the sample of children.

    a What questions would you ask the investigator to clarify the problem?

    b Assuming any answers you like to the questions you asked, suggest a design for the survey.

    c How will your survey design affect the regression analysis of the data? How do you propose to analyze the data? Are there other explanatory variables that you would suggest to the investigator?

3 Use the data in file spanish.dat (see Exercise 5 of Chapter 5). Let domain 1 consist of students who are planning a trip to a Spanish-speaking country in the next year and domain 2 consist of the students who are not planning such a trip. We are interested in whether the mean vocabulary score (y) differs in the two domains. The population domain mean in domain 1 is $\bar{y}_{U1}$ and the population domain mean in domain 2 is $\bar{y}_{U2}$. Using regression, estimate $\bar{y}_{U1} - \bar{y}_{U2}$ and give a 95% CI. Is there evidence that the domain means differ?

B. Working with Survey Data

4 Use the data in anthrop.dat for this problem.

a Construct a population from the 3000 observations in anthrop.dat in which the 1000 individuals with the highest value of y have been removed. Now take an SRS of size 200 from the remaining 2000 individuals, and plot the data along with the ordinary least squares regression line. How does this line compare to the population regression line?

b Repeat (a), but use as the population the 2000 individuals with the lowest value of x.

c Is there a difference in the regression equations in (a) and (b)? Explain, and relate your findings to the model in (11.1).

5 Use the data in nybight.dat (see Exercise 18 of Chapter 3) for this problem. Using the 1974 data, estimate the coefficients in a straight line regression model predicting weight of the catch from the number of fish caught. Give standard errors for your estimates. Be sure to plot the data!

6 Perform a model-based analysis for the setting in Exercise 5. Be sure to examine the residuals and postulate an appropriate variance structure for the model.

7 Repeat Exercise 5 for predicting number of species caught from the surface temperature.

8 Repeat Exercise 6 for predicting number of species caught from the surface temperature.

9 Use the data in teachers.dat (described in Exercise 15 of Chapter 5) for this problem.

a Estimate the coefficients in a straight line regression model predicting *preprmin* from *size*. Give standard errors for your estimates. Is there evidence that the two variables are related? (Be sure to plot the data!)

b Perform a model-based analysis of the same data. Be sure to examine the residuals and postulate an appropriate variance structure for the model.

10 Use the data in books.dat (described in Exercise 8 of Chapter 5) for this problem.

a Plot *replace* vs. *purchase* for the raw data.

b Plot *replace* vs. *purchase* using the sampling weights.

c Using a design-based approach, estimate the regression equation for predicting *replace* from *purchase*, along with its standard error. How many df would you use in constructing a CI for the slope?

11 For the situation in Exercise 10, postulate a model for the variance structure. Using your model, estimate the slope of the regression line predicting replace from purchase. How do your estimate and its standard error compare with your answers in Exercise 10?

12 Use your data set from Exercise 13 of Chapter 3 for this problem. Using the weights, fit a regression model predicting *acres92* from *largef92*. Give a standard error for the estimated slope. Now ignore the sampling design, and calculate the ordinary least

squares estimate of the slope. Do your point estimates differ? Explain why or why not by examining plots of the data.

13   Lush (1945, p. 95) discussed different estimates of heritability for milk fat percentage in dairy cattle herds. Heritability is defined to be the percentage of variability in fat percentage that is attributable to differences in the heredity of different individuals; the remainder of the variability is attributed to differences in environment. He noted that when the herd was treated as an SRS, the estimate of heritability was about 0.8; when fat percentage for daughters was regressed on fat percentage for dams, and where each dam was represented by only one record, the estimate of heritability decreased to below 0.3.

From a sampling perspective, why are these estimates so different? Discuss how you would analyze the full herd data from both a design-based and a model-based perspective.

14   Using the data in nhanes.dat, fit a straight line regression model predicting $y = $ triceps skinfold (variable *bmxtri*) from $x = $ body mass index (variable *bmxbmi*). You plotted these data in Exercise 15 of Chapter 7. Give a 95% CI for the slope, and calculate R^2 for these data. Draw your regression line on the plot.

15   Using the data in nhanes.dat, fit a straight line regression model predicting $y = $ waist circumference (variable *bmxwaist*) from $x = $ thigh circumference (variable *bmxthicr*). You plotted these data in Exercise 16 of Chapter 7. Give a 95% CI for the slope, and calculate R^2 for these data. Draw your regression line on the plot.

16   Using the data in ncvs2000.dat, fit a logistic regression model predicting whether a person is a victim of violent crime from age and sex. Is a quadratic term needed for age?

C. Working with Theory

17   *Comparison of domain means.* Suppose the population may be divided into two groups, with respective sizes N_1 and N_2 and population means $\bar{y}_{1U}$ and $\bar{y}_{2U}$. The overall population mean is $\bar{y}_U = (N_1\bar{y}_{1U} + N_2\bar{y}_{2U})/N$, with $N = N_1 + N_2$. Let $x_i = 1$ if observation unit i is in group 1, and $x_i = 0$ if it is in group 2. The weight for observation unit i is w_i.

Show that $B_1 = \bar{y}_{1U} - \bar{y}_{2U}$ and $B_0 = \bar{y}_{2U}$. Also show that

$$\hat{B}_1 = \frac{\displaystyle\sum_{i \in S} w_i x_i y_i}{\displaystyle\sum_{i \in S} w_i x_i} - \frac{\displaystyle\sum_{i \in S} w_i(1 - x_i)y_i}{\displaystyle\sum_{i \in S} w_i(1 - x_i)} = \hat{\bar{y}}_1 - \hat{\bar{y}}_2$$

and $\hat{B}_0 = \hat{\bar{y}}_2$.

18   Consider the SRS data in file uneqvar.dat.

a   Plot y vs. x.

b   Find the fitted regression line under the assumption of equal variances.

c   Calculate $\hat{V}_M(\hat{\beta}_1)$ and $\hat{V}_L(\hat{\beta}_1)$. How do they compare?

19   Show that (11.10) is equivalent to (11.6) and (11.7) for straight-line regression.

20 (Requires linear algebra and calculus.) The linearization estimator of $V(\hat{\mathbf{B}})$ can be found by the method outlined in Section 9.1 (see Shah et al., 1977). However, the calculations are easier using the Demnati–Rao (2004) method discussed in Exercise 23 of Chapter 9. Show (11.11) using the Demnati–Rao method. HINT: Use the fact (see Harville, 1997, p. 307) that if $\mathbf{F}$ is a nonsingular matrix whose entries are functions of u, then

$$\frac{\partial \mathbf{F}^{-1}}{\partial u} = -\mathbf{F}^{-1} \frac{\partial \mathbf{F}}{\partial u} \mathbf{F}^{-1}.$$

21 (Requires linear algebra and calculus.) Use the Demnati–Rao (2004) method discussed in Exercise 23 of Chapter 9 to estimate $V(\hat{t}_{y\text{GREG}})$ for $\hat{t}_{y\text{GREG}}$ defined in (11.20). HINT: Use the matrix differentiation result given in Exercise 20.

22 *Plotting residuals in regression models with complex survey data.* In a design-based framework for inference, the regression coefficients $\hat{\mathbf{B}}$ estimate the population values $\mathbf{B}$. Inferences such as CIs depend on the inclusion probabilities in the sampling design and thus do not depend on model assumptions. Design-based inferences about the finite-population regression parameters of a bad model are valid. Nevertheless, as discussed in Section 11.4, we often are interested in an underlying theoretical model and want to assess how well the population regression model fits. We can plot residuals versus predicted values, incorporating the weights, using methods in Section 7.4.2. Plot the residuals versus predicted values for the regression model in Example 11.6. What do you see in your plot? How would you change the model?

23 *Regression diagnostics for complex survey data.* Jenney (2005) and Li and Valliant (2006) independently developed regression diagnostics methods for complex survey data. The **leverages** for the population values are the diagonal elements of the matrix

$$\mathbf{H} = \mathbf{X}_U (\mathbf{X}_U^T \mathbf{X}_U)^{-1} \mathbf{X}_U^T,$$

so that the leverage of unit i in the population is

$$h(\mathcal{U})_i = \mathbf{x}_i^T (\mathbf{X}_U^T \mathbf{X}_U)^{-1} \mathbf{x}_i.$$

The leverage of an observation is a measure of the distance from $\mathbf{x}_i$ to the means of the set of explanatory variables (Kutner et al., 2005). Using the weights, define the leverage of unit i in the sample as

$$h(\mathcal{S})_i = w_i \mathbf{x}_i^T \left(\sum_{j \in \mathcal{S}} w_j \mathbf{x}_j \mathbf{x}_j^T \right)^{-1} \mathbf{x}_i.$$

a Show that $\sum_{i \in \mathcal{S}} h(\mathcal{S})_i = p$, the number of parameters in the regression model.

b Calculate the leverage, using the weights, for each observation in file anthuneq.dat, used in Section 11.2. Which points have the highest values of leverage? Does your assessment of the high-leverage points change if you do not use the weights?

24 The diagnostic statistic DFFITS (Belsley et al., 1980) is often used to assess influential observations in a data set. A complex survey version of DFFITS can be calculated using the survey-weighted leverage $h(\mathcal{S})_i$ from Exercise 23 along with the residual for observation i, $e_i = y_i - \hat{y}_i$:

$$\text{DFFITS} = \frac{h(\mathcal{S})_i e_i}{1 - h(\mathcal{S})_i} \frac{1}{\text{SE}(\hat{y}_i)},$$

where

$$\hat{V}(\hat{y}_i) = \hat{V}(\mathbf{x}_i^T \hat{\mathbf{B}}) = \mathbf{x}_i^T \hat{V}(\hat{\mathbf{B}}) \mathbf{x}_i.$$

Calculate DFFITS for each observation in file anthuneq.dat.

25 (Requires theory of linear models.) *Local polynomial regression with survey data.* Section 7.4.2 discussed using smoothed trend lines in bivariate plots of survey data (see Korn and Graubard, 1998 and Bellhouse and Stafford, 2001). We posit a model $y_i = g(x_i) + \varepsilon_i$, where the second derivative of g is continuous, and estimate the underlying smooth function $g(x)$ by sliding a kernel window along the data and fitting a straight line (or higher-order polynomial) to the weighted data in that window. As with density estimation, briefly discussed in Section 7.4.1, the kernel function K is a symmetric density function such as the normal kernel function $K_N(t) = \exp(-t^2/2)/\sqrt{2\pi}$ or the quadratic kernel function $K_Q(t) = \frac{3}{4}(1-t^2)$ for $|t| < 1$. Since the data are from a complex survey, the weights used in fitting the local regression include the survey weights as well as the kernel weights. Let $x_1, \ldots, x_n$ and $y_1, \ldots, y_n$ denote the observations in the sample. For local linear regression, the function g at a point t is estimated by:

$$\hat{g}(t) = [1 \ 0](\mathbf{X}_t^T \mathbf{W}_t \mathbf{X}_t)^{-1} \mathbf{X}_t^T \mathbf{W}_t \mathbf{y},$$

where

$$\mathbf{X}_t = \begin{bmatrix} 1 & t - x_1 \\ \vdots & \vdots \\ 1 & t - x_n \end{bmatrix}, \qquad \mathbf{y} = \begin{bmatrix} y_1 \\ \vdots \\ y_n \end{bmatrix},$$

and

$$\mathbf{W}_t = \text{diag}\left[\frac{w_1}{b} K\left(\frac{t - x_1}{b}\right), \ldots, \frac{w_n}{b} K\left(\frac{t - x_n}{b}\right)\right].$$

Calculate a local linear regression function for the NHANES data, using $y = $ triceps skinfold and $x = $ body mass index.

26 (Requires theory of linear models.) Suppose the "true" model describing the relation between x and y is

$$Y_i | x_i = \beta_0 + \beta_1 x_i + \varepsilon_i,$$

where the ε_i are independently generated from a $N(0, \sigma_i^2)$ distribution. Let Σ be a matrix with diagonal entries $\sigma_1^2, \sigma_2^2, \ldots, \sigma_n^2$. What is the covariance matrix for the ordinary least squares parameter estimators? How does this relate to the discussion of different estimators of the variance in Section 11.2.2?

TABLE   11.1
(a) Population Counts and (b) Sample Sizes for Exercise 30.

(a) Population Counts

	Age Group	
	≤ 25	>25
F	50	100
M	250	900

(b) Sample Sizes

	Age Group	
	≤ 25	>25
F	15	5
M	26	14

27 The coefficient of determination R^2 is often reported for regression analyses. For a straight line regression, the finite population quantity R^2 is defined to be

$$R^2 = \frac{B_1 \sum_{i=1}^{N} (x_i - \bar{x}_U)(y_i - \bar{y}_U)}{\sum_{i=1}^{N} (y_i - \bar{y}_U)^2}.$$

a Show that R^2 is the square of the population correlation coefficient R defined in (4.1).

b Write R^2 as a function of population totals.

c Give an estimator $\hat{R}^2$ of R^2 for data from a complex sample, using weights.

28 Fienberg (1980) says, "we know of no justification whatsoever for applying standard multivariate methods to weighted data . . . the automatic insertion of a matrix of sample-based weights into a weighted least-squares analysis is more often than not misleading, and possibly even incorrect." Which approach to regression inference does Fienberg advocate? What is your reaction?

29 Assuming a model

$$y_i = \beta_0 + \beta_1 x_i + \varepsilon_i$$

with $V(\varepsilon_i) = \sigma^2$, what is the generalized regression estimator of t_y?

30 Suppose the population counts for a cross-classification by age and gender are given in Table 11.1(a). Table 11.1(b) gives the sample sizes in each group for a sample from the population. The sampling weight for each observation is $w_i = 20$.

a Find the poststratification weight adjustments g_i for observations in each of the four cells. Show that poststratification adjustments must always be positive, but they can be less than one.

b Now find weight adjustments g_i based on the model

$$y_i = \beta_0 + \beta_1 x_i + \beta_1 z_i + \varepsilon_i, \quad V_M(\varepsilon_i) = \sigma^2,$$

where $x_i = 1$ if observation i is female and 0 otherwise, and $z_i = 1$ if observation i is 25 or younger and 0 otherwise. Do the weight adjustments have to be positive in this model?

D. Projects and Activities

31   *Trucks.*   Use the data in Exercise 34 of Chapter 3 and Exercise 27 of Chapter 7 for this exercise.

    a   Fit a straight line model predicting $y = $ *miles_annl* from $x = $ model year (*adm_modelyear*). Give a 95% CI for the slope.

    b   How well does this model fit the data?

    c   What other variables in the data set might be useful for predicting y? Fit a multiple regression model predicting y using x variables of your choice.

32   *Baseball data.*   Use the data from Exercise 29 in Chapter 7. What variables in the data do you think might be useful for predicting log(*salary*)? Fit a multiple regression model predicting log(*salary*) from these variables.

33   *IPUMS exercises.*

    a   Regress *inctot* on covariates of your choice using your sample from Exercise 38 of Chapter 5. Write a paragraph interpreting the results of your analysis.

    b   Perform a logistic regression predicting whether a person is in the labor force (variable *labforce*) from covariates of your choice.

34   *Activity for course project.*   Using the survey you chose in Exercise 31 of Chapter 7, use regression methods to predict a response of interest from covariates in the data. If the survey has no continuous responses, use logistic regression to predict a binary response. Make sure you plot the data appropriately.

12

Two-Phase Sampling

Nearly the whole of the states have now returned their census. I send you the result, which as far as founded on actual returns is written in black ink, and the numbers not actually returned, yet pretty well known, are written in red ink. Making a very small allowance for omissions, we are upwards of four millions; and we know in fact that the omissions have been very great.

—Thomas Jefferson, letter to David Humphreys, August 23, 1791

Sometimes, you would like to use stratification, unequal-probability sampling, or ratio estimation to increase the precision of your estimator, but the sampling frame lacks information on useful auxiliary variables. For example, suppose you want to sample businesses with probability proportional to income but do not have income information in the sampling frame. Or you want to estimate the total timber volume that has been cut in the forest by measuring the total volume in a sample of truckloads of logs. Timber volume in a truck is related to the weight of the truckload, so you would expect to gain precision by using ratio estimation with $y_i = $ timber volume in truck i and $x_i = $ weight of truck i. But the ratio estimator $\hat{t}_{yr} = t_x \hat{t}_y / \hat{t}_x$ requires that the total weight for all truckloads be known, and weighing every truck in the population is impractical.

Two-phase sampling, also called **double sampling**, provides a solution. Two-phase sampling, introduced by Neyman (1938), is useful when the variable of interest y is relatively expensive to measure, but a correlated variable x can be measured fairly easily and used to improve the precision of the estimator of t_y. It may also be used to adjust for nonresponse, to sample rare populations, or to improve the sampling frame. We discuss some of these applications later in the chapter.

Suppose the population has N observation units. The sample is taken in two phases:

1 *Phase I sample.*   Take a probability sample of $n^{(1)}$ units, called the phase I sample. Measure the auxiliary variables **x** for every unit in the phase I sample. In the survey of businesses, you could take a simple random sample (SRS) of tax records and record the reported income for each business in the sample. For measuring timber volume, you could weigh a sample of trucks selected either with

an SRS or with probability proportional to estimated timber volume. The phase I sample is generally relatively large (and can be large because the auxiliary information is inexpensive to obtain), and should provide accurate information about the distribution of the **x**'s.

2 *Phase II sample.* Now act as though the phase I sample is a population and select a probability sample of size $n^{(2)}$ from the phase I sample. Measure the variables of interest for each unit in the subsample, called the phase II sample. Since you are treating the phase I sample as the population from which the phase II sample is drawn, you may use the auxiliary information gathered in phase I when designing the phase II sample. You might select the businesses to be contacted with probability proportional to the income measured in the phase I sample. Alternatively, you might use the income information to stratify the businesses in the phase I sample and then contact a randomly selected subset of the businesses in each income stratum to obtain the desired information on variables such as total expenses. You could select the truckloads on which timber volume is to be measured with probability proportional to weight, or you could use the information in the phase I sample to obtain a better estimate of total weight and use ratio estimation. In each case, the *y* variables are relatively expensive to measure, but *y* is correlated with *x*.

Two-phase sampling can save time and money if the auxiliary information is relatively inexpensive to obtain and if having that auxiliary information can increase the precision of the estimates for quantities of interest.

EXAMPLE **12.1** Stockford and Page (1984) used two-phase sampling to estimate the percentage of Vietnam-era veterans in U.S. Veterans Administration (VA) hospitals who actually served in Vietnam.

The 1982 VA Annual Patient Census (APC) included a random sample of 20% of the patients in VA hospitals. The following question was included: "If period of service is 'Vietnam era,' was service in Vietnam?" with answer categories "yes," "no," and "not available." The answers to the question were obtained from patients' medical records. But the response from medical records could be inaccurate for several reasons: (1) The medical record classification was largely self-reported, and the patient may not have been able to recall the location of service due to medical problems, or may have been confused about the definition of Vietnam service (some pilots whose duty station was officially recorded as Thailand flew missions over Vietnam); (2) a patient might misstate Vietnam service because he or she thought the answer might affect VA benefits; or (3) errors might be made in recording the response in the medical record. In addition, a large number of patients had "not available" for the answer. Thus, the answer to the question on Vietnam service in the APC survey was unsatisfactory for estimating the percentage of Vietnam-era veterans in VA hospitals who served in Vietnam.

Stockford and Page checked the military records for a stratified subsample of the hospitalized veterans to determine the true classification of Vietnam service. The information in the original survey was used for the stratification, as different percentages with Vietnam service were expected in the "yes", "no," and "not available" groups in the APC survey. Military records for all of the patients in the "not available" stratum were checked. It was expected that the within-stratum variances would be relatively low in the "yes" and "no" strata—even though the APC survey data are inaccurate, you would expect a higher percentage of "yes" respondents to have served in Vietnam than "no" respondents—and military records for a 10% subsample were checked for each of those two strata.

The results for the question "Was service in Vietnam?" were as follows:

APC Group	APC Survey Classification	Subsample Size	Vietnam Service in Subsample
Yes	755	67	49
No	804	72	11
Not available	505	505	211
Total	2064	644	271

As expected, the percentage of veterans with Vietnam service differed for the three groups: Of the veterans with a "yes" response to the APC survey question, 73% actually served in Vietnam, compared with 15% for the "no" group and 42% for the veterans for which the information was not available. ∎

EXAMPLE 12.2   Two-phase sampling is often used in forestry surveys. Aerial photographs are available for the region of interest, and points are systematically distributed across the photographs. Areas around the points are inspected on the photographs and classified by land class: forest land, unproductive forest land, nonforest land, and water. A phase I sample of points is then drawn from the grid, with a higher sampling fraction for grid points classified as forest land than those classified as nonforest land. Areas in the phase I sample are examined more closely to classify them by stand size and density. Then, a subsample is taken of the points in the phase I sample, and ground measurements such as land use, volume, and mortality taken; the percentage of area that is forest from the phase II ground sample may differ somewhat from the photo estimate in phase I, and ratio estimation can be used in the phase II sample to increase the precision of the estimator. ∎

EXAMPLE 12.3   We have already seen two-phase sampling used in nonresponse adjustment, in Section 8.3. A probability sample is taken from the population; the sampled units are then divided into the two strata of respondents and nonrespondents. Then a subsample is taken of the nonrespondents. The phase I sample is the original probability sample. The variable

$$x_i = \begin{cases} 1 & \text{if observation } i \text{ responds} \\ 0 & \text{if observation } i \text{ is a nonrespondent} \end{cases}$$

is observed for everyone in the phase I sample. Then the information about x_i is used in the phase II sample. The variable of interest y_i is observed for all observations with $x_i = 1$; a subsample is taken for observations with $x_i = 0$. ∎

What is the difference between a *two-phase* design, discussed in this chapter, and the *two-stage* designs discussed in Chapters 5 and 6? In two-phase sampling, the phase I sample is used to collect inexpensive auxiliary information on the sampling units; this information is then used to improve the efficiency of the phase II sample design. In two-stage sampling, different sizes of sampling units are collected at the two stages. All primary sampling units (psus) selected at stage 1 of a two-stage design are subsampled. In a two-phase design, it is possible that some psus sampled in the phase I sample will not be represented at all in the phase II sample. A two-stage design might have hospitals as psus at the first stage, and then subsample patients as ssus at the second stage. The final two-stage sample contains patients from each psu selected in the first stage. A two-phase sample of hospitals might take a probability sample of hospitals in phase I, then divide the hospitals in the phase I sample into strata based on number of heart attack patients. In phase II, a stratified random subsample of the hospitals would be selected using the stratification information from phase I.

12.1
Theory for Two-Phase Sampling

A general framework for two-phase sampling is given in Särndal and Swensson (1987) and Legg and Fuller (2009). Let $\mathcal{S}^{(1)}$ denote the phase I sample; the units selected for the sample are determined by the random variables

$$Z_i = \begin{cases} 1 & \text{if unit } i \text{ is in the phase I sample} \\ 0 & \text{if unit } i \text{ is not in the phase I sample.} \end{cases}$$

Let $w_i^{(1)}$, for $i \in \mathcal{S}^{(1)}$, be the sampling weights for the phase I sample: $w_i^{(1)} = 1/[P(Z_i = 1)]$. We observe a vector of auxiliary characteristics $\mathbf{x}_i = (x_{i1}, x_{i2}, \ldots, x_{ik})^T$ for each observation unit in the phase I sample. Using the theory developed in earlier chapters, we can estimate the population total for auxiliary variable j as

$$\hat{t}_{x_j}^{(1)} = \sum_{i \in \mathcal{S}^{(1)}} w_i^{(1)} x_{ij} = \sum_{i=1}^{N} Z_i w_i^{(1)} x_{ij}.$$

Now, indicate membership in the phase II sample $\mathcal{S}^{(2)}$ by the random variable

$$D_i = \begin{cases} 1 & \text{if unit } i \text{ is in the phase II sample} \\ 0 & \text{if unit } i \text{ is not in the phase II sample.} \end{cases}$$

The probability that a unit is in the phase II sample depends on whether it is in the phase I sample and also may depend on auxiliary information collected in the phase I sample; we denote this dependence by writing $P(D_i = 1 \mid \mathbf{Z})$, where $\mathbf{Z}$ is the vector $(Z_1, Z_2, \ldots, Z_N)$. Thus, when we find an expectation conditional on $\mathbf{Z}$, we are treating the information from the phase I sample as known. The subsampling weights for the final, phase II sample also depend on which units were selected to be in the phase I sample.

$$w_i^{(2)} = w_i^{(2)}(\mathbf{Z}) = \begin{cases} \dfrac{1}{P(D_i = 1 \mid \mathbf{Z})} & \text{if } Z_i = 1 \\ \\ 0 & \text{if } Z_i = 0. \end{cases}$$

An analog of the Horvitz-Thompson estimator for two-phase sampling is

$$\hat{t}_y^{(2)} = \sum_{i \in \mathcal{S}^{(2)}} w_i^{(1)} w_i^{(2)} y_i = \sum_{i=1}^{N} Z_i D_i w_i^{(1)} w_i^{(2)} y_i. \tag{12.1}$$

Kott and Stukel (1997) call (12.1) the double expansion estimator; it "expands" the weight on y_i by the product of the two sampling weights.

We use the following device to find properties of the estimator in (12.1). The phase II sample is selected by treating the phase I sample as the population, so we can find properties of the subsample relative to phase I using standard methods. Define

$$\hat{t}_y^{(1)} = \sum_{i \in \mathcal{S}^{(1)}} w_i^{(1)} y_i = \sum_{i=1}^{N} Z_i w_i^{(1)} y_i.$$

Now, we do not know what $\hat{t}_y^{(1)}$ is, because we only observe the y_i's in the phase II sample. But $\hat{t}_y^{(1)}$ serves as the "population total" estimated in phase II—if we knew y_i for all units in the phase I sample, we would estimate t_y by $\hat{t}_y^{(1)}$. Treating the phase I sample as known, we have

$$E[\hat{t}_y^{(2)}|\mathbf{Z}] = \sum_{i=1}^{N} Z_i w_i^{(1)} w_i^{(2)} y_i E[D_i \mid \mathbf{Z}] = \sum_{i=1}^{N} Z_i w_i^{(1)} y_i = \hat{t}_y^{(1)}.$$

Then, using successive conditioning (see Section A.4),

$$E[\hat{t}_y^{(2)}] = E\{E[\hat{t}_y^{(2)} \mid \mathbf{Z}]\} = E\left[\sum_{i=1}^{N} Z_i w_i^{(1)} y_i\right] = t_y. \tag{12.2}$$

Also, from Property A.4 in Section A.4,

$$V(\hat{t}_y^{(2)}) = V(E[\hat{t}_y^{(2)} \mid \mathbf{Z}]) + E(V[\hat{t}_y^{(2)} \mid \mathbf{Z}]) = V(\hat{t}_y^{(1)}) + E(V[\hat{t}_y^{(2)} \mid \mathbf{Z}]). \tag{12.3}$$

The first term is the variance that would be obtained if y_i had been observed for every observation in $\mathcal{S}^{(1)}$; the second term is the additional variance from subsampling in phase II. Consequently, the variance from two-phase sampling is always larger than if we measured y on every unit in the phase I sample of $n^{(1)}$ units. We hope, though, that if y is related to $\mathbf{x}$, the second term in (12.3) will be smaller than the variance of an estimator of t_y from a sample of size $n^{(2)}$ that does not use the auxiliary information in the design.

12.2
Two-Phase Sampling with Stratification

In two-phase sampling with stratification, information on a stratification variable is selected in phase I. That information is then used to select a stratified sample (the phase II sample) from the phase I sample. For simplicity, assume that an SRS is taken in phase I, and that stratified random sampling is used in phase II. Särndal et al. (1992, Chapter 9) give a more general treatment, allowing unequal-probability sampling for either phase. Define $\mathcal{S}^{(1)}$, $\mathcal{S}^{(2)}$, Z_i, and D_i as in Section 12.1. If an SRS of size n is taken in phase I, then $P(Z_i = 1) = n/N$.

The observation units are divided among H strata, but we do not know stratum membership for a unit until it is selected in phase I. In the population, however, stratum h has N_h units (assume N_h is unknown) and $N = \sum_{h=1}^{H} N_h$ (assume N is known). Let

$$x_{ih} = \begin{cases} 1 & \text{if unit } i \text{ is in stratum } h \\ 0 & \text{if unit } i \text{ is not in stratum } h. \end{cases}$$

Observe x_{ih}, $h = 1, \ldots, H$, for each unit in the phase I sample. The number of units in the phase I sample that belong to stratum h is a random variable:

$$n_h = \sum_{i=1}^{N} Z_i x_{ih}.$$

Now take a simple random subsample of size m_h in stratum h; m_h may depend on the first phase of the sampling. The subsamples in different strata are selected independently, given the information in the phase I sample. With random subsampling,

$$P(D_i = 1 \mid \mathbf{Z}) = Z_i \sum_{h=1}^{H} x_{ih} \frac{m_h}{n_h}.$$

Although $P(D_i = 1 \mid \mathbf{Z})$ is written as a sum, all but one of the x_{ih}'s $(h = 1, \ldots, H)$ will equal zero because each unit belongs to exactly one stratum, so that $P(D_i = 1 \mid \mathbf{Z}) = Z_i m_h / n_h$ for unit i determined to be in stratum h. The sampling weight for a phase II unit in stratum h is $w_i^{(2)} = n_h / m_h$; in general, $w_i^{(2)} = Z_i \sum_{h=1}^{H} x_{ih} n_h / m_h$.

The two-phase-sampling stratified estimator of the population total is

$$\begin{aligned}
\hat{t}_{\text{str}}^{(2)} &= \sum_{i=1}^{N} Z_i D_i w_i^{(1)} w_i^{(2)} y_i \\
&= \sum_{i=1}^{N} Z_i D_i \frac{N}{n} \left(\sum_{h=1}^{H} \frac{n_h}{m_h} x_{ih} \right) y_i \\
&= N \sum_{h=1}^{H} \frac{n_h}{n} \bar{y}_h^{(2)},
\end{aligned} \tag{12.4}$$

where $\bar{y}_h^{(2)} = \sum_{i \in S^{(2)}} x_{ih} y_i / m_h$ is the average of the phase II units in stratum h. We showed in (12.2) that $E[\hat{t}_{\text{str}}^{(2)}] = t_y$. The corresponding estimator of the population mean is

$$\hat{\bar{y}}_{\text{str}}^{(2)} = \frac{1}{N} \sum_{i \in S^{(2)}} w_i^{(1)} w_i^{(2)} y_i = \sum_{h=1}^{H} \frac{n_h}{n} \bar{y}_h^{(2)}. \tag{12.5}$$

Recall that a stratified random sampling estimator of the population total from (3.1) is

$$\hat{t}_{\text{str}} = N \sum_{h=1}^{H} \frac{N_h}{N} \bar{y}_h;$$

the two-phase-sampling estimator simply substitutes n_h / n for N_h / N.

The variance is also computed conditionally using (12.3):

$$
\begin{aligned}
V\left(\hat{t}_{\text{str}}^{(2)}\right) &= V\left(E\left[\hat{t}_{\text{str}}^{(2)} \mid \mathbf{Z}\right]\right) + E\left(V\left[\hat{t}_{\text{str}}^{(2)} \mid \mathbf{Z}\right]\right) \\
&= V\left(\hat{t}_y^{(1)}\right) + N^2 E\left(V\left[\sum_{h=1}^{H} \frac{n_h}{n}\bar{y}_h^{(2)} \mid \mathbf{Z}\right]\right) \\
&= N^2\left(1 - \frac{n}{N}\right)\frac{S_y^2}{n} + N^2 E\left[\sum_{h=1}^{H} \left(\frac{n_h}{n}\right)^2 \left(1 - \frac{m_h}{n_h}\right)\frac{s_h^{2(1)}}{m_h}\right].
\end{aligned}
$$

(12.6)

The first term is the variance from the phase I SRS; the second term is the additional variance resulting from the subsampling in phase II. Here, $S_y^2 = \sum_{i=1}^{N}(y_i - \bar{y}_U)^2/(N-1)$ is the population variance of the y's;

$$
s_h^{2(1)} = \frac{1}{n_h - 1}\sum_{i \in \mathcal{S}^{(1)}} x_{ih}(y_i - \bar{y}_h^{(1)})^2
$$

would be the sample variance of the y_i's in stratum h in the phase I sample if we observed all of them. The second term in (12.6) is left as an expectation because n_h and m_h are random variables.

Rao (1973) estimates the variance in two-phase sampling with stratification as

$$
\begin{aligned}
\hat{V}\left(\hat{t}_{\text{str}}^{(2)}\right) &= N(N-1)\sum_{h=1}^{H}\left(\frac{n_h - 1}{n - 1} - \frac{m_h - 1}{N - 1}\right)\frac{n_h}{n}\frac{s_h^{2(2)}}{m_h} \\
&\quad + \frac{N^2}{n-1}\left(1 - \frac{n}{N}\right)\sum_{h=1}^{H}\frac{n_h}{n}\left(\bar{y}_h^{(2)} - \bar{y}_{\text{str}}^{(2)}\right)^2,
\end{aligned}
$$

(12.7)

where

$$
s_h^{2(2)} = \frac{1}{m_h - 1}\sum_{i \in \mathcal{S}^{(2)}} x_{ih}\left(y_i - \bar{y}_h^{(2)}\right)^2
$$

is the sample variance of the y_i's in stratum h (see Exercise 11). If we can ignore the finite population corrections (fpcs),

$$
\hat{V}(\bar{y}_{\text{str}}^{(2)}) \approx \sum_{h=1}^{H}\frac{n_h - 1}{n - 1}\frac{n_h}{n}\frac{s_h^{2(2)}}{m_h} + \frac{1}{n-1}\sum_{h=1}^{H}\frac{n_h}{n}\left(\bar{y}_h^{(2)} - \bar{y}_{\text{str}}^{(2)}\right)^2.
$$

(12.8)

EXAMPLE 12.4 Let's apply these results to the data in Example 12.1. Because $\bar{y}_h^{(2)} = \hat{p}_h$ is a proportion, $s_h^{2(2)} = m_h\hat{p}_h(1 - \hat{p}_h)/(m_h - 1)$. The statistics from the phase II sample are as follows:

Stratum	n_h	m_h	$\hat{p}_h$	$s_h^{2(2)}$
Yes	755	67	0.7313	0.1995
No	804	72	0.1528	0.1313
Not available	505	505	0.4178	0.2437
Total	2064	644		

The estimated percentage of Vietnam-era VA hospital patients who served in Vietnam is, from (12.5),

$$\hat{\bar{y}}_{str}^{(2)} = \left(\frac{755}{2064}\right)(0.7313) + \left(\frac{804}{2064}\right)(0.1528) + \left(\frac{505}{2064}\right)(0.4178) = 0.4293.$$

The phase I sample is an SRS with $n/N = 0.2$, so the fpc should be included in the variance estimate. Calculating the terms in (12.7),

$$\sum_{h=1}^{H} \left(\frac{n_h - 1}{n - 1} - \frac{m_h - 1}{N - 1}\right) \frac{n_h}{n} \frac{s_h^{2(2)}}{m_h} = 0.000391 + 0.000271 + 0.0000231 = 0.000686,$$

and

$$\frac{1}{n-1}\left(1 - \frac{n}{N}\right) \sum_{h=1}^{H} \frac{n_h}{n}(\bar{y}_h^{(2)} - \hat{\bar{y}}_{str}^{(2)})^2 = 1.29 \times 10^{-5} + 1.16 \times 10^{-5} + 1.24 \times 10^{-8}$$

$$= 0.0000245.$$

Thus, $\hat{V}(\hat{\bar{y}}_{str}^{(2)}) = 0.000686 + 0.0000245 = 0.00071$, and $SE(\hat{\bar{y}}_{str}^{(2)}) = 0.027$.

Was two-phase sampling more efficient here? Had an SRS of size 644 been taken directly from the records, and had $\hat{p} = 0.429$ been observed, the standard error would have been $SE(\hat{p}) = 0.019$, which is actually smaller than the standard error from the two-phase sampling design. If you look at the individual terms in the variance estimates, you can see why two-phase sampling did not increase efficiency in this example. All of the phase I units in the "not available" stratum were subsampled, giving a very low value of $s_h^{2(2)}/m_h$ for that stratum. But the sample sizes in the other two strata were too small, leading to relatively large contributions to the overall variance from those two strata.

Suppose proportional allocation had been used in the phase II sample instead and that the same sample proportions had been observed. Then, you would subsample 236 records in the "yes" stratum, 251 records in the "no" stratum, and 157 records in the "not available" stratum. In that case, if the sample proportions remained the same, the standard error from the two-phase sample would have been 0.017, a modest decrease from the standard error of an SRS of size 644. But proportional allocation does not make the most efficient use of the phase I information. More savings would have been achieved if some sort of optimal allocation had been used (see Exercise 9). ■

Ideally, you would use the information about stratum membership from phase I to have a more efficient sampling design in phase II. This usually means using optimal allocation in the stratified phase II sample. For example, in a survey to study total sales of manufacturing firms, you might obtain total revenue from the tax records (x) for a sample of manufacturing firms. Then you could use that tax information to stratify the phase I sample by the reported revenue, and take higher sampling fractions in phase II for the strata with higher revenue in the tax records.

A screening survey is a special case of a two-phase sample using stratification. The U.S. National Immunization Survey (Smith et al., 2005) collects a phase I sample using random digit dialing. The households in the phase I sample are divided into two strata: (1) households with children 19–35 months old, and (2) households with no children between 19 and 35 months old. Since the goal of the survey is to estimate vaccination rates for children in the 19–35 month age group, no households in stratum 2 are included in the phase II sample. The parent or guardian in stratum 1 households is asked to consent for information to be obtained from the child's vaccination providers, and those providers are asked about the child's immunizations. Nonresponse and other nonsampling errors in the phase II sample require weighting adjustments.

EXAMPLE 12.5 McNamee (2003) discusses the use of two-phase sampling to estimate disease prevalence. In the first phase an inexpensive, but not completely accurate, method is used to classify persons as having the disease or not. The second phase is a more accurate test for the disease. For example, the phase I survey might ask people whether they have diabetes, and divide the respondents into stratum 1, persons who say they have diabetes, and stratum 2, persons who say they do not have diabetes. But some persons with diabetes are unaware that they have it. You therefore need to subsample both strata in the phase II sample, which evaluates persons through a medical examination, to guarantee that diabetics who are unaware they have diabetes can be included in the sample. Although we expect a smaller fraction of the persons in stratum 2 to have diabetes, compared with the fraction in stratum 1 who have diabetes, the characteristics of persons with diabetes in stratum 2 might be quite different from those in stratum 1. ∎

12.3
Ratio and Regression Estimation in Two-Phase Samples

The stratified two-phase sampling design in Section 12.2 uses the auxiliary information collected from the phase I sample in the design of the phase II sample. Alternatively, or in addition, the information about the auxiliary variables x can be used in the estimator, through ratio and regression estimation.

12.3.1 Two-Phase Sampling with Ratio Estimation

Suppose that x, a variable thought to be highly correlated with y, can be measured inexpensively in the phase I sample. Define $\mathcal{S}^{(1)}$, $\mathcal{S}^{(2)}$, Z_i, and D_i as in Section 12.1. The auxiliary variable x_i is measured for each observation in the phase I sample; from that sample, we may estimate the population total $t_x = \sum_{i=1}^{N} x_i$ by

$$\hat{t}_x^{(1)} = \sum_{i \in \mathcal{S}^{(1)}} w_i^{(1)} x_i = \sum_{i=1}^{N} Z_i w_i^{(1)} x_i.$$

Now select the phase II subsample and measure y_i on units in the subsample. From the phase II sample $\mathcal{S}^{(2)}$, we can calculate $\hat{t}_y^{(2)}$ using (12.1) and

$$\hat{t}_x^{(2)} = \sum_{i \in \mathcal{S}^{(2)}} w_i^{(1)} w_i^{(2)} x_i = \sum_{i=1}^{N} Z_i D_i w_i^{(1)} w_i^{(2)} x_i.$$

Then,

$$\hat{t}_{yr}^{(2)} = \hat{t}_x^{(1)} \frac{\hat{t}_y^{(2)}}{\hat{t}_x^{(2)}} = \hat{t}_x^{(1)} \hat{B}^{(2)}. \tag{12.9}$$

Note that this estimator is very similar to the ratio estimator in (4.2); we use $\hat{t}_x^{(1)}$ from the phase I sample instead of the unknown quantity t_x.

Using linearization,

$$\hat{t}_{yr}^{(2)} \approx t_y + \frac{t_x}{t_x}(\hat{t}_y^{(2)} - t_y) + \frac{t_y}{t_x}(\hat{t}_x^{(1)} - t_x) - \frac{t_y t_x}{t_x^2}(\hat{t}_x^{(2)} - t_x).$$

Then,

$$
\begin{aligned}
V(\hat{t}_{yr}^{(2)}) &\approx V\left[\hat{t}_y^{(2)} + \frac{t_y}{t_x}(\hat{t}_x^{(1)} - \hat{t}_x^{(2)})\right] \\
&= V\left\{E\left[\hat{t}_y^{(2)} + \frac{t_y}{t_x}(\hat{t}_x^{(1)} - \hat{t}_x^{(2)}) \,\Big|\, \mathbf{Z}\right]\right\} + E\left\{V\left[\hat{t}_y^{(2)} + \frac{t_y}{t_x}(\hat{t}_x^{(1)} - \hat{t}_x^{(2)}) \,\Big|\, \mathbf{Z}\right]\right\} \\
&= V[\hat{t}_y^{(1)}] + E\left[V\left(\hat{t}_y^{(2)} - \frac{t_y}{t_x}\hat{t}_x^{(2)} \,\Big|\, \mathbf{Z}\right)\right] \\
&= V[\hat{t}_y^{(1)}] + E\left[V(\hat{t}_d^{(2)} \mid \mathbf{Z})\right],
\end{aligned}
$$

where $d_i = y_i - (t_y/t_x)x_i$. Thus, the variance of the two-phase ratio estimator is the variance that would be calculated for $\hat{t}_y^{(1)}$ if we observed y_i for every unit in the phase I sample, plus an extra term involving the variance of the residuals from the ratio model.

If an SRS of $n^{(1)}$ units is taken for phase I and an SRS of $n^{(2)}$ units is taken in phase II, then

$$V(\hat{t}_{yr}^{(2)}) \approx N^2 \left(1 - \frac{n^{(1)}}{N}\right) \frac{S_y^2}{n^{(1)}} + N^2 \left(1 - \frac{n^{(2)}}{n^{(1)}}\right) \frac{S_d^2}{n^{(2)}}, \tag{12.10}$$

where $d_i = y_i - Bx_i$ and $S_d^2 = \sum_{i=1}^{N} d_i^2/(N-1)$, and

$$\hat{V}(\hat{t}_{yr}^{(2)}) = N^2 \left(1 - \frac{n^{(1)}}{N}\right) \frac{s_y^2}{n^{(1)}} + N^2 \left(1 - \frac{n^{(2)}}{n^{(1)}}\right) \frac{s_e^2}{n^{(2)}}, \tag{12.11}$$

where $s_y^2 = \sum_{i \in \mathcal{S}^{(2)}} (y_i - \bar{y}^{(2)})^2/(n^{(2)} - 1)$ and $s_e^2 = \sum_{i \in \mathcal{S}^{(2)}} (y_i - \hat{B}^{(2)}x_i)^2/(n^{(2)} - 1)$, is an approximately unbiased estimator of $V(\hat{t}_{yr}^{(2)})$ (see Exercise 12). Another estimator of the variance is given in Exercise 13.

EXAMPLE **12.6**   Suppose, for the population in agpop.dat sampled in Examples 2.5 and 4.2, that we do not know the value of $x_i = acres87$, the acreage devoted to farms in 1987, or of t_x before sampling. To use x as auxiliary information through two-phase ratio estimation, we take an SRS of size 400, measure *acres87* on every unit in this phase I sample, and then take an SRS of size 30 from the phase I sample to serve as the phase II sample. We measure $y = acres92$ on the units in the phase II sample. We then employ ratio estimation to estimate t_y, using $\hat{t}_x^{(1)}$ to estimate the unknown auxiliary population total t_x. Using the SAS code on the website, we calculate

$$\hat{t}_{yr}^{(2)} = \hat{t}_x^{(1)} \frac{\hat{t}_y^{(2)}}{\hat{t}_x^{(2)}} = \hat{t}_x^{(1)} \frac{\bar{y}^{(2)}}{\bar{x}^{(2)}} = 1,002,814,347 \left(\frac{322,385}{335,444} \right) = 963,774,784.$$

From (12.11),

$$\hat{V}(\hat{t}_{yr}^{(2)}) = N^2 \left(1 - \frac{n^{(1)}}{N} \right) \frac{s_y^2}{n^{(1)}} + N^2 \left(1 - \frac{n^{(2)}}{n^{(1)}} \right) \frac{s_e^2}{n^{(2)}}$$

$$= 3000^2 \left(1 - \frac{400}{3000} \right) \frac{112,160,218,976}{400} + 3000^2 \left(1 - \frac{30}{400} \right) \frac{1,908,426,448}{30}$$

$$= 2.7 \times 10^{15}.$$

Thus, an approximate 95% confidence interval (CI) for t_y is $963,774,784 \pm 2.05 \times \sqrt{2.7 \times 10^{15}} = [856,924,072, 1,070,625,496]$. The corresponding CI for $\bar{y}_U$ is $[285,641, 356,875]$ with $\hat{\bar{y}}_r^{(2)} = 321,258$. The widths of these CIs are comparable to those of the SRS of size 300 in Example 2.10, even though here y_i was measured only on the 30 units in the phase II sample. If measuring x is inexpensive relative to measuring y, the high correlation between x and y makes two-phase sampling with ratio estimation very efficient. ∎

12.3.2 Generalized Regression Estimation in Two-Phase Sampling

We can also use a two-phase version of the generalized regression (GREG) estimator of Section 11.7. The two-phase GREG estimator takes the form

$$\hat{t}_{yGREG}^{(2)} = \hat{t}_y^{(2)} + (\hat{\mathbf{t}}_{\mathbf{x}}^{(1)} - \hat{\mathbf{t}}_{\mathbf{x}}^{(2)})^T \hat{\mathbf{B}}^{(2)}, \tag{12.12}$$

where

$$\hat{\mathbf{B}}^{(2)} = \left(\sum_{i \in \mathcal{S}^{(2)}} w_i^{(1)} w_i^{(2)} \frac{1}{\sigma_i^2} \mathbf{x}_i \mathbf{x}_i^T \right)^{-1} \sum_{i \in \mathcal{S}^{(2)}} w_i^{(1)} w_i^{(2)} \frac{1}{\sigma_i^2} \mathbf{x}_i y_i \tag{12.13}$$

and the constants σ_i^2 are determined by the analyst. In (12.12), the estimator is calibrated to the estimated population totals of $\mathbf{x}$ from phase I, $\hat{\mathbf{t}}_{\mathbf{x}}^{(1)}$ (see Exercise 15).

The estimator in (12.12) may be written using a modification of the weights. Analogously to (11.22), let

$$g_i = 1 + (\hat{\mathbf{t}}_{\mathbf{x}}^{(1)} - \hat{\mathbf{t}}_{\mathbf{x}}^{(2)})^T \left(\sum_{i \in \mathcal{S}^{(2)}} w_i^{(1)} w_i^{(2)} \frac{1}{\sigma_i^2} \mathbf{x}_i \mathbf{x}_i^T \right)^{-1} \frac{1}{\sigma_i^2} \mathbf{x}_i.$$

Then,

$$\hat{t}_{yGREG}^{(2)} = \sum_{i \in \mathcal{S}^{(2)}} w_i^{(1)} w_i^{(2)} g_i y_i.$$

We again use Property A.4 in Section A.4 to find $V\left(\hat{t}_{y\mathrm{GREG}}^{(2)}\right)$:

$$V\left(\hat{t}_{y\mathrm{GREG}}^{(2)}\right) = V\left[E\left(\hat{t}_{y\mathrm{GREG}}^{(2)} \mid \mathbf{Z}\right)\right] + E\left[V\left(\hat{t}_{y\mathrm{GREG}}^{(2)} \mid \mathbf{Z}\right)\right]. \tag{12.14}$$

Since the GREG estimator is approximately unbiased, if the sizes of the phase I and phase II samples are sufficiently large, then

$$V\left[E\left(\hat{t}_{y\mathrm{GREG}}^{(2)} \mid \mathbf{Z}\right)\right] \approx V\left(\hat{t}_y^{(1)}\right),$$

where $V(\hat{t}_y^{(1)})$ is the variance of the estimator $\hat{t}_y^{(1)} = \sum_{i \in \mathcal{S}^{(1)}} w_i^{(1)} y_i$ of the population total we would have if we had been able to measure y on every unit in the phase I sample. By linearization, the conditional variance in the second term of (12.14) is

$$V\left(\hat{t}_{y\mathrm{GREG}}^{(2)} \mid \mathbf{Z}\right) = V\left(\sum_{i \in \mathcal{S}^{(2)}} w_i^{(1)} w_i^{(2)} d_i^{(1)} \,\middle|\, \mathbf{Z}\right)$$

where $d_i^{(1)} = y_i - \mathbf{x}_i^T \hat{\mathbf{B}}^{(1)}$ and $\hat{\mathbf{B}}^{(1)} = \left(\sum_{i \in \mathcal{S}^{(1)}} w_i^{(1)} \frac{1}{\sigma_i^2} \mathbf{x}_i \mathbf{x}_i^T\right)^{-1} \sum_{i \in \mathcal{S}^{(1)}} w_i^{(1)} \frac{1}{\sigma_i^2} \mathbf{x}_i y_i$.

Särndal et al. (1992, Chapter 9) estimate the two terms in (12.14) separately. The conditional variance in the second term is unbiased for its expectation, so $E\left[V\left(\hat{t}_{y\mathrm{GREG}}^{(2)} \mid \mathbf{Z}\right)\right]$ may be estimated by

$$\hat{V}\left(\hat{t}_{y\mathrm{GREG}}^{(2)} \mid \mathbf{Z}\right) = \hat{V}\left(\sum_{i \in \mathcal{S}^{(2)}} w_i^{(1)} w_i^{(2)} e_i \,\middle|\, \mathbf{Z}\right),$$

with $e_i = y_i - x_i \hat{B}^{(2)}$ substituted for the unknown values $d_i^{(1)}$. If the phase I sampling design is an SRS of $n^{(1)}$ units, the first term in (12.14) may be estimated by

$$\hat{V}(\hat{t}_y^{(1)}) = N^2 \left(1 - \frac{n^{(1)}}{N}\right) \frac{\hat{S}_y^2}{n^{(1)}},$$

with

$$\hat{S}_y^2 = \frac{1}{n^{(2)} - 1} \sum_{i \in \mathcal{S}^{(2)}} (y_i - \bar{y}^{(2)})^2.$$

Estimating the first term in (12.14) is more challenging for complex designs; Exercise 16 presents an estimator for this situation.

EXAMPLE 12.7   Barnett et al. (2001) describe an application of two-phase sampling to an accounting problem. The auditor has access to a large phase I SRS of transactions. Each transaction in the phase I sample has been checked by internal auditors, who record errors they find. A small random subsample is taken of the phase I transactions; each transaction in the phase II SRS is examined by an external auditor. If the internal and external auditor disagree on a transaction, the external auditor is assumed to be correct. If the internal auditors are largely correct, then regression estimation can be used, either through ratio estimation or poststratification into classes based on types of errors, to greatly increase the precision of the amounts of errors in the population of transactions. ∎

12.4
Jackknife Variance Estimation for Two-Phase Sampling

As we have seen, the formulas for variance estimators are complicated in two-phase sampling. Fortunately, in many cases we can use resampling methods to estimate variances. In this section, we describe jackknife variance estimators for two-phase sampling, studied by Rao and Shao (1992), Rao and Sitter (1995, 1997), Kott and Stukel (1997), Sitter (1997), and Kim et al. (2006). The jackknife method presented in Section 9.3.2 needs to be modified for two-phase sampling because y is observed only for units selected in phase II. We mimic the sampling design, including the two phases of sampling, in the resamples. The jackknife estimates the with-replacement variance; it is a good approximation to the without-replacement variance if the sampling fractions are small.

Suppose the phase I design is an SRS of size $n^{(1)}$ and the phase II design is an SRS of size $n^{(2)}$. Consider the ratio estimator of t_y in Section 12.3.1,

$$\hat{t}_{yr}^{(2)} = \hat{t}_x^{(1)} \frac{\hat{t}_y^{(2)}}{\hat{t}_x^{(2)}}.$$

When we delete unit j in the phase I sample, we obtain

$$\hat{t}_{yr(j)}^{(2)} = \hat{t}_{x(j)}^{(1)} \frac{\hat{t}_{y(j)}^{(2)}}{\hat{t}_{x(j)}^{(2)}},$$

where $\hat{t}_{x(j)}^{(1)}$, $\hat{t}_{y(j)}^{(2)}$, and $\hat{t}_{x(j)}^{(2)}$ are calculated using the jackknife weights as described in the next paragraph. Then,

$$\hat{V}_{\text{JK}}(\hat{t}_{yr}^{(2)}) = \sum_{j \in \mathcal{S}^{(1)}} \frac{n^{(1)} - 1}{n^{(1)}} \left[\hat{t}_{yr(j)}^{(2)} - \hat{t}_{yr}^{(2)} \right]^2.$$

When both samples are SRSs, $w_i^{(1)} = N/n^{(1)}$ for the phase I sample and $w_i^{(2)} = n^{(1)}/n^{(2)}$ for the phase II sample. The jackknife weights for phase I are constructed exactly like those in Section 9.3.2. When unit j from the phase I sample is deleted, the modified weight for phase I is

$$w_{i(j)}^{(1)} = \begin{cases} 0 & \text{if } i = j \\ \dfrac{n^{(1)}}{n^{(1)} - 1} w_i^{(1)} & \text{if } i \neq j. \end{cases}$$

The modified weight for phase II depends on whether the unit deleted from the phase I sample is in the phase II sample or not:

$$w_{i(j)}^{(2)} = \begin{cases} 0 & \text{if } i = j \text{ and } j \in \mathcal{S}^{(2)} \\ \dfrac{n^{(1)} - 1}{n^{(2)} - 1} & \text{if } i \neq j \text{ and } j \in \mathcal{S}^{(2)} \\ \dfrac{n^{(1)} - 1}{n^{(2)}} & \text{if } j \notin \mathcal{S}^{(2)} \end{cases}$$

Using the jackknife weights,

$$\hat{t}_{x(j)}^{(1)} = \sum_{i \in \mathcal{S}^{(1)}} w_{i(j)}^{(1)} x_i = N \frac{n^{(1)} \bar{x}^{(1)} - x_j}{n^{(1)} - 1}$$

$$\hat{t}_{x(j)}^{(2)} = \sum_{i \in \mathcal{S}^{(2)}} w_{i(j)}^{(1)} w_{i(j)}^{(2)} x_i = \begin{cases} N \dfrac{n^{(2)} \bar{x}^{(2)} - x_j}{n^{(2)} - 1} & \text{if } j \in \mathcal{S}^{(2)} \\ \hat{t}_x^{(2)} & \text{if } j \notin \mathcal{S}^{(2)} \end{cases}$$

and

$$\hat{t}_{y(j)}^{(2)} = \sum_{i \in \mathcal{S}^{(2)}} w_{i(j)}^{(1)} w_{i(j)}^{(2)} y_i = \begin{cases} N \dfrac{n^{(2)} \bar{y}^{(2)} - y_j}{n^{(2)} - 1} & \text{if } j \in \mathcal{S}^{(2)} \\ \hat{t}_y^{(2)} & \text{if } j \notin \mathcal{S}^{(2)} \end{cases}$$

For two-phase sampling with stratification, suppose that an SRS of size $n^{(1)}$ is taken in phase I and a stratified random sample is taken in phase II, where the sampling fractions for the strata are specified before the phase I sample is collected. Consider the estimator $\hat{t}_{\text{str}}^{(2)}$ in (12.4). Define the jackknife replicate, deleting unit j, by

$$\hat{t}_{\text{str}(j)}^{(2)} = \sum_{i \in \mathcal{S}^{(2)}} w_{i(j)}^{(1)} w_{i(j)}^{(2)} y_i,$$

where

$$w_{i(j)}^{(1)} = \begin{cases} 0 & \text{if } j = i \\ \dfrac{N}{n^{(1)} - 1} & \text{if } j \neq i, \end{cases} \qquad w_{i(j)}^{(2)} = \begin{cases} \dfrac{n_h}{m_h} & \text{if } x_{ih} = 1, x_{jh} \neq 1 \\ \dfrac{n_h - 1}{m_h} & \text{if } x_{ih} = 1, x_{jh} = 1, j \notin \mathcal{S}^{(2)} \\ \dfrac{n_h - 1}{m_h - 1} & \text{if } x_{ih} = 1, x_{jh} = 1, j \in \mathcal{S}^{(2)}. \end{cases}$$

Then,

$$\hat{V}_{\text{JK}}(\hat{t}_{\text{str}}^{(2)}) = \sum_{j \in \mathcal{S}^{(1)}} \frac{n^{(1)} - 1}{n^{(1)}} \left[\hat{t}_{\text{str}(j)}^{(2)} - \hat{t}_{\text{str}}^{(2)} \right]^2.$$

See Kim et al. (2006) for jackknife variance estimators for other designs and estimators. As always, the jackknife estimates the with-replacement variance—in two-phase sampling, both phases are assumed to be sampled with replacement.

12.5
Designing a Two-Phase Sample

Two-phase sample designs require all the considerations of one-phase samples, plus the additional decision of how many resources to devote to each phase. A two-phase sample is more complicated than a one-phase sample; before you use one, study the relative costs and make sure a two-phase sample really will be more efficient. The two-phase design for the veterans survey in Examples 12.1 and 12.4 was actually less efficient than a one-phase design would have been. A two-phase sample uses resources to measure x on units that are not subsampled in phase II—resources that could alternatively be used to measure y on additional units. If x and y are strongly related, then the information in the phase I sample improves the efficiency of data collection. But if x and y are not related—for example, if x is the last digit of a student's telephone number and y is the student's grade point average—then the resources used to measure x are essentially wasted; you would be better off if you just sampled y directly.

Deming (1977) discusses issues to be considered when deciding whether to use a two-phase sample. A two-phase design adds complexity for both administering and analyzing the survey. It also can increase respondent burden, since in many cases respondents need to be contacted twice. If a two-phase design is used to identify persons with a certain characteristic such as diabetes, persons ultimately selected for the phase II sample may first be asked to answer a questionnaire for phase I and then be asked to participate in a medical examination for phase II. However, the two-phase sample has the advantage that it can give useful information about the screening method used in phase I.

12.5.1 Two-Phase Sampling with Stratification

Consider the situation in which phase I is an SRS and phase II is a stratified random sample. Efficiency gains for two-phase sampling arise when more observations are subsampled in strata with large variance, large values of N_h, or low cost. Rao (1973) proposes letting $m_h = v_h n_h$ for stratum h, with v_h, $h = 1, \ldots, H$, being constants to be determined before sampling. Let $c^{(1)}$ be the cost to sample a unit in the SRS taken for phase I and to determine its stratum membership. Let c_h be the cost of measuring y for a unit in stratum h in phase II. Assume the total cost will be a linear function, with

$$C = c^{(1)} n^{(1)} + \sum_{h=1}^{H} c_h m_h. \tag{12.15}$$

The total cost C varies from sample to sample, since the m_h's are only determined after the phase I sample is taken. The expected cost, however, is

$$E[C] = c^{(1)} n^{(1)} + n^{(1)} \sum_{h=1}^{H} c_h v_h W_h \tag{12.16}$$

where $W_h = N_h/N$. With v_h fixed, we can write $V(\hat{\bar{y}}_{\text{str}}^{(2)})$ from (12.6) as:

$$V(\hat{\bar{y}}_{\text{str}}^{(2)}) = S_y^2 \left(\frac{1}{n^{(1)}} - \frac{1}{N} \right) + \frac{1}{n^{(1)}} \sum_{h=1}^{H} W_h S_h^2 \left(\frac{1}{v_h} - 1 \right).$$

Then $V(\hat{\bar{y}}_{\text{str}}^{(2)})$ is minimized, subject to the constraint in (12.16), when

$$v_{h,\text{opt}} = \sqrt{\frac{c^{(1)} S_h^2}{c_h \left(S^2 - \sum_{j=1}^{H} W_j S_j^2 \right)}} \tag{12.17}$$

(see Exercise 17). If $v_{h,\text{opt}} > 1$ for a stratum h, then set $v_{h,\text{opt}} = 1$ and recalculate the other values. With a predetermined expected cost C^*, the phase I sample should have size

$$n_{\text{opt}}^{(1)} = \frac{C^*}{c^{(1)} + \sum_{h=1}^{H} c_h W_h v_{h,\text{opt}}}.$$

If $0 < v_{h,\text{opt}} < 1$ for $h = 1, \ldots, H$ and the optimal allocation is used, then, as shown in Exercise 18, the two-phase sample has variance

$$V_{\text{opt}}(\hat{\bar{y}}_{\text{str}}^{(2)}) = \frac{1}{C^*} \left[\sum_{h=1}^{H} W_h S_h \sqrt{c_h} + \sqrt{c^{(1)}} \sqrt{S_y^2 - \sum_{h=1}^{H} W_h S_h^2} \right]^2 - \frac{S_y^2}{N}. \quad (12.18)$$

Finding the optimal sample sizes for two-phase sampling with stratification requires estimates of the within-stratum variances S_h^2, similarly to the optimal allocation in stratified sampling discussed in Section 3.4.2. In addition, since the stratum sizes are unknown, the values of W_h and S^2 must also be estimated or guessed.

We can compare the variance achieved by a two-phase stratified sample with optimal allocation with the variance that we would get if we measured y on a one-phase sample with the same cost. For simplicity, assume $c_h = c^{(2)}$ for $h = 1, \ldots, H$. This is a reasonable cost structure for two-phase studies used to estimate disease prevalence, for example, in which all persons sampled in phase II are given the same medical examination. If, instead of taking a two-phase sample, we took an SRS in one phase with the same cost C^*, we could sample $n' = C^*/c^{(2)}$ units. Then, if S_y^2/N and $(1 - n'/N)$ are negligible, the ratio of the two-phase variance with optimal allocation to the one-phase variance with the same expected cost is approximately

$$\frac{V_{\text{opt}}(\hat{\bar{y}}_{\text{str}}^{(2)})}{V_{\text{SRS}}(\bar{y})} \approx \left[\sum_{h=1}^{H} W_h \frac{S_h}{S_y} + \sqrt{\frac{c^{(1)}}{c^{(2)}}} \sqrt{\frac{S_y^2 - \sum_{h=1}^{H} W_h S_h^2}{S_y^2}} \right]^2. \quad (12.19)$$

Thus, a two-phase sample with stratification is more efficient than an SRS for estimating $\bar{y}_U$ if the within-strata variances S_h^2 are small relative to S_y^2 and if the cost to sample phase I units is smaller than the cost to sample phase II units.

For two-phase sampling with stratification, n_h, the number of units in the phase I sample that are in stratum h, is a random variable. If we select a different phase I sample from the population, it is likely that we will get a different value for n_h. It is possible that some phase I samples will have $n_h = 0$ for one or more strata. In that case, we cannot subsample that stratum in phase II, and y will not be measured on any unit in stratum h. The estimator in (12.4) is unbiased for the population total only if we assume $n_h > 0$ for all strata. Similarly, we need a subsample size of at least 2 in each stratum to estimate the variance within the stratum. We thus want to design a two-phase sample so that $P(n_h = 0)$ is extremely small. If an SRS is taken at phase I and $n^{(1)}/N$ is small, then $P(n_h = 0) \approx (1 - N_h/N)^{n^{(1)}}$, so $P(n_h = 0)$ is small if $n^{(1)}$ is large or N_h/N is not close to 0. Strata should thus be formed so that all strata are large enough to have very high probability of being represented in the phase I sample.

12.5.2　Optimal Allocation for Ratio Estimation

How should the sample be allocated to phase I and phase II if ratio estimation is to be used? Suppose an SRS is taken in each of phase I and phase II and that the total cost of the two-phase sample is $C = c^{(1)}n^{(1)} + c^{(2)}n^{(2)}$. In Exercise 19, you will show that the variance of $\hat{t}_{yr}^{(2)}$, given in (12.10), is minimized subject to a fixed cost C when

$$\frac{n^{(2)}}{n^{(1)}} = v = \sqrt{\frac{c^{(1)}S_d^2}{c^{(2)}(S_y^2 - S_d^2)}}, \qquad (12.20)$$

where S_d^2 is the population variance of the residuals $d_i = y_i - Bx_i$. Consequently, for a fixed cost C, the optimal phase I sample size is

$$n^{(1)} = \frac{C}{c^{(1)} + vc^{(2)}}$$

and the optimal phase II sample size is

$$n^{(2)} = \frac{C - n^{(1)}c^{(1)}}{c^{(2)}}.$$

The optimal sample sizes can often be estimated using results from a preliminary survey or prior work. If x and y are highly correlated, we expect S_d^2 to be small relative to S_y^2. In that situation, it makes sense to measure x on a large phase I sample, particularly if the cost of measuring x is small, and use a relatively small phase II sample to estimate B.

12.6
Chapter Summary

Two-phase sampling can increase precision of estimators of t_y for a fixed budget if there exist auxiliary variables **x** such that (1) the cost of measuring **x** is low compared with the cost of measuring y and (2) the auxiliary variables are correlated with y. The auxiliary information collected in the phase I sample can be used to improve the efficiency of the phase II sampling design, as when the auxiliary information **x** collected at phase I is used to stratify the phase II sample. Alternatively, or additionally, the information in the phase I sample can be used through ratio or regression estimation.

Key Terms

Phase I sample: A sample selected from a population on which auxiliary variables **x** are measured.

Phase II sample: A subsample selected from the phase I sample on which the variable of interest y is measured.

Two-phase sampling: A sampling design in which a preliminary (phase I) sample is selected from the population, and then a subsample (phase II sample) is selected from the phase I sample.

For Further Reading

Watson (1937) presents an early example of two-phase sample for regression. Neyman (1938) developed theory for two-phase sampling with stratification. See Cochran (1977) for more discussion on two-phase sampling with simple random samples; Särndal et al. (1992, Chapter 9) and Legg and Fuller (2009) give a theoretical development for general probability sampling designs. Hidiroglou et al. (2009) develop Sen-Yates-Grundy-type variance estimators for two-phase samples. Rao and Sitter (1995) and Kim et al. (2006) present jackknife variance estimators for two-phase sampling. Armstrong et al. (1993) give an example of two-phase sampling for tax records. Hidiroglou (2001) discusses non-nested two-phase sampling designs.

12.7
Exercises

A. Introductory Exercises

1 A health official takes a two-phase sample to estimate the prevalence of diabetes in a population. In phase I, an SRS of size 1000 is taken from the population of size 100,000, and each individual is asked demographic information and whether he or she has diabetes. It is known that some demographic groups are more at risk for diabetes than others; in addition, the self report of diabetes may be inaccurate. Therefore, each individual in the phase II sample is given a medical exam to determine diabetes status. The phase I sample is divided into 4 strata. Stratum h has n_h observations in the phase I sample and m_h observations in the phase II sample. After the medical exam, r_h persons in stratum h of the phase II sample were determined to have diabetes.

Stratum	n_h	m_h	r_h
High risk group and reports diabetes	241	96	86
High risk group and does not report diabetes	113	45	17
Low risk group and reports diabetes	174	35	29
Low risk group and does not report diabetes	472	47	8

Estimate the total number of persons with diabetes in the population, along with its standard error.

2 Data mining methods in statistics are used to discover relationships among variables in very large data sets (Hastie et al., 2001). A company, for example, has databases of all its financial transactions. A very small fraction of these transactions involve fraud, but fraudulent transactions are expensive for the company. Discovering whether a transaction is fraudulent requires an investigation, so the company can only determine whether transactions are fraudulent for a small sample. Discuss how two-phase sampling might be used to improve prediction of fraudulent transactions. (Chen et al., 2002, discuss using two-phase sampling in data mining, but with purposive sampling at phase II rather than probability sampling.)

B. Working with Survey Data

3 Bart and Earnst (2002) describe the use of two-phase sampling with ratio estimation to estimate the density of nesting birds. The phase I sample, selected from the 2130 plots in the region of interest, is conducted using a rapid search method involving bird sightings to obtain an approximate count of birds in each phase I plot. Then, a subsample of 12 of the phase I sample plots are surveyed using an intensive method to obtain a more accurate count of the number of nests in each plot. In the intensive method, a surveyor visits a plot for several hours over a period of days and searches for nests and other indications of territorial males in the plot. In this setting, $x_i =$ number of nests counted in plot i using the rapid method, and $y_i =$ number of nests counted in plot i using the intensive method. Using the data in file shorebirds.dat, which were generated using summary statistics from Bart and Earnst (2002), estimate the total number of nests using the two-phase ratio estimator. Give the standard error of your estimate.

4 Dunn et al. (1999) discuss issues in analyzing two-phase data to estimate prevalence of psychiatric disorders. Participants in a phase I sample were given the General Health Questionnaire (GHQ) and classified into three strata based on their GHQ score. The stratification was used to take a stratified random sample of 250 persons for the phase II sample; the Composite International Diagnostic Interview (CIDI), considered to be a more accurate diagnostic tool, was administered to each person in the phase II sample. The CIDI score was used to classify the phase II sample members as having at least one psychiatric disorder (case) or having no psychiatric disorder (non-case). The results are given in the following table. The counts of cases and non-cases are from the phase II sample.

Stratum	n_h	m_h	Non-case	Case
GHQ $\leq$ 3 (low)	1049	60	33	27
GHQ $=$ 4, 5 (medium)	237	48	14	34
GHQ $\geq$ 6 (high)	272	142	23	119
Total	1558	250		

a Calculate the phase II sampling weight $w_i^{(2)}$ for each stratum.

b Use the two-phase sample to estimate the percentage of persons with at least one psychiatric disorder, along with its standard error. Since we do not know the population size N, use a relative phase I weight of $w_i^{(1)} = 1$.

5 Dunn et al. (1999) also classified the phase II sample by gender. In the following table, the entries in columns 3–7 are the counts from the phase II sample in the categories: Male Non-Case (MNC), Male Case (MC), Female Non-Case (FNC), and Female Case (FC).

Stratum	n_h	m_h	MNC	MC	FNC	FC
GHQ $\leq$ 3 (low)	1049	60	16	8	17	19
GHQ $=$ 4, 5 (medium)	237	48	9	8	5	26
GHQ $\geq$ 6 (high)	272	142	15	28	8	91

a   Estimate the percentages of persons in each cell of a 2×2 contingency table classified by gender and case/non-case. Find the standard error of each entry in the table.

b   Find the design effect for each cell proportion $\hat{p}_{ij}$ and marginal proportion ($\hat{p}_{i+}$ and $\hat{p}_{+j}$) in the table.

c   Use the Rao–Scott method (Section 10.3.2) to test $H_0 : p_{ij} = p_{i+}p_{+j}$.

6   Ismail et al. (2002) report results of a two-phase survey to estimate prevalence of psychiatric disorders in Gulf War veterans. A random sample of the $N = 53,462$ Gulf War veterans was administered the SF-36 questionnaire, a 36-question survey on health. Respondents who scored below 72.2 on physical functioning subscale were defined as disabled (stratum 1, with $n_1 = 406$); respondents who scored above 72.2 were defined as not disabled (stratum 2, with $n_2 = 3047$). A random subsample of 111 veterans was taken from stratum 1, and a random subsample of 98 veterans was taken from stratum 2. The 209 veterans in the phase II sample were evaluated by psychiatrists; the counts in the table below give the number of veterans who are determined to have any alcohol related disorder (Alcohol), any sleep disorder (Sleep), or any psychiatric related disorder (Psych). A veteran can be in more than one of these categories.

Stratum	n_h	m_h	Alcohol	Sleep	Psych
Disabled	406	111	8	20	27
Not disabled	3047	98	10	17	12

Estimate the total number of Gulf War veterans with an alcohol related disorder, and give a 95% CI.

7   Repeat Exercise 6 for sleep disorders.

8   Repeat Exercise 6 for any psychiatric disorder.

9   Use the results of Section 12.5 to determine an optimal allocation for a follow-up survey similar to that in Example 12.1. Assume that the relative costs are $c^{(1)} = 1$ and $c_h = 20$ for $h = 1, 2, 3$. Use the data in Example 12.1 to estimate quantities such as W_h and S_h^2. How does your allocation differ from the one used? From proportional allocation?

C. Working with Theory

10   (Requires probability.) Suppose the phase I sample is an SRS of size $n^{(1)}$, and the phase II subsample is an SRS of size $n^{(2)}$, with $n^{(2)} < n^{(1)}$. Show that

$$V(\hat{t}_y^{(2)}) = N^2 \left(1 - \frac{n^{(2)}}{N}\right) \frac{S_y^2}{n^{(2)}}$$

the same variance that would result if a SRS of size $n^{(2)}$ were taken directly.

11   *Estimating the variance in two-phase sampling for stratification.* Show that (12.7) is an approximately unbiased estimator of $V\left(\hat{t}_{ystr}^{(2)}\right)$ in large samples. HINT: Use the result derived from Table 3.3 in Chapter 3 that

$$S^2 = \left[\sum_{h=1}^{H} (N_h - 1)S_h^2 + \sum_{h=1}^{H} N_h(\bar{y}_{hU} - \bar{y}_U)^2 \right] / (N - 1).$$

12   (Requires probability.) For two-phase sampling with ratio estimation (Section 12.3.1), suppose the phase I sample is an SRS of size $n^{(1)}$, and the phase II sample is an SRS of fixed size $n^{(2)}$.

   a   Show that $P(Z_i = 1) = n^{(1)}/N$, and $P(D_i = 1 \mid \mathbf{Z}) = Z_i n^{(2)}/n^{(1)}$.

   b   Show that (12.10) gives the approximate variance of $\hat{t}_{yr}^{(2)}$.

   c   Let $e_i = y_i - \hat{B}^{(2)}x_i$ and let s_y^2 and s_e^2 be the sample variances of the y_i's and the e_i's from the phase II sample,

$$s_y^2 = \frac{1}{n^{(2)} - 1} \sum_{i \in \mathcal{S}^{(2)}} (y_i - \bar{y}^{(2)})^2 \quad \text{and} \quad s_e^2 = \frac{1}{n^{(2)} - 1} \sum_{i \in \mathcal{S}^{(2)}} e_i^2.$$

   Show that (12.11) is an approximately unbiased estimator of $V(\hat{t}_{yr}^{(2)})$.

13   Rao and Sitter (1995) propose an alternative linearization variance estimator for the situation in Exercise 12,

   a   Using part (b) of Exercise 12, show that

$$V(\hat{t}_{yr}^{(2)}) \approx N^2 \left(1 - \frac{n^{(1)}}{N} \right) \frac{2BS_{xd} + B^2 S_x^2}{n^{(1)}} + N^2 \left(1 - \frac{n^{(2)}}{N} \right) \frac{S_d^2}{n^{(2)}}$$

   where $S_{xd} = \sum_{i=1}^{N} (x_i - \bar{x}_U)d_i/(N - 1)$. HINT: Write

$$S_y^2 = \frac{1}{N - 1} \sum_{i=1}^{N} (y_i - \bar{y}_U)^2 = \frac{1}{N - 1} \sum_{i=1}^{N} (y_i - Bx_i + Bx_i - B\bar{x}_U)^2.$$

   b   Show that

$$\hat{V}_2(\hat{t}_{yr}^{(2)}) = N^2 \left(1 - \frac{n^{(1)}}{N} \right) \frac{2\hat{B}^{(2)}s_{xe} + \hat{B}^{(2)}s_x^{2(1)}}{n^{(1)}} + N^2 \left(1 - \frac{n^{(2)}}{N} \right) \frac{s_e^2}{n^{(2)}},$$

   is an approximately unbiased estimator of $V(\hat{t}_{yr}^{(2)})$, where

$$s_{xe} = \frac{1}{n^{(2)} - 1} \sum_{i \in \mathcal{S}^{(2)}} (x_i - \bar{x}^{(2)})e_i, \quad s_x^{2(1)} = \frac{1}{n^{(1)} - 1} \sum_{i \in \mathcal{S}^{(1)}} (x_i - \bar{x}^{(1)})^2.$$

14   *Demnati–Rao (2004) linearization variance estimator in two-phase sampling.* The linearization variance estimator presented in Exercise 23 of Chapter 9 can be extended to two-phase sampling. Let θ be the population quantity of interest, and define the estimator $\hat{\theta}$ to be a function of the vectors of sampling weights for the phase I and phase II samples and the population values:

$$\hat{\theta} = g(\mathbf{w}^{(1)}, \mathbf{w}, \mathbf{x}_1, \mathbf{x}_2, \dots, \mathbf{x}_m, \mathbf{y}_1, \mathbf{y}_2, \dots, \mathbf{y}_k),$$

where $\mathbf{w}^{(1)} = (w_1^{(1)}, \dots, w_N^{(1)})^T$ with $w_i^{(1)}$ the phase I sampling weight of unit i ($w_i^{(1)} = 0$ if i is not in the phase I sample), $\mathbf{w} = (w_1, \dots, w_N)^T$ with w_i the final sampling weight of unit i in the phase II sample ($w_i = w_i^{(1)}w_i^{(2)}$ if $i \in \mathcal{S}^{(2)}$ and $w_i = 0$ if $i \notin \mathcal{S}^{(2)}$), $\mathbf{x}_j$

is the vector of population values for the jth auxiliary variable (measured in phase I), and $\mathbf{y}_j$ is the vector of population values for the jth response variable (measured in phase II). Now let

$$z_i^{(1)} = \frac{\partial g(\mathbf{w}^{(1)}, \mathbf{w}, \mathbf{x}_1, \mathbf{x}_2, \dots, \mathbf{x}_m, \mathbf{y}_1, \mathbf{y}_2, \dots, \mathbf{y}_k)}{\partial w_i^{(1)}}$$

and

$$z_i^{(2)} = \frac{\partial g(\mathbf{w}^{(1)}, \mathbf{w}, \mathbf{x}_1, \mathbf{x}_2, \dots, \mathbf{x}_m, \mathbf{y}_1, \mathbf{y}_2, \dots, \mathbf{y}_k)}{\partial w_i}.$$

Then,

$$\hat{V}_{\mathrm{DR}}(\hat{\theta}) = \hat{V}\left(\sum_{i \in \mathcal{S}^{(1)}} w_i^{(1)} z_i^{(1)} + \sum_{i \in \mathcal{S}^{(2)}} w_i z_i^{(2)} \right).$$

a Consider the two-phase ratio estimator in (12.9). We can write

$$\hat{t}_{yr}^{(2)} = \sum_{i \in \mathcal{S}^{(1)}} w_i^{(1)} x_i \frac{\displaystyle\sum_{i \in \mathcal{S}^{(2)}} w_i y_i}{\displaystyle\sum_{i \in \mathcal{S}^{(2)}} w_i x_i},$$

where $w_i = w_i^{(1)} w_i^{(2)}$. Show that the Demnati-Rao linearization variance estimator is

$$\hat{V}_{\mathrm{DR}}\left(\hat{t}_{yr}^{(2)}\right) = \hat{V}\left[\sum_{i \in \mathcal{S}^{(1)}} w_i^{(1)} x_i \hat{B}^{(2)} + \frac{\hat{t}_x^{(1)}}{\hat{t}_x^{(2)}} \sum_{i \in \mathcal{S}^{(2)}} w_i (y_i - \hat{B}^{(2)} x_i) \right].$$

b Suppose that the phase I sample is an SRS of size $n^{(1)}$ and the phase II sample is an SRS of size $n^{(2)}$. What is $\hat{V}_{\mathrm{DR}}\left(\hat{t}_{yr}^{(2)}\right)$ for this case?

15 Show that if the estimator in (12.12) is applied to any of the auxiliary variables in $\mathbf{x}$, then $\hat{t}_{\mathbf{x}\mathrm{GREG}}^{(2)} = \hat{\mathbf{t}}_{\mathbf{x}}^{(1)}$.

16 (Requires probability.) Suppose the phase I sample is an unequal-probability sample of observations. If we observed y_i for every unit in $\mathcal{S}^{(1)}$, we could use the Horvitz–Thompson estimator of the variance of the Horvitz–Thompson estimator in (6.22) to estimate the first term in (12.14):

$$\hat{V}_{\mathrm{HT}}^{(1)}(\hat{t}_y^{(1)}) = \sum_{i \in \mathcal{S}^{(1)}} \sum_{k \in \mathcal{S}^{(1)}} \frac{\pi_{ik}^{(1)} - \pi_i^{(1)} \pi_k^{(1)}}{\pi_{ik}^{(1)}} \frac{y_i}{\pi_i^{(1)}} \frac{y_k}{\pi_k^{(1)}},$$

where $\pi_i^{(1)} = P(Z_i = 1)$ and $\pi_{ik}^{(1)} = P(Z_i Z_k = 1)$ for $i \neq k$ and $\pi_{ii}^{(1)} = P(Z_i = 1)$. We need an estimator of $V(\hat{t}_y^{(1)})$, however, that depends only on the y values in the phase II sample. Let $\pi_{ik}^{(2)} = P(D_i D_k = 1 \mid \mathbf{Z}) > 0$. Show that

$$\hat{V}_{\mathrm{HT}}(\hat{t}_y^{(1)}) = \sum_{i \in \mathcal{S}^{(2)}} \sum_{k \in \mathcal{S}^{(2)}} \frac{\pi_{ik}^{(1)} - \pi_i^{(1)} \pi_k^{(1)}}{\pi_{ik}^{(1)} \pi_{ik}^{(2)}} \frac{y_i}{\pi_i^{(1)}} \frac{y_k}{\pi_k^{(1)}}$$

is an unbiased estimator of $V(\hat{t}_y^{(1)})$.

17   (Requires calculus.) *Optimal allocation for two-phase sampling with stratification.* Suppose phase I is an SRS and phase II is a stratified random sample, and that the total cost for the sample is given in (12.15), where $c^{(1)}$ is the cost to sample a unit in phase I and c_h is the cost to sample a unit in stratum h in phase II. Let $v_h = m_h/n_h$, $h = 1, \ldots, H$ be the proportion of phase I units in stratum h to be sampled in phase II.

   a   Show that the expected cost is (12.16).

   b   Show that $V(\hat{\bar{y}}_{str}^{(2)})$ is minimized, subject to the constraint in (12.16), when v_h is given in (12.17). HINT: Use Lagrange multipliers.

18   Show that when the optimal allocation is used, the variance of $\hat{\bar{y}}_{str}^{(2)}$ for two-phase sampling with stratification is given by (12.18).

19   (Requires calculus.) Show that if an SRS of size $n^{(1)}$ is taken in phase I, and an SRS of size $n^{(2)}$ is taken in phase II, then taking the ratio $n^{(2)}/n^{(1)}$ in (12.20) minimizes the variance in (12.10) for a fixed cost C.

20   This exercise is based on results in McNamee (2003) on the use of two-phase sampling to estimate disease prevalence. An inexpensive, but possibly inaccurate, screening test for the disease is given in the phase I sample, an SRS of size $n^{(1)}$. Let $x_i = 1$ if person i tests positive on the screening test and $x_i = 0$ if person i tests negative on the screening test. Persons are then classified into stratum 1 ($x_i = 0$) and stratum 2 ($x_i = 1$). The persons sampled in phase II are given a test for the presence of the disease that, for purposes of this exercise, is assumed to be 100% accurate: The phase II response is $y_i = 1$ if person i has the disease and 0 otherwise. We can write the population values in a contingency table:

		Screening Test		
		Negative	Positive	
Disease	No	C_{11}	C_{12}	C_{1+}
present?	Yes	C_{21}	C_{22}	C_{2+}
		$C_{+1} = N_1$	$C_{+2} = N_2$	N

We wish to estimate $p = \bar{y}_U = C_{2+}/N$ from the two-phase sample; $p_1 = C_{21}/N_1$ and $p_2 = C_{22}/N_2$ are the proportions with the disease in strata 1 and 2, respectively.

   a   Epidemiologists often use the concepts of specificity and sensitivity to assess a test for a disease, with

$$S_1 = \text{Specificity} = P(\text{test is negative} \mid \text{disease absent}) = \frac{C_{11}}{C_{1+}}$$

and

$$S_2 = \text{Sensitivity} = P(\text{test is positive} \mid \text{disease present}) = \frac{C_{22}}{C_{2+}}.$$

Show that

$$\frac{N_1}{N}p_1 = (1 - S_2)p, \qquad \frac{N_1}{N}(1 - p_1) = (1 - p)S_1,$$

$$\frac{N_2}{N}p_2 = pS_2, \qquad \frac{N_2}{N}(1 - p_2) = (1 - p)(1 - S_1).$$

b   Suppose that the optimal allocation is used (see Section 12.5.1) and that $0 < v_{h,opt} < 1$ for $h = 1, 2$. Using (12.19) and part (a), show that

$$\frac{V_{opt}(\hat{p}_{str}^{(2)})}{V_{SRS}(\hat{p})} \approx \left[\sqrt{(1 - S_2)S_1} + \sqrt{S_2(1 - S_1)} + R\sqrt{\frac{c^{(1)}}{c^{(2)}}} \right]^2 ,$$

where R is the population Pearson correlation coefficient between x and y, given in (4.1). HINT: For the second term, first show that $RS_y = p(S_2 - W_2)/\sqrt{W_1 W_2}$.

c   Calculate the ratio of variances in (b) when $S_1 = S_2$ and $R = \min\{S_1 + S_2 - 0.9, 0.95\}$, for $S_1 \in \{0.5, 0.6, 0.7, 0.8, 0.9, 0.95\}$ and $c^{(1)}/c^{(2)} \in \{0.0001, 0.01, 0.1, 0.5, 1\}$. Display your results in a table. For which settings would you recommend two-phase sampling to estimate disease prevalence?

21   *Inverse Sampling.*   Hinkins et al. (1997) (also see Rao et al., 2003) note that in some situations one might want to apply a statistical procedure developed for an SRS to data from a complex survey, but the stratification and clustering in the survey make direct application inappropriate. They propose an inverse sampling algorithm to create a subsample from the complex survey that is an SRS from the population, essentially by inverting the procedure used to draw the complex sample. The procedure can be repeated multiple times.

Suppose that the complex survey is a stratified random sample. The population stratum sizes are $N_1, \ldots, N_H$ and the sample sizes are $n_1, \ldots, n_H$. It would be possible for all the observations in an SRS from the population to be in one stratum, so the maximum possible size of the subsample is $m = \min\{n_1, \ldots, n_H\}$. Use the hypergeometric distribution to generate subsampling sizes $m_1, \ldots, m_H$ from the strata, with $\sum_{h=1}^{H} m_h = m$, where

$$P(M_1 = m_1, M_2 = m_2, \ldots, M_H = m_H) = \frac{1}{\binom{N}{m}} \binom{N_1}{m_1} \binom{N_2}{m_2} \cdots \binom{N_H}{m_H} .$$

In stratum h, select an SRS of m_h of the n_h sampled units to be in $\mathcal{S}^{(2)}$.

a   Show that the probability that any subset of m units in the population is selected as the sample through this procedure (first taking a stratified random sample of size n, then using inverse sampling to select a subsample of size m) is $\binom{N}{m}^{-1}$.

b   Use inverse sampling to select a subsample of size 21 from the stratified random sample in file agstrat.dat, first discussed in Example 3.2.

22   *Ranked set sampling.*   McIntyre (1952) proposed ranked set sampling as a method of improving precision of estimates by a method related to two-phase sampling. Stokes (1980) and Patil (2002) recommend ranked set sampling for situations in which the response of interest, y, may be difficult to measure but it is easy to estimate an approximate ranking of the units in a sample, either visually or by using a correlated variable. In McIntyre's application $y_i =$ pasture yield of field i. Measuring y_i requires mowing and weighing—a time-consuming process. An expert, however, can assess and rank a small number of fields from lowest to highest yield by visual inspection, which is much less effort.

To implement ranked set sampling, select k independent SRSs, each of size k. Rank each of the k samples from low to high, using either judgment or a correlated easy-to-measure variable. (This ranking must be done without knowing any values of y_i.) Then select the smallest unit of the first sample for measurement of y, the second smallest unit of the second sample, and so on until the largest unit of sample k is selected. Repeat this procedure until m replicates are obtained. At the end of the process mk^2 units have been ranked and y has been measured on a sample of $n = mk$ units. Let $\bar{y}_j$ be the mean of the y-values measured in replicate $j, j = 1, \ldots, m$, and let $\hat{\bar{y}}_{RSS} = \frac{1}{m} \sum_{j=1}^{m} \bar{y}_j$.

We illustrate the method with a small example using the data in agpop.dat (Husby et al., 2005, have a similar example using NHANES data). We want to estimate the population total for $y = acres92$. We take mk SRSs, each of size k, using $m = 10$ and $k = 4$ (typically, k is relatively small to allow an expert to rank the elements). We rank each of the mk SRSs using the correlated variable $acres87$. Here are the values of $acres87$ for the 4 samples in the first replicate (SAS code used to obtain the samples is on the website).

		Observation			
Sample 1:	x	119,956	144,986	108,861	302,659
	Rank	2	3	1	4
Sample 2:	x	351,106	294,551	80,104	226,954
	Rank	4	3	1	2
Sample 3:	x	241,276	253,421	702,173	412,225
	Rank	1	2	4	3
Sample 4:	x	529,964	823,729	355,973	121,119
	Rank	3	4	2	1

We then measure y on the third unit in Sample 1 (which has rank 1), the fourth unit in Sample 2 (rank 2), the fourth unit in Sample 3 (rank 3), and the second unit in Sample 4 (rank 4), obtaining the y values 106,206, 246,038, 379,044, and 783,715. Note that since y is highly correlated with x, these four y values are forced to be spread out.

The same procedure is repeated for the remaining nine replicates. We obtain

$\bar{y}_1$	$\bar{y}_2$	$\bar{y}_3$	$\bar{y}_4$	$\bar{y}_5$
378,750.75	51,841	280,658.5	187,791.5	175,436.75

$\bar{y}_6$	$\bar{y}_7$	$\bar{y}_8$	$\bar{y}_9$	$\bar{y}_{10}$
446,398.5	1,092,582.5	499,146	350,570	665,457.75

Consequently, $\hat{\bar{y}}_{RSS} = \frac{1}{m} \sum_{j=1}^{m} \bar{y}_j = 412{,}863.33$.

a   How is ranked set sampling similar to two-phase sampling? How does it differ?

b   Argue that if the ranking is based on an auxiliary variable x, and the value of y itself is not used in the ranking, then $\hat{\bar{y}}_{RSS}$ is an unbiased estimator of the population mean.

c   Show that if the *m* replicates are selected independently, then

$$\hat{V}(\hat{\bar{y}}_{\text{RSS}}) = \frac{1}{m(m-1)} \sum_{j=1}^{m} (\bar{y}_j - \hat{\bar{y}}_{\text{RSS}})^2$$

is an unbiased estimator of $V(\hat{\bar{y}}_{\text{RSS}})$. (HINT: See Section 9.2.) Using this method, we obtain $\text{SE}(\hat{\bar{y}}_{\text{RSS}}) = 93{,}943.21$ for our sample.

d   The properties of ranked set sampling require that the ranking in each of the *mk* samples be done using the same method. What might go wrong in practice? How might a ranker who knows the sampling procedure produce bias in the estimates?

D. Projects and Activities

23   *Forest data.*   Treat the data in forest.dat, described in Exercise 36 of Chapter 2, as the population for this problem. Suppose you are interested in estimating the total number of cells with cover type 1 (spruce/fir). Select an SRS of 5000 records. Determine the elevation of each cell, and form two strata: (1) elevation less than 3000 m and (2) elevation greater than or equal to 3000 m. Now select a stratified random sample of 500 of the records in your phase I sample, and use the subsample to estimate the total number of cells with spruce/fir. You may want to use a small pilot sample to estimate the optimal allocation of sample sizes in the two strata.

24   *Ranked set sampling using the forest data.*   Use the ranked set sampling procedure described in Exercise 22 with the forest data. We expect the variables $x = Hillshade\_9am$ and $y = Hillshade\_3pm$ to be negatively correlated.

a   Draw an SRS of size 25 from the data. Find $\bar{y}$ and $\hat{V}(\bar{y})$ using formulas from Chapter 2.

b   Draw 100 SRSs of size 4 from the forest data, and arrange these as 25 sets of 4 SRSs. Rank the *x* variables in each SRS, and choose one *y* value from each SRS using the ranked values. Find $\hat{\bar{y}}_{\text{RSS}}$ and $\hat{V}(\hat{\bar{y}}_{\text{RSS}})$ for your data.

c   How do the two estimated variances compare? Does ranked set sampling improve precision for this example?

13

Estimating Population Size

I caught a large number of fishes in the neighbourhood of Suez. I passed a copper ring through their tails, and threw them back into the sea. Some months later, on the coast of Syria, I caught some of my fish ornamented with the ring.

—Jules Verne, *Twenty Thousand Leagues Under the Sea*

13.1
Capture–Recapture Estimation

EXAMPLE 13.1 Suppose we want to estimate N, the number of fish in a lake. One method is as follows: Catch and mark 200 fish in the lake, then release them. Allow the marked and released fish to mix with the other fish in the lake. Then, take a second, independent sample of 100 fish. Suppose that 20 of the fish in the second sample are marked. Then, assuming that the population of fish has not changed between the two samples and that each catch gives a simple random sample (SRS) of fish in the lake, estimate that 20% of the fish in the lake are marked, and, therefore, that the 200 fish tagged in the original sample represent approximately 20% of the population of fish. The population size N is then estimated to be approximately 1000. ∎

This method for estimating the size of a population is called two-sample **capture–recapture estimation**. Other names sometimes used are tag- or mark-recapture, multiple record system, the Petersen (1896) method, or the Lincoln (1930) index. The method relies on the following assumptions:

1 The population is *closed*—no fish enter or leave the lake between the samples. This means that N is the same for each sample.

2 Each sample of fish is an SRS from the population. This means that each fish is equally likely to be chosen in a sample—it is not the case, for example, that smaller or less healthy fish are more likely to be caught. Also, there are no "hidden fish" in the population that are impossible to catch.

3   The two samples are independent. The marked fish from the first sample become re-mixed in the population, so that the marking status of a fish is unrelated to the probability that the fish is selected in the second sample. Also, fish included in the first sample do not become "trap-shy" or "trap-happy"—the probability that a fish will be caught in the second sample does not depend on its capture history.

4   Fish do not lose their markings, and marked fish can be identified as such. Water-soluble paint, for example, would not be a good choice for marking material.

In this simple form, capture–recapture is a special case of ratio estimation of a population total, and results from Chapter 4 may be used when the samples and population are large. Let n_1 be the size of the first sample, n_2 the size of the second sample, and m the number of marked fish caught in the second sample. In Example 13.1, $n_1 = 200$, $n_2 = 100$, $m = 20$, and we used the estimator $\hat{N} = n_1 n_2/m$. To see how this estimator fits into the framework of Chapter 4, let

$$y_i = 1 \quad \text{for every fish in the lake}$$

and

$$x_i = \begin{cases} 1 & \text{if fish } i \text{ is marked} \\ 0 & \text{if fish } i \text{ is not marked.} \end{cases}$$

Then estimate $N = t_y = \sum_{i=1}^{N} y_i$ by $\hat{t}_{yr} = t_x \hat{B}$, where $t_x = \sum_{i=1}^{N} x_i = n_1$ and $\hat{B} = \bar{y}/\bar{x} = n_2/m$. This ratio estimator,

$$\hat{N} = \hat{t}_{yr} = \frac{n_1 n_2}{m}, \tag{13.1}$$

is also the maximum likelihood estimator (see Exercises 13 and 14). Applying (4.10) to the second SRS and ignoring the fpc, $s_e^2 = n_2(n_2 - m)/[m(n_2 - 1)]$ and

$$\hat{V}(\hat{N}) = t_x^2 \hat{V}(\hat{B}) = \left(\frac{n_1 n_2}{m}\right)^2 \frac{n_2 - m}{m(n_2 - 1)} \approx \frac{n_1^2 n_2(n_2 - m)}{m^3}.$$

For the data in Example 13.1, $\hat{V}(\hat{N}) = 40{,}000$.

Being a ratio estimator, though, $\hat{N}$ is biased, and the bias can be large in wildlife applications with small sample sizes. Indeed, it is possible for the second sample to consist entirely of unmarked animals, making the estimate in (13.1) infinite. Chapman (1951) proposes the less biased estimator

$$\tilde{N} = \frac{(n_1 + 1)(n_2 + 1)}{m + 1} - 1. \tag{13.2}$$

A variance estimator for $\tilde{N}$ (Seber, 1970) is

$$\hat{V}(\tilde{N}) = \frac{(n_1 + 1)(n_2 + 1)(n_1 - m)(n_2 - m)}{(m + 1)^2(m + 2)}. \tag{13.3}$$

The estimators in (13.2) and (13.3) are often used in wildlife applications. For the fish data, $\tilde{N} = (201)(101)/21 - 1 = 966$, and $\hat{V}(\tilde{N}) = 30{,}131$.

Many researchers have constructed confidence intervals (CIs) for the population size using either $\hat{N} \pm 1.96\sqrt{\hat{V}(\hat{N})}$ or $\tilde{N} \pm 1.96\sqrt{\hat{V}(\tilde{N})}$. These are not entirely satisfactory, however, because both require that $\hat{N}$ or $\tilde{N}$ be approximately normally distributed, and the normal distribution may not be a good approximation to the distribution of $\hat{N}$ or $\tilde{N}$ for small populations and samples. We'll discuss CIs in Section 13.1.2; first, however, let's look at another approach for these data that will be useful in developing CIs.

13.1.1 Contingency Tables for Capture–Recapture Experiments

Fienberg (1972) suggests viewing capture–recapture data in an incomplete contingency table. For the data in Example 13.1, the table is as follows:

		In Sample 2? Yes	No	
	Yes	20	180	200
In Sample 1?				
	No	80	?	?
		100	?	N

In general, if x_{ij} is the observed count in cell (i,j), the contingency table looks as follows. An asterisk indicates that we do not observe that cell.

		In Sample 2? Yes	No	
	Yes	$x_{11}(=m)$	x_{12}	$x_{1+}(=n_1)$
In Sample 1?				
	No	x_{21}	x_{22}^*	x_{2+}^*
		$x_{+1}(=n_2)$	x_{+2}^*	x_{++}^*

The expected counts are:

		In Sample 2? Yes	No	
	Yes	m_{11}	m_{12}	m_{1+}
In Sample 1?				
	No	m_{21}	m_{22}^*	m_{2+}^*
		m_{+1}	m_{+2}^*	$m_{++}^* = N$

To estimate the expected counts, we use $\hat{m}_{11} = x_{11}$, $\hat{m}_{12} = x_{12}$, and $\hat{m}_{21} = x_{21}$. If presence in sample 1 is independent of presence in sample 2, then the odds of being in sample 2 are the same for marked fish as for unmarked fish: $m_{11}/m_{12} = m_{21}/m_{22}$. Consequently, under independence, the estimated count in the cell of fish not included in either sample is

$$\hat{m}_{22} = \frac{\hat{m}_{12}\hat{m}_{21}}{\hat{m}_{11}} = \frac{x_{12}x_{21}}{x_{11}},$$

and

$$\hat{N} = \hat{m}_{11} + \hat{m}_{12} + \hat{m}_{21} + \hat{m}_{22} = \frac{x_{+1}x_{1+}}{x_{11}} = \frac{n_1 n_2}{m}.$$

The estimator $\hat{N}$ is calculated based on the assumption that the two samples are independent; unfortunately, this assumption cannot be tested because only three of the four cells of the contingency table are observed.

13.1.2 Confidence Intervals for N

In many applications of capture–recapture, CIs have been constructed using $\hat{N} \pm 1.96\sqrt{\hat{V}(\hat{N})}$ or $\tilde{N} \pm 1.96\sqrt{\hat{V}(\tilde{N})}$. If we use the first interval for the data in Example 13.1, $\hat{V}(\hat{N}) = 40,000$, and an asymptotic 95% CI would be $1000 \pm 1.96(200) = [608, 1392]$. The CI using the normal approximation and $\tilde{N}$ is [626, 1306]. Unfortunately, CIs based on the assumption that $\hat{N}$ or $\tilde{N}$ follow a normal distribution often have poor coverage probability in small samples because the distribution of $\hat{N}$ and $\tilde{N}$ is actually quite skewed, as you will see in Exercise 17. In general, we do not recommend using these CIs.

An additional shortcoming of CIs based on the normal distribution can occur in small samples. For example, suppose that $n_1 = 30$, $n_2 = 20$, and $m = 15$. Then $\hat{N} = (30)(20)/15 = 40$, and $\hat{V}(\hat{N}) = 26.7$. Using a normal approximation to the distribution of $\hat{N}$ results in the CI [30, 50]. The lower bound of 30 is silly, however; a total of 35 distinct animals were observed in the two samples, so we know that N must be at least 35.

Cormack (1992) discusses using the Pearson or likelihood ratio chi-square test for independence to construct a CI. Using this method, we fill in the missing observation x_{22} by some value u and perform a chi-square test for independence on the artificially completed data set. The 95% CI for m_{22} is then all values of u for which the null hypothesis of independence for the two samples would not be rejected at the 0.05 level. For the data in Example 13.1, let's try the value $u = 600$. With this value, the "completed" contingency table is

		In Sample 2? Yes	In Sample 2? No	
	Yes	20	180	200
In Sample 1?				
	No	80	600	680
		100	780	880

We can easily perform Pearson's chi-square test for independence on this table, obtaining a p-value of 0.49. As $0.49 > 0.05$, the value 600 would be inside the 95% CI for u, and the value 880 would be inside the 95% CI for N. Setting u equal to 1500, though, gives p-value 0.0043, so 1500 is outside the 95% CI for u, and 1780 is thus outside the 95% CI for N. Continuing in this manner, we find that values of u between 430 and 1198 are the only ones that result in p-value > 0.05, so [430, 1198] is a 95% CI for m_{22}. The corresponding CI for N is obtained by adding the number of observed animals in the other cells, 280, to the endpoints of the CI for m_{22}, resulting in the interval [710, 1478].

The likelihood ratio test may be used in similar manner, by including in the CI all values of u for which the p-value from the likelihood ratio test exceeds 0.05. Using the computer code given on the website, we find that values of u between 437 and 1233 give a likelihood ratio p-value exceeding 0.05. The CI for N, using the likelihood ratio test, is then [717, 1513].

Another alternative for CIs is to use the bootstrap (Buckland, 1984). To apply the bootstrap here, resample from the observed individuals in the second sample. Take R samples of size 100 with replacement from the 20 tagged and 80 untagged fish we observed. Calculate $\hat{N}^*$ for each of the R resamples, and find the 2.5 and 97.5 percentage points of the R values. With $R = 999$, the 95% CI is the 25th and 975th values from the ordered list of the $\hat{N}^*$, [714, 1538].

Note that all three of these CIs resulting from Pearson's chi-square test, the likelihood ratio chi-square test, and the bootstrap are similar, but all differ from the CIs based on the asymptotic normality of $\hat{N}$ or $\tilde{N}$.

13.1.3 Using Capture–Recapture on Lists

Capture–recapture estimation is not limited to estimating wildlife populations. It can also be used when the two samples are lists of individuals, provided that the assumptions for the method are met. Suppose you want to estimate the number of statisticians in the United States, and obtain membership lists from the American Statistical Association (ASA) and the Institute for Mathematical Statistics (IMS). Every statistician either is or is not a member of the ASA, and either is or is not a member of the IMS. (Of course, there are other worthy statistical organizations, but for simplicity let's limit the discussion here to these two.) Then n_1 is the number of ASA members, n_2 the number of IMS members, and m is the number of persons on both lists. We can estimate the number of statisticians using $\hat{N} = n_1 n_2 / m$, exactly as if statisticians were fish. The assumptions for this estimate are as above, but with slightly different implications than in wildlife settings:

1   The population is closed. In wildlife surveys, this assumption may not be met because animals often die or migrate between samples. When treating lists as the samples, though, we can usually act as though the population is closed if the lists are from the same time period.

2   Each list provides an SRS from the population of statisticians. This assumption is more of a problem; it implies that the probability of belonging to ASA is the same for all statisticians, and the probability of belonging to IMS is the same for all statisticians. It does not allow for the possibility that a group of statisticians may refuse to belong to either organization, or for the possibility that subgroups of statisticians may have different probabilities of belonging to an organization.

3   The two lists are independent. Here, this means that the probability that a statistician is in ASA does not depend on his or her membership in IMS. This assumption is also often not met—it may be that statisticians tend to belong to only one organization, and therefore that ASA members are less likely to belong to IMS than non-ASA members.

4 Individuals can be matched on the lists. This sounds easy, but often proves surprisingly difficult. Is J. Smith on List 1 the same person as Jonquil Smith on List 2? Larsen and Rubin (2001) describe some of the problems that can occur when you try to link records.

An important application of capture–recapture methods is estimating undercoverage in a census. In this setting, sample 1 is the census and sample 2 is an independent probability sample taken from the population.

EXAMPLE 13.2 The U.S. Census Bureau tries to enumerate everyone in the decennial census. Inevitably, however, persons are missed, leading population estimates from the census to underestimate the true population count. Moreover, it is thought that the undercount rate is not uniform; the undercount is thought to be greater for inner city areas and minority groups, and varies among different regions of the United States. As Congressional Representatives, billions of dollars of federal funding, and other resources are apportioned based on census results, it is important that the population counts be accurate. Capture–recapture estimation, called *dual system estimation* in this context, has been used since 1950 to evaluate the coverage of the U.S. decennial census. Fienberg (1992) gives a bibliography for dual-system estimation.

Hogan (1993) describes the Post-Enumeration Survey (PES) used in the 1990 U.S. census. The general principles are the same for other census years: Citro et al. (2004, Chapters 5–6) describe procedures that were used to assess coverage of the census in 2000. Kostanich et al. (2004, Chapter 5–6) describe plans for the Census Coverage Measurement program to be used for the 2010 census. A similar procedure, called the Reverse Record Check, is used in Canada. Two samples are taken. The P-sample is taken directly from the population, independently of the census, and is used to estimate number of persons missed by the census. The E-sample is taken from the census enumeration itself, and is used to estimate errors in the census such as nonexistent persons or duplicates.

Separate population estimates are derived for each poststratum, where the population is poststratified by region, race, ownership of dwelling unit, age, and other variables. Poststrata are used because it is hoped that assumption 2 of equal recapture probabilities is approximately satisfied within each poststratum; we know it is not satisfied for the population as a whole because of the differential undercount rates in the census. The population table for a poststratum is as follows:

		In Census Enumeration?		
		Yes	No	
In PES?	Yes	N_{11}	N_{12}	N_{1+}
	No	N_{21}	N_{22}^*	N_{2+}^*
		N_{+1}	N_{+2}	N

Then,

$$\hat{N} = \frac{\hat{N}_{+1}\hat{N}_{1+}}{\hat{N}_{11}}.$$

The estimates $\hat{N}_{1+}$ and $\hat{N}_{11}$ are from the P-sample: $\hat{N}_{1+}$ is the estimate of the poststratum total, using weights, from the P-sample, and $\hat{N}_{11}$ is a weighted estimate of matches between the P-sample and the census enumeration. Here, $\hat{N}_{+1}$ is not the actual count from the census, but is the census count adjusted using the E-sample to remove duplicates and fictitious persons. Many sample sizes in poststrata were small, leading to large variances for the estimates of population count, so the estimates were smoothed and adjusted using regression models.

The assumptions above need to be met for dual-system estimation to give a better estimate of the population than the original census data. It is hoped that assumption 2 holds within the poststrata. Assumption 3 is also of some concern, though, as the P-sample also has nonresponse. Persons missed in the census may also be missed in the P-sample. Another concern is the ability to match persons in the P-sample to persons in the census. Because P-sample persons not matched are assumed to have been missed by the census, errors in matching persons in the two samples can lead to biases in the population estimates. In the 2000 Census Accuracy and Coverage Evaluation, the initial evaluation missed a substantial number of duplicate Census enumerations; this was corrected in a revised evaluation (see U.S. Census Bureau, 2003b). ∎

13.2
Multiple Recapture Estimation

The assumptions for the two-sample capture–recapture estimators described in Section 13.1 are strong: The population must be closed and the two random samples independent. Moreover, these assumptions cannot be tested, because we observe only three of the four cells in the contingency table—we need all four cells to be able to test independence of samples.

More complicated models may be fit if $K > 2$ random samples are taken, and especially if different markings are used for individuals caught in the different samples. With fish, for example, the left pectoral fin might be marked for fish caught in the first sample, the right pectoral fin marked for fish caught in the second sample, and a dorsal fin marked for fish caught in the third sample. A fish caught in Sample 4 that had markings on the left pectoral fin and dorsal fin, then, would be known to have been caught in Sample 1 and Sample 3, but not Sample 2.

Schnabel (1938) first discussed how to estimate N when K samples are taken. She found the maximum likelihood estimator of N to be the solution to

$$\sum_{i=1}^{K} \frac{(n_i - r_i)M_i}{N - M_i} = \sum_{i=1}^{K} r_i,$$

where n_i is the size of sample i, r_i is the number of recaptured fish in sample i, and M_i is the number of tagged fish in the lake when sample i is drawn.

If individual markings are used, we can also explore issues of immigration or emigration from the population, and test some of the assumptions of independence.

EXAMPLE **13.3** Domingo-Salvany et al. (1995) used capture–recapture to estimate the prevalence of opiate addiction in Barcelona, Spain. One of their data sets consisted of three samples from 1989: (1) a list of opiate addicts from emergency rooms (E list), (2) a list of persons who started treatment for opiate addiction during 1989, reported to the Catalonia Information System on Drug Abuse (T list), and (3) a list of heroin overdose deaths registered by the forensic institute in 1989 (D list). A total of 2864 distinct persons were on the three lists. Persons on the three lists were matched, with the following results:

		In D list?			
		Yes		No	
		In T list?		In T list?	
		Yes	No	Yes	No
In E list?	Yes	6	27	314	1728
	No	8	69	712	?

It is unclear whether these data will fulfill the assumptions for the two-sample capture–recapture method. The assumption of independence among the samples may not be met—if treatment is useful, treated persons are less likely to appear in one of the other samples. In addition, persons on the death list are much less likely to subsequently appear on one of the other lists; the closed population assumption is also not met because one of the samples is a death list. Nevertheless, an analysis using the imperfectly met assumptions can provide some information on the number of opiate addicts. Because there are more than two samples, we can assess the assumptions of independence among different samples by using loglinear models. There is one assumption, though, that we can *never* test: The missing cell follows the same model as the rest of the data. ∎

If three samples are taken, the expected counts are:

		In Sample 3?			
		Yes		No	
		In Sample 2?		In Sample 2?	
		Yes	No	Yes	No
In Sample 1?	Yes	m_{111}	m_{121}	m_{112}	m_{122}
	No	m_{211}	m_{221}	m_{212}	m_{222}^*

Loglinear models were discussed in Section 10.4. The saturated model for three samples is

$$\ln m_{ijk} = \mu + \alpha_i + \beta_j + \gamma_k + (\alpha\beta)_{ij} + (\alpha\gamma)_{ik} + (\beta\gamma)_{jk} + (\alpha\beta\gamma)_{ijk}.$$

This model cannot be fit, however, as it requires eight degrees of freedom (df) and we only have seven observations. The following models may be fit, with α referring to the E list, β referring to the T list, and γ referring to the D list.

1 *Complete independence.*

$$\ln m_{ijk} = \mu + \alpha_i + \beta_j + \gamma_k.$$

This model implies that presence on any of the lists is independent of presence on any of the other lists. The independence model must always be adopted in two-sample capture–recapture.

2 *One list is independent of the other two.*

$$\ln m_{ijk} = \mu + \alpha_i + \beta_j + \gamma_k + (\alpha\beta)_{ij}.$$

Presence on the E list is related to the probability that an individual is on the T list, but presence on the D list is independent of presence on the other lists. There are three versions of this model: the other two substitute $(\alpha\gamma)_{ik}$ or $(\beta\gamma)_{jk}$ for $(\alpha\beta)_{ij}$.

3 *Two samples are independent given the third.*

$$\ln m_{ijk} = \mu + \alpha_i + \beta_j + \gamma_k + (\alpha\beta)_{ij} + (\alpha\gamma)_{ik}.$$

Three models of this type exist; the other two substitute either $(\alpha\beta)_{ij} + (\beta\gamma)_{ik}$ or $(\alpha\gamma)_{ij} + (\beta\gamma)_{ik}$ for $(\alpha\beta)_{ij} + (\alpha\gamma)_{ik}$. Presence on the death and treatment lists are conditionally independent given the E list status—once we know that a person is on the emergency room list, knowing that he or she is on the death list gives us no additional information about the probability that he or she will be on the treatment list.

4 *All two-way interactions.*

$$\ln m_{ijk} = \mu + \alpha_i + \beta_j + \gamma_k + (\alpha\beta)_{ij} + (\alpha\gamma)_{ik} + (\beta\gamma)_{jk}.$$

This model will always fit the data perfectly: It has the same number of parameters as there are cells in the contingency table.

Unfortunately, in none of these models can we test the hypothesis that the missing cell follows the model. But at least we can examine hypotheses of pairwise independence among the samples. For the addiction data, the following loglinear models were fit from the data, using the SAS PROC CATMOD code on the website (any loglinear model program that finds estimates using maximum likelihood will work):

Model	G^2	df	p-value	$\hat{m}_{222}$	$\hat{N}$	95% CI
1 Independence	1.80	3	0.62	3,967	6,831	[6,322, 7,407]
2a E*T	1.09	2	0.58	4,634	7,499	[5,992, 9,706]
2b E*D	1.79	2	0.41	3,959	6,823	[6,296, 7,425]
2c T*D	1.21	2	0.55	3,929	6,793	[6,283, 7,373]
3a E*T, E*D	0.19	1	0.67	6,141	9,005	[5,921, 16,445]
3b E*T, T*D	0.92	1	0.34	4,416	7,280	[5,687, 9,820]
3c E*D, T*D	1.20	1	0.27	3,918	6,782	[6,253, 7,388]
4 E*T, E*D, T*D	—	0	—	7,510	10,374	[4,941, 25,964]

Here, G^2 is the likelihood ratio test statistic (deviance) for that model. Somewhat surprisingly, the model of independence fits the data well. The predicted cell counts under model 1, complete independence, are:

		In D list?			
		Yes		No	
		In T list?		In T list?	
		Yes	No	Yes	No
In E list?	Yes	5.1	28.3	310.8	1730.7
	No	11.7	64.9	712.4	3966.7

These predicted cell counts lead to the estimate

$$\hat{N} = 2864 + 3967 = 6831$$

if the model of independence is adopted. The values of $\hat{N}$ for the other models are calculated similarly, by estimating the value in the missing cell from the model and adding that estimate to the known total for the other cells, 2864.

We can use an inverted likelihood ratio test (Cormack, 1992) to construct a CI for N using any of the models. A 95% CI for the missing cell consists of those values u for which a 0.05-level hypothesis test of $H_0 : m_{222} = u$ would not be rejected for the loglinear model adopted. Let $G^2(u)$ be the likelihood-ratio test statistic (deviance) for the completed table with u substituted for the missing cell, let t be the total of the seven observed cells, and let $\hat{u}$ be the estimate of the missing cell using that loglinear model. Cormack shows that the set

$$\left\{ u : G^2(u) - G^2(\hat{u}) + \ln\left(\frac{u}{t+u}\right) - \ln\left(\frac{\hat{u}}{t+\hat{u}}\right) < q_1(\alpha) \right\},$$

where $q_1(\alpha)$ is the percentile of the χ_1^2 distribution with right-tail area α, is an approximate $100(1-\alpha)\%$ CI for m_{222}. We give a computer program for calculating Cormack's CI on the website. This CI is conditional on the model selected and does not include uncertainty associated with the choice of model. Cormack also discusses extending the inverted Pearson chi-square test for goodness of fit, which produces a similar interval. Buckland and Garthwaite (1991) discuss using the bootstrap to find CIs for multiple recapture using loglinear models; they incorporate the model-selection procedure into each bootstrap iteration.

For these data, the point estimate and CI appear to rely heavily on the particular model fit, even though all seem to fit the observed cells. Note that the estimate $\hat{N}$ is larger and the CIs much wider for models including the E*T interaction, even though that interaction is not statistically significant. The good fit of the independence model is somewhat surprising because you would not expect the assumptions for independence to be satisfied. In addition, the population is not closed, but we have little information on migration in and out of the population.

13.3
Chapter Summary

Multiple samples from a population may be used to estimate its size. In the simplest form, two independent SRSs are taken and the number of population units found in both SRSs is used to estimate the population size. If the two samples are not independent—in particular, if individuals in the first sample are more likely to also appear in the second sample—then $\hat{N}$ calculated assuming independence is likely to underestimate the population size N.

Some forms of dependence can be assessed if three or more samples are taken. In that case, loglinear models can be fit to the data and used to predict the value of the missing cell.

Key Terms

Capture–recapture estimation: A method for estimating population size in which two independent samples are taken and the overlap used to estimate N.

Dual-system estimation: A form of capture–recapture estimation used to estimate undercount in a population census.

For Further Reading

In this chapter, we have just presented an introduction to estimating population size, under the assumption that the population is closed. Much other research has been done in capture–recapture estimation, including models for populations with births, deaths, and migrations; good sources for further reading are Seber (1982), Pollock (1991), and the review papers by the International Working Group for Disease Monitoring and Forecasting (1995a, 1995b). Chao et al. (2001) summarize recent research in capture–recapture methods used to estimate disease prevalence and provide links to S-Plus programs.

13.4
Exercises

A. Introductory Exercises

1   Suppose that an SRS of 500 fish is caught from a lake; each is marked and released, and the fish are allowed to remix with the other fish in the lake. A second sample of 300 fish has 120 marked fish. Estimate the total number of fish in the lake, along with a 95% CI.

B. Working with Survey Data

2   Investigators in the Wisconsin Department of Natural Resources (1993) used capture–recapture to estimate the number of fishers in the Monico Study Area in Wisconsin.

a In the first study, 7 fishers were captured between August 11, 1981 and January 31, 1982. Twelve fishers were captured between February 1 and February 19, 1982; of those 12, 4 had also been captured in the first sample. Give an estimate of the total number of fishers in the area, along with a 95% CI.

b In the second study, 16 fishers were captured between September 28, 1982 and October 31, 1982, and 19 fishers were captured between November 1 and November 17, 1982. Eleven of the 19 fishers in the second sample had also been caught in the first sample. Give an estimate of the total number of fishers in the area, along with a 95% CI.

c What assumptions are you making to calculate these estimates? What do these assumptions mean in terms of fisher behavior and "catchability"?

3 Alexander et al. (1997) apply capture–recapture methods to estimate the number of Mead's milkweed plants in a tract in Kansas. In some years, Mead's milkweed plants do not produce aboveground parts; in addition, if they are in dense vegetation and are not flowering, they are difficult to observe. Thus, a census of observed plants in any given year is likely an undercount. From the first two years of observation, 15 plants were observed in year 1 but not in year 2, 12 plants were observed in year 2 but not in year 1, and 33 plants were observed in both years. Estimate the total number of plants in the tract, along with a 95% CI.

4 Bellemain et al. (2005) relied on moose hunters in Norway to collect fecal samples from brown bears. Each sample was genotyped, and the number of distinct individuals was found in each of 2001 and 2002. In 2001, 311 unique genotypes were obtained (134 males and 177 females). In 2002, the procedure was repeated and 239 unique genotypes were obtained (106 males and 133 females). 165 of the individuals sampled in 2001 were also sampled in 2002.

a Fifty-six bears in the area in 2001 had also been followed with radio transmitters; 36 of these bears were represented in the 311 genotypes from the 2001 feces samples. Estimate the number of bears in 2001, along with a 95% CI.

b In 2002, 57 bears had radio transmitters, and 28 of them were among the 239 genotypes from the 2002 feces samples. Estimate the number of bears in 2002, along with a 95% CI.

c Estimate the number of bears, along with a 95% CI, treating the samples from 2001 and 2002 as independent SRSs (and ignoring the radio transmitter data). What assumptions are needed to use capture–recapture estimators of population size?

5 Domingo-Salvany et al. (1995) also used capture–recapture on the emergency room survey by dividing the list into four samples according to trimester (TR). The following data are from Table 1 of their paper:

	TR1 yes TR2 yes	TR1 yes TR2 no	TR1 no TR2 yes	TR1 no TR2 no
TR3 yes, TR4 yes	29	35	35	96
TR3 yes, TR4 no	48	58	80	400
TR3 no, TR4 yes	25	77	50	376
TR3 no, TR4 no	97	357	312	?

Fit loglinear models to these data. Which model do you think is best? Use your model to estimate the number of persons in the missing cell, and construct a 95% CI.

6 Chao et al. (2001) report data on an outbreak of Hepatitis A virus among students of a college. Investigators want to estimate N, the total number of students with Hepatitis A. Cases were reported from three sources: (1) a serum test conducted by the Institute of Preventive Medicine (P list), (2) local hospital records from the National Quarantine Service (Q list), and (3) records collected by epidemiologists (E list). The following table gives the counts from the three sources:

P list?	Q list?	E list?	Count
no	no	yes	63
no	yes	no	55
no	yes	yes	18
yes	no	no	69
yes	no	yes	17
yes	yes	no	21
yes	yes	yes	28

a Suppose that only the P list and Q list had been collected, with $n_1 = 135, n_2 = 122$, and $m = 49$. Calculate $\hat{N}$, Chapman's estimate $\tilde{N}$, and the standard error for each estimate.

b Fit loglinear models to the data. Using the deviance, evaluate the fit of these models. Is there evidence that the lists are dependent?

7 Cochi et al. (1989) recorded data on congenital rubella syndrome from two sources. The National Congenital Rubella Syndrome Registry (NCRSR) obtained data through voluntary reports from state and local health departments. The Birth Defects Monitoring Program (BDMP) obtained data from hospital discharge records from a subset of hospitals. Below are data from 1970 to 1985, from the two systems:

Year	NCRSR	BDMP	Both
1970	45	15	2
1971	23	3	0
1972	20	6	2
1973	22	13	3
1974	12	6	1
1975	22	9	1
1976	15	7	2
1977	13	8	3
1978	18	9	2
1979	39	11	2
1980	12	4	1
1981	4	0	0
1982	11	2	0
1983	3	0	0
1984	3	0	0
1985	1	0	0

a   The authors state that the NCRSR and the BDMP are independent sources of information. Do you think that is plausible? What about the other assumptions for capture–recapture?

b   Use Chapman's estimate (13.2) to find $\tilde{N}$ for each year for which you can calculate the estimate. What estimate will you use for the years in which Chapman's estimate cannot be calculated?

c   Now aggregate the data for all the years, and estimate the total number of cases of congenital rubella syndrome between 1970 and 1985. How does your estimate from the aggregated data compare with the sum of the estimates from (b)? Which do you think is more reliable?

d   Is there evidence of a decline in congenital rubella syndrome? Provide a statistical analysis to justify your answer.

8   Frank (1978) reports on the following experiment to estimate the number of minnows in a tank. The first two samples used a minnow trap to catch fish, while the third used a net to catch the fish. Minnows trapped in the first sample were marked by clipping their caudal fin, and minnows trapped in the second sample were marked by clipping the left pectoral fin.

Sample 1?	Sample 2?	Sample 3?	Number of Fish
yes	yes	yes	17
yes	no	yes	28
no	yes	yes	52
no	no	yes	234
yes	yes	no	80
yes	no	no	223
no	yes	no	400

Which loglinear model provides the best fit to these data? Using that model, estimate the total number of fish, and provide a 95% CI.

9   In the experiment in Exercise 8, what does it mean in terms of fish behavior if there is an interaction between presence in sample 1 and presence in sample 2? Between presence in sample 1 and presence in sample 3?

10   Egeland et al. (1995) use capture–recapture to estimate the total number of fetal alcohol syndrome cases among Alaska natives born between 1982 and 1989. Two sources of cases were used: thirteen cases identified by private physicians, and 45 cases identified by the Indian Health Service (IHS). Eight cases were on both lists.

a   Estimate the total number of fetal alcohol syndrome cases. Give a 95% CI for your estimate, using either the inverted chi-square test or the bootstrap method.

b   The capture–recapture estimate relies on the assumption that the two sources of data are independent—that is, a child on the IHS list has the same probability of appearing on the private physicians list as a child not on the IHS list. Do you think this assumption will hold here? Why or why not? What advice would you give the investigators if they were concerned about independence?

c Suppose that children who are seen by private physicians are less likely to be seen by the IHS. Is $\hat{N}$ then likely to underestimate or to overestimate the number of children with fetal alcohol syndrome? Explain.

C. Working with Theory

11 Note that in (13.1), $\hat{N} = n_1/\hat{p}$, where $\hat{p}$ is the sample proportion of individuals in the second sample that are tagged. Use the linearization method of Chapter 9 to find an estimator of $V(\hat{N})$.

12 The distribution of $\hat{N}$ in (13.1) is often not approximately normal. The distribution of $\hat{p} = m/n_2$, however, is often close to normality, and CIs for $\hat{p}$ are easily constructed. For the data in Example 13.1, find a 95% CI for $\hat{p}$. How can you use that interval to obtain a CI for $\hat{N}$? How does the resulting CI compare with others we calculated? Is the interval symmetric about $\hat{N}$?

13 (Requires mathematical statistics.) In a lake with N fish, n_1 of them tagged, the probability of obtaining m recaptured and $n_2 - m$ previously uncaught fish in a simple random sample of size n_2 is

$$\mathcal{L}(N \mid n_1, n_2) = \frac{\binom{n_1}{m}\binom{N - n_1}{n_2 - m}}{\binom{N}{n_2}}.$$

The maximum likelihood estimator $\hat{N}$ of N is the value which maximizes $\mathcal{L}(N)$—it is the value that makes the observed value of m appear most probable if we know n_1 and n_2. Find the maximum likelihood estimator of N. HINT: When is $\mathcal{L}(N) \geq \mathcal{L}(N - 1)$?

14 (Requires mathematical statistics.) *Maximum likelihood estimation of N in large samples.* Suppose that n_1 of the N fish in a lake are marked. An SRS of n_2 fish is then taken, and m of those fish are found to be marked. Assume that N, n_1, and n_2 are all "large." Then the probability that m of the fish in the sample are marked is approximately:

$$\mathcal{L}(N) = \binom{n_2}{m}\left(\frac{n_1}{N}\right)^m \left(1 - \frac{n_1}{N}\right)^{n_2 - m}.$$

a Show that $\hat{N} = n_1 n_2/m$ is the maximum likelihood estimator of N.

b Using maximum likelihood theory, show that the asymptotic variance of $\hat{N}$ is approximately $N^2(N - n_1)/(n_1 n_2)$.

15 (Requires calculus.) For the situation in Exercise 14, suppose the cost of catching a fish is the same for each fish in the first and second samples, and you have enough resources to catch a total of $n_1 + n_2 = C$ fish altogether. If N and C are known and $C < N$, what should n_1 and n_2 be to minimize the variance in Exercise 14(b)?

16 (Requires probability.)

a For Chapman's estimator $\tilde{N}$ in (13.2), let X be the random variable denoting the number of marked individuals in the second sample. What is the probability distribution of X?

b Show that $E[\tilde{N}] = N$ if $n_2 \geq N - n_1$.

17   Suppose the lake has N fish, and n_1 of them are marked. A sample of size n_2 is then drawn from the lake. Choose three values of N, n_1, and n_2. Approximate the distribution of $\hat{N}$ by drawing 1000 different samples of size n_2 from the population of N units and drawing a histogram of the $\hat{N}$ that result from the different samples. Repeat this for other values of N, n_1, and n_2. When does the histogram appear approximately normally distributed?

D. Projects and Activities

18   Try out the two-sample capture–recapture method to estimate the total number of popcorn kernels or dried beans in a package, or to estimate the total number of coins in a jar. Describe fully what you did, and give the estimate of the population size along with a 95% CI for N. How did you select the sizes of the two samples?

19   Repeat the preceding exercise, using three samples and loglinear models. Would you expect the model of complete independence to fit well? Does it?

14

Rare Populations and Small Area Estimation

Housework can't kill you, but why take a chance?[1]

—Phyllis Diller

EXAMPLE 14.1 The bestselling book *The Millionaire Mind* (Stanley, 2000) uses data from a survey of millionaire households in the United States. The population of millionaires is difficult to sample because no list of all millionaires exists. A simple random sample (SRS) from the U.S. population is likely to be inefficient because only a small fraction of American households have net worth over a million dollars; most returned surveys in an SRS will contain few members of the population of interest. Stanley had the sampling frame prepared by estimating the proportion of millionaire households in each census block group of the United States. He then stratified the block groups by estimated proportion of millionaires, and took an SRS of block groups within each stratum having at least 30% estimated millionaire households. A total of 5063 households were selected in those neighborhoods to receive the questionnaire; 1001 questionnaires were returned, and 733 came from households reporting a net worth of at least one million dollars. The sample selected did not cover the entire population of millionaires in the United States, since households in block groups with fewer than 30% estimated millionaire households were not included in the sampling frame. ∎

In this chapter, we discuss two situations for designing surveys. The first relates to Example 14.1: How to design a survey to sample units that belong to a rare population. A population can be rare in several ways. The number of individuals belonging to the rare population may be very small; snow leopards are a rare population simply because there are not very many of them. Or there may be a large number of individuals, but they form only a small fraction of the population. Millionaires, for example, are reasonably plentiful (one estimate has about 3.5 millionaires in the United States) but comprise a small percentage of the U.S. population (U.S. Census Bureau, 2007, Table 697). An SRS of persons in the United States, therefore, will yield few millionaires. Moreover, millionaires tend to be highly clustered, so that

---

[1] Although this is a wonderful quote, unfortunately housework *can* kill you; the Vital Statistics at www.cdc.gov tabulate deaths by accident in the house.

many geographic primary sampling units (psus) may have few, if any, millionaires. If we had a list of all millionaires in the United States, it would be quite easy to select a probability sample of them. For many rare populations, however, no such list exists; indeed, for some rare populations such as persons with Alzheimer's disease, it may be very difficult to determine membership in the population because persons may be unaware they have the disease. The challenge is to obtain a sufficiently large probability sample of the rare population for the desired accuracy while controlling costs.

In many surveys, we want to estimate quantities for many subpopulations, for example, to estimate the unemployment rate for every county in the United States. If we only wanted to estimate the unemployment rate in one county, we could design a survey with large sample size in that county. But a survey with sufficiently large sample size in every county will have unacceptably large cost. Instead, we would like to estimate county unemployment rates using an existing national survey on unemployment. Such a survey will likely have small sample sizes (or perhaps even no observations) in some counties. A sample of 60,000 households may give accurate estimates of the national unemployment rate, but the sample might have only a few observations in Larimer County—so few that an estimate of the unemployment rate in Larimer County that uses only the observations in the sample will have an unacceptably large margin of error. Larimer County, in this example, is a *small area* (also called a *small domain*); the population or land area of Larimer County may be large, but the sample size in the county is small. Section 14.2 explores models that may be used to improve accuracy of estimates for small areas.

14.1
Sampling Rare Populations

Sometimes you would like to investigate characteristics of a population that is difficult to find, or that is dispersed widely in the target population. For example, relatively few people are victims of violent crime in a given year, but you may want to obtain information about the population of violent crime victims. In an epidemiology survey, you may want to estimate the incidence of a rare disease, and to make sure you have enough persons with the disease in your sample to analyze how the persons with the disease differ from persons without the disease.

One possibility, of course, is to take a very large sample. That is done in the National Crime Victimization Survey (NCVS), which is used to estimate victimization rates. Because the NCVS was intended to estimate victimization rates for many different types of victimizations and to investigate households' victimization experiences over time, it was designed to be approximately self-weighting. The sample size for domestic violence victims, however, is very small. The NCVS would need to be prohibitively expensive to remain a self-weighting survey and still give sufficient sample sizes for all types of crime victims. In this section, we describe survey designs that have been proposed for estimating the prevalence of a rare characteristic or estimating quantities of interest for a rare population; several are based on concepts we have already discussed in this book.

Nonresponse can be an especial hazard in surveys of rare populations. If population members with the rare characteristic are more likely to be nonrespondents than members without the rare characteristic, estimates of prevalence will be biased. In some health surveys, the characteristic itself can lead to nonresponse—a survey of cancer patients may have nonresponse because the illness prevents persons from responding. It is therefore important to try to minimize nonresponse for any survey of a rare population.

14.1.1   Stratified Sampling with Disproportional Allocation

Sometimes strata can be constructed so that the rare characteristic is much more prevalent in one of the strata (say, in stratum 1). Then a stratified sample in which the sampling fraction is higher in stratum 1 can give a more accurate estimate of the prevalence of the rare characteristic in the general population. The higher sampling fraction in stratum 1 also increases the domain sample size for population members with the rare characteristic. The National Maternal and Infant Health Survey (MIHS), discussed in Example 11.1, sampled a higher fraction of records from low-birth-weight infants to ensure an adequate sample size of such infants.

Disproportional stratified sampling may work well when the allocation is efficient for all items of interest. For example, in the MIHS, a major concern was low-birth-weight infants, who have many more health problems. But disproportional stratification may not be helpful for all items of interest in other surveys. Having higher sampling fractions in census block groups thought to have high proportions of millionaires is sensible for estimating characteristics of millionaires. The design is not as efficient for estimating characteristics of persons who work at home, a rare population that is not necessarily concentrated in those block groups.

EXAMPLE **14.2**   Edwards et al. (2005) use models to construct strata for sampling rare lichen species in Washington, Oregon, and northern California. The rare species were uncommon in the pilot sample, so they fit classification tree models to predict presence of four common lichen species that frequently occur with the rare lichen species of interest. Each of the common lichen species was detected on at least 120 of the 840 sites sampled in a lichen air quality study, giving sufficient information to build models predicting presence of each species of common lichen from variables such as slope, aspect, precipitation, temperature, and relative humidity. Using data collected in a second sample, Edwards et al. (2005) estimate that a disproportional stratified sample based on the classification tree models would result in a 1.2- to 5-fold gain in sampling efficiency for four of the rare lichen species.   ∎

14.1.2   Two-Phase Sampling

Two-phase sampling methods were discussed in Chapter 12 as a way of using stratification when the information needed to form the strata is not available before sampling. To sample a rare population, we would like to stratify on the variable that indicates whether individuals belong to the population or not. Screen the phase I sample units to determine whether they have the rare characteristic or not. Then subsample all (or a high sampling fraction) of the units with the rare characteristic for the phase II sample. If the screening technique is completely accurate, use the phase I sample to estimate prevalence of the rare characteristic and the phase II sample to estimate other quantities for the rare population.

What if the screening technique is not completely accurate? If sampling arctic regions for presence of walruses, it is possible that you will not see walruses in some of the sectors from the air because the walruses are under the ice. Asking persons whether they have diabetes will not always produce an accurate response because persons do not always know whether they have it. As Deming (1977) points out, placing a person with diabetes in the "no-diabetes" stratum is more serious than placing a person without diabetes in the "diabetes" stratum: If only the "diabetes" stratum is subsampled, it is likely that the persons without diabetes who have been erroneously placed in that stratum will be discovered, while the error for a diabetic misclassified into the "no-diabetes" stratum will not be found. One possible solution is to broaden the screening criterion so that it encompasses all units that might have the rare characteristic. Another solution is to subsample both strata in phase II, but to use a much higher sampling fraction in the "likely to have diabetes" stratum.

You may want to use a different two-phase design for estimating characteristics of rare population members than for estimating prevalence of the rare population. Exercise 20 of Chapter 12 presented optimal sampling strategies for using a two-phase sample to estimate prevalence of a disease.

14.1.3   Unequal-Probability Sampling

To oversample individuals with the rare characteristic, we can create a model for the inclusion probabilities based on related characteristics. This is similar to disproportional stratified sampling, except that the unequal probabilities may be used directly as well as in stratification. Hoeting et al. (2000) developed a model for predicting the presence or absence of a species from satellite data. The model gives a predicted probability that the species is present for each pixel in the satellite image. The predicted probabilities may then be used to form strata or to specify inclusion probabilities π_i.

The Mitofsky–Waksberg method for random digit dialing, discussed in Example 6.12, can be used to sample rare populations that are clustered. In a survey of millionaires, census block groups can be treated as clusters. A probability sample of block groups is drawn (the probability sampling design should rely on stratification as well). Select one household from each cluster; if it is a millionaire household, then sample additional households in that cluster. This procedure samples clusters with probability proportional to the number of millionaire households.

14.1.4   Multiple Frame Surveys

Even though you may not have a list of all of the members of the rare population, you may have some incomplete sampling frames that contain a high percentage of units with the rare characteristic. You can sometimes combine these incomplete frames, omitting duplicates, to construct a complete sampling frame for the population. Alternatively, you can select samples independently from the frames, then combine sample estimates from the incomplete frames (and, possibly, a complete frame) to obtain general population estimates. This **multiple frame survey** approach was pioneered by Hartley (1962).

FIGURE **14.1**

Examples of dual frame surveys. In (a), frame *A* is complete and frame *B* is incomplete. In (b), both frames are incomplete.

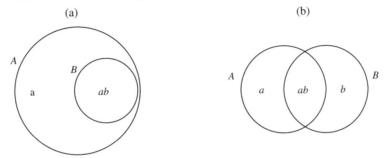

Suppose you would like to estimate characteristics of persons with Alzheimer's disease in the noninstitutionalized population. Since many users of adult day care centers have Alzheimer's, you would expect that a sample of adult day care centers would yield a higher percentage of persons with Alzheimer's than a general population survey. But not all persons with Alzheimer's attend an adult day care center. Thus, you might have two sampling frames: frame *A*, which is the sampling frame for the general population survey, and frame *B*, which is the sampling frame for adult day care centers. All persons in frame *B* are presumed to also be in the frame for the general population survey, so the design in Figure 14.1(a) has two domains: *ab*, which consists of persons in frame *A* and also in frame *B*, and *a*, which consists of persons in frame *A* but not in frame *B*. In other situations, both frames are incomplete, leading to three domains as in Figure 14.1(b): domain *a*, consisting of persons in frame *A* but not frame *B*; domain *b*, with persons in frame *B* but not frame *A*; and domain *ab*, consisting of persons in both frames.

To estimate population quantities from the general dual frame survey depicted in Figure 14.1(b), determine the domain membership of each sampled person. Estimate the population total $t = \sum_{i=1}^{N} y_i$ by $\hat{t}_a + \hat{t}_{ab} + \hat{t}_b$, where $\hat{t}_a$, $\hat{t}_{ab}$, and $\hat{t}_b$ estimate the population totals in domains *a*, *ab*, and *b*, respectively. A variety of estimators can be used to estimate the two domain totals; some of these are summarized in Lohr and Rao (2000) and Lohr (2009). Exercise 3 gives Hartley's (1962) estimator for the survey depicted in Figure 14.1(b).

EXAMPLE **14.3** The National Survey of Veterans in 2001 (Choudhry et al., 2002) used a dual frame survey to sample the target population of veterans living in private households. Frame *A* was a random digit dialing (RDD) frame that covered the population of telephone households. Frame *A* included all veterans living in telephone households, but many households contacted through the RDD survey contained no veterans. Frame *B* was a list of veterans constructed from the Veterans Administration Healthcare Enrollment, and Compensation and Pension files. Everyone in frame *B* was eligible for the survey so frame *B* was less expensive to sample than frame *A*, but frame *B* did not include everyone in the target population. The dual frame survey thus accorded with Figure 14.1(a). It combined complete coverage of the telephone household population from the frame *A* survey and lower cost of sampling from the frame *B* survey. ∎

EXAMPLE 14.4 Iachan and Dennis (1993) described the use of multiple frames to sample the homeless population in Washington, D.C. Four frames were used: (1) homeless shelters, (2) soup kitchens, (3) encampments such as vacant buildings and locations under bridges, and (4) streets, sampled by census blocks. Although the union of the frames should include more of the homeless population than a single frame, it will not include all homeless persons.

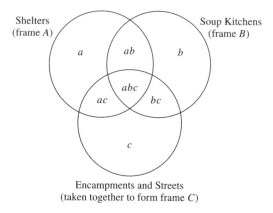

Shelters
(frame *A*)

Soup Kitchens
(frame *B*)

a   *ab*   *b*

abc

ac   *bc*

c

Encampments and Streets
(taken together to form frame *C*)

Membership in more than one frame was estimated by asking survey respondents whether they had been or expected to be in soup kitchens, in shelters, or on the street in the 24-hour period of sampling. ∎

14.1.5 Network or Multiplicity Sampling

In a household survey such as the NCVS, each household provides information only on victimizations that have occurred to members of that household. In a network sample to study crime victimization (see Czaja and Blair, 1986; Sudman et al., 1988 for the general method), each household in the population is linked to other units in the population; the sampled household can also provide information on units linked to it (called the network for that household). For example, the network of a household might be defined to be the adult siblings of adult household members.

Suppose a probability sample of households is taken. Define $\mathcal{G}_i$ to be the network for unit i in the probability sample. Suppose household 1 has adults John and Mary. Then, if networks are formed using the sibling rule, $\mathcal{G}_1$ consists of John, Mary, John's adult siblings (Suzy and Fred), and Mary's adult sibling (Mark). John is asked about crime incidents that occurred to him, Suzy, and Fred; Mary is asked about incidents that occurred to her and Mark. John's (or Suzy's or Fred's) response can be included up to three times in the sample: if John's household is selected, Suzy's household is selected, or Fred's household is selected. Mark's or Mary's information has only two chances of inclusion, if Mark's or Mary's household is chosen in the sample. An only child is included only if his or her household is selected in the probability sample.

The **multiplicity** of individual k is the number of links leading to that individual. Let $\omega_k = 1/(\text{multiplicity of person } k)$ be the multiplicity weight for person k in the population of interest. In our example, John, Suzy, and Fred each have multiplicity weight 1/3, and Mary and Mark each have multiplicity weight 1/2. Let y_k be an indicator variable for whether person k was a victim of crime. Estimate the total number of crime victims by

$$\hat{t}_{y,\text{net}} = \sum_{i \in \mathcal{S}} w_i \sum_{k \in \mathcal{G}_i} \omega_k y_k. \tag{14.1}$$

This estimator and its variance are derived in Exercise 4.

Network sampling can reduce the sampling variability of the estimated prevalence of a rare characteristic because it can provide more information per sampled individual. Czaja et al. (1986) found that network sampling provided greater precision for estimating prevalence of cancer cases. There are, however, additional possibilities for error in network sampling. If John is selected in the initial sample, he must report: (1) his value of the response y_i, (2) the response for each person in John's network (y_k for persons k linked to John), and (3) the number of population units linked to each person k in John's network (the multiplicity for person k in John's network).

John will probably give the correct multiplicity for his siblings. But with other linking rules, John's report of multiplicity for units in his network may be inaccurate—if John's network consists of students who are in class with him, John may not know the number of other classes taken by his classmates. Also, John might not report the correct value of the response for persons in his network. John might not be aware of criminal victimizations experienced by his or her siblings and give an inaccurate count. Social desirability of responses is also an issue. John may know which of his siblings have cancer, but may not know that one of them is a substance abuser.

14.1.6 Snowball Sampling

Snowball sampling is based on the premise that members of the rare population know one another. To take a snowball sample of homeless persons, you would locate a few homeless persons. Ask each of those persons to identify other homeless persons for your sample, ask the new persons in your sample to identify additional homeless persons, and so on, until a desired sample size is attained. Snowball sampling can create a fairly large sample of a rare population, but in general does not produce a probability sample; strong modeling assumptions are needed to generalize results from a snowball sample to the population. Although snowball sampling can identify members of a rare population who would be difficult to find with other designs, the resulting sample is often far from an SRS. Persons with many connections in the population of interest are more likely to be included in the sample than persons with few connections. Isolated persons may not be reachable at all. Respondent-driven sampling methods (Heckathorn, 1997; Salganik and Heckathorn, 2004) use information about network connections in the sample to weight the sample units.

14.1.7 Sequential Sampling

In sequential sampling, observations or psus are sampled one or a few at a time, and information from previously drawn psus can be used to modify the sampling design for subsequently selected psus. In one method dating back to Stein (1945) and Cox (1952), an initial sample is taken, and results from that sample are used to estimate the additional sample size necessary to achieve a desired precision. If it is desired that the sample contain a certain number of members from the rare population, the initial sample could be used to obtain a preliminary estimate of prevalence, and that estimate of prevalence used to estimate the necessary size of the second sample. After the second sample is collected, it is combined with the initial sample to obtain estimates for the population. A sequential sampling scheme generally needs to be accounted for in the estimation; in Cox's method, for example, the sample variance obtained after combining the data from the initial and second samples is biased downward (Lohr, 1990). Lai (2001) reviews history and uses of sequential methods.

Adaptive cluster sampling (Thompson, 1990) assumes that the rare population is aggregated—caribou are in herds, an infectious disease is concentrated in regions of the country, or artifacts are clustered at specific sites of an archaeological dig. An initial probability sample of psus (often quadrats, in wildlife applications) is selected. For each psu in the initial sample, measure a response such as the number of caribou in the psu. If the number of caribou in psu i exceeds a predetermined value c, then add neighbors of psu i to the sample. Count the number of caribou in each of the neighboring units and add the neighbors of any of those units with more than c caribou to the sample. Continue the procedure until none of the neighbors has more than c caribou. The adaptive nature of the sampling scheme needs to be accounted for when estimating population quantities—if you estimate caribou density by (number of caribou observed)/(number of psus sampled) from an adaptive cluster sample, your estimate of caribou density will be far too high. Thompson and Seber (1996), Thompson and Collins (2002), and Turk and Borkowski (2005) describe various approaches for adaptive cluster sampling and give references to other work.

14.2
Small Area Estimation

In most surveys, estimates are desired not only for the population as a whole, but also for subpopulations (domains). We discussed estimation in domains in Section 4.2 for SRSs and showed that estimating domain means is a special case of ratio estimation because the sample size in the domain varies from sample to sample. But we noted that if the sample size for the domain in an SRS was large enough, we could essentially act as though the sample size was fixed for inference about the domain mean.

In complex surveys with many domains, estimation is not quite that simple. One worry is that the sample size for a given domain will be too small to provide a useful estimate. The NCVS, for example, gives reliable information on the incidence of different types of criminal victimizations in the United States as a whole. However, if you are interested in estimating the violent crime rate at the state level for the purpose of allocating federal funds for additional police officers, the sample sizes for some states are so small that direct estimates of the violent crime rates for those states are of very little use. You might conjecture, though, that crime rates are similar in neighboring states with similar characteristics, and use information from other states to improve the estimate of violent crime rate for the state with a small sample size. You could also incorporate information on crime rate from other sources, such as police statistics, to improve your estimate.

Similarly, the National Assessment of Educational Progress (NAEP; see Example 11.10) data collected on students in New York may be sufficient for estimating eighth grade mathematics achievement for students in the state, but not for a direct assessment of mathematics achievement in individual cities such as Rochester. The survey data from Rochester, though, can be combined with estimates from other cities and with school administrative data (scores on other standardized tests, for example, or information about mathematics instruction in the schools) to produce an estimate of eighth grade mathematics achievement for Rochester that we hope has smaller mean squared error.

Small area estimation techniques, in which estimates are obtained for domains with small sample sizes, have in recent years been the focus of intense research in statistics. Rao (2003) describes small area estimation methods and gives a bibliography for further reading. Here, we summarize some of the proposed approaches. Let $a_{id} = 1$ if observation unit i is in domain d and 0 otherwise. In this section, the quantities of interest are the domain totals $t_d = \sum_{i=1}^{N} a_{id} y_i$, the domain sizes $N_d = \sum_{i=1}^{N} a_{id}$, and the domain means $\bar{y}_{Ud} = t_d / N_d$, for domains $d = 1, \ldots, D$.

14.2.1 Direct Estimators

A direct estimator of t_d depends only upon the sampled observations in domain d:

$$\hat{t}_d = \sum_{i \in S} w_i a_{id} y_i. \tag{14.2}$$

The estimated domain totals in (14.2) satisfy the following **additive property**: If domains d_1 and d_2 are mutually exclusive, and if domain d_3 is the union of domains d_1 and d_2, then

$$\hat{t}_{d_1} + \hat{t}_{d_2} = \hat{t}_{d_3}.$$

The additive property is desirable since we would like the estimated numbers of people without health insurance in each demographic group to sum to the estimated number of people without health insurance in the population.

The domain mean is estimated by

$$\hat{\bar{y}}_d = \frac{\sum\limits_{i \in S} w_i a_{id} y_i}{\sum\limits_{i \in S} w_i a_{id}}. \tag{14.3}$$

Because $\hat{\bar{y}}_d$ is a ratio, the variance is estimated using linearization (see Example 9.2) as

$$\hat{V}(\hat{\bar{y}}_d) = \frac{1}{\hat{N}_d^2} \hat{V} \left[\sum_{i \in S} w_i a_{id}(y_i - \hat{\bar{y}}_d) \right]. \tag{14.4}$$

The approximation to the variance is valid if the expected sample size in the domain is sufficiently large. Section 11.3 discussed comparing domain means using regression.

Warning: In an SRS, if you create a new data set that consists solely of sampled observations in domain d and then apply the standard variance formula, your variance estimator is approximately unbiased. **Do not** adopt this approach for estimating the variance of domain means in complex samples. A sampled psu may contain no observations in domain d; if you eliminate such psus and then apply the standard variance formula, you may underestimate the variance (see Exercises 5 and 8). Survey software such as SAS PROC SURVEYMEANS calculates the variance correctly when you specify domains.

In practice, the sample size in domain d may be so small that the variance of $\hat{\bar{y}}_d$ is unacceptably large. Some domains of interest may have no observations at all so that a direct estimator cannot be calculated. The next sections describe methods that may be used to estimate domain mean and totals in these cases.

14.2.2 Synthetic and Composite Estimators

Assume that we know some quantity associated with t_d for each domain d. For estimating violent crime victimization rates, we might use t_{ud} = total amount of violent crime in domain d obtained from police reports. Then, if the ratios t_d/t_{ud} are similar in different domains, and if each ratio is similar to the ratio of population totals t_y/t_u, then a simple form of synthetic estimator

$$\hat{t}_d(\text{syn}) = \frac{\hat{t}_y}{\hat{t}_u} t_{ud}$$

may be more accurate than $\hat{t}_d$ in (14.2). Certainly the variance of $\hat{t}_d(\text{syn})$ will be relatively small, since $(\hat{t}_y/\hat{t}_u)$ is estimated from the entire sample and is expected to be precise. If the ratios are not homogeneous, however—if, for example, the proportion of violent crime victimizations reported to the police varies greatly from domain to domain—the synthetic estimator may have large bias.

You can also use synthetic estimation in subsets of the population, and then combine the synthetic estimators for each subset. For estimating violent crime victimization in small areas, you could divide the population into different age-race-gender classes. Then you could find a synthetic estimate of the total violent crime victimization in domain d for each age-race-gender class, and sum the estimates for the age-race-gender classes to estimate the total violent crime victimizations in small area d. It is hoped that the ratios (violent crime victimizations in domain d for age-race-gender class c from NCVS)/(violent crime victimizations in domain d for age-race-gender class c from police reports) are more homogeneous than the ratios t_d/t_{ud}.

The direct estimator is unbiased but may have large variance; the synthetic estimator has smaller variance but may have large bias. They may be combined to form a composite estimator:

$$\hat{t}_d(\text{comp}) = \alpha_d \hat{t}_d(\text{dir}) + (1 - \alpha_d)\hat{t}_d(\text{syn})$$

for $0 \leq \alpha_d \leq 1$. The relative optimal weights α_d depend on the relative variance and bias of the direct and synthetic estimators, but one possible solution has α_d related to the sample size in domain d. If few units are observed in domain d, α_d will be close to zero and more reliance will be placed on the synthetic estimator.

14.2.3 Model-Based Estimators

In a model-based approach, a superpopulation model is used to predict values in domain d. The model "borrows strength" from the data in related domains, or incorporates auxiliary information from administrative data or other surveys. The models can often be used to determine the weights α_d in a composite estimator. Mixed models, described in Section 11.5, are often used in small area estimation.

The Fay–Herriot model (Fay and Herriot, 1979) is commonly used when a vector of auxiliary information $\mathbf{x}_d$ is available for each domain $d = 1, \ldots, D$. We wish to estimate the population domain mean $\theta_d = \bar{y}_{Ud}$. Let $\hat{\bar{y}}_d$ be a direct estimator of θ_d from the survey with variance $V(\hat{\bar{y}}_d) = \psi_d$. Assume that

$$\hat{\bar{y}}_d = \theta_d + e_d,$$

where $e_d \sim N(0, \psi_d)$ and $e_1, \ldots, e_D$ are independent. Also assume that the population domain means θ_d are related to the covariates $\mathbf{x}_d$ through the model

$$\theta_d = \mathbf{x}_d^T \boldsymbol{\beta} + v_d,$$

where $v_1, \ldots, v_D$ are independent $N(0, \sigma_v^2)$ random variables. Combining the two models, we have

$$\hat{\bar{y}}_d = \mathbf{x}_d^T \boldsymbol{\beta} + v_d + e_d, \tag{14.5}$$

which includes the error term e_d from the direct estimator as well as the error v_d from the model that is assumed to hold for the population domain means. If ψ_d and σ_v^2 are known, then the best linear unbiased predictor of θ_d is

$$\tilde{\theta}_d = \alpha_d \hat{\bar{y}}_d + (1 - \alpha_d)\mathbf{x}_d^T \tilde{\boldsymbol{\beta}},$$

where $\alpha_d = \sigma_v^2/(\sigma_v^2 + \psi_d)$ and $\tilde{\boldsymbol{\beta}}$ is the weighted least squares estimator of $\boldsymbol{\beta}$. The estimator $\tilde{\theta}_d$ thus depends more heavily on the direct estimator $\hat{\bar{y}}_d$ when $\psi_d = V(\hat{\bar{y}}_d)$ is small; it depends more heavily on the predicted value from the regression model $\mathbf{x}_d^T \tilde{\boldsymbol{\beta}}$ when ψ_d is large. In practice, σ_v^2 must be estimated from the data and the estimator $\hat{\sigma}_v^2$ used to estimate α_d by $\hat{\alpha}_d$.

The Fay–Herriot model is an example of an *area-level model* for small area estimation; it includes quantities that describe the domain as a whole. In the U.S. Small Area Income and Poverty Estimates program (www.census.gov/did/www/saipe), the estimated poverty rate for a county is a weighted average of the direct estimate from a survey (since 2006, the American Community Survey is used; before that, the Current Population Survey was used) and the predicted value from a regression equation using auxiliary information from tax records, food stamp programs, and other sources. Each covariate is at the domain level; the covariates include the number of food stamp participants in the county and the number of Internal Revenue Service exemptions on tax returns with adjusted gross income below the poverty threshold (Bell et al., 2007). In a county with a large sample size, $\tilde{\theta}_d$ is very close to the estimator from the survey. Logarithms are taken of the variables so that their distribution is closer to a normal distribution.

A *unit-level model* requires knowledge of covariate values for each individual in the survey. In the NAEP, if Y_{dj} is the mathematics achievement of student j in domain

d in the population, you might postulate a model such as

$$Y_{dj} = \beta_0 + x_{dj}\beta_1 + u_d + \varepsilon_{dj},$$

where $u_d \sim N(0, \sigma_u^2)$ and $\varepsilon_{dj} \sim N(0, \sigma_\varepsilon^2)$ for $d = 1, \ldots, D$ and student j in domain d, assuming all random variables u_d and ε_{dj} are independent (Battese et al., 1988). The student-level covariate x_{dj} (we included one covariate for simplicity, but of course several covariates could be included) could come from administrative records, for example, the student's score on an achievement test given to all students in the state or the student's grades in mathematics classes. Assume that the population mean of x in domain d, $\bar{x}_{Ud} = \frac{1}{N_d} \sum_{j=1}^{N_d} x_{dj}$, is known. Then, if the domain sample size n_d is small relative to the population size N_d, it can be shown that the best linear unbiased predictor of the modeled domain mean $\mu_d = \beta_0 + \bar{x}_{Ud}\beta_1 + u_d$ is

$$\tilde{\mu}_d = \tilde{\beta}_0 + \bar{x}_{Ud}\tilde{\beta}_1 + \gamma_d \left(\hat{\bar{y}}_d - \tilde{\beta}_0 - \hat{\bar{x}}_d \tilde{\beta}_1 \right),$$

where $\gamma_d = \sigma_u^2 / (\sigma_u^2 + \sigma_\varepsilon^2/n_d)$ and $\tilde{\beta}_0$ and $\tilde{\beta}_1$ are the best linear unbiased estimators of β_0 and β_1. The predictor depends more on the direct estimator $\hat{\bar{y}}_d$ if $V_M(\hat{\bar{y}}_d) = \sigma_\varepsilon^2/n_d$ is small; otherwise, it depends more on the predicted value from the regression at the population domain mean of x. Rao (2003) describes unit-level models and other models commonly used in small area estimation.

An indirect estimator, whether synthetic, composite, or model-based, is essentially an exercise in predicting missing data. Indirect estimators are thus highly dependent on the model used to predict the missing data—the synthetic estimator, for example, assumes that the ratios are homogeneous across domains. When possible, the model assumptions should be checked empirically; one method for exploring validity of the model assumptions is to pretend that some of the data you have is actually not available, and to compare the indirect estimator with the direct estimator computed with all the data.

14.3
Chapter Summary

Rare populations present special challenges for sampling, since many standard sampling designs yield few units in the rare population. Several designs discussed in previous chapters can be used to increase the number of rare population units in the sample. Auxiliary information associated with the rare characteristic can be used to design a stratified sample with disproportional allocation. If such auxiliary information is not known in advance, a two-phase sampling design can collect inexpensive screening information in the phase I sample, and then collect the detailed survey information in phase II.

Multiple frame surveys, in which independent probability samples are selected from sampling frames whose union is assumed to include the entire rare population, can greatly reduce the cost of a survey of a rare population. One frame might cover the entire population, while other frames might be incomplete yet inexpensive to sample. By sampling from multiple frames, you achieve the cost savings from sampling lists of rare population members, while also having complete population coverage by sampling from a complete frame.

Network, snowball, and adaptive cluster samples use connections among population members to increase the efficiency of the sampling design. In network sampling, persons in a probability sample are asked about themselves as well as persons defined to be in their network, for example, their adult siblings. A snowball sample often begins with a convenience sample of persons in the rare population, who are then asked to provide contact information for other persons in the rare population. In adaptive cluster sampling, the responses of an initial probability sample are used to select neighboring units for inclusion.

Small area estimation methods rely on auxiliary information and models to obtain estimators of population quantities in domains in which the sample size is too small for a direct estimator to be reliable.

Key Terms

Adaptive cluster sampling: A sequential sampling design in which estimates from the first units selected for the sample are used to determine inclusion probabilities for subsequent units.

Multiple frame survey: A survey in which independent samples are taken from two or more sampling frames that are thought to include the whole population.

Network sampling: A sampling method in which a probability sample is taken from a population and each sampled unit provides information on itself and on units in its network.

Rare population: A subpopulation that is uncommon relative to the whole population.

Small area: A subpopulation for which the sample size is small.

For Further Reading

Kalton and Anderson (1986), Sudman et al. (1988), Kalton (2003), and Christman (2009) review methods for sampling rare populations. The book edited by Thompson (2004) describes methods for sampling rare species in wildlife applications. Rao (2003) reviews methods for small area estimation, with applications to estimating poverty, unemployment, disease prevalence, and census undercounts.

14.4
Exercises

A. Introductory Exercises

1 What designs would you consider for sampling each of the following rare populations?

 a Alumni of your university who are currently working as engineers.

 b Persons who are caregivers for a household member who has Alzheimer's.

 c Households with children aged 18–36 months.

d Muslims in Canada; the 2001 Canadian Census estimated that there were approximately 600,000 Muslims in Canada in 2001 (about 2% of the population).

e Businesses that emit benzene. Some types of businesses—for example, gas stations—are thought to be likely to emit benzene. For other businesses, the benzene emissions are unknown, but it is thought that if one business is found to emit benzene, it is likely that other businesses in the same industry and area emit benzene as well.

C. Working with Theory

2 (Requires calculus.) Kalton and Anderson (1986) consider disproportional stratified random sampling for estimating the mean of a characteristic y_i in a rare population. Let $r_i = 1$ if person i is in the rare population and 0 otherwise. Stratum 1 contains N_1 persons, M_1 of whom are in the rare population; stratum 2 contains N_2 persons, with M_2 persons in the rare population. We wish to estimate the population mean $\bar{y}_{Ud} = \sum_{i=1}^{N} r_i y_i/(M_1 + M_2)$ using a stratified random sample of n_1 persons in stratum 1 and n_2 persons in stratum 2.

a Suppose $A = M_1/(M_1 + M_2)$ is known. Let $\hat{\bar{y}}_d = A\bar{y}_1 + (1 - A)\bar{y}_2$, where $\bar{y}_1$ and $\bar{y}_2$ are the sample means of the rare population members in strata 1 and 2, respectively. Show that, if you ignore the finite population corrections (fpcs) and if the sampled number of persons in the rare population in each stratum is sufficiently large, then

$$V(\hat{\bar{y}}_d) \approx \frac{A^2 S_1^2}{n_1 p_1} + \frac{(1 - A)^2 S_2^2}{n_2 p_2},$$

where S_j^2 is the the variance of y for the rare population members in stratum j and $p_j = M_j/N_j$ for $j = 1, 2$.

b Suppose that $S_1^2 = S_2^2$ and that the cost to sample each member of the population is the same. Let $f_2 = n_2/N_2$ be the sampling fraction in stratum 2, and write the sampling fraction in stratum 1 as $f_1 = kf_2$. Show that the variance in (a) is minimized for a fixed sample size n when $k = \sqrt{p_1/p_2}$.

3 (Requires calculus.) Consider the dual frame survey in Figure 14.1(b) in which independent probability samples are taken from frames A and B. Suppose that all three domains are nonempty. Let $\mathcal{S}^A$ denote the sample from frame A, with inclusion probabilities $\pi_i^A = P(i \in \mathcal{S}^A)$ and sampling weights $w_i^A = 1/\pi_i^A$. Corresponding quantities for frame B are $\mathcal{S}^B$, π_i^B, and w_i^B. Let $\delta_i = 1$ if unit i is in domain ab and 0 otherwise. Then $\hat{t}_a^A = \sum_{i \in \mathcal{S}^A} w_i^A(1 - \delta_i)y_i$ and $\hat{t}_b^B = \sum_{i \in \mathcal{S}^B} w_i^B(1 - \delta_i)y_i$ estimate the domain totals t_a and t_b, respectively. There are two independent estimators of the population total in the intersection domain ab: $\hat{t}_{ab}^A = \sum_{i \in \mathcal{S}^A} w_i^A \delta_i y_i$ and $\hat{t}_{ab}^B = \sum_{i \in \mathcal{S}^B} w_i^B \delta_i y_i$.

a Let $\theta \in [0, 1]$. Show that

$$\hat{t}_{y,\theta} = \hat{t}_a^A + \theta \hat{t}_{ab}^A + (1 - \theta)\hat{t}_{ab}^B + \hat{t}_b^B$$

is an unbiased estimator of $t_y = \sum_{i=1}^{N} y_i$ with

$$V(\hat{t}_{y,\theta}) = V\left[\hat{t}_a^A + \theta \hat{t}_{ab}^A\right] + V\left[(1 - \theta)\hat{t}_{ab}^B + \hat{t}_b^B\right].$$

b Show that $V(\hat{t}_{y,\theta})$ is minimized when

$$\theta = \frac{V(\hat{t}^B_{ab}) + \text{Cov}\,(\hat{t}^B_b, \hat{t}^B_{ab}) - \text{Cov}\,(\hat{t}^A_a, \hat{t}^A_{ab})}{V(\hat{t}^A_{ab}) + V(\hat{t}^B_{ab})}.$$

4 The estimator in Exercise 22 of Chapter 6 for indirect sampling can be applied to network sampling (Lavallée, 2007) to give the estimator in Sirken (1970). In the context of network sampling, $\mathcal{U}^A$ is the sampling frame population for the initial sample and $\mathcal{U}^B$ is the population of interest, with M elements. The links ℓ_{ik} define the networks: $\ell_{ik} = 1$ if person k in $\mathcal{U}^B$ is in the network of unit i in $\mathcal{U}^A$. Thus, $L_k = \sum_{i=1}^N \ell_{ik}$ is the multiplicity for person k.

a Show that $\hat{t}_{y,\text{net}}$ in Equation (14.1) equals the estimator $\hat{t}_y$ given in Exercise 22(a) of Chapter 6. Consequently, $\hat{t}_{y,\text{net}}$ is an unbiased estimator of t_y.

b Suppose that $\mathcal{U}^A = \mathcal{U}^B$ is a population of N persons, and the sample from $\mathcal{U}^A$, $\mathcal{S}^A$, is an SRS of size n. Let $y_k = 1$ if person k has the rare characteristic and 0 otherwise. Find $V(\hat{t}_{y,\text{net}})$.

c How does the variance in (b) compare with the variance of $\hat{t}_y = \sum_{i \in \mathcal{S}^A} \frac{N}{n} y_i$, which uses only information from $\mathcal{S}^A$?

5 Consider a stratified sample in which an SRS of n_h psus is selected from the population of N_h psus in stratum h, for $h = 1, \ldots, H$. We wish to estimate the mean of domain d.

a Find $\hat{V}(\hat{\bar{y}}_d)$ using linearization.

b Now suppose that a data analyst creates a new data set by deleting observations that are not in domain d. If you (incorrectly) act as though this is the full data set, what is the estimated variance of $\hat{\bar{y}}_d$?

c Show that the estimators of the variance in (a) and (b) are unequal if some sampled psus have no observations in domain d. The correct variance estimator is given in (a) and (14.4).

6 Estevao and Särndal (1999) and Hidiroglou and Patak (2004) study the use of auxiliary information in domain estimation, which can reduce the variance of the direct domain estimator $\hat{t}_d$ in Section 14.2.1.

a If the population total for an auxiliary variable x, t_x, is known, we may use the ratio estimator

$$\hat{t}_{dr1} = \hat{t}_d \frac{t_x}{\hat{t}_x}.$$

If the sample size in domain d is sufficiently large to use linearization, what is $\hat{V}(\hat{t}_{dr1})$? Does $\hat{t}_{dr1}$ have the additive property?

b If we know the population total of x for each domain d, with $t_{xd} = \sum_{i=1}^N a_{id} x_i$, then we can use a domain-specific ratio estimator

$$\hat{t}_{dr2} = \hat{t}_d \frac{t_{xd}}{\hat{t}_{xd}}.$$

What is $\hat{V}(\hat{t}_{dr2})$? Does $\hat{t}_{dr2}$ have the additive property?

7 (Requires calculus.) Consider the Fay–Herriot model in (14.5). Suppose that ψ_d, σ_v^2, and $\boldsymbol{\beta}$ are known.

a   Let

$$\tilde{\theta}_d(a) = a\hat{\bar{y}}_d + (1-a)\mathbf{x}_d^T\boldsymbol{\beta}$$

with $a \in [0, 1]$. Show that, under the model in (14.5), $E_M[\tilde{\theta}_d(a) - \theta_d] = 0$ for any $a \in [0, 1]$.

b   Show that $V_M[\tilde{\theta}_d(a) - \theta_d]$ is minimized when $a = \alpha_d$ and that $V_M[\tilde{\theta}_d(\alpha_d) - \theta_d] = \alpha_d \psi_d$. Consequently, under the model, $V_M[\tilde{\theta}_d(\alpha_d) - \theta_d] \le V_M[\hat{\bar{y}}_d - \theta_d]$.

D. Projects and Activities

8   Construct a population with 20 strata. Each stratum has 8 psus and each psu has 3 secondary sampling units (ssus), so that the population has a total of 480 ssus. Observation j of psu i in stratum h has $y_{hij} = h$, for $h = 1, \ldots, 20$, so that all observations in stratum 1 have the value 1, all observations in stratum 2 have the value 2, and so on. Within each stratum, all observations in psus 1–4 are in domain 1, and all observations in psus 5–8 are in domain 2.

a   Select a one-stage stratified sample from the population by selecting an SRS of two psus from each stratum and including all ssus within the selected psus in the sample. Your sample should have 120 observations. Estimate the population mean for each domain along with its standard error.

b   Repeat (a) for a second stratified sample, selected independently (i.e., use a different random seed). Compare the domain means from this sample with those from (a). Do the domain means vary from sample to sample?

c   Now create a new data set for your sample in (a) that consists only of observations in domain 1, by deleting all the observations in domain 2. What is the estimated domain mean from this data set? What is the standard error using this data set, and why is it incorrect?

9   *Forest data.*   Cells with primary cover-type cottonwood/willow form a rare population in the forest data. What methods discussed in this chapter might be used to sample the cottonwood/willow cells? Which do you think will be most efficient?

10   *Activity for course project.*   Are there rare populations of interest for the survey you studied in Exercise 31 of Chapter 7? If so, what design features were used in the survey to sample members of the rare population?

15

Survey Quality

DUDLEY. Do you think you've learned from your mistakes?
PETER. Oh, yes, I've learned from my mistakes and I'm sure I could repeat them exactly.

—Peter Cook and Dudley Moore, *Good Evening: A Comedy-revue in Two Acts*

EXAMPLE 15.1 The American Community Survey (U.S. Census Bureau, 2005) is the largest continuing sample survey in the history of the United States. Each year, approximately 3 million questionnaires are mailed out to households across the United States. Of course, a survey of this scale requires a great deal of planning and development, and potential inaccuracies in the data need to be resolved before the survey is launched. For national estimates of quantities such as unemployment and household size, the sampling error of the survey will be very small. But other sources of error such as nonresponse and undercoverage are important. It is thus crucial in planning such a survey that all errors be considered in the design. ∎

Throughout this book, we have concentrated on designing surveys that will produce accurate and timely statistics. Chapters 2–7 discussed survey designs that could be used to control sampling error for estimating population means and totals. Chapters 9–14 outlined other methods for analyzing data from complex surveys. Chapters 1 and 8 discussed nonsampling errors that can arise in surveys.

In Chapters 2–7 we assumed that there are no nonsampling errors, and the only reason that survey estimates differ from population quantities is that a sample was taken rather than a census. In many surveys, the margin of error reported is based entirely on the sampling error; nonsampling errors are sometimes acknowledged in the text, but generally are not included in the reported measures of uncertainty.

Dalenius (1977, p. 21) referred to the practice of reporting only sampling error and ignoring other sources of error as "'strain at a gnat and swallow a camel'; this characterization applies especially to the practice with respect to the accuracy: the sampling error plays the role of the gnat, sometimes malformed, while the nonsampling error plays the role of the camel, often of unknown size and always of unwieldy shape."

In this chapter, we explore approaches to survey design and analysis that consider the whole camel. Much of the early inspiration for this approach came from W. Edwards Deming, who in addition to writing one of the first books on survey sampling (Deming, 1950) was also one of the leaders in developing quality improvement methods for industry after World War II (Boardman, 1994). Not surprisingly, Deming was one of the earliest writers to consider factors that might affect the quality of survey estimates. Deming (1944) discussed survey errors due to interviewer variability, survey mode, questionnaire design, sampling variability, nonresponse, and other sources now considered to be part of total survey error.

Quality in surveys draws on many ideas from Deming's work on quality improvement (Deming, 1986). Biemer and Lyberg (2003) define survey quality as "fitness for use." While somewhat vague, this definition recognizes the multiple purposes of survey data. Eurostat (2000) considers quality to encompass seven dimensions:

1 *Relevance of statistical concept.* The statistics collected must meet user needs.

2 *Accuracy of estimates.* Estimates should be close to the true values of population quantities.

3 *Timeliness.* Results need to be disseminated quickly to be useful. Indeed, as argued in Chapter 1, one reason for taking a survey rather than conducting a census is that the survey can be completed much more rapidly.

4 *Accessibility and clarity of information.* Particularly in official statistics, data and data products must be accessible to users, and sufficient documentation should be provided to enable users to interpret the results.

5 *Comparability.* Many surveys such as the National Crime Victimization Survey (NCVS) have a purpose of comparing estimates over time; such surveys must be conducted so that these comparisons are meaningful. When survey results are to be compared for different countries, care must be taken to ensure the concepts being measured are interpreted the same way in different countries and that appropriate methodologies are used to ensure comparable results (Harkness et al., 2003).

6 *Coherence.* Common definitions and standards should be used when data come from several sources.

7 *Completeness.* The data collector should be able to provide statistics for all domains identified by the community of data users.

While all of these quality dimensions may be important in different contexts, we argue that data accuracy is the most important aspect of data quality. Timely, coherent, comparable statistics are of little use if they are wildly inaccurate. As defined in Chapter 2, an estimator $\hat{\theta}$ of a population quantity θ is **accurate** if it is close to the true value of the quantity being estimated, that is, if $\text{MSE}[\hat{\theta}] = E[(\hat{\theta} - \theta)^2]$ is small. We can consider the **total survey error** (Andersen et al., 1979) to be the sum of five main sources of error:

$$\text{total survey error} = \text{coverage error} + \text{nonresponse error} + \text{measurement error} + \text{processing error} + \text{sampling error}.$$

Lessler and Kalsbeek (1992) and Linacre and Trewin (1993) emphasize the concept of **total survey design**: You should design a survey to reduce errors in general, not just sampling errors. Of course, to design a survey to minimize all the error, you need to know what the major error components are. If you know that most of the error in survey estimates is caused by coverage problems, then you can devote resources to improving the coverage. If you know that the coding is highly accurate, then you do not need to devote as many resources to improving the quality of the coding procedures.

Total survey design calls for an interdisciplinary approach. The areas of expertise needed to study and reduce sources of error include statistical theory of complex surveys, design of experiments, statistical process control, mixed models, cognitive psychology, management, and ethnography.

15.1
Coverage Error

As discussed in Chapter 1, coverage is the percentage of the population of interest that is included in the sampling frame. A mismatch between the target population and the sampling frame can cause coverage bias. Most common is undercoverage, where the sampling frame misses part of the population. If the target population mean for all N units in the population is $\bar{y}_U$, let $\bar{y}_{UF}$ be the mean for the N_F units in the sampling frame, and $\bar{y}_{UN}$ be the mean for the $N - N_F$ units in the target population but not in the sampling frame. The bias due to undercoverage is then

$$\bar{y}_{UF} - \bar{y}_U = \frac{N - N_F}{N}(\bar{y}_{UF} - \bar{y}_{UN}). \qquad (15.1)$$

The bias is thus low if (1) the population means are approximately the same for the covered and noncovered units in the population, that is, $\bar{y}_{UF} \approx \bar{y}_{UN}$, or (2) the coverage rate, N_F/N, is high.

15.1.1 Measuring Coverage and Coverage Bias

Estimating undercoverage or bias caused by undercoverage is, in general, difficult. If it were easy to identify and reach units missed by the sampling frame, those units would have been included when the frame was constructed. By definition, undercoverage is external to the survey and thus information external to the survey must be used to assess it.

You can sometimes tell if there has been undercoverage or nonresponse by comparing survey estimates of demographic characteristics with known values of those characteristics for the population. If your estimated number of 18- to 24-year-old males from the survey is much lower than the total number of 18- to 24-year-old males from a census, then there is likely undercoverage of that subpopulation. But demographic counts that match do not necessarily mean that you have full coverage. A sampling frame for an e-mail survey may have equal numbers of men and women, but it will lack both men and women without e-mail addresses.

The coverage rate can sometimes be estimated using information from other studies or external records. For example, undercoverage in a survey of households with infants might be assessed by taking a sample of recent birth certificates and checking whether the households are in the sampling frame. The U.S. Census Bureau (2002b) reports that about 70% of eligible voters were registered to vote in November 2000, so a survey using voter registration lists as a sampling frame for eligible voters would have an estimated coverage of 70%.

Election polls present many challenges for constructing sampling frames and assessing coverage. The target population for a pre-election poll is persons who will vote in the election, but no one knows in advance exactly who those persons will be. A sampling frame of registered voters will include many persons who do not vote on election day. A sampling frame of persons who voted in the last election will miss new voters. Many pre-election polls in the United States use models to predict who is likely to vote, taking into account voting history and other information. Most current polls are conducted by telephone and thus do not include nontelephone households.

EXAMPLE **15.2** The U.S. Census is intended to count every person in the United States; in a sense, its mission is to obtain complete and accurate coverage of the country. It is thus essential that the coverage be assessed. In the 1980, 1990, and 2000 censuses, *postenumeration surveys* were used to estimate the degree of undercoverage and duplicate records. Some of the methods used for these surveys were described in Example 13.2.

Mulry (2004) describes components of error in the 2000 census and in the surveys used to evaluate its coverage. The census itself has undercoverage (from households that are not contacted), nonresponse (from households that do not return their form, or that omit persons living in the household from the form), and duplicate records (from persons listed twice, for example, a college student counted at his residence in the college city and also listed on his parents' form). The evaluation survey also has undercoverage and nonresponse. Persons who move between census day and the postenumeration survey are at different residences for the two surveys; while persons who move into the survey areas are counted correctly, the persons who move out of the survey areas must be estimated. Both census and the postenumeration survey have measurement error, since some persons report an erroneous residence for census day. Matching records from the census and postenumeration survey is also subject to error. ∎

Network or snowball sampling (see Sections 14.1.5 and 14.1.6) can sometimes be used to estimate coverage rate, particularly in surveys of rare populations. Rothbart et al. (1982) found that a network sample of Vietnam veterans gave improved coverage of Vietnam veterans from minority groups.

The methods discussed so far involve estimating the coverage rate, N_F/N, in (15.1). The second factor of the undercoverage bias in (15.1), $\bar{y}_{UF} - \bar{y}_{UN}$, depends on the mean value for units not in the sampling frame. Estimating $\bar{y}_{UN}$ requires data from the uncovered part of the population, which in general must be obtained from an external source. Large government surveys can sometimes be used to estimate coverage bias on responses related to responses of interest. The American Community Survey, for example, includes telephone and nontelephone households. It could thus be used to estimate the bias for some responses from an educational survey that is conducted by telephone. If the telephone and nontelephone households in the American Community Survey have significantly different proportions of college graduates, then you would expect the educational survey that excludes nontelephone households to have bias for estimating the proportion of college graduates and related items.

15.1.2 Coverage and Survey Mode

The mode of survey administration (in-person, telephone, mail, e-mail, fax, Internet, and so on) exerts great influence on the coverage properties, and choice of mode should be influenced in part by the coverage that can be obtained. Dillman et al. (2009) provide an excellent discussion of coverage issues in sample surveys. Other considerations for choice of mode, such as response rate and accuracy of responses for various modes, are discussed in Chapter 7.

Area frames usually have the highest coverage. An area frame is constructed by selecting a sample of geographical primary sampling units (psus) from the region of interest. Field investigators construct a list of housing units for the psus selected for the sample, and a probability sample of housing units is selected from the list in each psu. Not surprisingly, area frames are also generally the most expensive to construct and sample. The nontelephone households in the sample often require in-person interviews.

The sampling frame for a mail or e-mail survey is a list of physical or e-mail addresses. The coverage of the frame depends on the completeness and accuracy of the list. Even if the frame contains everyone in the target population, the addresses may be wrong because persons may have moved or changed their e-mail addresses. E-mail surveys often work well for surveys in a university or organization in which everyone uses e-mail; for other populations, they exclude persons who do not have e-mail or who never check their e-mail accounts. Mail and e-mail surveys carry risks that the questionnaire will not reach the intended respondent. A mail survey might be discarded by another household member; an e-mail survey might be deleted by a spam filter.

Telephone surveys may use list frames constructed from directories or random digit dialing (see Example 6.12). List frames for telephone surveys, like lists of addresses for mail or e-mail surveys, may be incomplete or have incorrect telephone numbers. Persons who move frequently are less likely to appear in the directory. At this writing, most U.S. telephone surveys include only households with landline telephones. Estimating the percentage of households who do not have a landline telephone is challenging. Tucker et al. (2007) used data from a Current Population Survey supplement to estimate that in 2004, 6% of U.S. households had a cellular telephone only and an additional 5.4% of households had no telephone service.[1] The households with only a cellular telephone, however, differed from those with landline service: They were more likely to be renters rather than owners of the housing unit and more likely to be one-person households. Adults aged 15–24 and unmarried adults were more likely to be in cellular-only households than their older or married counterparts. There may be additional nonresponse in a random digit dialing (RDD) survey if households with landlines and cellular telephones primarily use their cellular phones.

Internet surveys are appealing because of their low cost, but obtaining good coverage of the target population is challenging. At this writing, the most reliable surveys that use the Internet to collect data select the sample from a mail, telephone, or area frame. They contact the sampled individuals through another mode and then ask them to submit survey responses through the Internet. Some survey organizations provide

---

[1]The proportion of households with cellular telephone service has increased since 2004; the National Center for Health Statistics reports updated estimates at www.cdc.gov/nchs.

computers and Internet access for persons in the sample who do not have them. These surveys thus include members of the population who do not have Internet access.

Unfortunately, careful Internet surveys that collect a probability sample are rare. It is difficult to assess coverage in other Internet surveys (Couper, 2000). Internet surveys in which website visitors volunteer to participate are untrustworthy and should not be used to estimate characteristics of a population. The coverage of such surveys is unknown because the sample consists of volunteers. Some of the volunteers may take the survey many times in an attempt to influence the results. Even if the sample matches the target population on demographic characteristics, or is weighted to match demographic characteristics through poststratification, it is likely that other characteristics will differ.

EXAMPLE 15.3 In April 2007 the city of Tempe, Arizona, arranged for a market research company to conduct a survey of Tempe residents to solicit opinions about a proposed neighborhood shuttle bus service. An announcement of the upcoming survey and a map of the proposed bus route was mailed to every address in the neighborhood in March, 2007. The market research company took a telephone survey of approximately 700 Tempe residents. Because the survey was done by telephone, the questionnaire started with screening questions to ensure the respondent lived in the area of interest. The respondent was then asked to refer to a map that had been mailed earlier and to answer questions about proposed bus routes shown on the map. Neighborhood residents who were not selected to be in the telephone sample were given the opportunity to respond to the survey over the Internet.

Telephone was a poor choice for the survey mode. The telephone survey required screening questions to exclude persons not in the neighborhood. Cell phones were not sampled, resulting in undercoverage in this neighborhood close to Arizona State University. The survey required respondents to refer to a map that had been mailed earlier, and it is likely that many respondents would not have ready access to that map when they were called, resulting in measurement error. In addition, city planners and neighborhood activists were interested in whether residents adjacent to the proposed bus route had different opinions than other residents. The telephone survey could not guarantee a sufficient sample size of residents along the route.

A mail survey would have been a much better choice. The city had already gone to the expense of mailing the map to all neighborhood residents. It could have easily selected a stratified sample (stratified by proximity to proposed route, with a higher sampling fraction for addresses close to the proposed route) of those addresses and included the survey in the envelopes mailed to the households in the stratified sample. The money saved by not taking a telephone sample could have been used for nonresponse follow-up. ∎

15.1.3 Improving Coverage

As with nonresponse, the best way to deal with undercoverage and overcoverage is to prevent it. Some options for improving coverage in the survey design are:

- Check the sampling frame to remove duplicates.
- Compare the sampling frame with external sources to check for members of the target population that are missing in the frame.

■ Choose a survey mode or modes that have high coverage of the target population.

■ Use a multiple frame survey. An area frame, though expensive to sample, often has good coverage of the population. Data from the area frame sample can then be combined with data from incomplete frames that are inexpensive to sample, as discussed in Section 14.1.4. Coverage can often be improved by combining samples from several incomplete frames; even if the union of the frames does not include the entire target population, the multiple frame survey will have better coverage than any of the frames taken singly. A dual frame survey with one sample from a frame of landline telephones and another sample from a frame of cellular telephones will miss nontelephone households, but will have better coverage than a sample of landline telephones that excludes persons who use a cellular telephone exclusively.

Poststratification, discussed in Section 8.5.2, can partially alleviate coverage bias, but, as with all after-the-fact adjustments for nonresponse or coverage errors, you do not know whether the adjustment truly compensates for coverage bias unless you obtain data on the persons not covered by the sampling frame.

15.2
Nonresponse Error

In Chapter 8, we looked at possible remedies for nonresponse that has already occurred. It is far better, of course to be able to prevent or reduce nonresponse before it occurs. The methods outlined in Section 8.2 can be used to reduce nonresponse at the survey design stage.

We recommend using the AAPOR (2008b) standards for reporting nonresponse rates. As with undercoverage, it is often challenging to assess the bias due to nonresponse. In some cases, you can obtain accurate data for nonrespondents from an external source such as a population register and use the external records to evaluate the bias due to nonresponse. In a health survey, you might be able to access the medical records of a subsample of nonrespondents and a subsample of respondents. You can then compare the respondents and nonrespondents on quantities given in the medical records. If the nonrespondents have significantly higher blood pressure than the respondents, they may differ from the respondents on key survey items as well.

Similar comparisons can be done if your sampling frame has substantial auxiliary information about each individual in the frame. A university administrator taking a survey of students can compare the grade point averages and majors of survey respondents with those of the survey nonrespondents. In addition to identifying potential nonresponse bias, the frame information can be used to construct weighting classes or impute values for nonrespondents.

Comparing survey estimates of demographic quantities to those of a census or large survey such as the American Community Survey can also indicate nonresponse problems. This can identify undercoverage or nonresponse for different demographic groups, and suggest weighting variables that might be useful for nonresponse adjustment. When comparing results with another survey, be careful that the same definitions and measurements are used. Your survey may have different estimates of unemployment than the American Community Survey because you define unemployment differently, or because your survey covers a different time period.

You can also compare persons who respond initially with those who respond after several attempts to reach them. You might speculate that nonrespondents are similar to persons who are reached only after great effort, and use the information from the callbacks to estimate the nonresponse bias. This is a big assumption, though, and not always well founded (Lin and Scheaffer, 1995).

Under a response propensity framework, nonresponse bias depends on the relation between the (unknown) response propensity ϕ_i of each unit and the variable of interest y_i (see Exercise 16 of Chapter 8). If ϕ_i and y_i are uncorrelated, then the bias incurred by estimating the population mean using the respondents only is approximately zero. A model adopted for nonresponse that estimates the response propensities accurately will also reduce the nonresponse bias. Unfortunately, we do not know whether the propensity to respond is correlated with the responses or whether the model for nonresponse is good because we have no data on the nonrespondents. We can, however, fit several models for nonresponse and investigate the sensitivity of the results to the modeling assumptions, as described in Little and Rubin (2002).

Groves (2006) concludes from a review of 30 empirical studies that nonresponse bias occurs but is not necessarily correlated with the nonresponse rate. Some studies have relatively high response rates and yet still have high bias, while other studies with lower response rates have low bias. In general, higher response rates are better and complete response is best of all. Paradoxically, though, sometimes efforts to increase response rates can also increase nonresponse bias. This occurs, for example, when the measures taken to increase response rates also increase the correlation between ϕ_i and y_i. An incentive given in a survey might increase the propensity of low-income persons to respond and thus result in more bias for estimates of income.

Much research (see Groves and Couper, 1998, and Groves et al., 2002, Chapters 1–17) has been done on why persons choose to respond to a survey and how surveys can be designed to increase cooperation. Cialdini (1984) identifies factors associated with willingness to respond to a survey:

1 *Reciprocation.* Will the potential respondent gain something by participating in the survey? The 2005 Census Test in Maricopa County advertised that your participation helps your community by making sure everyone is counted. Informational booklets describing how the survey data will be used can motivate some persons to respond. Incentives can increase survey cooperation in some instances; Singer (2002) reviews experiments on the effectiveness of incentives for increasing response rates and concludes that monetary incentives are most effective in surveys for which persons have few other motivations to participate.

2 *Authority.* Persons are often more likely to provide responses to a survey if it is issued by a recognized authority. University faculty members may be more likely to respond to a survey sent by the university president than a survey distributed by a graduate student. The U.S. Census Bureau (Griffin et al., 2003) sent one group of potential respondents a "mandatory" letter saying that participation in the survey is "required by law. We are conducting this survey under the authority of Title 13, United States Code, sections 141–193, and 221." Another group was sent a "voluntary" letter saying "Your participation in the survey is important;

however, you may decline to answer any or all questions." The mail response rate was more than 20 percentage points higher with the "mandatory" letter than with the "voluntary" letter.

3 *Consistency.* Once someone is persuaded to participate in a survey, that person is likely to continue and perhaps participate in other surveys.

4 *Scarcity.* The scarcity heuristic is related to reciprocation: A potential respondent who believes that the opportunity to participate in the survey is reserved for the select few may be more likely to respond.

5 *Social validation.* Potential respondents may be more likely to participate if they believe others do so.

6 *Liking.* Potential respondents may be more amenable to participation if they like the interviewer.

Persons choose to participate in a survey for many different reasons, so a flexible approach to soliciting responses is helpful. Different survey introductions may work better with some subsets of the population. Skilled interviewers use a variety of approaches to persuade persons to respond to a survey. We know that some nonresponse will occur despite the best efforts of the survey designer. Thus, it is valuable to have additional information in the sampling frame—not just for adjusting the estimates for nonresponse after the data are collected, but for giving interviewers additional information to use when recruiting respondents.

As mentioned in Chapter 8, different survey modes tend to have different response rates. Hox and deLeeuw (1994) found, in their review of studies comparing response rates, that in-person surveys typically obtain the highest response rates, telephone surveys the second highest response rates, and mail surveys the lowest. Tourangeau et al. (2000) report more item nonresponse in self-administered questionnaires than in questionnaires administered by interviewers. Some surveys have obtained increased response rates by offering potential respondents a choice of response mode.

Several modes may be used for nonresponse follow-up. The American Community Survey conducts interviews using three modes. The initial surveys are sent out by mail; this is the least expensive form of data collection at present. The following month, households that did not respond to the mail survey are contacted by telephone. In the third month, in-person interviews are conducted with a subsample of households that did not respond to the mail survey or the telephone survey. Households who respond by different modes may have different characteristics, however. For example, households that respond by mail may be more likely to own their homes and have a household head who is white non-Hispanic (Citro et al., 2004, pp. 101–102).

15.3
Measurement Error

In Chapters 2–7 we assumed that y_i, a characteristic of interest on unit i, is a fixed quantity measured without error. When there is measurement error, however, y_i is not the true characteristic of interest for unit i. Instead, there is some underlying value μ_i, and y_i is a measurement of μ_i taken from the survey. For example, suppose that the characteristic of interest is $\mu_i =$ the true amount that household i spent on medical care between March and June of last year. The response provided by the household, y_i, is not necessarily equal to μ_i. The question may be worded confusingly so that the respondent omits some medical expenses (perhaps omitting over-the-counter medications); the respondent may forget some expenses or include expenses from July; characteristics of the interviewer may lead some respondents to give inaccurate answers (perhaps excluding expenses they are embarrassed by); or other circumstances may lead to y_i differing from μ_i. Measurement error is the difference between the response y_i provided by a survey respondent and the true value of the response, μ_i. Estimating measurement error, like coverage and nonresponse bias, requires additional information and modeling.

Often, as for measuring the amount spent on medical care between March and June, μ_i is a fixed value that could be found exactly using specific definitions of "medical care" and "household." Demographic characteristics such as age or ethnicity, physical measurements such as body mass index, behavioral variables such as number of visits to doctors, and monetary variables can be thought of as having a true underlying value μ_i that could be determined if the measuring instruments were precise enough. In other cases, the true characteristic of interest may not have a precise physical meaning, as when a consumer confidence survey asks you whether you think you will be better off, worse off, or about the same financially a year from now. Although it would be possible to compare your financial status 12 months from now with your financial status now, that is not the point of the survey—the survey researchers want to know how optimistic or confident you are about your short-term financial future. Psychometricians call a possibly unobservable underlying characteristic, in this case consumer confidence, a **construct**, and attempt to approximate the construct through items that can be measured. It is rarely possible for survey questions to correspond exactly to certain underlying constructs, however, which is the reason for the advice in Section 1.5 to report the actual questions asked when summarizing results from a survey.

The survey instrument, the interviewer, and the respondent all can contribute to measurement error. To reduce measurement error due to the survey instrument, follow Bradburn's (2004) Law for Questionnaires: "Ask what you want to know, not something else." Bradburn's Law, while eminently sensible, can be challenging to implement and Bradburn reviews recent research by linguists, psychologists, and statisticians on reducing measurement error when constructing a questionnaire. Presser et al. (2004) describe methods that can be used to test and evaluate questionnaires.

Interviewers can contribute to response variability and bias, and interviewer effect varies with different modes of data collection. Interviewers can often increase response rates and improve accuracy by explaining questions to respondents. But some respondents may give a more socially desirable response to a survey conducted by an interviewer than to a self-administered survey, and may report that they exercise more, and gamble less, than they actually do. Some interviewers may prompt a respondent toward a particular response. Extreme interviewer effects may occur when interviewers *falsify* the data by changing responses or fabricating entire interviews. The American Association of Public Opinion Research website (www.aapor.org/pdfs/falsification.pdf) provides guidelines on detecting and minimizing interviewer falsification.

Respondents may deliberately or inadvertently provide inaccurate information to a survey. Respondents to the NCVS may forget about criminal victimizations that have occurred to them or, if reporting for another person, be unaware of that person's experiences with criminal victimization. A respondent may choose not to report an incident of domestic violence to the survey, particularly if the perpetrator is present during the interview.

15.3.1 Measuring and Modeling Measurement Error

Biemer and Stokes (1991) review models that have been proposed for measurement error. Suppose that T replications of the measurement of unit i could be taken, and let y_{it} be the value of the tth replicate measurement on unit i. A simple additive model for the measurement error is

$$y_{it} = \mu_i + \beta_i + \varepsilon_{it}, \tag{15.2}$$

where β_i is a fixed bias for respondent i and ε_{it} is a random variable representing unexplained sources of measurement error. In the simplest model, the ε_{it}'s are assumed to be independent random variables with mean 0 and variance σ_i^2. Define $\bar{\mu} = \sum_{i=1}^{N} \mu_i/N$ and $V(\mu_i) = (N-1)^{-1} \sum_{i=1}^{N} (\mu_i - \bar{\mu})^2$. The assumptions of this model imply that all conditions remain the same for replicate measurements, and that there are no carryover effects for multiple responses of the same person.

If μ_i is the true characteristic of interest, the survey measurement y_{it} should be as close to μ_i as possible. For the model in (15.2), β_i and σ_i^2 should both be close to zero. In psychometrics, two concepts called **validity** and **reliability** are used to assess this closeness. Validity deals with the correlation between a survey item and the true score μ_i. Many types of validity have been proposed (Groves, 1989); we define theoretical validity to be the correlation between the true score and its observed value: theoretical validity $= \text{Corr}(y_i, \mu_i)$.

Sometimes you can find the true value μ_i by checking external records. In a survey asking the question "Did you vote in the election on September 13, 2005?" you may be able to check the voting records to determine whether the person actually voted (of course, you must have an accurate way to link persons from the survey to the voting records for this to work). Then $\mu_i = 1$ if the person is listed as voting in the voting records and 0 otherwise; $y_i = 1$ if the person responds that he or she voted. The validity of the question is estimated by the correlation between y_i and μ_i. If there is no external source of the true value, you can sometimes estimate validity by other methods such as looking at the correlations among answers to closely related questions.

Note that validity is not the same thing as unbiasedness or accuracy. Suppose μ_i is weight of person i, and the scale has negligible variability but erroneously adds 5 kg to every measurement. Then $\text{Corr}(y_i, \mu_i) \approx 1$ but $E[y_i|\mu_i] = \mu_i + 5$; in an SRS, $\bar{y}$ will overestimate the true mean weight of the population. In general, you need an external source of information to be able to evaluate measurement bias.

Reliability deals with variability of responses under repeated measurements. If all the values of σ_i^2 are equal to σ^2 in the model in (15.2),

$$\text{Reliability} = \frac{V(\mu)}{\sigma^2 + V(\mu)} = \frac{\text{variance of true values}}{\text{variance of values reported to the survey}}. \tag{15.3}$$

If the reliability is 1, then $\sigma^2 = 0$, that is, respondent i gives exactly the same answer over repeated trials. If the answers of respondent i are highly variable over repeated trials, then the reliability is low.

Cronbach's alpha (Cronbach, 1951) is often used to estimate reliability when multiple questions are used to assess the same construct:

$$\alpha = \frac{k\bar{r}}{1 + (k - 1)\bar{r}},$$

where k is the number of items, and $\bar{r}$ is the average of the pairwise correlations of the items. If α is close to one, then there is high reliability. High reliability can occur, however, when the questionnaire is constructed so that answers to one question affect answers to another. It is possible for all questions to be highly consistent yet for none of them to measure the true construct of interest.

Hansen et al. (1961) and Kish (1962) proposed methods for studying errors due to interviewers. Kish (1962) proposed considering the interviewers to be randomly selected from a population of possible interviewers, so that a random or mixed effects model (see Section 11.5) would be reasonable for examining the effect of interviewer variability on the overall variability of estimators. He noted that the measurement error component cannot be distinguished from the sampling error unless replicate measurements are taken from the respondents.

We can add an interviewer term to the basic measurement error model in (15.2). Let y_{ijt} be the response given by respondent i to interviewer j on replicate t:

$$y_{ijt} = \mu_i + \beta_i + b_j + \varepsilon_{ijt} \tag{15.4}$$

where b_j is the systematic effect of interviewer j. We assume that $E_M(b_j) = 0$, $V_M(b_j) = \sigma_b^2$, $E_M(\varepsilon_{ijt}) = 0$, $V_M(\varepsilon_{ijt}) = \sigma_i^2$, and that all of the b_j's and ε_{ijt}'s are uncorrelated. The model assumes that any respondent asked a question by interviewer j is likely to deviate from the true value by an amount b_j that is intrinsic to interviewer j. For example, in a health survey, perhaps Fred has a tendency to take blood pressure readings just a little below the true value. Then every person examined by Fred will have a blood pressure reading that is slightly too low, and respondents examined by Fred will tend to be more similar to each other than respondents selected at random. Or, in a victimization survey, respondents may tend to find an interviewer more sympathetic and be more likely to tell him or her about victimizations. That interviewer would tend to have more reported victimizations than other interviewers. The variability due to interviewers can be estimated using standard methods for mixed models (Demidenko, 2004).

The model in (15.4) can be expanded by including interaction effects between interviewers and respondents; for example, it might be thought that female respondents will report a different number of criminal victimizations to a female interviewer than to a male interviewer. Terms for mode effects can be added in a mixed-mode survey.

Mahalanobis (1946) proposed interpenetrating subsampling for estimating interviewer effects. The basic idea is the same as for estimating the variance of systematic sampling (Section 5.5): Assign each interviewer a random subsample of the interviews. Often in surveys, interviewers are assigned according to convenience; for example, an interviewer might be assigned to all households in a psu, which confounds the effect of the psu with the effect of the interviewer. In interpenetrating subsampling with an SRS, interviewers are assigned households at random.

15.3.2 Reducing Measurement Error

The first step in reducing measurement error is to estimate its prevalence and identify the main sources. If the largest component of measurement error is interviewer variability, then more standardized interview procedures may reduce the variance component. If respondents misinterpret questions, then better questions should be written and tested.

We recommend collecting data using randomized experiments to estimate components of variability and likely sources of bias. Hartley and Rao (1978) and Hartley and Biemer (1981) give designs that can be used to estimate interviewer variability from surveys. Scott and Davis (2001) consider interviewer effects for binary data. Randomized experiments, conducted before a survey is implemented, can compare versions of questionnaires, alternative field procedures, methods of interviewer training, and almost any other factor affecting survey quality.

Fowler (1991) provides advice for reducing interviewer-related measurement error:

- *Test your questions.* Interview potential respondents to see if they interpret the questions as you intend.

- *Write clear questions.* If a respondent does not know how to answer a question, the interviewer is likely to have more influence on the response. In a self-administered survey, unclear questions can lead to more variability or bias in the responses. Open-ended questions may be more susceptible to interviewer effects than closed questions.

- *Write procedures for administering the survey that will reduce errors.*

- *Hire good interviewers.*

- *Provide training and supervision for interviewers so they act consistently.* Interviewers should read the questions exactly as written, and should not indicate that one response is preferred over another. An interviewer should have a professional and neutral demeanor.

- *Give interviewers a reasonable workload.* Deming (1986) argues that assigning numerical quotas to workers decreases quality: An industrial worker who is required to make 130 parts per hour cannot pay attention to the quality of the part. Cannell et al. (1977) found similar effects for survey interviewers: Interviewers with high assignments had more errors in responses.

- *Apply quality improvement principles to the interviewing process.* Montgomery (2008) describes quality improvement methods for many settings. Reducing measurement error in surveys fits nicely into this framework.

15.4
Sensitive Questions

15.4.1   Nonresponse and Measurement Error

Many surveys involve questions that persons might view as sensitive. The American Community Survey asks a respondent to report his or her income in the past 12 months from eight different possible sources of income (wages, alimony and child support, interest income, and other sources). The National Household Survey on Drug Abuse asks respondents about their use of marijuana and cocaine. Some respondents view such questions as intrusive; others may fear that providing accurate information may expose them to penalties (for example, they may fear that reporting their true income on a survey may lead to penalties for underpayment of income taxes). Some persons may protect their personal information by refusing to respond to the survey or to specific items, while others may give inaccurate answers to sensitive questions.

Reputable survey takers promise respondents that their answers will be kept confidential. Respondents to the American Community Survey are assured that "Your data are confidential under Title 13, United States Code, Sections 9 and 214. Title 13 specifies that the Census Bureau can use the information provided by individuals for statistical purposes only and cannot publish or release information that would identify any individual. Instead, data are released as profiles of groups of individuals within broad geographic areas" (U.S. Census Bureau, 2003a). Singer (2003) reports that persons who said they were concerned about confidentiality were less likely to return their Census forms by mail, although other factors such as age had a higher association with nonresponse than did confidentiality concerns. The American Statistical Association Privacy, Confidentiality, and Data Security website (www.amstat.org, under Committees) provides links to numerous resources on assuring and protecting confidentiality of data. Even if promises of confidentiality do not influence the response rate, they are an ethical obligation to the respondents.

There is much evidence that many people simply do not provide accurate answers to sensitive questions. Tourangeau et al. (2000, Chapter 9) summarize studies in which record checks indicate underreporting of certain behaviors. Urine samples often contradict persons who say they do not use illegal drugs; counts of abortions from abortion clinics far exceed estimates of the total number of abortions from surveys. There is also overreporting of behaviors that many deem socially desirable: Studies comparing self-reports of voting with actual voting records show that many people say they voted when they actually did not (Presser, 1990).

As with coverage, mode of administration can have a great effect on responses to sensitive questions (see Tourangeau and Smith, 1996, and Kreuter et al., 2008). Many studies report that higher percentages of people say they have used illegal drugs when they fill out the questionnaire themselves than when the questionnaire is administered by an interviewer (Tourangeau et al., 2000, p. 295). Some surveys on sensitive topics use computer-assisted self-administration, where the respondent types answers directly onto the computer. The questions are displayed on-screen and are also played through a recording. An interviewer may be in the room to answer questions, but the interviewer does not see the responses typed into the computer (Newman et al., 2002).

15.4.2 Randomized Response

Sometimes you want to conduct a survey asking very sensitive questions, such as "Do you use cocaine?", "Have you ever shoplifted?", or "Did you understate your income on your tax return?"

These are all questions that "yes" respondents could be expected to lie about. A question form that encourages truthful answers but makes people comfortable is desired. Horvitz et al. (1967), in a variation of Warner's (1965) original idea, suggested using two questions: the sensitive question and an innocuous question. A randomizing device (such as a coin flip) determines which question the respondent should answer. If a coin flip is used as the randomizing device, the respondent might be instructed to answer the question "Did you use cocaine in the past week?" if the coin is heads, and "Is the second hand on your watch between 0 and 30?" if the coin is tails. The interviewer does not know whether the coin was heads or tails, and hence does not know which question is being answered. It is hoped that the randomization, and the knowledge that the interviewer does not know which question is being answered, will encourage respondents to tell the truth if they have used cocaine in the past week.

The randomizing device can be anything, but it must have known probability P that the person is asked the sensitive question and probability $1 - P$ that the person is asked the innocuous question. Other forms of randomized response are described in Fox and Tracy (1986). The key to randomized response is that the probability that the person responds yes to the innocuous question, p_I, is known. We want to estimate p_S, the proportion responding yes to the sensitive question. If everyone answers the questions truthfully, then

$$\phi = P(\text{respondent replies yes})$$
$$= P(\text{yes} \mid \text{asked sensitive question})P(\text{asked sensitive question})$$
$$\quad + P(\text{yes} \mid \text{asked innocuous question})P(\text{asked innocuous question})$$
$$= p_S P + p_I(1 - P).$$

Let $\hat{\phi}$ be the estimated proportion of "yesses" from the sample. Since both P and p_i are known, p_S may be estimated by

$$\hat{p}_S = \frac{\hat{\phi} - (1 - P)p_I}{P}. \qquad (15.5)$$

Then the estimated variance of $\hat{p}_S$ is

$$\hat{V}(\hat{p}_S) = \frac{\hat{V}(\hat{\phi})}{P^2}.$$

The penalty for randomized response appears in the factor $1/P^2$ in the estimated variance. If $P = 1/3$, for example, the variance is nine times as great as it would have been had everyone in the sample been asked the sensitive question and responded truthfully. The larger P is, the smaller the variance of $\hat{p}_S$. But if P is too large, respondents may think that the interviewer will know which question is being answered. Some respondents may think that only a $P = 0.5$ is "fair" and that no other probabilities exist when choosing among two items.

EXAMPLE 15.4 An SRS of high school seniors is selected. Each senior in the sample is presented with a card containing the following two questions:

Question 1: Have you ever cheated on an exam?
Question 2: Were you born in July?

We know from birth records that $p_I = 0.085$. Suppose the randomizing device is a spinner, with $P = 1/5$. Of the 800 people surveyed, 175 say yes to whichever question the spinner indicated they should answer. Then $\hat{\phi} = 175/800$. Because this is an SRS,

$$\hat{V}(\hat{\phi}) = \hat{\phi}(1 - \hat{\phi})/(n - 1) = 0.0002139.$$

Thus,

$$\hat{p}_S = \frac{175/800 - (4/5)(0.085)}{1/5} = 0.75375,$$

and $\hat{V}(\hat{p}_S) = (0.0002139)/(1/5)^2 = 0.0053.$ ∎

Before using randomized response methods in your survey, you should test the method to see if the extra complication does indeed increase compliance and reduce bias. Danermark and Swensson (1987) found that randomized response methods worked well for estimating drug use in schools and appeared to reduce response bias. Duffy and Waterton (1988), however, concluded that randomized response methods were not helpful in their survey to estimate incidence of various alcohol-related problems in Edinburgh, Scotland. They compared response rates and responses for a randomized response group with those for a group asked the questions directly, and found that the randomized response group had a lower response rate and lower estimated proportion of persons who had drunk more than the legal limit immediately before driving a car. Randomized response did, however, increase the complexity of the interviews, and interviewers reported that many persons were confused by the method.

15.5
Processing Error

Data entry error occurs when an answer given by a respondent differs from that entered into the database. Before computer-assisted interviewing became common, a frequent source of errors was a clerk typing the responses into the database. Statistical process control methods can be used to reduce errors due to data transfer and coding. Mudryk et al. (2001) describe the methods used to monitor and reduce errors from character recognition software used to capture the data from scanned survey questionnaires in the Canadian Census of Agriculture.

Coding errors can occur when responses are recorded. Open-ended questions are particularly prone to coding error, since someone must make a decision about how to classify a response. A person, when asked about why he or she patronizes a certain restaurant, might say that it is because the restaurant has good food and is cheap—two different responses. The coder must then decide what response to enter for the person.

Data editing can also introduce errors. Most survey organizations edit data files to remove internal inconsistencies and correct obvious errors (an individual with age 103 listed as living with his or her parents probably represents a coding error). Some organizations also impute values for missing data. Public-use data files are often edited to protect confidentiality of respondents' data—observations may be swapped from one location to another or some responses may be modified (see Doyle et al., 2001, for methods used to protect confidentiality). Editing, in general, removes errors introduced in other stages of data collection. Over-editing, however, can introduce additional errors. Granquist and Kovar (1997) report that "… it was not until a demographer 'discovered' that wives are on average two years younger than their husbands that the edit rule which performed this exact imputation was removed from the Canadian Census system!"

15.6
Total Survey Quality

It can happen that reducing one source of error actually increases another, and little is known about how this might work in practice. For example, heroic efforts by interviewers might result in some of the die-hard nonrespondents providing answers for the survey. But there is no guarantee that those responses are accurate—persons may make up data to stop the calls. Similarly, it is not clear how use of incentives to increase survey response may affect other aspects of accuracy.

Much research is still needed on how to estimate and improve survey quality. Biemer and Lyberg (2003) recommend a holistic approach to survey design, considering all possible sources of errors at the design stage. Most of this book has focused on methods for reducing and estimating sampling error using a design-based approach. The study of error sources involves proposing and fitting stochastic models for the sources or error; commonly, mixed models are used that incorporate terms describing bias and different sources of variability. A multivariate approach is needed since most surveys have multiple responses and errors may be correlated among different responses. The models can quickly become very complex as more terms are added for interactions of error sources and relationships between bias and variance so that the analyst must be careful not to overfit the data. Brick et al. (1994) found that some of the proposed models for studying sources of error in the U.S. Survey of Recent College Graduates were too complicated to fit without oversimplifying the assumptions and instead adopted a less structured approach.

Marker and Morganstein (2004) describe the use of statistical quality improvement methods in survey organizations. Quality improvement programs require that every person in the organization be committed to, and rewarded for, survey quality. As was stated in Chapter 8, the best time to reduce sampling and nonsampling errors in a survey is at the design stage.

The quality of the survey should be communicated to data users. Kasprzyk and Giesbrecht (2001) recommend that even abbreviated reports on survey results should contain the following:

- Information about the data set, including whether it was based on a probability sample
- Sources of sampling and nonsampling error
- Total in-scope sample size
- Unit nonresponse rates
- A reference to more detailed information about data collection
- A contact for more information

Some organizations, claiming that sampling error is a small part of total survey error, have returned to taking convenience samples, using opt-in Internet panels of respondents. Langer (2009) points out that such polls have "multiple methodological challenges": Some of them create potential biases by offering financial rewards to persons who volunteer to take surveys, and all of them live outside the framework of design-based inference. Probability samples are not perfect, and are subject to nonresponse and other nonsampling errors. But they remove many possible sources of bias, including the possibility that advocacy groups will bias a poll by encouraging their members to participate. The arguments put forward by some proponents of inexpensive convenience samples that their results agree with those from organizations that take probability samples do not prove the quality of their surveys. After all, the *Literary Digest* Survey discussed in Example 1.1 was accurate for several years—until it wasn't. As Groves (2006, p. 670) says:

> Probability sampling offers measurable sampling errors and unbiased estimates when 100 percent response rates are obtained. There is no such guarantee with low response rate surveys. Thus, within the probability sampling paradigm, high response rates are valued. Unfortunately, the alternative research designs for descriptive statistics, most notably volunteer panels, quota samples from large compilations of personal data records, and so forth, require even more heroic assumptions to derive the unbiased survey estimates.

Statistical sampling is a relatively young field, with many dating its origin as a modern discipline to Kiaer (1897). The discipline has been spurred by societal needs as well as technological developments. Over the past 100 years, survey researchers have developed methods for probability sampling, nonresponse and undercoverage adjustment, measurement error models, designed experiments for improving sample design and reducing nonsampling errors, computer-intensive inference, small area estimation, sampling rare populations, and many other applications. You can now solve some of the challenges of the next 100 years. As Gertrude Cox (1957) said in her 1956 presidential address to the American Statistical Association, "We are surrounded with ever widening horizons of thought, which demand that we find better ways of analytical thinking. We must recognize that the observer is part of what he observes and that the thinker is part of what he thinks. We cannot passively observe the statistical universe as outsiders, for we are all in it."

15.7
Chapter Summary

Total survey error is the sum of five components: coverage error, nonresponse error, measurement error, processing error, and sampling error. The main concern with undercoverage and nonresponse is bias. Sampling error produces variability in the estimates. Measurement and processing error have both bias and variance aspects.

Reports of survey results should include an assessment of errors. Quality improvement methods can be used to control errors throughout the survey-taking process. Designed experiments are useful for improving survey quality.

Key Terms

Total survey design: A philosophy of survey design for minimizing nonsampling as well as sampling errors.

Total survey error: Sum of all sampling and nonsampling errors in the survey.

For Further Reading

The book *Introduction to Survey Quality* (Biemer and Lyberg, 2003) provides a comprehensive guide to sources of errors in surveys, and what to do about them. Lessler and Kalsbeek (1992) discuss nonsampling errors and emphasize designing surveys to minimize all types of errors. The books edited by Lyberg et al. (1997), deLeeuw et al. (2008), and Pfeffermann and Rao (2009a) each contain several chapters on improving quality in survey data collection and on choice of survey mode and interviewer training. The Federal Committee on Statistical Methodology (2001) summarizes best practices and methods used by U.S. government statistical agencies to measure and report sources of errors in surveys. Groves and Couper (1998) provide a thorough treatment of nonresponse errors in household surveys. Groves et al. (2009) present methods for improving survey quality at the design stage.

Designed experiments and methods used for quality improvement in industry are also useful for improving survey quality. The books on experimental design and quality control by Oehlert (2000), Juran and Godfrey (2000), Ryan (2000), and Montgomery (2008), while not specific for survey operations, give principles that should be much more widely used when designing surveys. Deming's (1986) book *Out of the Crisis* is an example-filled guide to quality improvement. Some useful references on quality improvement in surveys are Biemer and Caspar (1994), Colledge and March (1993), González (1994), Marker and Morganstein (2004), and the AAPOR (2008a) guidelines for best practices in surveys. The quality guidelines from Statistics Canada (2003) describe procedures for assuring quality in all steps of survey development and implementation. Scheuren and Alvey (2008) review the history and methods used in exit polling, with an emphasis on survey quality.

15.8
Exercises

A. Introductory Exercises

1   The National Do Not Call Registry allows U.S. residents to prohibit telemarketers from calling the registered numbers. Conkey (2005) reports on a telephone survey conducted by the Customer Care Alliance. Respondents who had signed up for the registry were asked about their satisfaction with the list; 51% said they are still getting calls they thought the registry was supposed to block. Discuss possible sources of measurement error in this survey.

2   A university wishes to estimate the proportion of its students who have used cocaine. Students were classified into one of three groups: undergraduate, graduate, or professional school (medical or law school), and were sampled randomly within the groups. Since there was some concern that students might be unwilling to disclose their use of cocaine to a university official, the following method was used. Thirty red balls, sixteen blue balls, and four white balls were placed in a box and mixed well. The student was then asked to draw one ball from the box. If the ball drawn was red, the person answered question (a). Otherwise question (b) was answered.

Question (a): Have you ever used cocaine?
Question (b): Is the ball you drew white?

The results are as follows:

Group	Undergraduates	Graduates	Professional
Total number of students in group	8972	1548	860
Number of students sampled	900	150	80
Number answering yes	123	27	27

Assuming that all responses were truthful, estimate the proportion of students who have used cocaine and report the standard error of your estimate. Compare this standard error with the standard error you would expect to have if you asked the sample students question (a) directly and if all answered truthfully.

Now suppose that all respondents answer truthfully with the randomized response method but 25% of those who have used cocaine deny the fact when asked directly. Which method gives an estimate of the overall proportion of students who have used cocaine with the smallest mean squared error?

C. Working with Theory

3   Kuk (1990) proposed the following randomized response method. Ask the respondent to generate two independent binary variables X_1 and X_2 with $P(X_1 = 1) = \theta_1$ and $P(X_2 = 1) = \theta_2$. The probabilities θ_1 and θ_2 are known. Now ask the respondent to tell you the value of X_1 if she is in the sensitive class, and X_2 if she is not in the sensitive class. Suppose the true proportion of persons in the sensitive class is p_S.

a What is the probability that the respondent reports 1?

b Using your answer to (a), give an estimator $\hat{p}_S$ of p_S. What conditions must θ_1 and θ_2 satisfy?

c What is $V(\hat{p}_S)$ if an SRS is taken?

4 (Requires linear models.) In Section 15.3.1, we discussed the reliability of a survey instrument. Suppose we measure each individual twice, under the same conditions. Let X_i be the score of person i on the first survey administration, and let Y_i be the score of person i on the second survey administration. Consider the following model: Assume $U_i \sim N(\mu, \sigma_S^2)$. Now let $X_i = U_i + R_{i1}$ and $Y_i = U_i + R_{i2}$, where R_{i1} and R_{i2} are independent $N(0, \sigma_R^2)$ random variables (and are also independent of U_i).

a Using matrices, find the distribution of $\begin{bmatrix} X_i \\ Y_i \end{bmatrix}$.

b What is the reliability of the test under this model?

c Find $E[Y \mid X = x]$ and $\text{Var}[Y \mid X = x]$.

D. Projects and Activities

5 Read Deming's (1944) article "On errors in surveys." What sources of error identified by Deming are still considered part of total survey error? Did Deming discuss any errors that are no longer relevant? What new sources of error have arisen since Deming wrote his article?

6 The goal of the National Comorbidity Survey Replication is to estimate the prevalence of mental disorders in the United States. Read the survey description by Kessler et al. (2004). What aspects of this survey might affect data quality? What design features were implemented to improve the quality of the survey?

7 One problem that has occurred in surveys on sexual behavior in the United States is that, typically, men report more opposite-sex sexual partners than women do. This has led some researchers to be skeptical of the data quality, since one would expect the total number of opposite-sex partners for men to equal the total number of opposite-sex partners for women. Read the article by Tourangeau and Smith (1996) on asking sensitive questions. What steps did the authors take to reduce measurement error in their study?

8 The websites fivethirtyeight.com and pollster.com provide commentary on the quality of polls in the United States. Read a recent entry and describe the aspects of survey quality discussed.

9 In Exercise 29 of Chapter 1, you volunteered to be in an online panel for a survey. If you were asked to participate in a survey, report on your experiences. What are the sources of error in the survey?

10 Read the section on survey design (pp. 181–188) from the Consumer Bankruptcy Study described in Exercise 29 of Chapter 3 (Warren and Tyagi, 2003). How did the investigators deal with the different components of total survey error? What might they have done differently?

11   Read the guidelines on statistical ethics by the American Statistical Association (1999) or the International Statistical Institute (2009). What specific recommendations in the guidelines apply to survey research?

12   Read the Code of Standards and Ethics for Survey Research by the Council of American Survey Research Organizations (2008). Give examples of how adhering to the standards in this code might improve the quality of survey data.

13   Return to the survey you critiqued in Exercise 27 of Chapter 1. What sources of error were reported? How might the quality of the survey have been improved?

14   *Activity for course project.* What methods were used to improve the quality of the survey you studied in Exercise 31 of Chapter 7? In your opinion, how effective were these methods? How could the survey have been improved?

Appendix A: Probability Concepts Used in Sampling

I recollect nothing that passed that day, except Johnson's quickness, who, when Dr. Beattie observed, as something remarkable which had happened to him, that he had chanced to see both No. 1, and No. 1000, of the hackney-coaches, the first and the last; "Why, Sir, (said Johnson,) there is an equal chance for one's seeing those two numbers as any other two." He was clearly right; yet the seeing of the two extremes, each of which is in some degree more conspicuous than the rest, could not but strike one in a stronger manner than the sight of any other two numbers."

—James Boswell, *The Life of Samuel Johnson*

The essence of probability sampling is that we can calculate the probability with which any subset of observations in the population will be selected as the sample. Most of the randomization theory results used in this book depend on probability concepts for their proof. In this appendix we present a brief review of some of the basic ideas used. The reader should consult a more comprehensive reference on probability, such as Ross (2006) or Durrett (1994), for more detail and for derivations and proofs.

Because all work in randomization theory concerns discrete random variables, only results for discrete random variables are given in this section. We use the results in Sections A.1–A.3 in Chapters 2–4, and the results in Section A.3–A.4 in Chapters 5 and 6.

A.1
Probability

Consider performing an experiment in which you can write out all of the outcomes that could possibly happen, but you do not know exactly which one of those outcomes will occur. You might flip a coin, or draw a card from a deck, or pick three names out of a hat containing 20 names. Probabilities are assigned to the different outcomes and to sets composed of outcomes (called **events**), in accordance with the likelihood that the events will occur. Let Ω be the **sample space**, the list of all possible outcomes. For flipping a coin, $\Omega = \{\text{heads, tails}\}$. Probabilities in finite sample spaces have three basic properties:

1 $P(\Omega) = 1$.

2 For any event A, $0 \le P(A) \le 1$.

3 If the events $A_1, \ldots, A_k$ are disjoint, then $P\left(\bigcup_{i=1}^{k} A_i\right) = \sum_{i=1}^{k} P(A_i)$.

   In sampling, we have a population of N units and use a probability sampling scheme to select n of those units. We can think of those N units as balls in a box labelled 1 through N in a box, and we draw n balls from the box. For illustration, suppose $N = 5$ and $n = 2$. Then we draw two labeled balls out of the box:

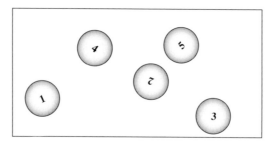

If we take a simple random sample (SRS) of one ball, each ball has an equal probability $1/N$ of being chosen as the sample.

A.1.1 Simple Random Sampling with Replacement

In a simple random sample with replacement (SRSWR), we put a ball back after it is chosen, so the same population is used on successive draws from the population. For the box with $N = 5$, there are 25 possible samples (a, b) in Ω, where a represents the first ball chosen and b represents the second ball chosen:

(1, 1)	(2, 1)	(3, 1)	(4, 1)	(5, 1)
(1, 2)	(2, 2)	(3, 2)	(4, 2)	(5, 2)
(1, 3)	(2, 3)	(3, 3)	(4, 3)	(5, 3)
(1, 4)	(2, 4)	(3, 4)	(4, 4)	(5, 4)
(1, 5)	(2, 5)	(3, 5)	(4, 5)	(5, 5)

   Since we are taking a random sample, each of the possible samples has the same probability, 1/25, of being the one chosen. When we take a sample, though, we usually do not care whether we chose unit 4 first and unit 5 second, or the other way around. Instead, we are interested in the probability that our sample consists of units 4 and 5 in either order, which we write as $\mathcal{S} = \{4, 5\}$. By the third property in the definition of a probability,

$$P(\{4, 5\}) = P[(4, 5) \cup (5, 4)] = P[(4, 5)] + P[(5, 4)] = \frac{2}{25}.$$

Suppose we want to find $P(\text{unit 2 is in the sample})$. We can either count that nine of the outcomes above contain 2, so the probability is 9/25, or we can use the **addition formula:**

$$P(A \cup B) = P(A) + P(B) - P(A \cap B). \tag{A.1}$$

Here, let $A = \{\text{unit 2 is chosen on the first draw}\}$ and let $B = \{\text{unit 2 is chosen on the second draw}\}$. Then,

$$P(\text{unit 2 is in the sample}) = P(A) + P(B) - P(A \cap B) = 1/5 + 1/5 - 1/25 = 9/25.$$

Note that, for this example,

$$P(A \cap B) = P(A) \times P(B).$$

That occurs in this situation because events A and B are **independent**, that is, whatever happens on the first draw has no effect on the probabilities of what will happen on the second draw. Independence of the draws occurs in finite population sampling when we sample with replacement.

A.1.2 Simple Random Sampling without Replacement

Most of the time, we sample without replacement because it is more efficient—if Heather is already in the sample, why should we use resources by sampling her again? If we plan to take an SRS (recall that SRS refers to a simple random sample without replacement) of our population with N balls, the ten possible samples (ignoring the ordering) are

$$
\begin{array}{ccccc}
\{1, 2\} & \{1, 3\} & \{1, 4\} & \{1, 5\} & \{2, 3\} \\
\{2, 4\} & \{2, 5\} & \{3, 4\} & \{3, 5\} & \{4, 5\}
\end{array}
$$

Since there are ten possible samples and we are sampling with equal probabilities, the probability that a given sample will be chosen is 1/10.

In general, there are

$$\binom{N}{n} = \frac{N!}{n!(N-n)!} \tag{A.2}$$

possible samples of size n that can be drawn without replacement and with equal probabilities from a population of size N, where

$$k! = k(k-1)(k-2)\cdots 1 \quad \text{and} \quad 0! = 1.$$

For our example, there are

$$\binom{5}{2} = \frac{5!}{2!(5-2)!} = \frac{5 \times 4 \times 3 \times 2 \times 1}{(2 \times 1)(3 \times 2 \times 1)} = 10$$

possible samples of size 2, as we found when we listed them.

Note that in sampling without replacement, successive draws are *not* independent. For this example,

$$P(\text{2 chosen on first draw, 4 chosen on second draw}) = \frac{1}{20}.$$

But P(2 chosen on first draw) = 1/5, and P(4 chosen on second draw) = 1/5, so P(2 chosen on first draw, 4 chosen on second draw) $\neq$ P(2 chosen on first draw) $\times$ P(4 chosen on second draw).

EXAMPLE A.1   Players of the Arizona State Lottery game "Fantasy 5" choose 5 numbers without replacement from the numbers 1 through 35. If the 5 numbers you choose match the 5 official winning numbers, you win $50,000. What is the probability you win $50,000? You could select a total of

$$\binom{35}{5} = \frac{35!}{5! \, 30!} = 324{,}632$$

possible sets of 5 numbers. But only

$$\binom{5}{5} = 1$$

of those sets will match the official winning numbers, so your probability of winning $50,000 is 1/324,632.

Cash prizes are also given if you match three or four of the numbers. To match four, you must select four numbers out of the set of five winning numbers, and the remaining number out of the set of 30 non-winning numbers, so the probability is

$$P(\text{match exactly 4 balls}) = \frac{\binom{5}{4}\binom{30}{1}}{\binom{35}{5}} = \frac{150}{324{,}632}. \qquad \blacksquare$$

EXERCISE A.1   What is the probability you match exactly 3 of the numbers? That you match at least one of the numbers?   ▪

EXERCISE A.2   *Calculating the sampling distribution in Example 2.4*
A box has eight balls; three of the balls contain the number 7. You select an SRS (without replacement) of size 4. What is the probability that your sample contains no 7s? Exactly one 7? Exactly two 7s?   ▪

A.2
Random Variables and Expected Value

A **random variable** is a function that assigns a number to each outcome in the sample space. Which number the random variable will actually assume is only determined after we conduct the experiment and depends on a random process: Before we conduct the experiment, we only know probabilities with which the different outcomes can occur. The set of possible values of a random variable, along with the probability with which each value occurs, is called the **probability distribution** of the random variable. Random variables are denoted by capital letters in this book to distinguish them from the fixed values y_i. If X is a random variable, then $P(X = x)$ is the probability that the random variable X takes on the value x. The quantity x is sometimes called a **realization** of the random variable X; x is one of the values that could occur if we performed the experiment.

EXAMPLE A.2 In the game "Fantasy 5," let X be the amount of money you will win from your selection of numbers. You win \$50,000 if you match all 5 winning numbers, \$500 if you match 4, \$5 if you match 3, and nothing if you match fewer than 3. Then the probability distribution of X is given in the following table:

x	0	5	500	50,000
$P(X = x)$	$\dfrac{320{,}131}{324{,}632}$	$\dfrac{4350}{324{,}632}$	$\dfrac{150}{324{,}632}$	$\dfrac{1}{324{,}632}$

∎

If you played "Fantasy 5" many, many times, what would you expect your average winnings per game to be? The answer is the **expected value** of X, defined by

$$E(X) = EX = \sum_x xP(X = x). \tag{A.3}$$

For "Fantasy 5,"

$$E(X) = \left(0 \times \frac{320{,}131}{324{,}632}\right) + \left(5 \times \frac{4350}{324{,}632}\right) + \left(500 \times \frac{150}{324{,}632}\right)$$

$$+ \left(50{,}000 \times \frac{1}{324{,}632}\right) = \frac{176{,}750}{324{,}632} = 0.45.$$

Think of a box containing 324,632 balls, in which 1 ball contains the number 50,000, 150 balls contain the number 500, 4350 balls contain the number 5, and the remaining 320,131 balls contain the number 0. The expected value is simply the average of the numbers written inside all the balls in the box. One way to think about expected value is to imagine repeating the experiment over and over again and calculating the long-run average of the results. If you play "Fantasy 5" many, many times, you would expect to win about 45 cents per game, even though 45 cents is not one of the possible realizations of X.

Variance, covariance, and the coefficient of variation are defined directly in terms of the expected value:

$$V(X) = E[(X - EX)^2] = \text{Cov}\,(X, X) \tag{A.4}$$

$$\text{Cov}\,(X, Y) = E[(X - EX)(Y - EY)] \tag{A.5}$$

$$\text{Corr}\,(X, Y) = \frac{\text{Cov}\,(X, Y)}{\sqrt{V(X)V(Y)}} \tag{A.6}$$

$$CV\,(X) = \frac{\sqrt{V(X)}}{E(X)}, \text{ for } E(X) \neq 0. \tag{A.7}$$

Expected value and variance have a number of properties that follow directly from the definitions above.

Properties of Expected Value

1   If g is a function, then $E[g(X)] = \sum_x g(x)P(X = x)$.

2   If a and b are constants, then $E(aX + b) = aE(X) + b$.

3   If X and Y are independent, then $E(XY) = (EX)(EY)$.

4   $\text{Cov}(X, Y) = E(XY) - (EX)(EY)$.

5   $\text{Cov}\left[\sum_{i=1}^{n}(a_i X_i + b_i), \sum_{j=1}^{m}(c_j Y_j + d_j)\right] = \sum_{i=1}^{n}\sum_{j=1}^{m} a_i c_j \text{Cov}(X_i, Y_j)$.

6   $V(X) = E(X^2) - (EX)^2$.

7   $V(X + Y) = V(X) + V(Y) + 2\text{Cov}(X, Y)$.

8   $-1 \leq \text{Corr}(X, Y) \leq 1$.

EXERCISE A.3   Prove properties 1 through 8 using the definitions in (A.3) through (A.7).   ∎

In sampling, we often use estimators that are ratios of two random variables. But $E[Y/X]$ usually does not equal EY/EX. To illustrate this, consider the following probability distribution for X and Y:

x	y	$\dfrac{y}{x}$	$P(X = x, Y = y)$
1	2	2	$\dfrac{1}{4}$
2	8	4	$\dfrac{1}{4}$
3	6	2	$\dfrac{1}{4}$
4	8	2	$\dfrac{1}{4}$

Then EY/EX = 6/2.5 = 2.4, but $E[Y/X]$ = 2.5. In this example, the values are close but not equal.

The random variable we use most frequently in this book is

$$Z_i = \begin{cases} 1 & \text{if unit } i \text{ is in the sample} \\ 0 & \text{if unit } i \text{ is not in the sample.} \end{cases} \tag{A.8}$$

This indicator variable tells us whether the ith unit is in the sample or not. In an SRS, n of the random variables $Z_1, Z_2, \ldots, Z_N$ will take on the value 1, and the remaining $N - n$ will be 0. For Z_i to equal 1, one of the units in the sample must be unit i, and the other $n-1$ units must come from the remaining $N-1$ units in the population, so

$$P(Z_i = 1) = P(i\text{th unit is in the sample})$$

$$= \frac{\binom{1}{1}\binom{N-1}{n-1}}{\binom{N}{n}}$$

$$= \frac{n}{N}. \tag{A.9}$$

Thus,

$$E[Z_i] = 0 \times P(Z_i = 0) + 1 \times P(Z_i = 1)$$

$$= P(Z_i = 1) = \frac{n}{N}.$$

Similarly, for $i \neq j$,

$$P(Z_i Z_j = 1) = P(Z_i = 1 \text{ and } Z_j = 1)$$

$$= P(i\text{th unit is in the sample and } j\text{th unit is in the sample})$$

$$= \frac{\binom{2}{2}\binom{N-2}{n-2}}{\binom{N}{n}}$$

$$= \frac{n(n-1)}{N(N-1)}.$$

Thus for $i \neq j$,

$$E[Z_i Z_j] = 0 \times P(Z_i Z_j = 0) + 1 \times P(Z_i Z_j = 1)$$

$$= P(Z_i Z_j = 1) = \frac{n(n-1)}{N(N-1)}.$$

EXERCISE A.4   Show that

$$V(Z_i) = \text{Cov}\,(Z_i, Z_i) = \frac{n(N-n)}{N^2}$$

and that, for $i \neq j$,

$$\text{Cov}\,(Z_i, Z_j) = -\frac{n(N-n)}{N^2(N-1)}. \qquad \blacksquare$$

The properties of expectation and covariance may be used to prove many results in finite population sampling. In Chapter 4, we use the covariance of $\bar{x}$ and $\bar{y}$ from an SRS. Let

$$\bar{x}_U = \frac{1}{N} \sum_{i=1}^{N} x_i, \qquad \bar{y}_U = \frac{1}{N} \sum_{j=1}^{N} y_j,$$

$$\bar{x} = \frac{1}{n} \sum_{i=1}^{N} Z_i x_i, \qquad \bar{y} = \frac{1}{n} \sum_{j=1}^{N} Z_j y_j,$$

and

$$R = \frac{\sum\limits_{i=1}^{N} (x_i - \bar{x}_U)(y_i - \bar{y}_U)}{(N-1)S_x S_y}.$$

Then,

$$\text{Cov}\,(\bar{x}, \bar{y}) = \left(1 - \frac{n}{N}\right) \frac{R S_x S_y}{n}. \qquad (A.10)$$

We use properties 5 and 6 of expected value, along with the results of Exercise A.4, to show (A.10):

$$\text{Cov}\,(\bar{x}, \bar{y}) = \frac{1}{n^2}\text{Cov}\left(\sum_{i=1}^{N} Z_i x_i, \sum_{j=1}^{N} Z_j y_j\right)$$

$$= \frac{1}{n^2}\sum_{i=1}^{N}\sum_{j=1}^{N} x_i y_j \text{Cov}\,(Z_i, Z_j)$$

$$= \frac{1}{n^2}\sum_{i=1}^{N} x_i y_i\, V(Z_i) + \frac{1}{n^2}\sum_{i=1}^{N}\sum_{j\neq i}^{N} x_i y_j \text{Cov}\,(Z_i, Z_j)$$

$$= \frac{1}{n}\frac{N-n}{N^2}\sum_{i=1}^{N} x_i y_i - \frac{1}{n}\frac{N-n}{N^2(N-1)}\sum_{i=1}^{N}\sum_{j\neq i}^{N} x_i y_j$$

$$= \frac{1}{n}\left[\frac{N-n}{N^2} + \frac{N-n}{N^2(N-1)}\right]\sum_{i=1}^{N} x_i y_i - \frac{1}{n}\frac{N-n}{N^2(N-1)}\sum_{i=1}^{N}\sum_{j=1}^{N} x_i y_j$$

$$= \frac{1}{n}\frac{N-n}{N(N-1)}\sum_{i=1}^{N} x_i y_i - \frac{1}{n}\frac{N-n}{N-1}\bar{x}_U \bar{y}_U$$

$$= \frac{1}{n}\frac{N-n}{N(N-1)}\sum_{i=1}^{N} (x_i - \bar{x}_U)(y_i - \bar{y}_U)$$

$$= \frac{1}{n}\left(1 - \frac{n}{N}\right) R S_x S_y.$$

EXERCISE A.5   Show that

$$\text{Corr}\,(\bar{x}, \bar{y}) = R. \quad \blacksquare \qquad (A.11)$$

A.3
Conditional Probability

In sampling without replacement, successive draws from the population are **dependent**: The unit we choose on the first draw changes the probabilities of selecting the other units on subsequent draws. When taking an SRS from our box of five balls in

Section A.1, each ball has probability 1/5 of being chosen on the first draw. If we choose ball 2 on the first draw and sample without replacement, then

$$P(\text{select ball 3 on second draw} \mid \text{select ball 2 on first draw}) = \frac{1}{4}.$$

(Read as "the conditional probability that ball 3 is selected on the second draw given that ball 2 is selected on the first draw equals 1/4.") Conditional probability allows us to adjust the probability of an event if we know that a related event occurred.

The **conditional probability** of A given B is defined to be

$$P(A \mid B) = \frac{P(A \cap B)}{P(B)}. \tag{A.12}$$

In sampling we usually use this definition the other way around:

$$P(A \cap B) = P(A \mid B)P(B). \tag{A.13}$$

If events A and B are independent—that is, knowing whether A occurred gives us absolutely no information about whether B occurred—then $P(A \mid B) = P(A)$ and $P(B \mid A) = P(B)$.

Suppose we have a population with 8 households (HHs) and 15 persons living in the households, as follows:

Household	Persons
1	1, 2, 3
2	4
3	5
4	6, 7
5	8
6	9, 10
7	11, 12, 13, 14
8	15

In a one-stage cluster sample, as discussed in Chapter 5, we might take an SRS of two households, then interview each person in the selected households. Then,

$$P(\text{select person 10}) = P(\text{select HH 6})\,P(\text{select person 10} \mid \text{select HH 6})$$

$$= \left(\frac{2}{8}\right)\left(\frac{2}{2}\right) = \frac{2}{8}.$$

In fact, for this example the probability that any individual in the population is interviewed is the same value, 2/8, because each household is equally likely to be chosen and the probability a person is selected is the same as the probability that the household is selected.

Suppose now that we take a two-stage cluster sample instead of a one-stage cluster sample, and we interview only one randomly selected person in each selected household. Then, in this example, we are more likely to interview persons living alone than those living with others:

$$P(\text{select person 4}) = P(\text{select HH 2})\,P(\text{select person 4} \mid \text{select HH 2})$$
$$= \left(\frac{2}{8}\right)\left(\frac{1}{1}\right) = \frac{2}{8},$$

but

$$P(\text{select person 12}) = P(\text{select HH 7})\,P(\text{select person 12} \mid \text{select HH 7})$$
$$= \left(\frac{2}{8}\right)\left(\frac{1}{4}\right) = \frac{2}{32}.$$

These calculations extend to multistage cluster sampling because of the general result

$$P(A_1 \cap A_2 \cap \cdots \cap A_k) = P(A_1 \mid A_2, \cdots, A_k)P(A_2 \mid A_3 \ldots, A_k) \ldots P(A_k). \quad \text{(A.14)}$$

Suppose we take a three-stage cluster sample of grade school students. First, we take an SRS of schools, then an SRS of classes within schools, then an SRS of students within classes. Then the event {Joe is selected in the sample} is the same as {Joe's school is selected ∩ Joe's class is selected ∩ Joe is selected} and we can find Joe's probability of inclusion by

$$P(\text{Joe in sample}) = P(\text{Joe's school is selected})$$
$$\times P(\text{Joe's class is selected} \mid \text{Joe's school is selected})$$
$$\times P(\text{Joe is selected} \mid \text{Joe's school and class are selected}).$$

If we sample 10% of the schools, 20% of classes within selected schools, and 50% of students within selected classes, then

$$P(\text{Joe in sample}) = (0.10)(0.20)(0.50) = 0.01.$$

A.4
Conditional Expectation

Conditional expectation is used extensively in the theory of cluster sampling. Let X and Y be random variables. Then, using the definition of conditional probability,

$$P(Y = y \mid X = x) = \frac{P(Y = y \cap X = x)}{P(X = x)}. \quad \text{(A.15)}$$

This gives the **conditional distribution** of Y given that $X = x$. The **conditional expectation** of Y given that $X = x$ simply follows the definition of expectation using the conditional distribution:

$$E(Y \mid X = x) = \sum_y yP(Y = y \mid X = x). \quad \text{(A.16)}$$

The **conditional variance** of Y given that $X = x$ is defined similarly:

$$V(Y \mid X = x) = \sum_y [y - E(Y \mid X = x)]^2 P(Y = y \mid X = x). \quad \text{(A.17)}$$

EXAMPLE A.3 Consider a box with two balls, A and B:

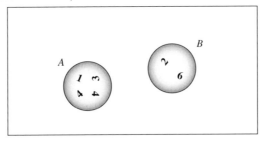

Choose one of the balls at random, then randomly select one of the numbers inside that ball. Let Y = the number that we choose and let

$$Z = \begin{cases} 1 & \text{if we choose ball A} \\ 0 & \text{if we choose ball B.} \end{cases}$$

Then,

$$P(Y = 1 \mid Z = 1) = \frac{1}{4},$$

$$P(Y = 3 \mid Z = 1) = \frac{1}{4},$$

$$P(Y = 4 \mid Z = 1) = \frac{1}{2},$$

and

$$E(Y \mid Z = 1) = \left(1 \times \frac{1}{4}\right) + \left(3 \times \frac{1}{4}\right) + \left(4 \times \frac{1}{2}\right) = 3.$$

Similarly,

$$P(Y = 2 \mid Z = 0) = \frac{1}{2}$$

and

$$P(Y = 6 \mid Z = 0) = \frac{1}{2},$$

so

$$E(Y \mid Z = 0) = \left(2 \times \frac{1}{2}\right) + \left(6 \times \frac{1}{2}\right) = 4.$$

In short, if we know that ball A is picked, then the conditional expectation of Y is the average of numbers in ball A since an SRS of size 1 is taken from the ball; the conditional expectation of Y given that ball B is picked is the average of the numbers in ball B. ∎

Note that $E(Y \mid X = x)$ is a function of x; call it $g(x)$. Define the **conditional expectation** of Y given X, $E(Y \mid X)$, to be $g(X)$, the same function but of the random variable instead. $E(Y \mid X)$ is a random variable and gives us the conditional expected value of Y for the general random variable X: for each possible value of x, the value $E(Y \mid X = x)$ occurs with probability $P(X = x)$.

EXAMPLE A.4   In Example A.3, we know the probability distribution of Z and can thus use the conditional expectations calculated to write the probability distribution of $E(Y \mid Z)$:

z	$E(Y \mid Z = z)$	Probability
0	4	$\dfrac{1}{2}$
1	3	$\dfrac{1}{2}$

■

In sampling, we need this general concept of conditional expectation largely so we can use the following properties of conditional expectation to find expected values and variances in cluster samples.

Properties of Conditional Expectation

1 $E(X \mid X) = X$.
2 $E[f(X)Y \mid X] = f(X)E(Y \mid X)$.
3 If X and Y are independent, then $E(Y \mid X) = E(Y)$.
4 $E(Y) = E[E(Y \mid X)]$.
5 $V[Y] = V[E(Y \mid X)] + E[V(Y \mid X)]$.

Conditional expectation can be confusing, so let's talk about what these properties mean. The interested reader should see Ross (2006) or Durrett (1994) for proofs of these properties.

1 $E(X \mid X) = X$. If we know what X is already, then we expect X to be X. The probability distribution of $E(X \mid X)$ is the same as the probability distribution of X.

2 $E[f(X)Y \mid X] = f(X)E(Y \mid X)$. If we know what X is, then we know X^2, or $\log X$, or any function $f(X)$ of X.

3 If X and Y are independent, then $E(Y \mid X) = E(Y)$. If X and Y are independent, then knowing X gives us no information about Y. Thus the expected value of Y, the average of all the possible outcomes of Y in the experiment, is the same no matter what X is.

4 $E(Y) = E[E(Y \mid X)]$. This property, called **successive conditioning**, and property 5 are the ones we use the most in sampling; we use them to find the bias and variance of estimators in cluster sampling. Successive conditioning simply says that if we take the weighted average of the conditional expected value of Y given that $X = x$, with weights $P(X = x)$, the result is the expected value of Y. You use successive conditioning every time you take a weighted average of a quantity over subpopulations: If a population has 60 women and 40 men, and if the average height of the women is 64 inches and the average height of the men is 69 inches, then the average height for the class is

$$(64 \times 0.6) + (69 \times 0.4) = 66 \text{ inches.}$$

In this example, 64 is the conditional expected value of height given that the person is a woman, 69 is the conditional expected value of height given that the person is a man, and 66 is the expected value of height for all persons in the population.

5 $V[Y] = V[E(Y \mid X)] + E[V(Y \mid X)]$. This property gives an easy way of calculating variances in two-stage cluster samples. It says that the total variability has two parts: (a) the variability that arises because $E(Y \mid X = x)$ varies with different values of x, and (b) the variability that arises because there can be different values of y associated with the same value of x. Note that, using property 6 of Expected Value in Section A.2,

$$V(Y \mid X) = E\{[Y - E(Y \mid X)]^2 \mid X\} = E[Y^2 \mid X] - [E(Y \mid X)]^2 \qquad \text{(A.18)}$$

and

$$\begin{aligned} V[E(Y \mid X)] &= E\left(\{E(Y \mid X) - E[E(Y \mid X)]\}^2\right) \\ &= E\left(\{E(Y \mid X) - E(Y)\}^2\right) \\ &= E\{[E(Y \mid X)]^2\} - [E(Y)]^2. \end{aligned} \qquad \text{(A.19)}$$

EXAMPLE A.5 Here's how conditional expectation properties work in Example A.3. Successive conditioning implies that

$$\begin{aligned} E(Y) &= E(Y \mid Z = 0)P(Z = 0) + E(Y \mid Z = 1)P(Z = 1) \\ &= \left(4 \times \frac{1}{2}\right) + \left(3 \times \frac{1}{2}\right) = 3.5. \end{aligned}$$

We can find the distribution of $V(Y \mid Z)$ using (A.18):

$$\begin{aligned} V(Y \mid Z = 0) &= E(Y^2 \mid Z = 0) - [E(Y \mid Z = 0)]^2 \\ &= \left(2^2 \times \frac{1}{2}\right) + \left(6^2 \times \frac{1}{2}\right) - (4)^2 = 4, \\ V(Y \mid Z = 1) &= E(Y^2 \mid Z = 1) - [E(Y \mid Z = 1)]^2 \\ &= \left(1^2 \times \frac{1}{4}\right) + \left(3^2 \times \frac{1}{4}\right) + \left(4^2 \times \frac{1}{2}\right) - (3)^2 = 1.5. \end{aligned}$$

These calculations give the following probability distribution for $V(Y \mid Z)$:

z	$V(Y \mid Z = z)$	Probability
0	4	$\frac{1}{2}$
1	1.5	$\frac{1}{2}$

Thus, using (A.19),

$$V[E(Y \mid Z)] = E\left\{[E(Y \mid Z) - E(Y)]^2\right\}$$
$$= [E(Y \mid Z = 0) - E(Y)]^2 P(Z = 0) + [E(Y \mid Z = 1) - E(Y)]^2 P(Z = 1)$$
$$= \left[(4 - 3.5)^2 \times \frac{1}{2}\right] + \left[(3 - 3.5)^2 \times \frac{1}{2}\right]$$
$$= 0.25.$$

Using the probability distribution of $V(Y \mid Z)$,

$$E[V(Y \mid Z)] = \left(4 \times \frac{1}{2}\right) + \left(1.5 \times \frac{1}{2}\right) = 2.75.$$

Consequently,

$$V(Y) = V[E(Y \mid Z)] + E[V(Y \mid Z)] = 0.25 + 2.75 = 3.00. \qquad \blacksquare$$

If we did not have the properties of conditional expectation, we would need to find the unconditional probability distribution of Y to calculate its expectation and variance—a relatively easy task for the small number of options in Example A.3 but cumbersome to do for general multistage cluster sampling.

EXERCISE A.6   Consider the box below, with 3 balls labelled 1, 2, and 3:

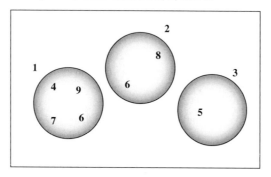

Suppose we take an SRS of one ball, then subsample an SRS of one number from the selected ball. Let Z represent the number of the ball chosen, and let Y represent the number we choose from the ball. Use the properties of conditional expectation to find $E(Y)$ and $V(Y)$. $\blacksquare$

References

Abelson, R. P., Loftus, E. F., and Greenwald, A. G. (1992). Attempts to improve the accuracy of self-reports of voting. In J. M. Tanur (Ed.), *Questions about survey questions: Meaning, memory, expression, and social interactions in surveys* (pp. 138–153). New York: Russell Sage Foundation.

Agresti, A. (2002). *Categorical data analysis, 2nd ed.* New York: Wiley.

Alexander, C. H. (1991). Comments on "The pros and cons of weighting versus non-weighting." In *Proceedings of the Section on Survey Research Methods, American Statistical Association,* 643–645.

Alexander, H. M., Slade, N. A., and Kettle, W. D. (1997). Application of mark-recapture models to estimation of the population size of plants. *Ecology, 78,* 1230–1237.

Alf, C., and Lohr, S. (2007). Sampling assumptions in introductory statistics classes. *The American Statistician, 61,* 71–77.

Altham, P. M. E. (1976). Discrete variable analysis for individuals grouped into families. *Biometrika, 63,* 263–269.

American Association of Public Opinion Research. (2008a). *Best practices for survey and public opinion research.* Lenexa, KS: AAPOR. Retrieved from www.aapor.org.

American Association of Public Opinion Research. (2008b). *Standard definitions: Final dispositions of case codes and outcome rates for surveys, 5th ed.* Lenexa, KS: AAPOR. Retrieved from www.aapor.org.

American Statistical Association. (1999). *Ethical guidelines for statistical practice.* Retrieved from www. amstat.org/committees/ethics.

Andersen, R., Kasper, J., Frankel, M. R., and associates. (1979). *Total survey error.* San Francisco: Jossey-Bass.

Armstrong, J., Block, C., and Srinath, K. P. (1993). Two-phase sampling of tax records for business surveys. *Journal of Business and Economic Statistics, 11,* 407–416.

Arnold, T. W. (1991). Intraclutch variation in egg size of American coots. *The Condor, 93,* 19–27.

Aye Maung, N. (1995). Survey design and interpretation of the British Crime Survey. In M. Walker (Ed.), *Interpreting crime statistics* (pp. 207–227). Oxford: Oxford University Press.

Azuma, D. L., Baldwin, J. A., and Noon, B. R. (1990). *Estimating the occupancy of spotted owl habitat areas by sampling and adjusting for bias.* U.S. Forest Service General Technical Report PSW-124. Berkeley, CA: Pacific Southwest Research Station, Forest Service, Department of Agriculture.

Babbie, E. R. (2007). *The practice of social research, 11th ed.* Belmont, CA: Wadsworth.

Barnett, V., Haworth, J., and Smith, T. M. F. (2001). A two-phase sampling scheme with applications to auditing or *sed quis custodiet ipsos custodes? Journal of the Royal Statistical Society, Series A, 164,* 407–422.

Bart, J., and Earnst, S. (2002). Double-sampling to estimate density and population trends in birds. *The Auk, 119*, 36–45.

Basow, S. A., and Silberg, N. T. (1987). Student evaluations of college professors: Are female and male professors rated differently? *Journal of Educational Psychology, 87*, 656–665.

Basu, D. (1971). An essay on the logical foundations of survey sampling, part 1. In V. P. Godambe and D. A. Sprott (Eds.), *Foundations of statistical inference* (pp. 203–242). Toronto: Holt, Rinehart & Winston.

Bates, N., Martin, E. A., DeMaio, T. J., and de la Puente, M. (1995). Questionnaire effects on measurements of race and Spanish origin. *Journal of Official Statistics, 11*, 433–459.

Battese, G. E., Harter, R. M., and Fuller, W. A. (1988). An error-components model for prediction of county crop areas using survey and satellite data. *Journal of the American Statistical Association, 83*, 28–36.

Beatty, P., and Herrmann, D. (2002). To answer or not to answer: Decision processes related to survey item nonresponse. In R. M. Groves, D. Dillman, J. Eltinge, and R. Little (Eds.), *Survey nonresponse* (pp. 71–85). New York: Wiley.

Beaumont, J.-F., and Alavi, A. (2004). Robust generalized regression estimation. *Survey Methodology, 30*, 195–208.

Beck, A. J., Kline, S. A., and Greenfeld, L. A. (1988). *Survey of youth in custody* (Tech. Rep. No. NCJ-113365). Washington, DC: Bureau of Justice Statistics.

Bedrick, E. J. (1983). Adjusted chi-squared tests for cross-classified tables of survey data. *Biometrika, 70*, 591–595.

Bell, W., Basel, W., Cruse, C., Dalzell, L., Maples, J., O'Hara, B., et al. (2007). *Use of ACS data to produce SAIPE model-based estimates of poverty for counties*. Washington, DC: U.S. Census Bureau.

Bellemain, E., Swenson, J. E., Tallmon, D., Brunberg, S., and Taberlet, P. (2005). Estimating population size of elusive animals with DNA from hunter-collected feces: Four methods for brown bears. *Conservation Biology, 19*, 150–161.

Bellhouse, D. R. (1984). A review of optimal designs in survey sampling. *The Canadian Journal of Statistics, 12*, 53–65.

Bellhouse, D. R., and Stafford, J. E. (1999). Density estimation from complex surveys. *Statistica Sinica, 9*, 407–424.

Bellhouse, D. R., and Stafford, J. E. (2001). Local polynomial regression in complex surveys. *Survey Methodology, 27*(2), 197–203.

Belsley, D. A., Kuh, E., and Welsch, R. E. (1980). *Regression diagnostics*. New York: Wiley.

Belson, W. A. (1981). *The design and understanding of survey questions*. Aldershot, Hants., England: Gower Publishing.

Berger, Y. G. (2004). A simple variance estimator for unequal probability sampling without replacement. *Journal of Applied Statistics, 31*, 305–315.

Bethlehem, J. (2002). Weighting nonresponse adjustments based on auxiliary information. In R. M. Groves, D. Dillman, J. Eltinge, and R. Little (Eds.), *Survey nonresponse* (p. 275–287). New York: Wiley.

Biderman, A. D., and Cantor, D. (1984). A longitudinal analysis of bounding, respondent conditioning, and mobility as sources of panel bias in the National Crime Survey. In *Proceedings of the Section on Survey Research Methods, American Statistical Association, 708–713*.

Biemer, P., and Caspar, R. (1994). Continuous quality improvement for survey operations: Some general principles and applications. *Journal of Official Statistics, 10*, 307–326.

Biemer, P. P., and Lyberg, L. E. (2003). *Introduction to survey quality*. New York: Wiley.

Biemer, P. P., and Stokes, S. L. (1991). Approaches to the modeling of measurement error. In P. P. Biemer, R. M. Groves, L. E. Lyberg, N. A. Mathiowetz, and S. Sudman (Eds.), *Measurement error in surveys* (pp. 487–516). New York: Wiley.

Binder, D. A. (1983). On the variances of asymptotically normal estimators from complex surveys. *International Statistical Review, 51*, 279–292.

Binder, D. A. (1996). Linearization methods for single phase and two phase samples: A cookbook approach. *Survey Methodology*, *22*, 17–22.

Binder, D. A., and Roberts, G. R. (2003). Design-based and model-based methods for estimating model parameters. In R. L. Chambers and C. J. Skinner (Eds.), *Analysis of survey data* (pp. 29–48). New York: Wiley.

Bisgard, K. M., Folsom, A. R., Hong, C. P., and Sellers, T. A. (1994). Mortality and cancer rates in nonrespondents to a prospective study of older women: 5-year follow-up. *American Journal of Epidemiology*, *139*, 990–1000.

Blackard, J. A. (1998). *Comparison of neural networks and discriminant analysis in predicting forest cover types*. Doctoral dissertation, Department of Forest Sciences, Colorado State University.

Boardman, T. J. (1994). The statistician who changed the world: W. Edwards Deming, 1900–1993. *The American Statistician*, *48*, 179–187.

Bowden, D. C., Anderson, A. E., and Medin, D. E. (1984). Sampling plans for mule deer sex and age ratios. *The Journal of Wildlife Management*, *48*, 500–509.

Box, G. E. P. (1979). Robustness in the strategy of scientific model building. In G. N. Wilkinson and R. L. Launer (Eds.), *Robustness in statistics* (pp. 201–236). New York: Academic Press.

Box, G. E. P., Hunter, W. G., and Hunter, J. S. (1978). *Statistics for experimenters: An introduction to design, data analysis, and model building*. New York: Wiley.

Brackstone, G. J., and Rao, J. N. K. (1979). An investigation of raking ratio estimators. *Sankhyā, Series C*, *41*, 97–114.

Bradburn, N. M. (2004). Understanding the question-answer process. *Survey Methodology*, *30*, 5–15.

Bradburn, N. M., and Sudman, S. (1979). *Improving interview method and questionnaire design: Response effects to threatening questions in survey research*. San Francisco: Jossey-Bass.

Breidt, F. J., and Opsomer, J. D. (2000). Local polynomial regresssion estimators in survey sampling. *The Annals of Statistics*, *28*, 1026–1053.

Brewer, K. R. W. (1963). Ratio estimation and finite populations: Some results deducible from the assumption of an underlying stochastic process. *The Australian Journal of Statistics*, *5*, 93–105.

Brewer, K. R. W. (1975). A simple procedure for sampling πpswor. *The Australian Journal of Statistics*, *17*, 166–172.

Brewer, K. R. W. (2002). *Combined survey sampling inference: Weighing Basu's elephants*. London: Arnold.

Brewer, K. R. W., and Donadio, M. E. (2003). The high entropy variance of the Horvitz-Thompson estimator. *Survey Methodology*, *29*, 189–196.

Brewer, K. R. W., and Hanif, M. (1983). *Sampling with unequal probabilities*. New York: Springer-Verlag.

Brewer, K. R. W., and Mellor, R. W. (1973). The effect of sample structure on analytical surveys. *The Australian Journal of Statistics*, *15*, 145–152.

Brick, J. M., Cahalan, M., Gray, L., and Severynse, J. (1994). *A study of selected nonsampling errors in the 1991 Survey of Recent College Graduates*. Technical Report NCES 95640. Washington, DC: U.S. Department of Education, National Center for Education Statistics.

Brick, J. M., and Tucker, C. (2007). Mitofsky-Waksberg: Learning from the past. *Public Opinion Quarterly*, *71*, 703–716.

Brier, S. S. (1980). Analysis of contingency tables under cluster sampling. *Biometrika*, *67*, 591–596.

Brogan, D. (2005). Sampling error estimation for survey data. In *Household sample surveys in developing and transition countries* (pp. 447–490). New York: United Nations Department of Social and Economic Affairs, unstats.un.org/unsd/HHsurveys/pdf/Chapter.21.pdf.

Buckland, S. T. (1984). Monte Carlo confidence intervals. *Biometrics*, *40*, 811–817.

Buckland, S. T., and Garthwaite, P. H. (1991). Quantifying precision of mark-recapture estimates using the bootstrap and related methods. *Biometrics, 47,* 255–268.

Burnard, P. (1992). Learning from experience: Nurse tutors' and student nurses' perceptions of experiential learning in nurse education: Some initial findings. *International Journal of Nursing Studies, 29,* 151–161.

Buskirk, T. D., and Lohr, S. L. (2005). Asymptotic properties of kernel density estimation with complex survey data. *Journal of Statistical Planning and Inference, 128,* 165–190.

Cable News Network (CNN). (2002). *Should the federal government legalize medical marijuana?* Retrieved July 2002 from www.cnn.com.

Calahan, D. (1989). The *Digest* poll rides again. *Public Opinion Quarterly, 53,* 129–133.

Cannell, C. F., Marquis, K. H., and Laurent, A. (1977). A summary of studies of interviewing methodology. In *Vital and health statistics, series 2, 69.* Washington, DC: U.S. Government Printing Office.

Canty, A. J., and Davison, A. C. (1999). Resampling-based variance estimation for labour force surveys. *The Statistician, 48,* 379–391.

Casady, R. J., and Lepkowski, J. M. (1993). Stratified telephone survey designs. *Survey Methodology, 19,* 103–113.

Casady, R. J., and Valliant, R. (1993). Conditional properties of post-stratified estimators under normal theory. *Survey Methodology, 19,* 183–192.

Casella, G., and Berger, R. L. (2002). *Statistical inference, 2nd ed.* Pacific Grove, CA: Duxbury Press.

Catlin, G., Ingram, S., and Hunter, L. (1988). The effect of CATI on data quality: A comparison of CATI and paper methods. *Proceedings of the Bureau of the Census Annual Research Conference, Vol. 4,* 291–299.

Chambers, J. M., Cleveland, W. S., Kleiner, B., and Tukey, P. A. (1983). *Graphical techniques for data analysis.* Belmont, CA: Duxbury Press.

Chambers, R. L., and Skinner, C. J. (Eds.). (2003). *Analysis of survey data.* New York: Wiley.

Chambless, L. E., and Boyle, K. E. (1985). Maximum likelihood methods for complex sample data: Logistic regression and discrete proportional hazards models. *Communications in Statistics: Theory and Methods, 14,* 1377–1392.

Chang, T. C., Lohr, S. L., and McLaren, C. G. (1992). Teaching survey sampling using simulation. *The American Statistician, 46,* 232–237.

Chao, A., Tsay, P. K., Lin, S.-H., Shau, W.-Y., and Chao, D.-Y. (2001). The applications of capture-recapture models to epidemiological data. *Statistics in Medicine, 20,* 3123–3157.

Chapman, D. G. (1951). Some properties of the hypergeometric distribution with applications to zoological sample censuses. *University of California Publications in Statistics, 1,* 131–160.

Chapman, D. W. (2005). Sample design for the FDIC's asset loss reserve project. In *Proceedings of the Section on Survey Research Methods, American Statistical Association,* 2839–2843.

Chen, B., Haas, P., and Scheuermann, P. (2002). A new two-phase sampling based algorithm for discovering association rules. In *Proceedings of the eighth ACM SIGKDD international conference on knowledge discovery and data mining,* 462–468.

Choudhry, G. H., Park, I., Kudela, M. S., and Helmick, J. C. (2002). *2001 National Survey of Veterans design and methodology: Final report.* Rockville, MD: Westat.

Chowdhury, S. R., and McGilchrist, C. A. (2001). Analysis of contingency tables with clustered observations. *Australian and New Zealand Journal of Statistics, 43,* 351–358.

Christian, C. A., and Kinney, A. (1999). The public impact of Hubble space telescope. Space Telescope Science Institute, oposite.stsci.edu/pubinfo/edugroup/PIHSTMono_102.pdf.

Christman, M. C. (2009). Sampling of rare populations. In D. Pfeffermann and C. R. Rao (Eds.), *Handbook of statistics, vol. 29A, Sample surveys: Design, methods and applications* (pp. 109–124). Amsterdam: North Holland.

Cialdini, R. B. (1984). *Influence: The new psychology of modern persuasion.* New York: Quill.

Citro, C. F., Cork, D. L., and Norwood, J. L. (Eds.). (2004). *The 2000 census: Counting under adversity.* Washington, DC: National Academies Press.

Cleveland, W. (1994). *Visualizing data.* Summit, NJ: Hobart Press.

Cochi, S. L., Edmonds, L. E., Dyer, K., Greaves, W. L., Marks, J. S., Rovira, E. Z., et al. (1989). Congenital rubella syndrome in the United States: 1970–1985. *American Journal of Epidemiology, 129,* 349–361.

Cochran, W. G. (1939). The use of analysis of variance in enumeration by sampling. *Journal of the American Statistical Association, 34,* 492–510.

Cochran, W. G. (1977). *Sampling techniques, 3rd ed.* New York: Wiley.

Cochran, W. G. (1978). Laplace's ratio estimator. In H. A. David (Ed.), *Contributions to survey sampling and applied statistics, in honor of H. O. Hartley* (pp. 3–10). New York: Academic Press.

Cohen, J. E. (1976). The distribution of the chi-squared statistic under clustered sampling from contingency tables. *Journal of the American Statistical Association, 71,* 665–670.

Colledge, M., and March, M. (1993). Quality management: Development of a framework for a statistical agency. *Journal of Business & Economic Statistics, 11,* 157–165.

Conkey, C. (2005). Do-not-call lists under fire. *The Wall Street Journal, September 28,* D1.

Converse, J. M. (1987). *Survey research in the United States: Roots and emergence, 1890–1960.* Berkeley: University of California Press.

Converse, J. M., and Presser, S. (1986). *Survey questions: Handcrafting the standardized questionnaire.* Beverly Hills, CA: Sage.

Cook, R. D., and Weisberg, S. (1994). *An introduction to regression graphics.* New York: Wiley.

Cormack, R. M. (1992). Interval estimation for mark-recapture studies of closed populations. *Biometrics, 48,* 567–576.

Cornfield, J. (1944). On samples from finite populations. *Journal of the American Statistical Association, 39,* 236–239.

Cornfield, J. (1951). Modern methods in the sampling of human populations. *American Journal of Public Health, 41,* 654–661.

Council of American Survey Research Organizations. (2008). *CASRO code of standards and ethics for survey research.* Retrieved from www.casro.org/codeofstandards.cfm.

Couper, M. P. (2000). Web surveys: A review of issues and approaches. *Public Opinion Quarterly, 64,* 464–494.

Courbois, J.-Y. P., and Urquhart, N. S. (2004). Comparison of survey estimates of the finite population variance. *Journal of Agricultural, Biological, and Environmental Statistics, 9,* 236–251.

Cox, D. R. (1952). Estimation by double sampling. *Biometrika, 39,* 217–227.

Cox, G. M. (1957). Statistical frontiers. *Journal of the American Statistical Association, 52,* 1–12.

Crewe, I. (1992). A nation of liars? Opinion polls and the 1992 election. *Parliamentary Affairs, 45,* 475–495.

Cronbach, L. J. (1951). Coefficient alpha and the internal structure of tests. *Psychometrika, 16,* 297–334.

Cullen, R. (1994). Sample survey methods as a quality assurance tool in a general practice immunisation audit. *New Zealand Medical Journal, 107,* 152–153.

Curtin, R., Presser, S., and Singer, E. (2005). Changes in telephone survey nonresponse over the past quarter century. *Public Opinion Quarterly, 69,* 87–98.

Czaja, R. F., and Blair, J. (1986). Using network sampling in crime victimization surveys. *Journal of Quantitative Criminology, 6,* 185–206.

Czaja, R. F., Snowden, C. B., and Casady, R. J. (1986). Reporting bias and sampling errors in a survey of a rare population using multiplicity counting rules. *Journal of the American Statistical Association, 81,* 411–419.

Dalenius, T. (1977). Strain at a gnat and swallow a camel: Or, the problem of measuring sampling and non-sampling errors. In *Proceedings of the Social Statistics Section, American Statistical Association,* 21–25.

Dalenius, T. E. (1981). The survey statistician's responsibility for both sampling and measurement errors. In D. Krewski, R. Platek, and J. N. K. Rao (Eds.), *Current topics in survey sampling* (pp. 17–29). New York: Academic Press.

D'Alessandro, U., Aikins, M. K., Langerock, P., Bennett, S., and Greenwood, B. M. (1994). Nationwide survey of bednet use in rural Gambia. *Bulletin of the World Health Organization, 72,* 391–394.

Danermark, B., and Swensson, B. (1987). Measuring drug use among Swedish adolescents: Randomized response versus anonymous questionnaires. *Journal of Official Statistics, 3,* 439–448.

David, I. P., and Sukhatme, B. V. (1974). On the bias and mean squared error of the ratio estimator. *Journal of the American Statistical Association, 69,* 464–466.

Davison, A. C., and Hinkley, D. V. (1997). *Bootstrap methods and their application.* Cambridge: Cambridge University Press.

deLeeuw, E. D., Hox, J. J., and Dillman, D. A. (Eds.). (2008). *International handbook of survey methodology.* New York: Erlbaum.

Demidenko, E. (2004). *Mixed models: Theory and applications.* New York: Wiley.

Deming, W. E. (1944). On errors in surveys. *American Sociological Review, 9,* 359–369.

Deming, W. E. (1950). *Some theory of sampling.* New York: Dover.

Deming, W. E. (1977). An essay on screening, or on two-phase sampling, applied to surveys of a community. *International Statistical Review, 45,* 29–37.

Deming, W. E. (1986). *Out of the crisis.* Cambridge, MA: Massachusetts Institute of Technology.

Deming, W. E., and Stephan, F. F. (1940). On a least squares adjustment of a sampled frequency table when the expected marginal totals are known. *Annals of Mathematical Statistics, 11,* 427–444.

Demnati, A., and Rao, J. N. K. (2004). Linearization variance estimators for survey data (with discussion). *Survey Methodology, 30,* 17–34.

Deville, J.-C. (1991). A theory of quota surveys. *Survey Methodology, 17,* 163–181.

Deville, J.-C., and Särndal, C.-E. (1992). Calibration estimators in survey sampling. *Journal of the American Statistical Association, 87,* 376–382.

DeVries, W., Keller, W., and Willeboordse, A. (1996). Reducing the response burden: Some developments in the Netherlands. *International Statistical Review, 64,* 199–213.

Dillman, D. A. (2008). The logic and psychology of constructing questionnaires. In E. D. deLeeuw, J. J. Hox, and D. A. Dillman (Eds.), *International handbook of survey methodology* (pp. 161–175). New York: Erlbaum.

Dillman, D. A., Clark, J. R., and Sinclair, M. D. (1995). How prenotice letters, stamped return envelopes and reminder postcards affect mailback response rates for census questionnaires. *Survey Methodology, 21,* 159–165.

Dillman, D. A., Smyth, J. D., and Christian, L. M. (2009). *Internet, mail, and mixed-mode surveys: The tailored design method, 3rd ed.* New York: Wiley.

Dobishinski, W. M. (1991). ASCAP/BMI primer. In *The musician's business and legal guide, 4th ed.* (pp. 176–221). Englewood Cliffs, NJ: Prentice Hall.

Domingo-Salvany, A., Hartnoll, R. L., Maquire, A., Suelves, J. M., and Anto, J. M. (1995). Use of capture-recapture to estimate the prevalence of opiate addiction in Barcelona, Spain, 1989. *American Journal of Epidemiology, 141,* 567–574.

Doyle, P., Lane, J. I., Theeuwes, J., and Zayatz, L. M. (Eds.). (2001). *Confidentiality, disclosure, and data access: Theory and practical application for statistical agencies.* Amsterdam: Elsevier.

Duce, R. A., Quinn, J. G., Olney, C. E., Piotrowicz, S. R., Ray, B. J., and Wade, T. L. (1972). Enrichment of heavy metals and organic compounds in the surface microlayer of Narragansett Bay, Rhode Island. *Science, 176,* 161–163.

Duffy, J. C., and Waterton, J. J. (1988). Randomised response vs. direct questioning: Estimating the prevalence of alcohol related problems in a field survey. *The Australian Journal of Statistics, 30,* 1–14.

DuMouchel, W. H., and Duncan, G. J. (1983). Using sample survey weights in multiple regression analyses of stratified samples. *Journal of the American Statistical Association, 78,* 535–543.

Dunn, G., Pickles, A., Tansella, M., and Vazquez-Barquero, J. (1999). Two-phase epidemiological surveys in psychiatric research. *British Journal of Psychiatry, 174*, 95–100.

Durbin, J. (1953). Some results in sampling theory when the units are sampled with unequal probabilities. *Journal of the Royal Statistical Society, Series B, 15*, 262–269.

Durrett, R. (1994). *The essentials of probability*. Belmont, CA: Duxbury Press.

Ebersole, S. (2000). Uses and gratifications of the web among students. *Journal of Computer-Mediated Communications, 6*. Retrieved from www.ascusc.org/jcmc/vol6/issue1/ebersole.html.

Edwards, T. C., Cutler, D. R., Zimmerman, N. E., Geiser, L., and Alegria, J. (2005). Model-based stratifications for enhancing the detection of rare ecological events. *Ecology, 86*, 1081–1090.

Efron, B. (1979). Bootstrap methods: Another look at the jackknife. *The Annals of Statistics, 7*, 1–26.

Efron, B. (1982). *The jackknife, the bootstrap and other resampling plans*. Philadelphia: SIAM.

Efron, B., and Tibshirani, R. (1993). *An introduction to the bootstrap*. London: Chapman & Hall.

Egeland, G. M., Perham-Hester, K. A., and Hook, E. B. (1995). Use of capture-recapture analyses in fetal alcohol syndrome surveillance in Alaska. *American Journal of Epidemiology, 141*, 335–341.

Einarsen, S., Matthiessen, S., and Skogstad, A. (1998). Bullying, burnout, and well-being among assistant nurses. *Journal of Occupational Health and Safety Australia and New Zealand, 14*, 563–568.

Elliott, M. R., and Little, R. J. A. (2000). Model-based alternatives to trimming survey weights. *Journal of Official Statistics, 16*(3), 191–209.

Eltinge, J. L., and Yansaneh, I. S. (1997). Diagnostics for formation of nonresponse adjustment cells, with an application to income nonresponse in the U.S. Consumer Expenditure Survey. *Survey Methodology, 23*, 33–40.

Estevao, V. M., and Särndal, C.-E. (1999). The use of auxiliary information in design-based estimation for domains. *Survey Methodology, 25*, 213–231.

Estevao, V. M., and Särndal, C.-E. (2006). Survey estimates by calibration on complex auxiliary information. *International Statistical Review, 74*, 127–147.

Eurostat. (2000). *Assessment of the quality in statistics*. Eurostat/A4/Quality/00/General Standard Report, April 4–5, Luxembourg.

Ezzati-Rice, T., and Murphy, R. S. (1995). Issues associated with the design of a national probability sample for human exposure assessment. *Environmental Health Perspectives, 103 (suppl. 3)*, 55–59.

Farkas, S., and Johnson, J. (1997). *Different drummers: How teachers of teachers view public education*. New York: Public Agenda.

Fay, R. E. (1985). A jackknifed chi-squared test for complex samples. *Journal of the American Statistical Association, 80*, 148–157.

Fay, R. E. (1989). Theory and application of replicate weighting for variance calculations. In *Proceedings of the Section on Survey Research Methods, American Statistical Association*, 212–217.

Fay, R. E. (1990). VPLX: Variance Estimates for Complex Samples. In *Proceedings of the Section on Survey Research Methods, American Statistical Association*, 266–271.

Fay, R. E. (1996). Alternative paradigms for the analysis of imputed survey data. *Journal of the American Statistical Association, 91*, 490–498.

Fay, R. E., and Herriot, R. A. (1979). Estimates of income for small places: An empirical Bayes application of James-Stein procedures to census data. *Journal of the American Statistical Association, 74*, 269–277.

Federal Committee on Statistical Methodology. (2001). *Measuring and reporting sources of error in surveys*. Statistical Policy Working Paper No. 31. Washington, DC: U.S. Office of Management and Budget. Retrieved from www.fcsm.gov/01papers/SPWP31_final.pdf.

Fellegi, I. P., and Holt, D. (1976). A systematic approach to automatic edit and imputation. *Journal of the American Statistical Association, 71*, 17–35.

Fienberg, S. E. (1972). The multiple recapture census for closed populations and incomplete 2^k contingency tables. *Biometrika, 59*, 591–603.

Fienberg, S. E. (1979). Use of chi-squared statistics for categorical data problems. *Journal of the Royal Statistical Society, Series B, 41*, 54–64.

Fienberg, S. E. (1980). The measurement of crime victimization: Prospects for panel analysis of a panel survey. *The Statistician, 29*, 313–350.

Fienberg, S. E. (1992). Bibliography on capture-recapture modelling with application to census undercount adjustment. *Survey Methodology, 18*, 143–154.

Fienberg, S. E., and Tanur, J. M. (1987). Experimental and sampling structures: Parallels diverging and meeting. *International Statistical Review, 55*, 75–96.

Fisher, R. A. (1925). *Statistical methods for research workers.* London: Oliver and Boyd.

Fisher, R. A. (1938). Presidential address. In *Proceedings of the Indian Statistical Conference.* Calcutta: Statistical Publishing Society.

Forman, S. L. (2004). *Baseball-reference.com—Major league statistics and information.* Retrieved November 2004 from www.baseball-reference.com.

Fowler, F. J. (1991). Reducing interviewer-related error through interviewer training, supervision, and other means. In P. P. Biemer, R. M. Groves, L. E. Lyberg, N. A. Mathiowetz, and S. Sudman (Eds.), *Measurement error in surveys* (pp. 259–278). New York: Wiley.

Fowler, F. J. (1995). *Improving survey questions: Design and evaluation.* Thousand Oaks, CA: Sage.

Fox, J. A., and Tracy, P. E. (1986). *Randomized response: A method for sensitive surveys.* Beverly Hills, CA: Sage.

Francisco, C. A., and Fuller, W. A. (1991). Quantile estimation with a complex survey design. *The Annals of Statistics, 19*, 454–469.

Frank, A. (1978). *The contingency table approach to mark-recapture population estimation.* University of Minnesota Department of Applied Statistics Plan B Paper. St. Paul, MN.

Frankovic, K. (2008). *Race, gender and bias in the electorate.* Retrieved August 2008 from www.cbsnews.com/stories/2008/03/17/opinions/pollpositons.

Fuller, W. A. (1975). Regression analysis for sample survey. *Sankhyā, Series C, 37*, 117–132.

Fuller, W. A. (1984). Least squares and related analyses for complex survey designs. *Survey Methodology, 10*, 97–118.

Fuller, W. A. (2002). Regression estimation for survey samples. *Survey Methodology, 28*, 5–23.

Fuller, W. A. (2009). *Sampling statistics.* Hoboken, NJ: Wiley.

Gabler, S. (1984). On unequal probability sampling: Sufficient conditions for the superiority of sampling without replacement. *Biometrika, 71*, 171–175.

Gelman, A. (2007). Struggles with survey weighting and regression modeling (with discussion). *Statistical Science, 22*, 153–164.

Gelman, A., and Carlin, J. B. (2002). Poststratification and weighting adjustments. In R. M. Groves, D. Dillman, J. Eltinge, and R. Little (Eds.), *Survey nonresponse* (pp. 289–302). New York: Wiley.

Giles, J. (2005). Internet encyclopaedias go head to head. *Nature, 438*, 900–901.

Gnanadesikan, R. (1997). *Statistical data analysis of multivariate observations.* New York: Wiley.

Gnap, R. (1995). *Teacher load in Arizona elementary school districts in Maricopa County.* Unpublished doctoral dissertation, Arizona State University.

González, M. E. (1994). Improving data quality awareness in the United States federal statistical agencies. *The American Statistician, 48*, 12–17.

Goren, S., Silverstein, L., and Gonzales, N. (1993). A survey of food service managers of Washington State boarding homes for the elderly. *Journal of Nutrition for the Elderly, 12*, 27–36.

Gower, A. R. (1979). Nonresponse in the Canadian Labour Force Survey. *Survey Methodology, 5*, 29–58.

Granquist, L., and Kovar, J. (1997). Editing of survey data: How much is enough? In L. Lyberg, P. Biemer, M. Collins, E. de Leeuw, C. Dippo, N. Schwarz, and D. Trewin (Eds.), *Survey measurement and process quality* (pp. 415–435). New York: Wiley.

Graybill, F. A. (1976). *Theory and application of the linear model*. North Scituate, MA: Duxbury Press.

Griffin, D. H., Raglin, D. A., Leslie, T. F., McGovern, P. D., and Broadwater, J. K. (2003). *Meeting 21st century demographic data needs—Implementing the American Community Survey*. Washington, DC: U.S. Census Bureau. Retrieved from www.census.gov/acs/Downloads/Report03.pdf.

Gross, S. T. (1980). Median estimation in sample surveys. In *Proceedings of the Section on Survey Research Methods, American Statistical Association,* 181–184.

Groves, R. M. (1989). *Survey errors and survey costs*. New York: Wiley.

Groves, R. M. (2006). Nonresponse rates and nonresponse bias in household surveys. *Public Opinion Quarterly, 70*, 646–675.

Groves, R. M., and Couper, M. (1998). *Nonresponse in household interview surveys*. New York: Wiley.

Groves, R. M., Dillman, D., Eltinge, J., and Little, R. (Eds.). (2002). *Survey nonresponse*. New York: Wiley.

Groves, R. M., Fowler, F. J., Couper, M. P., Lepkowski, J., Singer, E., and Tourangeau, R. (2009). *Survey methodology, 2nd ed.* New York: Wiley.

Hájek, J. (1960). Limiting distributions in simple random sampling from a finite population. *Publications of the Mathematical Institute of the Hungarian Academy of Sciences, 5,* 361–371.

Hájek, J. (1964). Asymptotic theory of rejection sampling with varying probabilities from a finite population. *Annals of Mathematical Statistics, 35,* 1491–1523.

Hand, D. J., Daly, F., Lunn, A. D., McConway, K. J., and Ostrowski, E. (1994). *A handbook of small data sets*. London: Chapman and Hall.

Hansen, M. H., and Hurwitz, W. N. (1943). On the theory of sampling from a finite population. *Annals of Mathematical Statistics, 14,* 333–362.

Hansen, M. H., and Hurwitz, W. N. (1946). The problem of non-response in sample surveys. *Journal of the American Statistical Association, 41,* 517–529.

Hansen, M. H., Hurwitz, W. N., and Bershad, M. A. (1961). Measurement errors in censuses and surveys. *Bulletin of the International Statistical Institute, 38,* 359–374.

Hansen, M. H., Hurwitz, W. N., and Madow, W. G. (1953). *Sample survey methods and theory. Volume 1: Methods and applications.* New York: Wiley.

Hansen, M. H., Madow, W. G., and Tepping, B. J. (1983). An evaluation of model-dependent and probability-sampling inferences in sample surveys. *Journal of the American Statistical Association, 78,* 776–793.

Hanurav, T. V. (1967). Optimum utilization of auxiliary information: πps sampling of two units from a stratum. *Journal of the Royal Statistical Society, Series B, 29,* 374–391.

Harkness, J., van de Vijver, F. J. R., and Mohler, P. P. (2003). *Cross-cultural survey methods*. New York: Wiley.

Hartley, H. O. (1946). Discussion of paper by F. Yates. *Journal of the Royal Statistical Society, 109,* 38–39.

Hartley, H. O. (1962). Multiple frame surveys. In *Proceedings of the Social Statistics Section, American Statistical Association,* 203–206.

Hartley, H. O., and Biemer, P. (1981). The estimation of nonsampling variance in current surveys. In *Proceedings of the Section on Survey Research Methods, American Statistical Association,* 257–262.

Hartley, H. O., and Rao, J. N. K. (1962). Sampling with unequal probabilities and without replacement. *Annals of Mathematical Statistics, 33,* 350–374.

Hartley, H. O., and Rao, J. N. K. (1978). Estimation of nonsampling variance components in sample surveys. In N. K. Namboodiri (Ed.), *Survey sampling and measurement* (pp. 35–44). New York: Academic Press.

Hartley, H. O., and Ross, A. (1954). Unbiased ratio estimators. *Nature, 174,* 270–271.

Harville, D. A. (1997). *Matrix algebra from a statistician's perspective*. New York: Springer.

Hastie, T., Tibshirani, R., and Friedman, J. (2001). *The elements of statistical learning*. New York: Springer.

Hayes, L. R. (2000). Are prices higher for the poor in New York City? *Journal of Consumer Policy, 23*, 127–152.

Heckathorn, D. D. (1997). Respondent driven sampling: A new approach to the study of hidden populations. *Social Problems, 44*, 174–199.

Hidiroglou, M. A. (2001). Double sampling. *Survey Methodology, 27*, 143–154.

Hidiroglou, M. A., Drew, J. D., and Gray, G. B. (1993). A framework for measuring and reducing nonresponse in surveys. *Survey Methodology, 19*, 81–94.

Hidiroglou, M. A., and Patak, Z. (2004). Domain estimation using linear regression. *Survey Methodology, 30*, 67–78.

Hidiroglou, M. A., Rao, J. N. K., and Haziza, D. (2009). Variance estimation in two-phase sampling. *Australian and New Zealand Journal of Statistics, 51*, 127–141.

Hinkins, S., Oh, H. L., and Scheuren, F. (1997). Inverse sampling design algorithms. *Survey Methodology, 23*, 11–21.

Hite, S. (1987). *Women and love: A cultural revolution in progress.* New York: Knopf.

Hoeting, J. A., Leecaster, M., and Bowden, D. (2000). An improved model for spatially correlated binary responses. *Journal of Agricultural, Biological, and Environmental Statistics, 5*, 102–114.

Hogan, H. (1993). The 1990 Post-Enumeration Survey: Operations and results. *Journal of the American Statistical Association, 88*, 1047–1060.

Holden, C. (2009). America's uncounted millions. *Science, 324*, 1008–1009.

Holt, D., and Elliot, D. (1991). Methods of weighting for unit non-response. *The Statistician, 40*, 333–342.

Holt, D., Scott, A. J., and Ewings, P. D. (1980). Chi-squared tests with survey data. *Journal of the Royal Statistical Society, Series A, 143*, 303–320.

Horvitz, D., Koshland, D., Rubin, D., Gollin, A., Sawyer, T., and Tanur, J. (1995). Pseudo-opinion polls: SLOP or useful data? *Chance, 8*, 16–25.

Horvitz, D. G., Shah, B. V., and Simmons, W. R. (1967). The unrelated question randomized response model. In *Proceedings of the Social Statistics Section, American Statistical Association*, 65–72.

Horvitz, D. G., and Thompson, D. J. (1952). A generalization of sampling without replacement from a finite universe. *Journal of the American Statistical Association, 47*, 663–685.

Hosmer, D. W., and Lemeshow, S. (2000). *Applied logistic regression, 2nd ed.* New York: Wiley.

Hox, J., and deLeeuw, E. (1994). A comparison of nonresponse in mail, telephone, and face-to-face surveys: Applying multilevel modeling to meta-analysis. *Quality and Quantity, 28*, 329–344.

Husby, C. E., Stasny, E. A., and Wolfe, D. A. (2005). An application of ranked set sampling for mean and median estimation using USDA crop production data. *Journal of Agricultural, Biological, and Environmental Statistics, 10*, 354–373.

Iachan, R., and Dennis, M. L. (1993). A multiple frame approach to sampling the homeless and transient population. *Journal of Official Statistics, 9*, 747–764.

International Statistical Institute. (2009). ISI declaration on professional ethics. *isi.cbs.nl/ethics0index.htm.*

International Working Group for Disease Monitoring and Forecasting. (1995a). Capture-recapture and multiple-record systems estimation I: History and theoretical development. *American Journal of Epidemiology, 142*, 1047–1058.

International Working Group for Disease Monitoring and Forecasting. (1995b). Capture-recapture and multiple-record systems estimation II: Application in human diseases. *American Journal of Epidemiology, 142*, 1059–1068.

Ismail, K., Kent, K., Brugha, T., Hotopf, M., Hull, L., Seed, P., et al. (2002). The mental health of UK gulf war veterans: Phase 2 of a two phase cohort study. *British Medical Journal, 325*, 576–579.

Jackson, K. W., Eastwood, I. W., and Wild, M. S. (1987). Stratified sampling protocol for monitoring trace metal concentrations in soil. *Soil Science, 143*, 436–443.

Jacoby, J., and Handlin, A. H. (1991). Non-probability sampling designs for litigation surveys. *Trademark Reporter, 81*, 169–179.

Jenney, B. (2005). *Regression diagnostics with complex survey data.* Unpublished M.S. applied project, Arizona State University, Tempe Arizona.

Jiang, J. (2007). *Linear and generalized linear mixed models and their applications.* New York: Springer.

Jinn, J.-H., and Sedransk, J. (1989). Effect on secondary data analysis of common imputation methods. *Sociological Methodology, 19*, 213–241.

Judkins, D. (1990). Fay's method for variance estimation. *Journal of Official Statistics, 6*, 223–240.

Juran, J. M., and Godfrey, A. B. (2000). *Juran's quality handbook, 5th ed.* New York: McGraw-Hill.

Kalton, G. (1983). Models in the practice of survey sampling. *International Statistical Review, 51*, 175–188.

Kalton, G. (2003). Practical methods for sampling rare and mobile populations. *Statistics in Transition, 6*, 491–501.

Kalton, G., and Anderson, D. W. (1986). Sampling rare populations. *Journal of the Royal Statistical Society, Series A, 149*, 65–82.

Kalton, G., and Kasprzyk, D. (1986). The treatment of missing survey data. *Survey Methodology, 12*, 1–16.

Karras, T. (2008). The disorder next door. *SELF*, May, 248–253.

Kasprzyk, D., and Giesbrecht, L. (2001). Reporting sources of error in U.S. federal government data collection programs. In L. Lyberg (Ed.), *Proceedings of the international conference on quality in official statistics* (p. 30.1). Stockholm: Statistics Sweden.

Kempthorne, O. (1952). *The design and analysis of experiments.* New York: Wiley.

Kessler, R. C., Berglund, P., Chiu, W. T., Demler, O., Heeringa, S., Hiripi, E., et al. (2004). The US National Comorbidity Survey Replication (NCS-R): Design and field procedures. *International Journal of Methods in Psychiatric Research, 13*, 69–92.

Kiaer, A. (1897). *The representative method of statistical surveys [1976 translation of the original Norwegian].* Oslo: Central Bureau of Statistics of Norway.

Kim, J. K., Navarro, A., and Fuller, W. A. (2006). Replication variance estimation for two-phase stratified sampling. *Journal of the American Statistical Association, 101*, 312–320.

Kinsley, M. (1981). The art of polling. *New Republic, 184*, 16–19.

Kirkman, E. E., Maxwell, J. W., and Rose, C. A. (2005). 2004 annual survey of the mathematical sciences. *Notices of the American Mathematical Society, 52*, 871–883.

Kish, L. (1962). Studies of interviewer variance for attitudinal variables. *Journal of the American Statistical Association, 57*, 92–115.

Kish, L. (1965). *Survey sampling.* New York: Wiley.

Kish, L. (1992). Weighting for unequal P_i. *Journal of Official Statistics, 8*, 183–200.

Kish, L. (1995). Methods for design effects. *Journal of Official Statistics, 11*, 55–77.

Kish, L., and Frankel, M. R. (1974). Inference from complex samples (with discussion). *Journal of the Royal Statistical Society, Series B, 36*, 1–37.

Kleppel, G. S., Madewell, S. A., and Hazzard, S. E. (2004). Responses of emergent marsh wetlands in upstate New York to variations in urban typology. *Ecology and Society, 5*, Retrieved from www.ecologyandsociety.org/vol9/iss5/art1.

Koch, G. G., Freeman, J., Daniel H., and Freeman, J. L. (1975). Strategies in the multivariate analysis of data from complex surveys. *International Statistical Review, 43*, 59–78.

Konijn, H. S. (1962). Regression analysis in sample surveys. *Journal of the American Statistical Association, 57*, 509–606.

Korn, E. L., and Graubard, B. I. (1995). Analysis of large health surveys: Accounting for the sampling design. *Journal of the Royal Statistical Society, Series A, 158*, 263–295.

Korn, E. L., and Graubard, B. I. (1998). Scatterplots with survey data. *The American Statistician, 52*, 58–69.

Korn, E. L., and Graubard, B. I. (1999). *Analysis of health surveys.* New York: Wiley.

Kosmin, B. A., and Lachman, S. P. (1993). *One nation under God: Religion in contemporary American society*. New York: Harmony Books.

Kostanich, D., Whitford, D., and Bell, W. R. (2004). Plans for measuring coverage of the 2010 U.S. Census. In *Proceedings of the Section on Survey Research Methods, American Statistical Association,* 1626–1635.

Kott, P. S. (1991). A model-based look at linear regression with survey data. *The American Statistician, 45,* 107–112.

Kott, P. S., and Stukel, D. M. (1997). Can the jackknife be used with a two-phase sample? *Survey Methodology, 23,* 81–89.

Kovar, J. G., Rao, J. N. K., and Wu, C. F. J. (1988). Bootstrap and other methods to measure errors in survey estimates. *Canadian Journal of Statistics, 16*(S), 25–45.

Krasilovsky, M. W., and Shemel, S. (2003). *The business of music, 9th ed.* New York: Watson-Guptill.

Krenzke, T. (1995). Reevaluating generalized variance model parameters for the National Crime Victimization Survey. In *Proceedings of the Section on Survey Research Methods, American Statistical Association,* 327–332.

Kreuter, F., Presser, S., and Tourangeau, R. (2008). Social desirability bias in CATI, IVR, and web surveys. *Public Opinion Quarterly, 72,* 847–865.

Krewski, D., and Rao, J. N. K. (1981). Inference from stratified samples: Properties of the linearization, jackknife and balanced repeated replication methods. *The Annals of Statistics, 9,* 1010–1019.

Kripke, D. F., Garfinkel, L., Wingard, D. L., Klauber, M. R., and Marler, M. R. (2002). Mortality associated with sleep duration and insomnia. *Archives of General Psychiatry, 59,* 131–136.

Kruuk, H., Moorhouse, A., Conroy, J. W. H., Durbin, L., and Frears, S. (1989). An estimate of numbers and habitat preferences of otters *lutra lutra* in Shetland, UK. *Biological Conservation, 49,* 241–254.

Kuk, A. Y. C. (1990). Asking sensitive questions indirectly. *Biometrika, 77,* 436–438.

Kutner, M. H., Nachtsheim, C. J., Neter, J., and Li, W. (2005). *Applied linear statistical models, 5th ed.* Boston: McGraw-Hill/Irwin.

Lahiri, D. B. (1951). A method of sample selection providing unbiased ratio estimates. *Bulletin of the International Statistical Institute, 33,* 133–140.

Lai, T. L. (2001). Sequential analysis: Some classical problems and new challenges (with discussion). *Statistica Sinica, 11,* 303–408.

Landers, A. (1976). If you had it to do over again, would you have children? *Good Housekeeping, 182* (June), 100–101, 215–216, 223–224.

Langer, G. (2009). *Beyond the plus or minus.* April 21. Retrieved May 2009 from www.abcnews.com.

Laplace, P. S. (1814). *Essai philosophique sur les probabilités.* Paris: MME VE Courcier, Imprimeur-Libraire pour les Mathématiques, quai des Augustins, no. 57. [An English translation was published by Dover in 1951.]

Larsen, M. D., and Rubin, D. B. (2001). Iterative automated record linkage using mixture models. *Journal of the American Statistical Association, 96,* 32–41.

Lavallée, P. (2007). *Indirect sampling.* New York: Springer.

Lavrakas, P. J., Shuttles, C. D., Steeh, C., and Fienberg, H. (2007). The state of surveying cell phone numbers in the United States: 2007 and beyond. *Public Opinion Quarterly, 71,* 840–854.

Legg, J. C., and Fuller, W. A. (2009). Two-phase sampling. In D. Pfeffermann and C. R. Rao (Eds.), *Handbook of statistics: Vol. 29A. Sample surveys: Design, methods and applications* (pp. 55–70). Amsterdam: North Holland.

Lehmann, E. L. (1999). *Elements of large-sample theory.* New York: Springer-Verlag.

Lehnen, R. G., and Skogan, W. G. (1981). *The National Crime Victimization Survey: Working papers.* Washington, DC: U.S. Department of Justice.

Lehtonen, R., and Pahkinen, E. (2004). *Practical methods for design and analysis of complex surveys, 2nd ed.* New York: Wiley.

Lenski, G., and Leggett, J. (1960). Caste, class, and deference in the research interview. *American Journal of Sociology, 65*, 463–467.

Lepkowski, J. M. (1988). Telephone sampling methods in the United States. In R. M. Groves, P. P. Biemer, L. E. Lyberg, J. T. Massey, W. L. Nicholls, and J. Waksberg (Eds.), *Telephone survey methodology* (pp. 73–98). New York: Wiley.

Lesser, V. M., and Kalsbeek, W. D. (1999). Nonsampling errors in environmental surveys. *Journal of Agricultural, Biological, and Environmental Statistics, 4*, 473–488.

Lessler, J. T., and Kalsbeek, W. (1992). *Nonsampling errors in surveys.* New York: Wiley.

Levy, P. S., and Lemeshow, S. (2008). *Sampling of populations: Methods and applications, 4th ed.* New York: Wiley.

Li, J., and Valliant, R. (2006). Influence analysis in linear regression with sampling weights. In *Proceedings of the Section on Survey Research Methods, American Statistical Association,* 3330–3337.

Lin, I.-F., and Schaeffer, N. (1995). Using survey participants to estimate the impact of non-participation. *Public Opinion Quarterly, 59*, 236–258.

Linacre, S. J., and Trewin, D. J. (1993). Total survey design—Application to a collection of the construction industry. *Journal of Official Statistics, 9*, 611–621.

Lincoln, F. C. (1930). Calculating waterfowl abundance on the basis of banding returns. *Circular of the U.S. Department of Agriculture, 118*, 1–4.

Link, H. C., and Hopf, H. A. (1946). *People and books: A study of reading and book-buying habits.* New York: Book Manufacturer's Institute.

Literary Digest. (1932). Roosevelt bags 41 states out of 48. *Literary Digest, 114* (November 5), 8–9.

Literary Digest. (1936a). "The *Digest*" presidential poll is on: Famous forecasting machine is thrown into gear for 1936. *Literary Digest, 122* (August 22), 3–4.

Literary Digest. (1936b). Landon, 1,293,669: Roosevelt, 972,897. *Literary Digest, 122* (October 31), 5–6.

Literary Digest. (1936c). What went wrong with the polls? *Literary Digest, 122* (November 14), 7–8.

Little, R. J. A. (1991). Inference with survey weights. *Journal of Official Statistics, 7*, 405–424.

Little, R. J. A. (2004). To model or not to model: Competing modes of inference for finite population sampling. *Journal of the American Statistical Association, 99*, 546–556.

Little, R. J. A., and Rubin, D. B. (2002). *Statistical analysis with missing data, 2nd ed.* New York: Wiley.

Lohr, S. L. (1990). Accurate multivariate estimation using triple sampling. *The Annals of Statistics, 18*, 1615–1633.

Lohr, S. L. (2001). Sample surveys: Model-based approaches. In N. J. Smelser and P. B. Baltes (Eds.), *International encyclopedia of the social & behavioral sciences* (pp. 13462–13467). New York: Elsevier.

Lohr, S. L. (2009). Multiple frame surveys. In D. Pfeffermann and C. R. Rao (Eds.), *Handbook of statistics: Vol. 29A. Sample surveys: Design, methods and applications* (pp. 71–88). Amsterdam: North Holland.

Lohr, S. L., and Liu, J. (1994). A comparison of weighted and unweighted analyses in the NCVS. *Journal of Quantitative Criminology, 10*, 343–360.

Lohr, S. L., and Rao, J. N. K. (2000). Inference from dual frame surveys. *Journal of the American Statistical Association, 95*, 271–280.

Lumley, T. (2000). *R survey functions.* www.r-project.org.

Lush, J. L. (1945). *Animal breeding plans.* Ames, IA: Iowa State College Press.

Lyberg, L., Biemer, P., Collins, M., deLeeuw, E., Dippo, C., Schwarz, N., et al. (1997). *Survey measurement and process quality.* New York: Wiley.

Lydersen, C., and Ryg, M. (1991). Evaluating breeding habitat and populations of ringed seals *Phoca hispida* in Svalbard fjords. *Polar Record, 27*, 223–228.

Macdonell, W. R. (1901). On criminal anthropometry and the identification of criminals. *Biometrika, 1*, 177–227.

Madow, W. G., Olkin, I., and Rubin, D. B. (Eds.). (1983). *Incomplete data in sample surveys (Vols. 1–3).* New York: Academic Press.

Mahalanobis, P. C. (1939). A sample survey of the acreage under jute in Bengal. *Sankhyā, 4,* 511–531.

Mahalanobis, P. C. (1946). Recent experiments in statistical sampling in the Indian Statistical Institute. *Journal of the Royal Statistical Society, 109,* 325–378.

Manly, B. F. J. (1997). *Randomization, bootstrap and Monte Carlo methods in biology, 2nd ed.* London: Chapman & Hall.

Marker, D. A., Judkins, D. R., and Winglee, M. (2002). Large-scale imputation for complex surveys. In R. M. Groves, D. Dillman, J. Eltinge, and R. Little (Eds.), *Survey nonresponse* (pp. 329–341). New York: Wiley.

Marker, D. A., and Morganstein, D. (2004). Keys to successful implementation of continuous quality improvement in a statistical agency. *Journal of Official Statistics, 20,* 125–136.

Martin, E., de la Puente, M., and Bennett, C. (2005). *The effects of questionnaire and content change on race data: Results of a replication of 1990 race and origin questions.* U.S. Census Bureau Survey Methodology Report 2005–05. Retrieved from www.census.gov/srd/papers/pdf/rsm2005-05.pdf.

Matei, A., and Tillé, Y. (2005). *The R sampling package.* cran.r-project.org/src/contrib/Descriptions/sampling.html.

Mayr, J., Gaisl, M., Purtscher, K., Noeres, H., Schimpl, G., and Fasching, G. (1994). Baby walkers—an underestimated hazard for our children? *European Journal of Pediatrics, 153,* 531–534.

McAuley, R. G., Paul, W. M., Morrison, G. H., Beckett, R. F., and Goldsmith, C. H. (1990). Five-year results of the peer assessment program of the College of Physicians and Surgeons of Ontario. *Canadian Medical Association Journal, 143,* 1193–1199.

McCarthy, P. J. (1966). Replication: An approach to the analysis of data from complex surveys. In *Vital and health statistics,* ser. 2, no. 14. Washington, DC.

McCarthy, P. J. (1969). Pseudo-replication: Half-samples. *Review of the International Statistical Institute, 37,* 239–264.

McCarthy, P. J. (1993). Standard error and confidence interval estimation for the median. *Journal of Official Statistics, 9,* 673–689.

McFarland, S. G. (1981). Effects of question order on survey responses. *Public Opinion Quarterly, 45,* 208–215.

McIlwee, J. S., and Robinson, J. G. (1992). *Women in engineering: Gender, power, and workplace culture.* Albany, NY: State University of New York Press.

McIntyre, G. A. (1952). A method of unbiased selective sampling, using ranked sets. *Australian Journal of Agricultural Research, 3,* 385–390.

McNamee, R. (2003). Efficiency of two-phase designs for prevalence estimation. *International Journal of Epidemiology, 32,* 1072–1078.

Mitofsky, W. (1970). Sampling of telephone households. Unpublished CBS News memorandum.

Mitofsky, W. J., and Edelman, M. (2002). Election night estimation. *Journal of Official Statistics, 18,* 165–179.

Montanari, G. E. (1987). Post-sampling efficient QR-prediction in large-scale surveys. *International Statistical Review, 55,* 191–202.

Montanari, G. E., and Ranalli, M. G. (2005). Nonparametric model calibration estimation in survey sampling. *Journal of the American Statistical Association, 100,* 1429–1442.

Montgomery, D. C. (2008). *Introduction to statistical quality control, 6th ed.* New York: Wiley.

Morton, H. C., and Price, A. J. (1989). *The ACLS survey of scholars: Final report of views on publications, computers, and libraries.* Washington, DC: University Press of America.

Mosteller, F., Hyman, H., McCarthy, P. J., Martis, E. S., and Truman, D. B. (1949). *The pre-election polls of 1948.* New York: Social Sciences Research Council.

Mudryk, W., Bougie, B., and Xie, H. (2001). Quality control of ICR data capture: 2001 Canadian Census of Agriculture. In L. Lyberg (Ed.), *Proceedings of the international conference on quality in official statistics.* Stockholm: Statistics Sweden, CD-20-2.pdf.

Mulry, M. H. (2004). Methodological lessons from Census 2000 coverage error measurement. In *Proceedings of the Section on Survey Research Methods, American Statistical Association,* 4066–4071.

Narain, R. D. (1951). On sampling without replacement with varying probabilities. *Journal of the Indian Society of Agricultural Statistics, 3,* 169–174.

Nathan, G., and Smith, T. M. F. (1989). The effect of selection on regression analysis. In C. J. Skinner, D. Holt, and T. M. F. Smith (Eds.), *Analysis of complex surveys* (pp. 149–163). New York: Wiley.

National Center for Health Statistics. (2005). *Analytic and reporting guidelines: The National Health and Nutrition Examination Survey (NHANES).* Hyattsville, MD: National Center for Health Statistics. Retrieved from www.cdc.gov/nchs/nhanes/nhanes.htm.

National Research Council. (2008). *Surveying victims: Options for the National Crime Victimization Survey* (Panel to Review the Programs of the Bureau of Justice Statistics., R. M. Groves, and D. L. Cork, Eds.). Committee on National Statistics and Committee on Law and Justices, Division of Behavioral and Social Sciences and Education. Washington, DC: National Academies Press.

Neter, J. (1978). How accountants save money by sampling. In J. M. Tanur (Ed.), *Statistics: A guide to the unknown.* San Francisco: Holden-Day.

Neter, J., Leitch, R. A., and Fienberg, S. E. (1978). Dollar unit sampling: Multinomial bounds for total overstatement and understatement errors. *Accounting Review, 53,* 77–93.

Newman, J. C., Des Jarlais, D. C., Turner, C., Gribble, J., Cooley, P., and Paone, D. (2002). The differential effects of face-to-face and computer interview modes. *American Journal of Public Health, 92,* 294–297.

Neyman, J. (1934). On the two different aspects of the representative method: The method of stratified sampling and the method of purposive selection. *Journal of the Royal Statistical Society, 97,* 558–606.

Neyman, J. (1938). Contribution to the theory of sampling human populations. *Journal of the American Statistical Association, 33,* 101–116.

Nolan, D., and Speed, T. (2000). *Stat labs: Mathematical statistics through applications.* New York: Springer.

Nusser, S. M., Carriquiry, A. L., Dodd, K. W., and Fuller, W. A. (1996). A semiparametric transformation approach to estimating usual daily intake distributions. *Journal of the American Statistical Association, 91,* 1440–1449.

O'Brien, L. A., Grisso, J. A., Maislin, G., LaPann, K., Krotki, K. P., Greco, P. J., et al. (1995). Nursing home residents' preferences for life-sustaining treatments. *Journal of the American Medical Association, 274,* 1775–1779.

Oehlert, G. W. (2000). *A first course in design and analysis of experiments.* New York: Freeman.

Office of Management and Budget. (2006). *Standards and guidelines for statistical surveys.* Retrieved from www.whitehouse.gov/omb/inforeg/statpolicy/standards_stat_surveys.pdf.

Oh, H. L., and Scheuren, F. J. (1983). Weighting adjustment for unit nonresponse. In W. G. Madow, I. Olkin, and D. B. Rubin (Eds.), *Incomplete data in sample surveys, Vol. 2* (pp. 143–184). New York: Academic Press.

Overton, W. S., and Stehman, S. V. (1995). The Horvitz-Thompson theorem as a unifying perspective for probability sampling: With examples from natural resource sampling. *The American Statistician, 49,* 261–268.

Patil, G. P. (2002). Ranked set sampling. In A. H. El-Shaarawi and W. W. Piegorsch (Eds.), *Encyclopedia of environmetrics, Vol. 3* (p. 1684–1690). Hoboken, NJ: Wiley.

Paulin, G. D., and Ferraro, D. L. (1994). Imputing income in the Consumer Expenditure Survey. *Monthly Labor Review, December,* 23–31.

Peart, D. (1994). *Impacts of feral pig activity on vegetation patterns associated with Quercus agrifolia on Santa Cruz Island, California.* Tempe, AZ: Ph.D. dissertation, Arizona State University.

Petersen, C. G. J. (1896). The yearly immigration of young plaice into the Limfjord from the German Sea. *Reports of the Danish Biological Station, 6,* 5–84.

Pfeffermann, D. (1993). The role of sampling weights when modeling survey data. *International Statistical Review*, *61*, 317–337.

Pfeffermann, D. (1996). The use of sampling weights for survey data analysis. *Statistical Methods in Medical Research*, *5*, 239–261.

Pfeffermann, D., and Holmes, D. J. (1985). Robustness considerations in the choice of a method of inference for regression analysis of survey data. *Journal of the Royal Statistical Society, Series A*, *148*, 268–278.

Pfeffermann, D., and Rao, C. R. (2009a). *Handbook of statistics: Vol. 29A. Sample surveys: Design, methods and applications*. Amsterdam: North Holland.

Pfeffermann, D., and Rao, C. R. (2009b). *Handbook of statistics: Vol. 29B. Sample surveys: Inference and analysis*. Amsterdam: North Holland.

Pfeffermann, D., Skinner, C. J., Holmes, D. J., Goldstein, H., and Rabash, J. (1998). Weighting for unequal selection probabilities in multilevel models. *Journal of the Royal Statistical Society, Ser. B*, *60*, 23–40.

Pincus, T. (1993). Arthritis and rheumatic diseases: What doctors can learn from their patients. In D. Goleman and J. Gurin (Eds.), *Mind/body medicine: How to use your mind for better health* (pp. 177–192). Yonkers, NY: Consumer Reports Books.

Platek, R. (1977). Some factors affecting non-response. *Survey Methodology*, *3*, 191–214.

Politz, A., and Simmons, W. (1949). An attempt to get the "not at homes" into the sample without callbacks. *Journal of the American Statistical Association*, *44*, 9–31.

Pollock, K. H. (1991). Modeling capture, recapture, and removal statistics for estimation of demographic parameters for fish and wildlife populations: Past, present, and future. *Journal of the American Statistical Association*, *86*, 225–238.

Potthoff, R. F., Manton, K. G., and Woodbury, M. A. (1993). Correcting for nonavailability bias in surveys by weighting based on number of callbacks. *Journal of the American Statistical Association*, *88*, 1197–1207.

Prentice, R. L., and Pyke, R. (1979). Logistic disease incidence models and case-control studies. *Biometrika*, *66*, 403–412.

Presser, S. (1990). Can changes in context reduce vote overreporting in surveys? *Public Opinion Quarterly*, *54*, 586–593.

Presser, S., Rothgeb, J. M., Couper, M. P., Lessler, J. T., Martin, E., Martin, J., et al. (2004). *Methods for testing and evaluating survey questionnaires*. New York: Wiley.

Quenouille, M. H. (1956). Notes on bias in estimation. *Biometrika*, *43*, 353–360.

R Development Core Team. (2008). *R: A language and environment for statistical computing*. Vienna: R Foundation for Statistical Computing.

Rabe-Hesketh, S., and Skrondal, A. (2006). Multilevel modelling of complex survey data. *Journal of the Royal Statistical Society, Series A*, *169*, 805–827.

Raghunathan, T. E., and Grizzle, J. E. (1995). A split questionnaire survey design. *Journal of the American Statistical Association*, *90*, 54-63.

Raj, D. (1968). *Sampling theory*. New York: McGraw-Hill.

Rao, J. N. K. (1963). On three procedures of unequal probability sampling without replacement. *Journal of the American Statistical Association*, *58*, 202–215.

Rao, J. N. K. (1973). On double sampling for stratification and analytical surveys. *Biometrika*, *60*, 125–133.

Rao, J. N. K. (1979a). On deriving mean square errors and their non-negative unbiased estimators in finite population sampling. *Journal of the Indian Statistical Association*, *17*, 125–136.

Rao, J. N. K. (1979b). Optimization in the design of sample surveys. In J. S. Rustagi (Ed.), *Optimizing methods in statistics: Proceedings of an international conference* (pp. 419–434). New York: Academic Press.

Rao, J. N. K. (1988). Variance estimation in sample surveys. In P. R. Krishnaiah and C. R. Rao (Eds.), *Handbook of statistics volume 6: Sampling* (pp. 427–447). New York: Elsevier Science.

Rao, J. N. K. (1994). Estimating totals and distribution functions using auxiliary information at the estimation stage. *Journal of Official Statistics*, *10*, 153–165.

Rao, J. N. K. (1996). On variance estimation with imputed survey data. *Journal of the American Statistical Association*, *91*, 499–506.

Rao, J. N. K. (1997). Developments in sample survey theory: An appraisal. *Canadian Journal of Statistics*, *25*, 1–21.

Rao, J. N. K. (2003). *Small area estimation*. New York: Wiley.

Rao, J. N. K. (2005). Interplay between sample survey theory and practice: An appraisal. *Survey Methodology*, *31*, 117–338.

Rao, J. N. K., Hartley, H. O., and Cochran, W. G. (1962). A simple procedure for unequal probability sampling without replacement. *Journal of the Royal Statistical Society, Series B*, *24*, 482–491.

Rao, J. N. K., and Scott, A. J. (1981). The analysis of categorical data from complex sample surveys: Chi-squared tests for goodness of fit and independence in two-way tables. *Journal of the American Statistical Association*, *76*, 221–230.

Rao, J. N. K., and Scott, A. J. (1984). On chi-squared tests for multiway contingency tables with cell proportions estimated from survey data. *The Annals of Statistics*, *12*, 46–60.

Rao, J. N. K., Scott, A. J., and Benhin, E. (2003). Undoing complex survey data structures: Some theory and applications of inverse sampling. *Survey Methodology*, *29*(2), 107–121.

Rao, J. N. K., Scott, A. J., and Skinner, C. J. (1998). Quasi-score tests with survey data. *Statistica Sinica*, *8*, 1059–1070.

Rao, J. N. K., and Shao, J. (1992). Jackknife variance estimation with survey data under hot deck imputation. *Biometrika*, *79*, 811–822.

Rao, J. N. K., and Sitter, R. R. (1995). Variance estimation under two-phase sampling with application to imputation for missing data. *Biometrika*, *82*, 453–460.

Rao, J. N. K., and Sitter, R. R. (1997). Variance estimation under stratified two-phase sampling with applications to measurement bias. In L. Lyberg, P. Biemer, M. Collins, E. de Leeuw, C. Dippo, N. Schwarz, and D. Trewin, (Eds.), *Survey measurement and process quality* (pp. 753–768). New York: Wiley.

Rao, J. N. K., and Thomas, D. R. (1988). The analysis of cross-classified categorical data from complex sample surveys. *Sociological Methodology*, *18*, 213–269.

Rao, J. N. K., and Thomas, D. R. (1989). Chi-squared tests for contingency tables. In C. J. Skinner, D. Holt, and T. M. F. Smith (Eds.), *Analysis of complex surveys* (pp. 89–114). New York: Wiley.

Rao, J. N. K., and Thomas, D. R. (2003). Analysis of categorical response data from complex surveys: An appraisal and update. In R. L. Chambers and C. J. Skinner (Eds.), *Analysis of survey data* (pp. 85–108). New York: Wiley.

Rao, J. N. K., and Wu, C. F. J. (1985). Inference from stratified samples: Second-order analysis of three methods for nonlinear statistics. *Journal of the American Statistical Association*, *80*, 620–630.

Rao, J. N. K., and Wu, C. F. J. (1987). Methods for standard errors and confidence intervals from sample survey data: Some recent work. *Bulletin of the International Statistical Institute*, *52*, 5–21.

Rao, J. N. K., and Wu, C. F. J. (1988). Resampling inference with complex survey data. *Journal of the American Statistical Association*, *83*, 231–241.

Rao, J. N. K., Wu, C. F. J., and Yue, K. (1992). Some recent work on resampling methods for complex surveys. *Survey Methodology*, *18*, 209–217.

Rässler, S., Rubin, D. B., and Schenker, N. (2008). Incomplete data: Diagnosis, imputation, and estimation. In E. D. deLeeuw, J. J. Hox, and D. A. Dillman (Eds.), *International handbook of survey methodology* (pp. 370–386). New York: Erlbaum.

Ravishanker, N., and Dey, D. K. (2002). *A first course in linear model theory*. Boca Raton, FL: CRC Press.

Remafedi, G., Resnick, M., Blum, R., and Harris, L. (1992). Demography of sexual orientation in adolescents. *Pediatrics*, *89*, 714–721.

Robbins, N. B. (2005). *Creating more effective graphs*. Hoboken, NJ: Wiley.

Roberts, G., Rao, J. N. K., and Kumar, S. (1987). Logistic regression analysis of sample survey data. *Biometrika*, *74*, 1–12.

Roberts, R. J., Sandifer, Q. D., Evans, M. R., Nolan-Ferrell, M. Z., and Davis, P. M. (1995). Reasons for non-uptake of measles, mumps, and rubella catch up immunisation in a measles epidemic and side effects of the vaccine. *British Medical Journal, 310,* 1629–1632.

Robinson, J. (1987). Conditioning ratio estimates under simple random sampling. *Journal of the American Statistical Association, 82,* 826–831.

Rosenbaum, P. R., and Rubin, D. B. (1983). The central role of the propensity score in observational studies for causal effects. *Biometrika, 70,* 41–55.

Ross, S. M. (2006). *A first course in probability, 7th ed.* Upper Saddle River, NJ: Pearson Prentice Hall.

Rothbart, G. S., Fine, M., and Sudman, S. (1982). On finding and interviewing the needles in the haystack: The use of multiplicity sampling. *Public Opinion Quarterly, 46,* 408–421.

Rothenberg, R. B., Lobanov, A., Singh, K. B., and Stroh, G. (1985). Observations on the application of EPI cluster survey methods for estimating disease incidence. *Bulletin of the World Health Organization, 63,* 93–99.

Roush, W. (1996). A census in which all Americans count. *Science, 274,* 713–714.

Royall, R. M. (1970). On finite population sampling theory under certain linear regression models. *Biometrika, 57,* 377–387.

Royall, R. M. (1976). The linear least-squares prediction approach to two-stage sampling. *Journal of the American Statistical Association, 71,* 657–664.

Royall, R. M. (1992a). The model based (prediction) approach to finite population sampling theory. In M. Ghosh and P. K. Pathak (Eds.), *Current issues in statistical inference: Essays in honor of D. Basu* (pp. 225–240). Hayward, CA: Institute of Mathematical Statistics.

Royall, R. M. (1992b). Robustness and optimal design under prediction models for finite populations. *Survey Methodology, 18,* 179–185.

Royall, R. M., and Eberhardt, K. R. (1975). Variance estimates for the ratio estimator. *Sankhyā, Series C, 37,* 43–52.

Rubin, D. B. (1985). The use of propensity scores in applied Bayesian inference. In J. M. Bernardo, M. H. DeGroot, D. V. Lindley, and A. F. M. Smith (Eds.), *Bayesian statistics 2* (pp. 463–472). New York: Elsevier.

Rubin, D. B. (1987). *Multiple imputation for nonresponse in surveys.* New York: Wiley.

Rubin, D. B. (1996). Multiple imputation after 18+ years. *Journal of the American Statistical Association, 91,* 473–489.

Rubin, D. B. (2004). The design of a general and flexible system for handling nonresponse in sample surveys. *The American Statistician, 58,* 298–302.

Rubin-Bleuer, S., and Kratina, I. S. (2005). On the two-phase framework for joint model and design-based inference. *Annals of Statistics, 33,* 2789–2810.

Ruggles, S. (1995). Sampling designs and sampling errors. *Historical Methods, 28,* 40–46.

Ruggles, S., Sobek, M., Alexander, T., Fitch, C. A., Goeken, R., Hall, P. K., et al. (2004). *Integrated public use microdata series: Version 3.0 [machine-readable database].* Retrieved from www.ipums/org/usa.

Russell, H. J. (1972). Use of a commercial dredge to estimate a hardshell clam population by stratified random sampling. *Journal of the Fisheries Research Board of Canada, 29,* 1731–1735.

Rust, K. F., and Rao, J. N. K. (1996). Variance estimation for complex surveys using replication techniques. *Statistical Methods in Medical Research, 5,* 283–310.

Ryan, A. S., Rush, D., Krieger, F. W., and Lewandowski, G. E. (1991). Recent declines in breastfeeding in the United States, 1984–1989. *Pediatrics, 88,* 719–727.

Ryan, T. P. (2000). *Statistical methods for quality improvement.* New York: Wiley.

Salganik, M. J., and Heckathorn, D. D. (2004). Sampling and estimation in hidden populations using respondent-driven sampling. *Sociological Methodology, 34,* 193–239.

Samuels, C. (1996). Full-time vs. part-time instructors. *Arizona AAUP Advocate, 45,* 1–3.

Sanzo, J. M., Garcia-Calabuig, M. A., Audicana, A., and Dehesa, V. (1993). Q fever: Prevalence of antibodies to *Coxiella Burnetii* in the Basque country. *International Journal of Epidemiology*, *22*, 1183–1188.

Särndal, C.-E. (1996). Efficient estimators with simple variance in unequal probability sampling. *Journal of the American Statistical Association*, *91*, 1289–1300.

Särndal, C.-E. (2007). The calibration approach in survey theory and practice. *Survey Methodology*, *33*, 99–119.

Särndal, C.-E., and Lundström, S. (2005). *Estimation in surveys with nonresponse*. Hoboken, NJ: Wiley.

Särndal, C.-E., and Swensson, B. (1987). A general view of estimation for two phases of selection with applications to two-phase sampling and nonresponse. *International Statistical Review*, *55*, 279–294.

Särndal, C.-E., Swensson, B., and Wretman, J. (1992). *Model assisted survey sampling*. New York: Springer-Verlag.

SAS Institute Inc. (2008). *SAS/STAT 9.2 user's guide*. Cary, NC: SAS Institute Inc.

Satterthwaite, F. E. (1946). An approximate distribution of estimates of variance components. *Biometrics*, *2*, 110–114.

Sauer, J. R., Hines, J. E., Gough, G., Thomas, I., and Peterjohn, B. G. (1997). *The North American Breeding Bird Survey, results and analysis* (Version 96.4). Laurel, MD: Patuxent Wildlife Research Center.

Saville, A. (1977). *Survey methods of appraising fishery resources*. FAO Fisheries Technical Paper no. 171. Rome: Food and Agriculture Organization of the United Nations.

Schei, B., and Bakketeig, L. S. (1989). Gynaecological impact of sexual and physical abuse by spouse: A study of a random sample of Norwegian women. *British Journal of Obstetrics and Gynaecology*, *96*, 1379–1383.

Schenker, N., Raghunathan, T. E., Chiu, P.-L., Makuc, D. M., Zhang, G., and Cohen, A. J. (2006). Multiple imputation of missing income data in the National Health Interview Survey. *Journal of the American Statistical Association*, 924–933.

Scheuren, F., and Alvey, W. (2008). *Elections and exit polling*. New York: Wiley.

Schnabel, Z. E. (1938). The estimation of the total fish population of a lake. *American Mathematical Monthly*, *45*, 348–352.

Schreuder, H. T., Sedransk, J., and Ware, K. D. (1968). 3-p sampling and some alternatives, I. *Forest Science*, *14*, 429–453.

Schuman, H., and Converse, J. M. (1971). The effects of black and white interviewers on black responses in 1968. *Public Opinion Quarterly*, *35*, 44–68.

Schuman, H., and Presser, S. (1981). *Questions and answers in attitude surveys: Experiments on question form, wording, and context*. New York: Academic Press.

Schuman, H., and Scott, J. (1987). Problems in the use of survey questions to measure public opinion. *Science*, *280*, 957–959.

Schwarz, N., and Sudman, S. (Eds.). (1996). *Answering questions: Methodology for determining cognitive and communicative processes in survey research*. San Francisco: Jossey-Bass.

Scott, A., and Davis, P. (2001). Estimating interviewer effects for survey responses. In *Proceedings of Statistics Canada Symposium 2001*. Ottawa, ON: Statistics Canada.

Scott, A. J., and Smith, T. M. F. (1969). Estimation in multi-stage surveys. *Journal of the American Statistical Association*, *64*, 830–840.

Scott, A. J. (2006). Population-based case control studies. *Survey Methodology*, *32*, 123–132.

Scott, A. J. (2007). Rao-Scott corrections and their impact. In *Proceedings of the Section on Survey Research Methods, American Statistical Association*, 3514–3518.

Scott, A. J., and Wild, C. J. (2003). Fitting logistic regression models in case-control studies with complex sampling. In R. L. Chambers and C. J. Skinner (Eds.), *Analysis of survey data* (pp. 109–121). New York: Wiley.

Scott, D. W. (1992). *Multivariate density estimation: Theory, practice, and visualization*. New York: Wiley.

Seber, G. A. F. (1970). The effects of trap response on tag recapture estimates. *Biometrics, 26,* 13–22.

Seber, G. A. F. (1982). *The estimation of animal abundance and related parameters, 2nd ed.* London: Charles Griffin.

Sen, A. R. (1953). On the estimate of the variance in sampling with varying probabilities. *Journal of the Indian Society of Agricultural Statistics, 5,* 119–127.

Senturia, Y. D., Christoffel, K. K., and Donovan, M. (1994). Children's household exposure to guns: A pediatric practice-based survey. *Pediatrics, 93,* 469–475.

Serdula, M., Mokdad, A., Pamuk, E., Williamson, D., and Byers, T. (1995). Effects of question order on estimates of the prevalence of attempted weight loss. *American Journal of Epidemiology, 142,* 64–67.

Shah, B. V., Holt, M. M., and Folsom, R. E. (1977). Inference about regression models from sample survey data. *Bulletin of the International Statistical Institute, 47,* 43–57.

Shao, J. (2003). Impact of the bootstrap on sample surveys. *Statistical Science, 18,* 191–198.

Shao, J., and Chen, Y. (1998). Bootstrapping sample quantiles based on complex survey data under hot deck imputation. *Statistica Sinica, 8,* 1071–1085.

Shao, J., and Tu, D. (1995). *The jackknife and bootstrap.* New York: Springer.

Shao, J., and Wu, C. F. J. (1992). Asymptotic properties of the balanced repeated replication method for sample quantiles. *The Annals of Statistics, 20,* 1571–1593.

Shen, J., and Hsieh, C.-L. (1999). Improving the professional status of teaching: Perspectives of future teachers, current teachers, and education professors. *Teaching and Teacher Education, 15,* 315–323.

Silva, P. L. D. N., and Skinner, C. J. (1997). Variable selection for regression estimation in finite populations. *Survey Methodology, 23,* 23–32.

Simonoff, J. S. (1996). *Smoothing methods in statistics.* New York: Springer.

Simonoff, J. S. (2006). *Analyzing categorical data.* New York: Springer.

Singer, E. (2002). The use of incentives to reduce nonresponse in household surveys. In R. M. Groves, D. Dillman, J. Eltinge, and R. Little (Eds.), *Survey nonresponse* (pp. 163–177). New York: Wiley.

Singer, E. (2003). The Eleventh Morris Hansen Lecture: Public perceptions of confidentiality. *Journal of Official Statistics, 19*(4), 333–341.

Siniff, D., and Skoog, R. O. (1964). Aerial censusing of caribou using stratified random sampling. *Journal of Wildlife Management, 28,* 391–401.

Sirken, M. (1970). Household surveys with multiplicity. *Journal of the American Statistical Association, 65,* 257–266.

Sitter, R. R. (1992). Comparing three bootstrap methods for survey data. *The Canadian Journal of Statistics, 20,* 135–154.

Sitter, R. R. (1997). Variance estimation for the regression estimator in two-phase sampling. *Journal of the American Statistical Association, 92,* 780–787.

Sitter, R. R., and Wu, C. F. J. (2001). A note on Woodruff confidence intervals for quantiles. *Statistics and Probability Letters, 52,* 353–358.

Skinner, C. J. (1989). Domain means, regression and multivariate analysis. In C. J. Skinner, D. Holt, and T. M. F. Smith (Eds.), *Analysis of complex surveys* (pp. 59–87). New York: Wiley.

Skinner, C. J., Holt, D., and Smith, T. M. F. (Eds.). (1989a). *Analysis of complex surveys.* New York: Wiley.

Skinner, C. J., Holt, D., and Smith, T. M. F. (1989b). General introduction. In C. J. Skinner, D. Holt, and T. M. F. Smith (Eds.), *Analysis of complex surveys* (pp. 1–20). New York: Wiley.

Smith, P. J., Hoaglin, D. C., and Battaglia, M. P. (2005). *Statistical methodology of the National Immunization Survey, 1994–2002* (Tech. Rep. No. 138). Washington, DC.

Smith, T. M. F. (1988). To weight or not to weight, that is the question. In J. M. Bernardo, M. H. DeGroot, D. V. Lindley, and A. F. M. Smith (Eds.), *Bayesian statistics 3* (pp. 437–451). Oxford: Clarendon Press [Oxford University Press].

Smith, T. M. F. (1994). Sample surveys 1975–1990; An age of reconciliation? (Discussion pp. 19–34). *International Statistical Review, 62,* 5–19.

Squire, P. (1988). Why the 1936 Literary Digest poll failed. *Public Opinion Quarterly*, *52*, 125–133.

Stanley, T. J. (2000). *The millionaire mind*. Kansas City, MO: Andrews McMeel.

Stasny, E. A. (1991). Hierarchical models for the probabilities of a survey classification and nonresponse: An example from the National Crime Survey. *Journal of the American Statistical Association*, *86*, 296–303.

Statistics Canada. (2003). *Statistics Canada quality guidelines*. Ottawa: Statistics Canada.

Steffey, D. L., Fienberg, S. E., and Sturgess, R. H. (2006). Statistical assessment of damages in breach of contract litigation. *Jurimetrics*, *46*, 129–138.

Stein, C. (1945). A two-sample test for a linear hypothesis whose power is independent of the variance. *Annals of Mathematical Statistics*, *37*, 36–50.

Stockford, D. D., and Page, W. F. (1984). Double sampling and the misclassification of Vietnam service. In *Proceedings of the Social Statistics Section, American Statistical Association,* 261–264.

Stokes, S. L. (1980). Estimation of variance using judgment ordered ranked set samples. *Biometrics*, *36*, 35–42.

Stokes, S. L., and Plummer, J. (2002). Using spreadsheet solvers in sample design. *Computational Statistics and Data Analysis*, *44*, 527–546.

Strunk, W., and White, E. B. (1959). *The elements of style*. New York: Macmillan.

Stuart, A. (1984). *The ideas of sampling, 3rd ed.* New York: Oxford University Press.

Student. (1908). On the probable error of the mean. *Biometrika*, *6*, 1–25.

Sudman, S., Bradburn, N., and Schwarz, N. (1995). *Thinking about answers: The application of cognitive processes to survey methodology*. San Francisco: Jossey-Bass.

Sudman, S., and Bradburn, N. M. (1982). *Asking questions*. San Francisco: Jossey-Bass.

Sudman, S., Sirken, M. G., and Cowan, C. D. (1988). Sampling rare and elusive populations. *Science*, *240*, 991–996.

Suessbrick, A., Schober, M. F., and Conrad, F. G. (2000). Different respondents interpret ordinary questions quite differently. In *Proceedings of the Section on Survey Research Methods, American Statistical Association,* 907–912.

Sugden, R. A., Smith, T. M. F., and Jones, R. P. (2000). Cochran's rule for simple random sampling. *Journal of the Royal Statistical Society, Series B*, *62*, 787–793.

Tanur, J. M. (Ed.). (1992). *Questions about questions: Inquiries into the cognitive bases of surveys*. New York: Russell Sage Foundation.

Tate, E. E. (1988). *Survey of radon in Minnesota homes*. Minneapolis: Minnesota Department of Health.

Taylor, B. M. (1989). *Redesign of the National Crime Survey*. Washington, DC: U.S. Department of Justice.

Teichman, J., Coltrin, D., Prouty, K., and Bir, W. (1993). A survey of lead contaminzation in soil along Interstate 880, Alameda County, California. *American Industrial Hygiene Association*, *54*, 557–559.

Theoharakis, V., and Skordia, M. (2003). How do statisticians perceive statistics journals? *The American Statistician*, *54*, 115–123.

Thomas, D. R. (1989). Simultaneous confidence intervals for proportions under cluster sampling. *Survey Methodology*, *15*, 187–201.

Thomas, D. R., Singh, A. C., and Roberts, G. R. (1996). Tests of independence on two-way tables under cluster sampling: An evaluation. *International Statistical Review*, *64*, 295–311.

Thomas, E. J., Studdert, D. M., Burstin, H. R., Orav, E. J., Zeena, T., Williams, E. J., et al. (2000). Incidence and types of adverse events and negligent care in Utah and Colorado. *Medical Care*, *38*, 261–271.

Thompson, M. E. (1997). *Theory of sample surveys*. London: Chapman & Hall.

Thompson, S. K. (1990). Adaptive cluster sampling. *Journal of the American Statistical Association*, *85*, 1050–1059.

Thompson, S. K. (2002). *Sampling, 2nd ed.* New York: Wiley.

Thompson, S. K., and Collins, L. M. (2002). Adaptive sampling in research on risk-related behaviors. *Drug and Alcohol Dependence*, *68*, S57–S67.

Thompson, S. K., and Seber, G. A. F. (1996). *Adaptive sampling*. New York: Wiley.

Thompson, W. L. (Ed.). (2004). *Sampling rare or elusive species: Concepts, designs, and techniques for estimating population parameters*. Washington DC: Island Press.

Thomsen, I., and Siring, E. (1983). On the causes and effects of nonresponse: Norwegian experience. In W. G. Madow and I. Olkin (Eds.), *Incomplete data in sample surveys* vol. 3 (pp. 25–59). New York: Academic Press.

Tillé, Y. (2006). *Sampling algorithms*. New York: Springer.

Tourangeau, R., Rips, L. J., and Rasinski, K. (2000). *The psychology of survey responses*. Cambridge: Cambridge University Press.

Tourangeau, R., and Smith, T. W. (1996). Asking sensitive questions: The impact of data collection mode, question format, and question context. *Public Opinion Quarterly*, *60*, 275–304.

Traugott, M. W. (1987). The importance of persistence in respondent selection for preelection surveys. *Public Opinion Quarterly*, *51*, 48–57.

Tremblay, V. (1986). Practical criteria for definition of weighting classes. *Survey Methodology*, *12*, 85–97.

Tucker, C., Brick, J. M., and Meekins, B. (2007). Household telephone service and usage patterns in the United States in 2004: Implications for telephone samples. *Public Opinion Quarterly*, *71*, 3–22.

Tucker, C., Lepkowski, J. M., and Piekarski, L. (2002). The current efficiency of list-assisted telephone sampling designs. *Public Opinion Quarterly*, *66*, 321–338.

Tukey, J. W. (1958). Bias and confidence in not-quite large samples [Abstract]. *Annals of Mathematical Statistics*, *29*, 614.

Tukey, J. W. (1968). Discussion of "Balanced repeated replications for analytical statistics" by L. Kish and M. Frankel. In *Proceedings of the Section on Government Statistics, American Statistical Association, 32*.

Turk, P., and Borkowski, J. J. (2005). A review of adaptive cluster sampling: 1990–2003. *Environmental and Ecological Statistics*, *12*, 55–94.

U.S. Census Bureau. (1994). *County and city data book: 1994*. Washington, DC: U.S. Census Bureau.

U.S. Census Bureau. (2002a). *Current population survey: Design and methodology* (Technical Paper No. 63RV). Washington, DC: U.S. Census Bureau. Retrieved from www.census.gov/prod/2002pubs/tp63rv.

U.S. Census Bureau. (2002b). *Voting and registration in the election of November 2000* (Current Population Reports No. P20-542). Washington, DC: U.S. Census Bureau.

U.S. Census Bureau. (2003a). *American Community Survey questions and answers*. Retrieved from www.census.gov/prod/www/abs/QandAbroch.pdf.

U.S. Census Bureau. (2003b). *Technical assessment of A.C.E. revision II*. Washington, DC: U.S. Census Bureau. Retrieved from www.census.gov/dmd/www/pdf/ACETechAssess.pdf.

U.S. Census Bureau. (2005). *American Community Survey web page*. www.census.gov/acs/.

U.S. Census Bureau. (2006a). *Current Population Survey: Design and methodology* (Technical Paper No. 66). Washington, DC: U.S. Census Bureau.

U.S. Census Bureau. (2006b). *Vehicle Inventory and Use Survey—Methods*. Washington, DC: U.S. Census Bureau. Retrieved from www.census.gov/svsd/www/vius/methods.html.

U.S. Census Bureau. (2007). *Statistical abstract of the United States: 2008 (127th ed.)*. Washington, DC: U.S. Census Bureau. Retrieved from www.census.gov/statab/.

U.S. Department of Justice. (1989). *Survey of youths in custody, 1987, United States computer file, conducted by Department of Commerce, Bureau of the Census, 2nd ICPSR ed.* Ann Arbor, MI: Inter-University Consortium for Political and Social Research.

U.S. Department of Justice. (1992). *Criminal victimization in the United States, 1990*. Ann Arbor, MI: Inter-University Consortium for Political and Social Research.

U.S. Department of Justice. (2002). *Criminal victimization in the United States, 2000*. Ann Arbor, MI: Inter-University Consortium for Political and Social Research.

U.S. Department of Justice. (2006). *National Crime Victimization Survey, 1992–2004, code-book.* Ann Arbor, MI: Inter-University Consortium for Political and Social Research.

U.S. Environmental Protection Agency. (1990a). *National pesticide survey: Summary results of EPA's national survey of pesticides in drinking water wells.* Washington, DC: U.S. Government Printing Office.

U.S. Environmental Protection Agency. (1990b). *National pesticide survey: Survey design.* Washington, DC: U.S. Government Printing Office.

U.S. Environmental Protection Agency. (2007). *A citizen's guide to radon.* Report no. 405-k-07-009. Washington, DC: U. S. Environmental Protection Agency. Retrieved from www.epa.gov/radon/pdfs/citizensguide.pdf.

Valliant, R. (1987). Generalized variance functions in stratified two-stage sampling. *Journal of the American Statistical Association, 82,* 499–508.

Valliant, R. (2002). Variance estimation for the general regression estimator. *Survey Methodology, 28,* 103–114.

Valliant, R. (2009). Model-based prediction of finite population totals. In D. Pfeffermann and C. R. Rao (Eds.), *Handbook of statistics: Vol. 29B. Sample surveys: Inference and analysis* (pp. 11–32). Amsterdam: North Holland.

Valliant, R., Dorfman, A. H., and Royall, R. M. (2000). *Finite population sampling and inference: A prediction approach.* New York: Wiley.

Vijayan, K. (1968). An exact πps sampling scheme: Generalization of a method of Hanurav. *Journal of the Royal Statistical Society, Series B, 30,* 556–566.

Wadsworth, J., Field, J., Johnson, A. M., Bradshaw, S., and Wellings, K. (1993). Methodology of the National Survey of Sexual Attitudes and Lifestyles. *Journal of the Royal Statistical Society, Series A, 156,* 407–421.

Waksberg, J. (1978). Sampling methods for random digit dialing. *Journal of the American Statistical Association, 73,* 40–46.

Wald, A. (1943). Tests of statistical hypotheses concerning several parameters when the number of observations is large. *Transactions of the American Mathematical Society, 54,* 426–482.

Wand, M. P., and Jones, M. C. (1995). *Kernel smoothing.* London: Chapman & Hall.

Warner, S. L. (1965). Randomized response: A survey technique for eliminating evasive answer bias. *Journal of the American Statistical Association, 60,* 63–69.

Warren, E., and Tyagi, A. W. (2003). *The two-income trap.* New York: Basic Books.

Watson, D. J. (1937). The estimation of leaf area in field crops. *Journal of Agricultural Science, 27,* 474–483.

Webb, W. B. (1955). The illusive phenomena in accident proneness. *Public Health Reports, 70,* 951.

Wilk, S. J., Morse, W. W., Ralph, D. E., and Azarovitz, T. R. (1977). *Fishes and associated environmental data collected in New York Bight, June 1974–June 1975.* NOAA Tech. Rep. No. NMFS SSRF-716. Washington, DC: U.S. Government Printing Office.

Wilson, J. R., and Koehler, K. J. (1991). Hierarchical models for cross-classified overdispersed multinomial data. *Journal of Business & Economic Statistics, 9,* 103–110.

Wisconsin Department of Natural Resources. (1993). *The fisher in Wisconsin.* Technical Bulletin no. 183. Madison, WI: Department of Natural Resources.

Wolter, K. M. (2007). *Introduction to variance estimation, 2nd ed.* New York: Springer.

Woodruff, R. S. (1952). Confidence intervals for medians and other position measures. *Journal of the American Statistical Association, 47,* 636–646.

Woodruff, R. S. (1971). A simple method for approximating the variance of a complicated estimate. *Journal of the American Statistical Association, 66,* 411–414.

Wright, J. (1988). The mentally ill homeless: What is myth and what is fact? *Social Problems, 35,* 182–191.

Wynia, W., Sudar, A., and Jones, G. (1993). Recycling human waste: Composting toilets as a remedial action plan option for Hamilton Harbour. *Water Pollution Research Journal of Canada, 28,* 355–368.

Yates, F. (1981). *Sampling methods for censuses and surveys, 4th ed.* New York: Macmillan.

Yates, F., and Grundy, P. M. (1953). Selection without replacement from within strata with probability proportional to size. *Journal of the Royal Statistical Society, Series B, 109,* 12–30.

Zehnder, G. W., Kolodny-Hirsch, D. M., and Linduska, J. J. (1990). Evaluation of various potato plant sample units for cost-effective sampling of Colorado potato beetle (*Coleoptera chrysomelidae*). *Journal of Economic Entomology, 83,* 428–433.

Zhang, F., Brick, J. M., Kaufman, S., and Walter, E. (1998). *Variance estimation of imputed survey data.* Working Paper no. 98–14. Washington, DC: U.S. Department of Education. Retrieved from nces.ed.gov/pubs98/9814.pdf.

Author Index

Subject Index